Lecture Notes in Artificial Intelligence   9330

Subseries of Lecture Notes in Computer Science

LNAI Series Editors

Randy Goebel
University of Alberta, Edmonton, Canada
Yuzuru Tanaka
Hokkaido University, Sapporo, Japan
Wolfgang Wahlster
DFKI and Saarland University, Saarbrücken, Germany

LNAI Founding Series Editor

Joerg Siekmann
DFKI and Saarland University, Saarbrücken, Germany

More information about this series at http://www.springer.com/series/1244

Manuel Núñez · Ngoc Thanh Nguyen
David Camacho · Bogdan Trawiński (Eds.)

Computational Collective Intelligence

7th International Conference, ICCCI 2015
Madrid, Spain, September 21–23, 2015
Proceedings, Part II

 Springer

Editors
Manuel Núñez
Universidad Complutense de Madrid
Madrid
Spain

Ngoc Thanh Nguyen
Wrocław University of Technology
Wrocław
Poland

David Camacho
Computer Science Department
Universidad Autónoma De Madrid
Madrid
Spain

Bogdan Trawiński
Wrocław University of Technology
Wrocław
Poland

ISSN 0302-9743 ISSN 1611-3349 (electronic)
Lecture Notes in Artificial Intelligence
ISBN 978-3-319-24305-4 ISBN 978-3-319-24306-1 (eBook)
DOI 10.1007/978-3-319-24306-1

Library of Congress Control Number: 2015948851

LNCS Sublibrary: SL7 – Artificial Intelligence

Springer Cham Heidelberg New York Dordrecht London

Printed on acid-free paper

Springer International Publishing AG Switzerland is part of Springer Science+Business Media
(www.springer.com)

Preface

This volume contains the proceedings of the 7th International Conference on Computational Collective Intelligence (ICCCI 2015), held in Madrid, Spain, September 21–23, 2015. The conference was co-organized by the Universidad Complutense de Madrid, Spain, the Universidad Autónoma de Madrid, Spain, and Wrocław University of Technology, Poland. The conference was run under the patronage of the IEEE SMC Technical Committee on Computational Collective Intelligence.

Following the successes of the 1st ICCCI (2009) held in Wrocław, Poland, the 2nd ICCCI (2010) in Kaohsiung, Taiwan, the 3rd ICCCI (2011) in Gdynia, Poland, the 4th ICCCI (2012) in Ho Chi Minh City, Vietnam, the 5th ICCCI (2013) in Craiova, Romania, and the 6th ICCCI (2014) in Seoul, South Korea, this conference continued to provide an internationally respected forum for scientific research in the computer-based methods of collective intelligence and their applications.

Computational Collective Intelligence (CCI) is most often understood as a sub-field of Artificial Intelligence (AI) dealing with soft computing methods that enable making group decisions or processing knowledge among autonomous units acting in distributed environments. Methodological, theoretical, and practical aspects of computational collective intelligence are considered as the form of intelligence that emerges from the collaboration and competition of many individuals (artificial and/or natural). The application of multiple computational intelligence technologies such as fuzzy systems, evolutionary computation, neural systems, consensus theory, etc., can support human and other collective intelligence, and create new forms of CCI in natural and/or artificial systems. Three subfields of the application of computational intelligence technologies to support various forms of collective intelligence are of special interest but are not exclusive: Semantic Web (as an advanced tool for increasing collective intelligence), social network analysis (as the field targeted to the emergence of new forms of CCI), and multiagent systems (as a computational and modeling paradigm especially tailored to capture the nature of CCI emergence in populations of autonomous individuals).

The ICCCI 2015 conference featured a number of keynote talks and oral presentations, closely aligned to the theme of the conference. The conference attracted a substantial number of researchers and practitioners from all over the world, who submitted their papers for the main track and 10 special sessions.

The main track, covering the methodology and applications of computational collective intelligence, included: knowledge integration; data mining for collective processing; fuzzy, modal, and collective systems; nature-inspired systems; language processing systems; social networks and Semantic Web; agent and multi-agent systems; classification and clustering methods; multi-dimensional data processing; web systems; intelligent decision making; methods for scheduling; and image and video processing.

The special sessions, covering some specific topics of particular interest, included: Collective Intelligence in Web Systems - Web Systems Analysis, Computational Swarm Intelligence, Cooperative Strategies for Decision Making and Optimization, Advanced Networking and Security Technologies, IT in Biomedicine, Collective Computational Intelligence in an Educational Context, Science Intelligence and Data Analytics, Computational Intelligence in Financial Markets, Ensemble Learning, and Big Data Mining and Searching.

We received in total 186 submissions. Each paper was reviewed by 2–4 members of the International Program Committee of either the main track or one of the special sessions. We selected the 110 best papers for oral presentation and publication in two volumes of the Lecture Notes in Artificial Intelligence series.

We would like to express our thanks to the keynote speakers, Francisco Herrera, B. John Oommen, Guy Theraulaz, and Jorge Ufano, for their world-class plenary speeches. Many people contributed towards the success of the conference. First, we would like to recognize the work of the Program Committee Co-chairs and special sessions organizers for taking good care of the organization of the reviewing process, an essential stage in ensuring the high quality of the accepted papers. The Workshops and Special Sessions Chairs deserve a special mention for the evaluation of the proposals and the organization and coordination of the work of the 10 special sessions. In addition, we would like to thank the PC members, of the main track and of the special sessions, for performing their reviewing work with diligence. We thank the Organizing Committee Chairs, Liaison Chairs, Publicity Chair, Special Issues Chair, Financial Chair, Web Chair, and Technical Support Chair for their fantastic work before and during the conference. Finally, we cordially thank all the authors, presenters, and delegates for their valuable contribution to this successful event. The conference would not have been possible without their support.

It is our pleasure to announce that the conferences of ICCCI series continue a close cooperation with the Springer journal Transactions on Computational Collective Intelligence, and the IEEE SMC Technical Committee on Transactions on Computational Collective Intelligence.

Finally, we hope and intend that ICCCI 2015 significantly contributes to the academic excellence of the field and leads to the even greater success of ICCCI events in the future.

September 2015

Manuel Núñez
Ngoc Thanh Nguyen
David Camacho
Bogdan Trawiński

ICCCI 2015 Conference Organization

General Chairs

Ngoc Thanh Nguyen	Wrocław University of Technology, Poland
Manuel Núñez	Universidad Complutense de Madrid, Spain

Steering Committee

Ngoc Thanh Nguyen	Wrocław University of Technology, Poland (Chair)
Piotr Jędrzejowicz	Gdynia Maritime University, Poland (Co-Chair)
Shyi-Ming Chen	National Taiwan University of Science and Technology, Taiwan
Adam Grzech	Wrocław University of Technology, Poland
Kiem Hoang	University of Information Technology, VNU-HCM, Vietnam
Lakhmi C. Jain	University of South Australia, Australia
Geun-Sik Jo	Inha University, South Korea
Janusz Kacprzyk	Polish Academy of Sciences, Poland
Ryszard Kowalczyk	Swinburne University of Technology, Australia
Toyoaki Nishida	Kyoto University, Japan
Ryszard Tadeusiewicz	AGH University of Science and Technology, Poland

Program Committee Chairs

David Camacho	Universidad Autónoma de Madrid, Spain
Costin Bădică	University of Craiova, Romania
Piotr Jędrzejowicz	Gdynia Maritime University, Poland
Toyoaki Nishida	Kyoto University, Japan

Organizing Committee Chairs

Jesús Correas	Universidad Complutense de Madrid, Spain
Sonia Estévez	Universidad Complutense de Madrid, Spain
Héctor Menéndez	Universidad Autónoma de Madrid, Spain

Liaison Chairs

Geun-Sik Jo	Inha University, South Korea
Attila Kiss	Eötvös Loránd University, Hungary
Ali Selamat	Universiti Teknologi Malaysia, Malaysia

Special Sessions and Workshops Chairs

Alberto Núñez   Universidad Complutense de Madrid, Spain
Bogdan Trawiński   Wrocław University of Technology, Poland

Special Issues Chair

Jason J. Jung   Chung-Ang University, South Korea

Publicity Chair

Antonio González   Bilbao Center for Applied Mathematics, Spain

Financial Chair

Mercedes G. Merayo   Universidad Complutense de Madrid, Spain

Web Chair

Luis Llana   Universidad Complutense de Madrid, Spain

Technical Support Chair

Rafael Martínez   Universidad Complutense de Madrid, Spain

Special Sessions

WebSys: Special Session on Collective Intelligence in Web Systems – Web Systems Analysis
Organizers: Kazimierz Choroś and Maria Trocan
CSI: Special Session on Computational Swarm Intelligence
Organizers: Urszula Boryczka, Mariusz Boryczka, and Jan Kozak
CSDMO: Special Session on Cooperative Strategies for Decision Making and Optimization
Organizers: Piotr Jędrzejowicz and Dariusz Barbucha
ANST: Special Session on Advanced Networking and Security Technologies
Organizers: Vladimir Sobeslav, Ondrej Krejcar, Peter Brida, and Peter Mikulecky
ITiB: Special Session on IT in Biomedicine
Organizers: Ondrej Krejcar, Kamil Kuca, Teodorico C. Ramalho, and Tanos C.C. Franca
ColEdu: Special Session on Collective Computational Intelligence in an Educational Context
Organizers: Danuta Zakrzewska and Marta Zorrilla
SIDATA: Special Session on Science Intelligence and Data Analytics.
Organizers: Attila Kiss and Binh Nguyen

CIFM: Special Session on Computational Intelligence in Financial Markets
Organizers: Fulufhelo Nelwamondo and Sharat Akhoury
EL: Special Session on Ensemble Learning
Organizers: Piotr Porwik and Michał Woźniak
BigDMS: Special Session on Big Data Mining and Searching
Organizers: Rim Faiz and Aymen Elkhlifi

Program Committee (Main Track)

Abulaish, Muhammad	Jamia Millia Islamia, Kingdom of Saudi Arabia
Bădică, Amelia	University of Craiova, Romania
Barbucha, Dariusz	Gdynia Maritime University, Poland
Bassiliades, Nick	Aristotle University of Thessaloniki, Greece
Bielikova, Maria	Slovak University of Technology in Bratislava, Slovakia
Byrski, Aleksander	AGH University Science and Technology, Poland
Calvo-Rolle, José Luis	University of A Coruña, Spain
Camacho, David	Universidad Autónoma de Madrid, Spain
Capkovic, Frantisek	Slovak Academy of Sciences, Slovakia
Ceglarek, Dariusz	Poznan High School of Banking, Poland
Chiu, Tzu-Fu	Aletheia University, Taiwan
Chohra, Amine	Paris-East University (UPEC), France
Choros, Kazimierz	Wrocław University of Technology, Poland
Colhon, Mihaela	University of Craiova, Romania
Czarnowski, Ireneusz	Gdynia Maritime University, Poland
Davidsson, Paul	Malmö University, Sweden
Do, Tien V.	Budapest University of Technology and Economics, Hungary
Elci, Atilla	Aksaray University, Turkey
Ermolayev, Vadim	Zaporozhye National University, Ukraine
Faiz, Rim	IHEC - University of Carthage, Tunisia
Florea, Adina Magda	Polytechnic University of Bucharest, Romania
Gaspari, Mauro	University of Bologna, Italy
Gónzalez, Antonio	Universidad Autónoma de Madrid, Spain
Ha, Quang-Thuy	Vietnam National University, Vietnam
Hoang, Huu Hanh	Hue University, Vietnam
Hong, Tzung-Pei	National University of Kaohsiung, Taiwan
Huang, Jingshan	University of South Alabama, USA
Hwang, Dosam	Yeungnam University, South Korea
Istrate, Dan	UTC, Laboratoire BMBI, France
Ivanovic, Mirjana	University of Novi Sad, Serbia
Jędrzejowicz, Joanna	University of Gdansk, Poland
Jozefowska, Joanna	Poznan University of Technology, Poland
Jung, Jason	Yeungnam University, South Korea
Kefalas, Petros	The University of Sheffield International Faculty - CITY College, Greece

Kim, Sang-Wook	Hanyang University, South Korea
Kisiel-Dorohinicki, Marek	AGH University of Science and Technology, Poland
Koychev, Ivan	University of Sofia "St. Kliment Ohridski", Bulgaria
Kozierkiewicz-Hetmanska, Adrianna	Wrocław University of Technology, Poland
Krejcar, Ondrej	University of Hradec Kralove, Czech Republic
Kulczycki, Piotr	Polish Academy of Science, Poland
Kuwabara, Kazuhiro	Ritsumeikan University, Japan
Leon, Florin	Technical University "Gheorghe Asachi" of Iasi, Romania
Li, Xiafeng	Texas A&M University, USA
Lu, Joan	University of Huddersfield, UK
Lughofer, Edwin	Johannes Kepler University Linz, Austria
Meissner, Adam	Poznan University of Technology, Poland
Menéndez, Hector	Universidad Autónoma de Madrid, Spain
Mercik, Jacek	Wrocław School of Banking, Poland
Michel, Toulouse	Université de Montréal, Canada
Moldoveanu, Alin	Polytechnic University of Bucharest, Romania
Mostaghim, Sanaz	Otto von Guericke University of Magdeburg, Germany
Nalepa, Grzegorz J.	AGH University of Science and Technology, Poland
Neri, Filippo	University of Napoli Federico II, Italy
Nguyen, Ngoc Thanh	Wrocław University of Technology, Poland
Nguyen, Linh Anh	University of Warsaw, Poland
Nguyen Trung, Duc	Yeungnam University, South Korea
Nguyen Tuong, Tri	Yeungnam University, South Korea
Núñez, Manuel	Universidad Complutense de Madrid, Spain
Núñez, Alberto	Universidad Complutense de Madrid, Spain
Ponnusamy, Ramalingam	Madha Engineering College, India
Precup, Radu-Emil	Polytechnic University of Timisoara, Romania
Quaresma, Paulo	Universidade de Evora, Portugal
Ratajczak-Ropel, Ewa	Gdynia Maritime University, Poland
Selamat, Ali	Universiti Teknologi Malaysia, Malaysia
Stoyanov, Stanimir	University of Plovdiv "Paisii Hilendarski", Bulgaria
Takama, Yasufumi	Tokyo Metropolitan University, Japan
Trawiński, Bogdan	Wrocław University of Technology, Poland
Treur, Jan	Vrije Universiteit Amsterdam, The Netherlands
Unold, Olgierd	Wrocław University of Technology, Poland
Wierzbowska, Izabela	GMU, Poland
Žagar, Drago	University of Osijek, Croatia
Zakrzewska, Danuta	Technical University of Lodz, Poland
Zamfirescu, Constantin-Bala	"Lucian Blaga" University of Sibiu, Romania
Zdravkova, Katerina	FINKI, FYR of Macedonia, Macedonia

Additional Reviewers

Abro, Altaf Hussain
Acs, Zoltan
Camacho, Azahara
Cañizares, Pablo C.
Eriksson, Jeanette
Hajas, Csilla

Hwang, Dosam
Kaczor, Krzysztof
Kluza, Krzysztof
Kontopoulos, Efstratios
Lócsi, Levente
Malumedzha, Tendani

Mollee, Julia
Molnár, Bálint
Ramírez-Atencia, Cristian
Rácz, Gábor
Salmeron, José L.
Thilakarathne, Dilhan

Contents – Part II

CSDMO: Special Session on Cooperative Strategies for Decision Making and Optimization

ANST: Special Session on Advanced Networking and Security Technologies

ITiB: Special Session on IT in Biomedicine

**ColEdu: Special Session on Collective Computational Intelligence
in Educational Context**

SIDATA: Special Session on Science Intelligence and Data Analytics

CIFM: Special Session on Computational Intelligence in Financial Markets

EL: Special Session on Ensemble Learning

BigDMS: Special Session on Big Data Mining and Searching

Contents – Part I

Ontologies and Information Extraction

WebSys: Special Session on Collective Intelligence in Web Systems - Web Systems Analysis

Optimization of 3D Rendering by Simplification of Complicated Scene for Mobile Clients of Web Systems

Tomas Marek and Ondrej Krejcar[✉]

Faculty of Informatics and Management, Center for Basic and Applied Research,
University of Hradec Kralove, Rokitanskeho 62 500 03, Hradec Kralove, Czech Republic
Tomas.Marek.5@uhk.cz, ondrej@krejcar.org

Abstract. Computer graphics in combination with mobile devices find much use in the fields of entertainment, education and data displaying. The amount of information that is possible to provide to the user depends greatly on the optimization of graphic chain in the development of given application. There is a large area of web based software solutions where the speed and fluency of visualization and rendering are the most important parameters of user satisfaction when using of mobile client application. The important element is simplification of the scene by removing objects that are not currently visible. The methods that can be used as a solution for these problems are frustum culling, occlusion culling and level of detail. The work describes implementation of the first of mentioned methods on the Android platform and it includes a comparison of results from different devices when displaying a complicated 3D scene.

Keywords: Mobile device · Engine · Graphics · Culling · Optimization

1 Introduction

Nowadays, the mobile devices replace every time more computers and laptops [17-20]. Their constantly increasing output enables development of application environment in which it is possible to find tools for solution of many tasks. Moreover, these devices are constantly available and ready to be used. This wide accessibility and output increases demand for better and smarter applications in many branches.

Consequently, the computer graphics comes in with the employment of interface Open Graphics Library (OpenGL) for Android and new Metal API (application programming interface) introduced with IOS8 [14]. The extent of employment is very wide, from the effects during the video recording and taking photographs [21], for example, in the article [1] described depth of field effect, to 3D talking avatar [2] presented in the conference APSIPA [13]. Significantly the most profitable and the most favourite category across the applications are games, nevertheless 2D and 3D graphics are finding application in other popular categories such as education or tools. In general, we talk about modules for professional programmes, data displaying or augmented reality. Even cheap mobile devices can display large medical data using volumetric rendering [3]. With this trend the mobile versions of big graphic engines,

© Springer International Publishing Switzerland 2015
M. Núñez et al. (Eds.): ICCCI 2015, Part II, LNCS 9330, pp. 3–12, 2015.
DOI: 10.1007/978-3-319-24306-1_1

such as Unity, CryEngine or Unreal engine are beginning to appear [17-20, 24-25]. For the developers and studies a questions comes to mind and that is: How should we start creating graphic content?

Freely available engines have some disadvantages that have to be considered. However, in order to have the possibility of exporting for mobile devices it is necessary to pay [16].

The incomparability of output of mobile devices and computers is in 2D and 3D graphic very evident. Saving resources in the places where they are not necessary leads to much bigger displaying possibilities and that way to a richer user's experience and competitiveness of the application. That goes for any device. Graphic effects are constantly being developed and with a bigger output it is possible to use more realistic physically based simulations. Dedication to optimization and to methods for lowering demands of the scene becomes a necessity.

2   Problem Definition

The basic characteristic in PC graphics is that better optimization enables displaying more content. The right idea about complicatedness of 3D scene can be acquired by using an example from gaming environment [4]. It can be thought about 3D as a world with dozens of rooms where each room contains other models and details. The complicatedness of the scene exceeds the possibility of the device for processing the whole content. There are animated models of figures moving between the rooms and some of them can be controlled by the user. Certain objects can be moved and destroyed, which is arranged using physical simulations. Everything is accompanied with sound effects to illustrate the atmosphere. In order to realize a vision like this it is convenient to prepare a tool that will facilitate and accelerate the work with the scene design. This kind of tool is called engine. Everything that a graphic framework can contain is described in figure (fig. 1) [16].

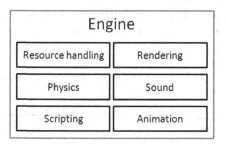

Fig. 1. Graphic engine

Resources represent data uploaded and processed by engine. That is object models, their textures, sounds but also shaders, animations, scripts and others. Models are a reflection of elements of the real world or designer's mind and are created in special rendering tools such as Blender or Autodesk 3ds Max. Models can be also called mesh data and they are a summary of all information needed for rendering of the complex scene. For data transmission different kinds of formats exist, for example OBJ [15], COLLADA [11], 3DS [12] and others [23].

However not even the most optimal use of suitable callings and loading data suffices for rendering big complicated scenes [16]. In that case it is necessary to invent methods that will focus on minimization of processed data.

3   New Solution

For identification of places where it is possible to apply further optimization it is convenient to understand the process of graphic content rendering [6]. OpenGL ES API is a tool for creating an image from 2D and 3D objects. The result of rendering is saved to framebuffer of the graphic card (visual output device). Objects are composed of graphic primitives that are the foundation of rendering OpenGL and they are set during calling glDraw. In order to define such a basic shape it is necessary to have a set of vertices where each vertex is a point in the space. Every vertex is connected to its characteristics that are transmitted with it and connection of these vertices defines the whole model. In figure (Fig. 2) there is demonstrated a simplified process of graphic chain.

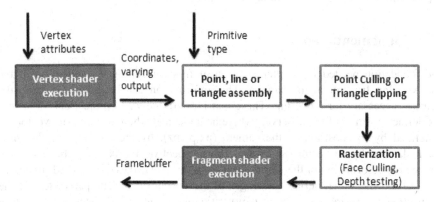

Fig. 2. Graphic chain

The results of vertex shader, which is run for every vertex, are the coordinates of vertex after transformations and possible varying variable. The next step, except for the automatic conversions for the following processes, is a creation of graphic primitives and cropping by viewing frustum. The next is rasterization (conversion to 2D image), activation of fragment shader and saving the figure to framebuffer. In case of triangle being partly out of the screen, some of his vertices are removed and for shape preservation new points are added on the edge of the viewing frustum. In the processing of the fragment (colour, depth, and varying data) it is sometimes necessary to read again the result of the previous record from framebuffer, for example when trying to resolve transparency. Before launching the fragment shader the testing of the depth happens and fragments that are lying behind other fragments are not processed. In process of rasterization, polygons are also being tested for their orientation. The triangle can lie with the front or back side to the camera. When it comes to the 3D objects, back side is oriented into the inside of the model and if it is fronting camera, it is most likely invisible.

When rendering a scene that contains a bigger amount of models, in the first place, each vertex is processed. However, the majority of the models are not visible. Except for the position of the model out of the visual field of the camera, the models can also overlap each other completely. During rasterization the areas that are hidden outside the objects are processed. In case that the camera is moved or rendering happens on random succession there are pointless passages through the fragment shader because the new fragment overlaps the old one. When it comes to the transparent objects it is also necessary to run fragment shader more times for one final fragment. All of the mentioned operations take certain time.

These problems can be solved by differently complicated methods and recommendations. The easiest way is skipping processing of reverse surfaces using setting front / back face culling to true. That way removal of these triangles from following processing is ensured after running the test. More complicated is the solution of the rendering succession where it is necessary to send to models to the graphic chain in the order from the closes to the camera. The transparency is better to avoid when it comes to problems as it was discussed in previous paper [16].

4   Implementation

The foundation needed for implementation of frustum culling is acquisition of the very viewing frustum of the camera. This information can be calculated from the matrixes necessary for rendering. These matrixes are view and projection. View matrix defines position of the observer, where he looks and where he is up (up vector). It is defined by the position of the camera (px,py,pz), by the direction of the view (dx,dy,dz) and by up vector (ux,uy,uz). The perspective matrix (projection) sets the perspective deformation of the world. Data from these matrixes can be used for acquisition of the shape that defines the edges of the visible scene. The parameters of the perspective matrix are field of view (fov), ratio (aspect ratio) and distance of the close and distant surface [10].

In figure (Fig. 3) [10], there are data of the pyramid shape of viewing frustum. The height and width of the planes can be calculated from the present data.

$$Hnear = 2 * \tan\left(\frac{fov}{2}\right) * nearDist$$

$$Wnear = Hnear * ratio$$

$$Hfar = 2 * \tan\left(\frac{fov}{2}\right) * farDist$$

$$Wfar = Hfar * ratio$$

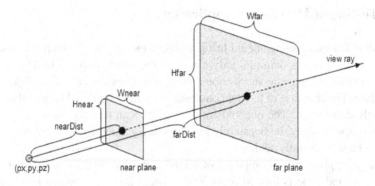

Fig. 3. Frustum planes [10].

From the acquired data and parameters of view matrix it is now possible to define the exact coordinates of vertices. For example the far centre and subsequently far top left.

$$farCenter = p \ (camera \ position) + d \ (view \ direction) * farDist$$

$$farTopLeft = farCenter + \left(upVector * \frac{Hfar}{2}\right) - \left(rightVector * \frac{Wfar}{2}\right)$$

The calculation of all the vertices will be done in a similar way. It is possible to use three points and a surface normal as a definition of each plane of the viewing frustum. This last datum can be acquired using cross product. Its result is perpendicular to both original vectors. One of the vectors can be up vector of the camera. The second one we can acquire as:

$$a = \left(nearCenter + rightVector * \frac{Wnear}{2}\right) - p$$

$$a = normalize(a)$$

$$normalRight = cross(up, a)$$

This way a normal vector of right plane of viewing frustum is acquired. All normals should be directed towards the inside of the shape to ease further processing. Apart from the stated geometrical process it is possible to calculate necessary data even in the space defined by the resulting matrix (VP) from view and projection (calculated in clip space). An arbitrary vertex $p = (x,y,z,1)$ after multiplication of VP matrix will be in clip space as $pc = (xc \ yc, zc, wc)$. This point is situated in homogeneous coordinates. After normalization (dividing by the wc figure) the result is $pcn = (nx,ny,nz)$. In this normalized space, the viewing frustum is defined using axes with the centre in the origin. Those create the volume of the shape of cube located from -1 to 1 on all axes. So that the point pcn is visible, all of its coordinates have to be between -1 and 1. In homogeneous coordinates the delimitations for the point pc are −wc and wc. From this data we can ascertain the location of the vertex with respect to the previously mentioned plane. Therefore, the vertex is on the right side of the left plane if −wc < xc. This method is described more in detail in paper [16].

It is possible to test another shapes this way but those require a calculations for more vertices.

5   Testing of Developed Application

The method frustum culling should bring a visible increase in the output in the situations where the camera monitors only a part of the whole scene. The testing is unfolded from this prerequisite and it should bring a clear evidence of the advantages of this method. The first part of testing constitutes of setting up the 3D scene that will be sufficiently demanding. The demands of the scene mean that for the usual rendering the process speed would be negatively influenced. By using frustum culling a higher speed can be subsequently achieved.

A sample application is composed of a landscape that forms a basis for rendered objects. The landscape is formed from 4096 vertices and it is covered in an ordinary skybox. This part of the scene is not influenced by cropping using viewing frustum. The influence of used method is tested on a model of a tree that has 1648 vertices. Part of this model is particle system for rendering details of branches with leaves. Each of the 99 trees is matched with a cover in a shape of a sphere. The sphere was chosen as the easiest and the least demanding shape for testing in relation to frustum. For a better illustration, the covers are smaller than models itself. The location of the covers is visualized using a cube with a size of the diameter of the sphere. In figure (Fig. 4), there is a view from the camera located towards the centre of the scene.

Fig. 4. Sample scene with vegetation and illustration of covers from the observer's view

The test is running on devices Huawei y300 and Samsung Galaxy S4. The parameters of these devices are stated in the table (Table 1).

Table 1. Used testing devices

Device	SoC	CPU	GPU	RAM
Samsung S4	Qualcomm Snapdragon 600 APQ8064T	Krait 300, 1900 MHz, Cores: 4	Qualcomm Adreno 320, 400 MHz, Cores: 4	2 GB, 600 MHz
Huawei y300	Qualcomm Snapdragon S4 Play MSM8225	ARM Cortex-A5, 1000 MHz, Cores: 2	Qualcomm Adreno 203	512 MB

Apart from the basic parameters it is good to include information about the resolution, mentioned in table number 2 that can have a considerable influence on the resulting output in the graphic applications.

Table 2. Resolution of the display on tested devices

Device	Width [pix]	Height [pix]
Samsung S4	1920	1080
Huawei y300	800	480

The main measured data is frame per second (FPS). FPS provides information about the speed of re-rendering. The relative minimum for preserving the continuity of the movement in the scene is 25 to 20 FPS. However, this value can be misleading for measuring the output. For example, 10 FPS deterioration has a different significance when the starting number is 100 FPS or 20 FPS. Frame per second is a calculated value from the speed of one scene rendering. So the value of 100 FPS means that one rendering lasts 0,001 s. Table number 3 shows a difference between the changes of 10 FPS deterioration when it has different starting FPS. The consequence is that a different load is needed for the 10 FPS change. When measuring the output, for that reason, we will also present data about the speed of rendering.

Table 3. Comparison of the speed change of image re-rendering entered in FPS in contrast to seconds

Initial re-rendering speed [FPS]	Re-rendering speed deterioration [FPS]	Initial re-rendering speed [s]	re-rendering speed after deterioration [s]	Difference / deterioration in re-rendering speed [s]
100	10	0,01	0,0111	0,001
20	10	0,05	0,1	0,05

The testing scene will be run twice in every device. First, the run will be recorded without frustum culling and for the second time it will be with employment of this method. In figures (Fig. 5 and 6), there is a ground plan of the scene and there is a visible difference in the picture with the employment of frustum culling.

Fig. 5. Sample scene without frustum culling in the ground plan

Fig. 6. Frustum culling obstructing rendering of the trees that are beyond the viewing frustum of the observer in figure (Fig. 4).

It is right to say that devices with Android have a firmly set refreshing frequency of 60 FPS. That means that it is not possible to reach higher figures. The demanding character of the scene should reduce the speed of re-rendering that way so that the benefit of frustum culling would be visible. It was discovered from the recorded data that using frustum culling reduces the number of the trees from 99 to 17 in average. Without including the landscape 17,2% of content is shown, which forms the visible part of the scene. The average figures from the testing are indicated in table number 4.

Table 4. Results of frustum culling testing

Device	Re-rendering speed without fr. c. [FPS]	Re-rendering speed with fr. c. [FPS]	Re-rendering speed without fr. c. [s]	Re-rendering speed with fr. c. [s]
Samsung S4	24,23	59,0	0,0413	0,0169
Huawei y300	13,86	25,19	0,0722	0,0397

The method frustum culling significantly helps to reduce the number of objects in the scene, and that way to reduce demands on its rendering. The result creates a possibility of creating a richer content without lowering FPS. A big subjective difference using this optimization was noted while using high-performance Samsung S4. However, Huawei also brought a significant improvement. Generally, frustum culling brings considerable rendering speed acceleration.

6   Conclusions

It was practically verified that using methods for rendering optimization allows to increase the detail of the scene and to reduce the load on the computing power even when using mobile devices. The applied frustum culling method successfully removes uselessly processed scene content and using this method has, in the end, an influence on user's experience. Due to the increasing output and users' base it represents a necessity for competitiveness and profitability of graphic applications. Considering faster development, smaller content and simpler publication of mobile applications in comparison with PS programmes, this branch is a significant possibility of achieving profits.

Acknowledgment. This work and the contribution were supported by project "SP-103-2015 - Smart Solutions for Ubiquitous Computing Environments" Faculty of Informatics and Management, University of Hradec Kralove, Czech Republic. Last but not least, we acknowledge the technical language assistance provided by Jirina Cancikova.

References

1. Wang, Q.S., Yu, Z., Rasmussen, C., Yu, J.Y.: Stereo vision-based depth of field rendering on a mobile device. In: Journal of Electronic Imaging **23**(2) (2014). doi:10.1117/1.JEI.23.2.023009
2. Lin, H.J., Jia, J., Wu, X.J., Cai, L.H.: Stereo talking android: an interactive, multimodal and real-time talking avatar application on mobile phones. In: Asia-Pacific Signal and Information Processing Association Annual Summit and Conference (APSIPA), pp. 1–4 (2013). doi:10.1109/APSIPA.2013.6694211
3. Hachaj, T.: Real time exploration and management of large medical volumetric datasets on small mobile devices-Evaluation of remote volume rendering approach. International Journal of Information Management **34**, 336–343 (2014). doi:10.1016/j.ijinfomgt.2013.11.005
4. Eberly, H.: 3D game engine architecture: engineering real-time applications with wild magic, ver. 1, p. 735. Morgan Kaufmann Publishers, Boston (2005)

5. Brothaler, K.: OpenGL ES 2 for Android: A Quick-Start Guide. The Pragmatic Programmers, Raleigh (2013). ISBN 978-193-7785-345
6. Khronos Group: OpenGL ES Common Profile Specification Version 2.0.25 (Full Specification) (2010). vyd. https://www.khronos.org/registry/gles/specs/2.0/es_full_spec_2.0.25.pdf (cit. November 14, 2014)
7. Level of Detail. Pixar (2014). http://renderman.pixar.com/view/level-of-detail (cit. December 18, 2014)
8. Level-of-detail Representation. In: New York University Computer Science. New York University (2001). http://cs.nyu.edu/~yap/classes/visual/01f/lect/l4/ (cit. November 16, 2014)
9. Gamedev.com, Pietari, L.: Geometry Culling in 3D Engines (2000). http://www.gamedev.net/page/resources/_/technical/graphics-programming-and-theory/geometry-culling-in-3d-engines-r1212 (cit. November 16, 2014)
10. Lighthouse3d.com. http://www.lighthouse3d.com (cit. December 06, 2014)
11. Khronos Group: Collada (2014). https://collada.org/ (cit. December 18, 2014)
12. Autodesk 3D Studio File Format Summary. FileFormat.info (2014). http://www.fileformat.info/format/3ds/egff.htm (cit. December 18, 2014)
13. Asia-Pacific Signal and Information Processing Association (2014). http://www.apsipa.org/ (cit. December 18, 2014)
14. IOS8. Apple Inc: Apple. https://www.apple.com/cz/ios/ (cit. December 18, 2014)
15. Wavefront OBJ File Format Summary. FileFormat.info (2014). http://www.fileformat.info/format/wavefrontobj/egff.htm (cit. December 18, 2014)
16. Marek, T., Krejcar, O.: Optimization of 3D rendering in mobile devices. In: Younas, M., Awan, I., Mecella, M. (eds.) MobiWIS 2015. LNCS, vol. 9228, pp. 37–48. Springer, Heidelberg (2015)
17. Behan, M., Krejcar, O.: Modern Smart Device-Based Concept of Sensoric Networks. EURASIP Journal on Wireless Communications and Networking **2013**(1), No. 155 (June 2013). doi:10.1186/1687-1499-2013-155
18. Krejcar, O., Jirka, J., Janckulik, D.: Use of Mobile Phone as Intelligent Sensor for Sound Input Analysis and Sleep State Detection. Sensors **11**(6), 6037–6055 (2011)
19. Krejcar, O.: Threading possibilities of smart devices platforms for future user adaptive systems. In: Pan, J.-S., Chen, S.-M., Nguyen, N.T. (eds.) ACIIDS 2012, Part II. LNCS, vol. 7197, pp. 458–467. Springer, Heidelberg (2012)
20. Behan, M., Krejcar, O.: Adaptive graphical user interface solution for modern user devices. In: Pan, J.-S., Chen, S.-M., Nguyen, N.T. (eds.) ACIIDS 2012, Part II. LNCS, vol. 7197, pp. 411–420. Springer, Heidelberg (2012)
21. Machacek, Z., Slaby, R., Hercik, R., Koziorek, J.: Advanced system for consumption meters with recognition of video camera signal. Elektronika Ir Elektrotechnika **18**(10), 57–60 (2012). ISSN: 1392-1215
22. Ozana, S., Pies, M., Hajovsky, R., Koziorek, J., Horacek, O.: Application of PIL approach for automated transportation center. In: Saeed, K., Snášel, V. (eds.) CISIM 2014. LNCS, vol. 8838, pp. 501–513. Springer, Heidelberg (2014)
23. Maresova, P., Halek, V.: Deployment of Cloud Computing in Small and Medium Sized Enterprises in the Czech Republic. E & M Ekonomie a Management **17**(4), 159–174 (2014)
24. Gantulga, E., Krejcar, O.: Smart access to big data storage – android multi-language offline dictionary application. In: Nguyen, N.-T., Hoang, K., Jędrzejowicz, P. (eds.) ICCCI 2012, Part I. LNCS, vol. 7653, pp. 375–384. Springer, Heidelberg (2012)
25. Benikovsky, J., Brida, P., Machaj, J.: Proposal of user adaptive modular localization system for ubiquitous positioning. In: Pan, J.-S., Chen, S.-M., Nguyen, N.T. (eds.) ACIIDS 2012, Part II. LNCS, vol. 7197, pp. 391–400. Springer, Heidelberg (2012)

Automatic Categorization of Shots in News Videos Based on the Temporal Relations

Kazimierz Choroś(⊠)

Department of Information Systems, Wrocław University of Technology,
Wybrzeże Wyspiańskiego 27, 50-370 Wrocław, Poland
kazimierz.choros@pwr.edu.pl

Abstract. The development of new methods and technologies of video indexing and retrieval is stimulated by the growing amount of digital video data stored in Internet video collections, TV shows archives, video-on-demand systems, personal video archives offered by Web services, etc. The videos very frequently offered in the Web by broadcast channels are news videos and sports news videos. Content-based indexing of videos is based on the automatic detection of a video structure. A video shot is the main structural video unit. Shots can be of different categories such as intro or final animation, chart or table shots, anchor, reporter, statement, or interview shots, and finally the most informative report shots. The temporal aggregation results in grouping of shots into scenes of a given category. The paper examines the usefulness of the temporal aggregation method to select report shots and non-report shots in a news video.

Keywords: Content-based video indexing · News videos · Video structures · Temporal aggregation · News shots categories · Digital video segmentation

1   Introduction

The great variety of approaches and methods of content-based video indexing are applied to an automatic processing of television broadcast. In television broadcast archives there is a huge number of news and TV sports news videos from last years as well as from the past. Many old analogue videos are digitized, new news videos are being stored every day. Very efficient methods of automatic content-based analyses of digital videos are strongly desirable. The main goal of content-based video indexing of broadcast news videos is to ensure an effective retrieval of special events or special people, of official statements or political polemics and commentaries, etc. Whereas, in the case of sports news the main purpose is to detect players and games, or to select reports of a given sports category. To achieve this purpose, the automatic categorization of sports events, i.e. the automatic detection of the sports disciplines of reported events should be provided. In consequence, the effective retrieval of sports news and of sports highlights of a given sports category such as the best or actual games, tournaments, matches, contests, races, cups, etc., special player behaviours or actions like penalties, jumps, or race finishes, etc. becomes possible.

M. Núñez et al. (Eds.): ICCCI 2015, Part II, LNCS 9330, pp. 13–23, 2015.
DOI: 10.1007/978-3-319-24306-1_2

In most approaches the content-based indexing method needs to analyze all frames of a video. Because such a procedure is extremely time consuming the indexing should be limited to the key-frames, to only one or to a few frames for every shot or for every video scene. The detection of a key-frame of a shot or of a scene requires very effective temporal segmentation methods.

The automatic detection of transitions between shots in a digital video with the purpose of temporal segmentation is relatively well managed and applied in practice. Unfortunately, the detection of scenes is not effectively carried out. A scene is usually defined as a group of consecutive shots sharing similar visual properties and having a semantic correlation – following the classical rule of unity of time, place, and action. The temporal aggregation method [1] detects player scenes taking into account only shot lengths. A player scene is a scene presenting the sports game, i.e. a given scene was recorded on the sports fields such as playgrounds, tennis courts, sports hall, swimming polls, ski jumps, etc. All other non-player shots and scenes usually re-corded in a TV studio such as commentaries, interviews, charts, tables, announcements of future games, discussions of decisions of sports associations, etc. are called studio shots or studio scenes. Studio shots are not useful for video categorization and therefore should be rejected. It was observed that the studio scenes may be even two thirds of sports news. This rejection of non-player scenes before starting content analyses creates a great opportunity to significantly reduce computing time and conduct these analyses more efficiently.

Generally different video genres have different editing style. The specific nature of videos has an important influence on the efficiency of temporal segmentation methods. In the experiments performed in [2] the efficiency of segmentation methods was analyzed for five different categories of movies: TV talk-show, documentary movie, animal video, action & adventure, and pop music video. It has been shown that the segmentation parameters should be suitable to the specificity of the videos.

TV news video editing is similar to that of TV sports news but shots are longer in average. Then the statements and commentaries can be more significant in news than in sports news because these statements are not spoken by anchorman but also for example by politicians or famous people. The detection of politicians is important and may be realized using for example face detection methods.

In this paper the usefulness of the temporal aggregation in detection of report and non-report shots in a news video is verified. The paper is organized as follows. The next section describes related work in the area of an automatic news and sports news shot categorization. The main idea of the temporal aggregation method is presented and the detection of pseudo-scenes using temporal aggregation is outlined in the third section. The forth section presents the experimental results of the detection of the main structural units of TV news obtained in the AVI Indexer. The final conclusions and the future research work areas are discussed in the last fifth section.

2   Related Work

Much research has been carried out in the area of automatic recognition of video content and of visual information indexing and retrieval [3–7]. Traditional textual techniques frequently applied for videos are not sufficient for nowadays video archive browsers. The effective methods of the automatic categorization of a huge amount of broadcast news videos – mainly sports news videos – would be highly desirable. Most of proposed methods require the detection of the structure of videos being indexed and the categorization of shots and scenes detected in videos [8].

Different criteria can be used for the shot categorization in indexing process, the most interesting criterion is the content of a shot. In news videos such shot categories can be defined as [9]: anchor shots, animation (intro), communication (static image of reporter), interview, reporter, maps, studio (discussion with a guest), synthetic (tables, charts, diagrams), whether, and of course report shots.

Other authors developed methods for sports shots classification to such classes as: court views, players and coach, player close-up views, audience views, and setting long views [10], but also as: long views, middle views, close-up views, out of fields views [11], as well as like: intro, headlines, player shots, studio shots [12]. In [13] all sports shots from all types of field video are classified basing on the perceived distance between the camera and the object presented in the shot. Fourteen different shot classes were defined: close up shot head with simple background, close up shot head complex background, close up shot head mixture background, close up shot waist up simple background, close up shot waist up complex background, close up shot waist up mixture background, short distance shot presenting player(s) simple background, short distance shot presenting player(s) complex background, short distance shot presenting player(s) mixture background, short distance shot presenting spectators, long distance shot presenting centre of the field, long distance shot presenting right side of the field, long distance shot presenting left side of the field, and long distance shot presenting spectators.

Anchor/non-anchor shots are frequently used as a starting point for the automatic recognition of a news or sports news video structure. Anchorperson shot detection is still a challenging and important stage of news video analysis and indexing. Recent years, many algorithms have been proposed to detect anchorperson shots. Because we observe the very high similarity between anchor shots (very static sequences of frames, small changes, the same repeated background) one of the approaches of an anchor shot detection is based on template matching. Whereas the other methods are based on different specific properties of anchor shots or on temporal analyses of shots. In the first group of methods a set of predefined models of an anchor should be defined and then, they are matched against all frames in a news video, in order to detect potential anchor shots. The second group of an anchor shot detection methods is mainly based on clustering. Unfortunately, the proposed methods are very time-consuming because they require complex analyses of a great number of video frames. The third approach based on temporal aggregation is very fast because only shot durations are analysed.

The high values of recall and precision for anchorperson detection have been obtained in the experiments on 10 news videos [14]. The news videos were firstly as usually segmented into shots by a four-threshold method. Then the key frames were extracted from each shot. The anchorperson detection was conducted from these key frames by using a clustering-based method based on a statistical distance of Pearson's correlation coefficient.

The new method presented in [15] can be also used for dynamic studio background and multiple anchorpersons. It is based on spatio-temporal slice analysis. This method proposes to extract two different diagonal spatio-temporal slices and divide them into three portions. Then all slices from two sliding windows obtained from each shot are classified to get the candidate anchor shots. And finally, the real anchor shots are detected using structure tensor. The experiments carried out on news programs of seven different styles confirmed the effectiveness of this method.

The algorithm described in [16] analyzes audio, frame and face information to identify the content. These three elements are independently processed during the cluster analysis and then jointly in a compositional mining phase. The temporal features of the anchorpersons for finding the speaking person that appears most often in the same scene are used to differentiate the role played by the detected people in the video. Significant values of precision and recall have been obtained in the experiments carried out for broadcast news coming from eight different TV channels.

A novel anchor shot detection method proposed in [17] detects an anchorperson cost-effectively by reducing the search space. It is achieved by using skin colour and face detectors, as well as support vector data descriptions with non-negative matrix factorization.

It is observed that the most frequent speaker is the anchorman [18]. An anchor speaks many times during the programme, so the anchorperson shots are distributed all along the programme timeline. This observation leads to the selection of the speaker who most likely is the anchorman. It is assumed that a speaker clustering process labels all the speakers present in the video and associates them to temporal segments of the content. However, there are some obvious drawbacks, because a shot with a reporter (interview shots) or with a politician (statement shots) frequently found in news can be erroneously recognized as an anchor shot.

Another observation in a large database [19] draws much attention to interview scenes. In many interview scenes an interviewer and an interviewee recursively appear. A technique called interview clustering method based on face similarity can be applied to merge these interview units.

In [20] a fast method of automatic detection of anchorperson shots has been presented. The method is useful for detection long duration shots such as anchor, reporter, interview, or any other statement shots.

Video analyses discussed in the related papers as well as in this research are the methods using visual features only. There are also audio-visual approaches analyzing not only visual information but also audio (see for example [21, 22]).

3   Temporal Segmentation and Aggregation in the AVI Indexer

The Automatic Video Indexer AVI [23] is a research system designed to develop new tools and techniques of automatic video content-based indexing for retrieval systems, mainly based on the video structure analyses [24] and using the temporal aggregation method [1]. The standard process of automatic content-based analysis and video indexing is composed of several stages. Usually it starts with a temporal segmentation resulting in the segmentation of a movie into small units called video shots. Shots can be grouped to make scenes, and then key-frame or key-frames for every scene can be selected for further analyses. In the case of TV sports news every scene is categorized using such strategies as: detection of playing fields, of superimposed text like player or team names, identification of player faces, detection of lines typical for a given playing field and for a given sports discipline, recognition of player and audience emotions, and also detection of sports objects specific for a given sports category. Whereas in the case of TV news scenes can be categorized basing on the people or place detection using face detection or object detection.

The detection of video scenes facilities the optimization of indexing process. The automatic categorization of news videos will be less time consuming if the analyzed video material is limited only to scenes the most adequate for content-based analyses like player scenes in TV sports news or official statements in TV news. The temporal aggregation method implemented in the AVI Indexer is applied for a video structure detection. The method detects and aggregates long anchorman shots. The shots are grouped into scenes basing on the length of the shots as a sufficient sole criterion.

The temporal aggregation method has two main advantages. First of all it detects player scenes, therefore the most informative parts of sports news videos. Then, it significantly reduces video material analyzed in content-based indexing of TV sports news because it permits to limit indexing process only to player scenes. Globally, the length of all player scenes is significantly lower than the length of all studio shots.

The temporal aggregation is specified by three values: minimum shot length as well as lower and upper limits representing the length range for the most informative shots. The values of these parameters should be determined taking into account specific editing style of a video and its high-level structure.

Formally, the temporal aggregation process is defined as follows [25]:

- single frame detected as a shot is aggregated to the next shot,
  if $(L(shot_i) == 1 \ [frame])$ then $L(shot_{i+1}) = L(shot_{i+1}) + 1;$
  $LS = LS - 1;$
  where $L(shot_i)$ is the length [measured in frames] of the detected shot i and $shot_{i+1}$
  is a next shot on a timeline and LS is a number of shots detected;
- very short shots should be aggregated till their aggregated length attains a certain value Min_Shot_Length,
  while ($(L(shot_i) < MIN\_Shot\_Length)$ and ($L(shot_{i+1}) < MIN\_Shot\_Length$)) do
    { $L(shot_i) = L(shot_i) + L(shot_{i+1});$
    $LS = LS - 1; }$

- all long consecutive shots should be aggregated because these shots seem to be useless in further content analyses and categorization of sports events,

   while ((L(shot$_i$) > MAX_Shot_Length) and (L(shot$_{i+1}$) > MAX_Shot_Length)) do

   { *L(shot$_i$) = L(shot$_i$) + L(shot$_{i+1}$);*

   LS = LS – 1; }

- after aggregation all shots of the length between two a priori defined maximum and minimum values should remain unchanged – these shots are very probably the most informative shots for further content-based analyses.

Very short shots including single frames are relatively very frequent. Generally very short shots of one or several frames are detected in case of dissolve effects or they are simply wrong detections. The causes of false detections may be different [26]. Most frequently it is due to very dynamic movements of players during the game, very dynamic movements of objects just in front of a camera, changes (lights, content) in advertising banners near the player fields, very dynamic movements of a camera during the game, light flashes during games or interviews. These extremely short shots resulting from temporal segmentation are joined with the next shot in a video. So, the first two steps of the temporal aggregation of shots also leads to the significant reduction of false cuts incorrectly detected during temporal segmentation.

4   Report and Non-report Shots in Temporally Aggregated News

The method of temporal aggregation has been applied in the experiments performed in the AVI Indexer. The temporal aggregation has been used with such parameters that only shots of the duration not lower than 45 frames (MIN_Shot_Length) and not greater than 305 frames (MAX_Shot_Length) have been not aggregated. These are shots of the length from 2 to 12 seconds ± 5 frames of tolerance.

Six editions of the TV News "Teleexpress" used in the experiments have been broadcasted in the first national Polish TV channel (TVP1). Their characteristics before and after temporal aggregation are presented in Table 1. The "Teleexpress" is broadcasted every day and is of 15 minutes. This TV program is mainly dedicated to young people. It is dynamically edited, it is very fast paced with very quickly uttered anchor comments. So, the dynamics of the "Teleexpress" News can be comparable to the dynamics of players scenes in TV sports news. However the number of topics and events reported in the news is usually much greater than in the sports news. The question is how the temporal aggregation method changes the temporal relations of shots, whether despite the fact that shots are aggregated their temporal characteristics enable us to predict a shot category.

Table 1. Characteristics before and after temporal aggregation of the six „Teleexpress" News videos broadcasted in March 2014 (03-03, 03-05, 03-06, 03-08, 03-09, and 03-11).

	Video 1	Video 2	Video 3	Video 4	Video 5	Video 6	Average
Length [sec.]	907	900	899	894	893	905	900
After temporal segmentation							
Total number of shots before aggregation	520	461	492	465	523	453	486
Number of shots of less than 15 frames	310	260	277	274	304	260	281
Number of shots of a single frame	261	201	222	205	223	181	216
Real number of anchor shots	13	12	13	14	13	13	13
After temporal aggregation							
Total number of shots after aggregation	176	178	173	159	197	169	175
Number of shots tipped to be reports (45<=length<=305)	161	164	157	144	182	153	160
Percentage of frames in the report shots in the video [%]	73	69	72	66	73	70	71
Number of long aggregated shots (>305 frames)	14	14	16	15	14	15	15
Number of incorrectly (two or more shots of different categories) aggregated shots	2	6	0	4	3	2	2.8

The aggregation is incorrect if the length of an aggregated shot becomes not adequate for its category, i.e. a report aggregated shot becomes so long that it can be treated as for example an anchor shot. The aggregation is also incorrect if two or more shots of different categories are aggregated, mainly if a report shot is aggregated with no report shot. It happens very rarely. Only 2.8 incorrectly aggregates shots in average have been observed (last row in Table 1). The most frequently it was a case of the aggregation of very short report shot with a subsequent anchor shot. Such a long aggregated shot would be then processed as an anchor shot, and thus it should not distort the results of content-based indexing.

The most important result is that after the application of the temporal aggregation the anchor shots are still the longest shots in news (Table 2). Although, it should be noticed that the statement, reporter, or interview shots are also at the beginning of the ranking of longest shots in news videos. They are almost as frequent (7 shots) in long shots as report shots (8 shots). So, the shot aggregation facilitates the detection of speaking person shots. Thus, the temporal analyses of aggregated shot makes it easy to select report shots and non-report shots. And this is a key problem in indexing.

Table 2. Analysis of long aggregated shots.

	Video 1	Video 2	Video 3	Video 4	Video 5	Video 6	Average
Number of aggregated long shots	14	14	16	15	14	15	14.67
Real number of all anchor shots	13	12	13	14	13	13	13.00
Number of anchor shots in long shots	12	11	11	13	12	11	11.67
Percentage of detected anchor shots [%]	92.31	91.67	84.62	92.86	92.31	84.62	89.77
Number of report shots	1	0	3	2	1	1	1.33
Percentage of report shots in long shots [%]	7.14	0	18.75	13.33	7.14	6.67	8.84

Between all 88 long aggregated shots there are 70 anchor shots and 7 other speaking people shots, eight report shots, one chart shot, and one final animation. The report shots represent only about 9 % of all aggregated long shots. To detect faster anchor shots it is desirable to reduce the video space by using temporal aggregation.

The structure of a news video is as follows: it starts with the intro animation, then several stories are presented and commented containing an anchor shot or shots followed by a sequence of report shots optionally enhanced by reporter, politician's statement, interview, or chart, table, diagram shots. Similarly to the began the news is finished with a final animation shot. The length of the intro and final animation is not constant, so after temporal segmentation at the beginning as well as at the end of a news video we can receive different sequence of shots. It is due to the fact that sometimes there is a fade from black to the intro part and the final animation usually fades away to a black frame. But we very often observed that the first frame of the intro or the last frame of the final animation are frozen, so the length of these parts of a news video are of an unpredictable duration.

The analyses of the lengths of both the 50 longest shots (Table 3) as well as of all shots (Table 4) in the tested videos clearly confirm that anchor shots are the longest parts of videos before and also after aggregation: 414 frames in average for the most longest shots in a news video and 341 frames in average for all shots. And at the same time report shots are the shortest parts of videos: 173 frames in average for the most longest shots in a news video and 92 frames in average for all shots.

Table 3. Number of shots of a given category and their average lengths among the 50 longest aggregated shots in each of the videos.

Shot categories	Video 1	Video 2	Video 3	Video 4	Video 5	Video 6	Averages for all videos
Intro	1 140	1 281	1 142	1 140	1 147	0 –	1 170
Anchor shots	11 385	10 441	13 381	11 447	10 409	11 420	11 414
Chart shots					2 272	1 225	1.5 249
Statement, Reporter, or Interview	7 220	6 218	10 249	4 225	5 191	11 230	7 222
Report	28 187	27 153	25 192	30 181	29 147	24 178	27 173
Final Animation	1 300	1 306	1 246	1 282	1 326	1 330	1 298

Table 4. Number of all shots of a given category and their average lengths among in the tested videos.

Shot categories	Video 1	Video 2	Video 3	Video 4	Video 5	Video 6	Averages for all videos
Intro	3 101	1 281	3 93	2 125	2 145	2 92	2.2 140
Anchor shots	11 385	13 292	16 345	13 385	14 310	15 329	13.7 341
Chart shots					2 272	1 225	1.5 249
Statement, Reporter, or Interview	7 220	6 218	11 236	5 203	7 163	12 218	8.0 210
Report	151 97	148 89	139 92	134 98	168 86	134 92	146 92
Final Animation	1 300	1 306	1 246	1 282	2 170	1 330	1.0 298

The most important is that the number of shots decreased almost three times. If the indexing process is based on the analysis of a single key frame for every shot in a video the processing time of indexing will be also decreased three times. This is a great advantage of the temporal aggregation.

5   Final Remarks

The methods of content-based video indexing are still being improved, new methods are still being proposed. Many of them adapted to the content analysis of news videos are based on video structure. The detection of news video structure and categorization of news video shots is very important. The key problem is to select report shots and non-report shots because usually different indexing strategies should be applied.

The temporal aggregation can successfully reduce the video space analyzed in content-base indexing without disturbing news shot categorization. Furthermore, the temporal aggregation can be also used to select report shots and non-report shots in news videos. The results of tests performed in the AVI Indexer have confirmed that the temporal aggregation facilitates the automatic parsing of video structure of news videos.

All candidate shots for non-report shots such as anchor shots as well as statement, reporter, or interview shots can be then analyzed using usually proposed approaches mainly based on face detection and person recognition. Whereas, the report shots demand a variety of methods based on different approaches.

References

1. Choroś, K.: Temporal aggregation of video shots in TV sports news for detection and categorization of player scenes. In: Bădică, C., Nguyen, N.T., Brezovan, M. (eds.) ICCCI 2013. LNCS, vol. 8083, pp. 487–497. Springer, Heidelberg (2013)
2. Choroś K., Gonet M.: Effectiveness of video segmentation techniques for different categories of videos. In: New Trends in Multimedia and Network Information Systems, pp. 34–45. IOS Press, Amsterdam (2008)
3. Truong, B.T., Venkatesh, S.: Video Abstraction: A Systematic Review and Classification. ACM Transactions on Multimedia Computing, Communications, and Applications (TOMM) 3(1), 1–37 (2007)
4. Money, A.G., Agius, H.: Video summarisation: a conceptual framework and survey of the state of the art. J. of Visual Communication and Image Representation 19, 121–143 (2008)
5. Hu, W., Xie, N., Li, L., Zeng, X., Maybank, S.: A survey on visual content-based video indexing and retrieval. IEEE Transactions on Systems, Man, and Cybernetics, Part C: Applications and Reviews 41(6), 797–819 (2011)
6. Del Fabro, M., Böszörmenyi, L.: State-of-the-art and future challenges in video scene detection: a survey. Multimedia Systems 19(5), 427–454 (2013)
7. Asghar, M.N., Hussain, F., Manton, R.: Video indexing: a survey. Int. J. of Computer and Information Technology 3(1), 148–169 (2014)
8. Kompatsiaris, Y., Mérialdo, B., Lian, S. (eds.): TV Content Analysis: Techniques and Applications. CRC Press, Boca Raton (2012)
9. Valdés, V., Martínez, J.M.: On-line video abstract generation of multimedia news. Multimedia Tools and Applications 59(3), 795–832 (2012)
10. Duan, L.Y., Xu, M., Tian, Q., Xu, C.S., Jin, J.S.: A unified framework for semantic shot classification in sports video. IEEE Transactions on Multimedia 7(6), 1066–1083 (2005)
11. Lang, C., Xu, D., Jiang, Y.: Shot type classification in sports video based on visual attention. In: Proc. of Int. Conf. on Computational Intelligence and Natural Computing (CINC 2009), vol. 1, pp. 336–339. IEEE (2009)
12. Choroś, K.: Automatic detection of headlines in temporally aggregated TV sports news videos. In: Proc. of the 8th Int. Symp. on Image and Signal Processing and Analysis (ISPA 2013). IEEE, pp. 147–152 (2013)
13. Kapela, R., McGuinness, K., O'Connor, N.E.: Real-time field sports scene classification using colour and frequency space decompositions. Journal of Real-Time Image Processing, 1–13 (2014)

14. Ji, P., Cao, L., Zhang, X., Zhang, L., Wu, W.: News videos anchor person detection by shot clustering. Neurocomputing **123**, 86–99 (2014)
15. Zheng, F., Li, S., Wu, H., Feng, J.: Anchor shot detection with diverse style backgrounds based on spatial-temporal slice analysis. In: Boll, S., Tian, Q., Zhang, L., Zhang, Z., Chen, Y.-P.P. (eds.) MMM 2010. LNCS, vol. 5916, pp. 676–682. Springer, Heidelberg (2010)
16. Broilo, M., Basso, A., De Natale, F.G.: Unsupervised anchorpersons differentiation in news video. In: Proc. of the 9th International Workshop on Content-Based Multimedia Indexing (CBMI), pp. 115–120. IEEE (2011)
17. Lee, H., Yu, J., Im, Y., Gil, J.M., Park, D.: A unified scheme of shot boundary detection and anchor shot detection in news video story parsing. Multimedia Tools and Applications **51**(3), 1127–1145 (2011)
18. Montagnuolo, M., Messina, A., Borgotallo, R.: Automatic segmentation, aggregation and indexing of multimodal news information from television and the Internet. Int. J. of Information Studies **1**(3), 200–211 (2010)
19. Dong, Y., Qin, G., Xiao, G., Lian, S., Chang, X.: Advanced news video parsing via visual characteristics of anchorperson scenes. Telecommunication Systems **54**(3), 247–263 (2013)
20. Choroś, K.: Automatic fast detection of anchorperson shots in temporally aggregated TV news videos. In: Nguyen, N.T., Trawiński, B., Kosala, R. (eds.) ACIIDS 2015. LNCS, vol. 9012, pp. 339–348. Springer, Heidelberg (2015)
21. El Khoury, E., Sénac, C., Joly, P.: Audiovisual diarization of people in video content. Multimedia Tools and Applications **68**(3), 747–775 (2014)
22. Qu, B., Vallet, F., Carrive, J., Gravier, G.: Content-based inference of hierarchical structural grammar for recurrent TV programs using multiple sequence alignment. In: Proc. of the IEEE International Conference on Multimedia and Expo., ICME, pp. 1–6 (2014)
23. Choroś, K.: Video structure analysis and content-based indexing in the automatic video indexer AVI. In: Nguyen, N.T., Zgrzywa, A., Czyżewski, A. (eds.) Advances in Multimedia and Network Information System Technologies. AISC, vol. 80, pp. 79–90. Springer, Heidelberg (2010)
24. Choroś, K.: Video structure analysis for content-based indexing and categorisation of TV sports news. Int. J. of Intelligent Information and Database Systems **6**(5), 451–465 (2012)
25. Choroś, K.: Automatic detection of headlines in temporally aggregated TV sports news videos. In: Proc. of the 8th International Symposium on Image and Signal Processing and Analysis (ISPA), pp. 147–152. IEEE (2013)
26. Choroś, K.: False and miss detections in temporal segmentation of TV sports news videos – causes and remedies. In: Zgrzywa, A., Choroś, K., Siemiński, A. (eds.) New Research in Multimedia and Internet Systems. Advances in Intelligent Systems and Computing 314, pp. 35–46. Springer, Heidelberg (2014)

Saliency-Guided Video Deinterlacing

Maria Trocan[1]([✉]) and François-Xavier Coudoux[2]

[1] Institut Superieur d'Electronique de Paris,
28 rue Notre Dame des Champs, Paris, France
maria.trocan@isep.fr
[2] IEMN (UMR CNRS 8520) Department OAE,
Valenciennes University, 59313 Valenciennes Cedex 9, France
francois-xavier.coudoux@univ-valenciennes.fr

Abstract. Video deinterlacing is a technique wherein the interlaced video format is converted into progressive scan format for nowadays display devices. In this paper a spatial saliency-guided motion compensated deinterlacing method is proposed: our algorithm classifies the field according to its texture and viewer's region of interest and adapts the motion estimation and compensation, as well as the saliency-guided interpolation in order to ensure high quality frame reconstruction. The experimental results show significant improvement of the proposed method over classical motion compensated and adaptive deinterlacing techniques.

1 Introduction

The process of deinterlacing involves converting a stream of interlaced frames within a video sequence to progressive frames [1], in order to ensure their playback on nowadays progressive devices. Deinterlacing requires the display device to buffer one or more fields and recombine them to a full progressive frame. There are various methods to deinterlace a video and each method produces its own artifacts, due to the temporal lack of information and the dynamics of the video sequence.

Spatial deinterlacers [2] use the information from the current field to interpolate the missing field lines. The most common types of spatial deinterlacing methods are line averaging and directional spatial interpolation. Edge-based line averaging is done by interpolation along the edge direction, by comparing the gradients of various directions. The interpolation accuracy of edge-based line averaging is increased by an efficient estimation of the directional spatial correlations of neighboring pixels. Usually, the spatial deinterlacing methods have low computational power. However, one disadvantage of spatial deinterlacing is that this class of methods is not optimal due to the fact that motion activity is not considered in interpolation; moreover, these kinds of algorithms fail to remove the flickering artifacts.

Motion adaptive methods use consecutive fields to analyze the characteristics of motion in order to choose the appropriate interpolation scheme. In such deinterlacers, dynamic areas are interpolated spatially and the static segments are interpolated temporally.

© Springer International Publishing Switzerland 2015
M. Núñez et al. (Eds.): ICCCI 2015, Part II, LNCS 9330, pp. 24–33, 2015.
DOI: 10.1007/978-3-319-24306-1_3

The best class of deinterlacers is given by the motion compensated ones [4]. In these schemes, the motion trajectory is estimated and the interpolation of the missing fields is done along the motion flow. However, motion compensated deinterlacers need massive computational resources. To reduce their complexity, block-based motion estimation is used at the expense of blocking artifacts and some unreliable motion information [6], which severely degrades the visual quality of the reconstructed video sequences.

In this paper, in order to reduce the blocking artifacts hence improving the Quality of Experience (QoE) of human viewers, we propose to use the block-based motion estimation on smooth areas, whether on highly textured areas optical-flow [5] pixel velocity is used. For improving the frame reconstruction quality, visual saliency-guided interpolation of the estimated temporal field is used. The use of visual saliency [3] as trigger for the spatio-temporal interpolator has two advantages: for non salient regions, no motion-estimation is performed, the areas being spatially interpolated, hence highly reducing the proposed deinterlacer complexity. The second advantage is the corollary of the first one: the computing resources, translated mainly into the motion estimation process, can be used entirely on the region of interest area.

In the sequel, the paper is organized as follows: Section 2 first introduces the notion of visual saliency and presents some existing saliency models. Then, Section 3 describes the proposed saliency-guided spatio-temporal video deinterlacing method. Some experimental results obtained with the proposed method for different video sequences are presented in Section 4. Finally, conclusions are drawn in Section 5.

2   Visual Saliency

Visual saliency is defined in [7] as *the distinct subjective perceptual quality which makes some items in the world stand out from their neighbors and immediately grab our attention*. The visual saliency process allows a human observer to specifically focus her/his attention on one or more visual stimuli into a scene depending on some semantic features like orientation, motion or color.

It constitutes one of the most important properties of the human visual system (HVS) with numerous applications in digital imaging applications including content-aware video coding, segmentation or image resizing [8][9]. In order to model human visual attention, several visual saliency models have been recently proposed in the literature [10][11][12]. Generally, these models allow computing a so-called visual saliency map as a topographically arranged map that represents visually salient parts, also called regions of interest (ROI), of a visual scene. Among the different existing saliency models, the one proposed by Itti et al. [7][13] is the most popular.

The Itti algorithm exploits three low-level semantic features of an image: color, orientation and intensity. These features are extracted from the image to establish feature maps. Finally, the saliency map is computed from these feature maps after normalization and pooling.

In [3], the authors propose a Graph-Based Visual Saliency (GBVS) model which improves the model developed by Itti et al. The GBVS model relies on a fully connected graph between feature maps at multiple spatial scales. It is shown that the GBVS model outperforms the Itti model in predicting human visual attention while viewing natural images.

3 Saliency-based Deinterlacing

The flow chart of the proposed algorithm is depicted in Figure 1. As the field interpolation model depends on the saliency map, the first step of our algorithm is given by the computation of the spatial saliency of the current field to be deinterlaced, using the graph-based visual saliency model proposed in [3]. The obtained saliency map, denoted in the following by S (i.e. depicted in Figure 2) and consisting of gray values $S(i,j) \in \{0..255\}$ will trigger, along with the texture type, the interpolation used for the current field. Equally, a Canny edge detector is applied on the current field and the edges mask C is obtained.

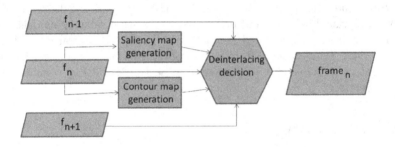

Fig. 1. Flow-chart diagram of proposed deinterlacing algorithm.

Further, the current field is partitioned into blocks of fixed size B^2, each block being categorized depending on its belonging to the salient region, as follows: the block b_n of size B^2 is said to be salient/important if the mean S_{b_n} of the entire collocated block s_n within the saliency map S, i.e.:

$$S_{b_n} = \Sigma_{i=1}^{B}\Sigma_{j=1}^{B}s_n(i,j)/B^2 \qquad (1)$$

is higher than a given threshold T_s; otherwise the block is classified as smooth.

Also, for each block b_n belonging to the current field f_n, its number of edges is derived as in eq.(2), by counting the amount of pixels on contours in the collocated block c_n in the mask field C, obtained with the Canny filter:

$$CE_{b_n} = \Sigma_{i=1}^{B}\Sigma_{j=1}^{B}c_n(i,j) \qquad (2)$$

where, CE_{b_n} is the number of identified edges in block b_n. The block b_n is classified as highly textured if CE_{b_n} is significant with respect to the blocksize B^2, i.e.:

(a)

(b)

Fig. 2. Saliency map obtained for (a) 10th frame of "Foreman" sequence, (b) 21st frame of "Salesman" sequence.

$$CE_{b_n} > T_b \qquad (3)$$

(T_b is a threshold depending on B^2), or smooth, if eq.(3) does not hold.

If the block b_n belongs to a salient region and its number of contours is significant (as in eq. (3)) optical flow based motion estimation is implemented, otherwise we use block-based estimation.

If the block b_n is determined as not belonging to a salient region, simple spatial 5-tap edge-line averaging techniques (Figure 3) are used to obtain the deinterlaced block $\hat{b}_n(i, j)$, i.e:

$$\hat{b}_n(i, j) = \frac{b_n(i - 1, j + x_0) + b_n(i + 1, j - x_0)}{2}, \qquad (4)$$

where the exact value of x_0 is given by the minimization:

$$|b_n(i-1, j+x_0) - b_n(i+1, j-x_0)| = \min_{x_0 \in \{-2,-1,0,1,2\}} |b_n(i-1, j+x_0) - b_n(i+1, j-x_0)|. \quad (5)$$

For the salient blocks, the motion vectors (MV) are obtained on the backward and forward directions for the current field, and applying either OF-based estimation proposed by Liu in [5], or simple block-based ME.

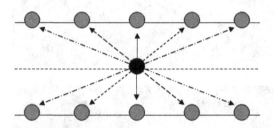

Fig. 3. 5-Tap Edge Line Average (ELA): the five interpolation directions are represented with different dashed lines.

We assume that the motion trajectory is linear, so the obtained forward motion vectors (MVs) are split into backward (MVB) and forward (MVF) motion vector fields for the current field f_n. As a block in f_n could have zero or more than one MVs passing through, the corresponding MV_n for the block $b_n \in f_n$ are obtained by the minimization of the Euclidean distance between b_n's center, $(y_{n,0}, x_{n,0})$, and the passing vectors MVs. In our minimization, we consider only the MVs obtained for the blocks in the neighborhood of the collocated block b_{n-1} in the left field f_{n-1} (thus a total of nine MVs, obtained for b_{n-1} and the blocks adjacent to $b_{n-1} \in f_{n-1}$, as these MVs are supposed to be the most correlated to the one in the current block, e.g., belonging to the same motion object).

If the motion vector MV corresponding to the collocated block $b_{n-1} \in f_{n-1}$ lies on the line:

$$\frac{y - y_{n-1,0}}{MV_y} = \frac{x - x_{n-1,0}}{MV_x} \quad (6)$$

where $(y_{n-1,0}, x_{n-1,0})$ is the center of b_{n-1} and MV_x, respectively MV_y, measures the displacement along the x, respectively y axis, the distances from the center $(y_{n,0}, x_{n,0})$ of the current block b_n to the MVs lines are obtained as:

$$D_{k \in \{1,...,9\}} = \frac{|MV_{k,x} y_{n,0} - MV_{k,y} x_{n,0} + MV_{k,y} x_{n-1,k} - MV_{k,x} y_{n-1,k}|}{\sqrt{MV_{k,x}^2 + MV_{k,y}^2}}. \quad (7)$$

MV^n is the closest motion vector to the current block b_n, if its corresponding distance to the center of b_n, $(y_{n,0}, x_{n,0})$, is minimal, i.e.: $D_n = min(D_{k \in \{1,...,9\}})$.

Hence, MV^n is generated for each block, containing the motion-estimation in the x and y directions for every pixel.

The forward and backward MVs for each block are obtained as:

$$MV_B^n = \frac{-MV^n}{2}, \quad MV_F^n = \frac{MV^n}{2}.$$ (8)

The backward prediction of b_n, denoted by $\mathcal{F}_{MV_B}^n$, is obtained as:

$$\mathcal{F}_{MV_B}^n(i,j) = b_{n-1}(i + MV_{By}^n, j + MV_{Bx}^n),$$ (9)

and the forward prediction of the current block, $\mathcal{F}_{MV_F}^n$, is obtained as:

$$\mathcal{F}_{MV_F}^n(i,j) = b_{n+1}(i + MV_{Fy}^n, j + MV_{Fx}^n),$$ (10)

The motion compensated block $\hat{b}^n$ to be further used for the deinterlacing of b_n is obtained as average of the backward, $\mathcal{F}_{MV_B}^n$, and forward, $\mathcal{F}_{MV_F}^n$, predictions:

$$\hat{b}^n(i,j) = \frac{\mathcal{F}_{MV_B}^n(i,j) + \mathcal{F}_{MV_B}^n(i,j)}{2}.$$ (11)

Finally, the deinterlaced block is found in a saliency-based motion-compensated manner, as:

$$\hat{b}_n(i,j) = \frac{b_n(i-1, j+x_0) + b_n(i+1, j-x_0) + s_n(i,j)\hat{b}^n(i,j)}{s_n(i,j) + 2},$$ (12)

s_n being the corresponding saliency value within the saliency map S, acting as a weight for the motion compensated interpolation, and x_0 is obtained by the edge-line minimization in (5).

4   Experimental Results

To objectively and comprehensively present the performance of the proposed deinterlacing approach, our method has been tested on several CIF-352 × 288 ("Foreman", "Hall", "Mobile", "Stefan" and "News") and QCIF-176×144 ("Carphone" and "Salesman") video sequences, which have been chosen for their different texture content and motion dynamics.

The selected video sequences were originally in progressive format. In order to generate interlaced content, the even lines of the even frames and the odd lines of the odd frames were removed, as shown in Figure 4. This way, objective quality measurements could be done, using the original sequences - progressive frames - as references.

In our experimental framework we have used $8 \times 8(B = 8)$ pixel blocks for a $16 \times 16(S = 16)$ search motion estimation window, for the salient blocks b_n having a small $CE_{b_n} < T_b$ number of contours. The parameter T_s for saliency detector was set up to 20, and the edge threshold T_b to 32 (eg, at least half of the block pixels are situated on contours).

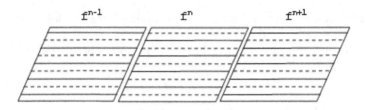

Fig. 4. Progressive to interlaced format frame conversion, by removing the dashed lines.

The tests were run on 50 frames for each sequence. The deinterlacing performance of our method is presented in terms of peak signal to-noise ratio (PSNR) computed on the luminance component. The efficiency of our proposed method -denoted in the followings by SGAD - is compared in Table I to Vertical Average (VA), Edge Line Average (ELA), Temporal Field Average (TFA), Adaptive Motion Estimation (AME) and Motion-Compensated Deinterlacing (MCD), which are the most common implementations in deinterlacing systems. Moreover, the proposed algorithm is compared to the work in [14], denoted by EPMC, [15] denoted by SMCD and the methods proposed in [16], [17] and [18] (these latter results are reported as in the corresponding references, NC denoting the non-communicated ones).

For visually showing the results of the proposed method, two deinterlaced frames are illustrated in Figure 5.

Table 1. PSNR RESULTS

	Foreman	Hall	Mobile	Stefan	News	Carphone	Salesman
VA	32.15	28.26	25.38	27.30	34.64	32.17	31.52
ELA	33.14	30.74	23.47	26.04	32.19	32.33	30.51
TFA	34.08	37.47	27.96	26.83	41.06	37.39	45.22
AME	33.19	27.27	20.95	23.84	27.36	29.63	28.24
MCD	35.42	34.23	25.26	27.32	35.49	33.55	33.16
$EPMC(S1)$	37.09	39.27	31.54	30.02	41.63	37.53	45.61
$EPMC(S2)$	37.18	39.08	30.56	30.11	39.44	37.55	42.28
[16]	33.77	NC	27.66	28.79	NC	NC	NC
[17]	NC	NC	NC	24.59	NC	NC	NC
[18]	33.93	38.79	24.67	26.88	NC	NC	NC
$SMCD(S1)$	37.52	39.71	30.41	31.77	41.85	37.59	45.95
$SMCD(S2)$	37.63	39.86	30.58	31.82	42.00	37.74	45.09
$SGAD$	**39.07**	**43.86**	**37.54**	**34.23**	**44.35**	**40.33**	**50.70**

As it can be seen in the presented results, our proposed method has an average PSNR gain of $\approx 4.5dBs$ with respect to a wild range of deinterlacers. Our framework has been implemented in Matlab (8.0.0.783 (R2012b)) and the

(a)

(b)

Fig. 5. Deinterlacing result for the (a) 10th frame of "Foreman" sequence, (b) 21th frame of "Salesman" sequence.

tests have been realized on a quad-core Intel-PC@4GHz. Due to the independent block-based processing, the proposed deinterlacing approach is prone to distributed/parallel implementation, thus highly reducing the computation time obtained with a sequential implementation. Moreover, as the proposed algorithm adapts the motion estimation in function of region's saliency, due to our used threshold T_s for motion computation, only $\approx 1/3$ of field regions are motion processed (as it can be seen in Figure 2). The parametrization allows thus to drastically decrease the complexity attached to motion-compensated schemes, by preserving its advantages where the user attention is focused.

5   Conclusion

In this paper, a spatial saliency-guided motion-compensated method for video deinterlacing is proposed. Our approach is an efficient deinterlacing tool, being able to adapt the interpolation method depending both of region of interest and its texture content. Experiments show that the proposed algorithm generates high quality results, having more than 4.5dBs PSNR gain, in average, compared to other deinterlacing approaches. Furthermore, the proposed method acknowledges the possibility of improving image quality and simultaneously reducing execution time, based on the saliency map.

References

1. Haan, G.D., Bellers, E.B.: Deinterlacing - An overview. Proceedings of the IEEE **86**(9), 1839–1857 (1998)
2. Atkins, C.B.: Optical image scaling using pixel classification. In: International Conference on Image Processing (2001)
3. Harel, J., Koch, C., Perona, P.: Graph-based visual saliency. In: Advances in Neural Information Processing Systems, pp. 545–552 (2006)
4. Liu, C.: Beyond Pixels: Exploring new Representations and applications for Motion Analysis, Doctoral Thesis, MIT, May 2009
5. Horn, B.K.P., Schunck, B.G.: Determining optical flow. Artificial Intelligence **17**, 185–203 (1981)
6. Trocan, M., Mikovicova, B., Zhanguzin, D.: An Adaptive Motion Compensated Approach for Video Deinterlacing. Multimedia Tools and Applications **61**(3), 819–837 (2011)
7. Itti, L., Koch, C., Niebur, E.: A Model of Saliency-based Visual Attention for Rapid Scene Analysis. IEEE Trans. on PAMI **20**(11), 1254–1259 (1998)
8. Itti, L.: Automatic Foveation for Video Compression Using a Neurobiological Model of Visual Attention. IEEE Trans. on Image Processing **13**(10), 1304–1318 (2004)
9. Rahtu, E., Kannala, J., Salo, M., Heikkilä, J.: Segmenting salient objects from images and videos. In: Daniilidis, K., Maragos, P., Paragios, N. (eds.) ECCV 2010, Part V. LNCS, vol. 6315, pp. 366–379. Springer, Heidelberg (2010)
10. Zhang, L., Tong, M., et al.: SUN: a Bayesian Framework for Saliency using Natural Statistics. Journal of Vision **9**(7), 1–20 (2008)
11. Lu, S., Lim, J.-H.: Saliency modeling from image histograms. In: Fitzgibbon, A., Lazebnik, S., Perona, P., Sato, Y., Schmid, C. (eds.) ECCV 2012, Part VII. LNCS, vol. 7578, pp. 321–332. Springer, Heidelberg (2012)
12. Seo, H.-J., Milanfar, P.: Static and Space-Time Visual Saliency Detection by Self-Resemblance. Journal of Vision **9**(12), 1–12 (2009)
13. Itti, L., Koch, C.: Computational Modeling of Visual Attention. Nature Reviews Neuroscience **2**(3), 194–203 (2001)
14. Zhanguzin, D., Trocan, M., Mikovicova, B.: An edge-preserving motion-compensated approach for video deinterlacing. In: IEEE/IET/BCS3rd International Workshop on Future Multimedia Networking, June 2010
15. Trocan, M., Mikovicova, B.: Smooth motion compensated video deinterlacing. In: 7th International Symposium on Image and Signal Processing and Analysis (ISPA), September 2011

16. Chen, Y., Tai, S.: True motion-compensated de-interlacing algorithm. IEEE Transactions on Circuits and Systems for Video Technology **19**(10), 1489–1498 (2009)
17. Wang, S.-B., Chang, T.-S.: Adaptive de-interlacing with robust overlapped block motion compensation. IEEE Transactions on Circuits and Systems for Video Technology **18**(10), 1437–1440 (2008)
18. Lee, G.G., Wang, M.J., Li, H.T., Lin, H.Y.: A motion-adaptive deinterlacer via hybrid motion detection and edge-pattern recognition. EURASIP Journal on Image and Video Processing (2008)

Towards a System for Cloud Service Discovery and Composition Based on Ontology

Rawand Guerfel[✉], Zohra Sbaï, and Rahma Ben Ayed

École Nationale d'Ingénieurs de Tunis, Université de Tunis El Manar,
BP. 37, Le Belvédère, 1002 Tunis, Tunisia
{Rawand.Guerfel,zohra.sbai,rahma.benayed}@enit.rnu.tn

Abstract. Cloud computing has become a widely used concept. It is characterized by its elasticity, its on demand services and the unlimited resources provided to end users. However, the increase of services number offered by different providers causes sometimes ambiguity problems. So, ontology must be used to relate between different services in order to facilitate their access. Many languages expressing ontology have been proposed namely RDF, OWL, etc. To access them, queries should be expressed using SPARQL. However, generally, users are not familiar with this query language. Therefore, there is a need to an interface allowing them to express their requirements without any expertise in SPARQL language. Besides, given the increase of complex service requirements, it is sometimes necessary to compose services to meet final user requirements.

This paper gives related works on Cloud service discovery based on ontology and composition and presents a system ensuring these procedures.

Keywords: Cloud computing · Services discovery · Cloud ontology · OWL · Cloud composition

1 Introduction

Nowadays, IT Services are facing many challenges namely: aging data centers, storage growth, application explosion, cost of ownership, globalisation and acquisitions. So, new revolutionary IT model is required which is Cloud computing.

Indeed, Cloud computing is an internet-based computing, where shared resources, softwares and informations are provided to computer and other devices on demand. It is characterised by its: as a service model in which everything is provided as a service, pay-per use and measured service, rapid elasticity, abstraction of implementation and underlying infrastructure to users and its shared resources. These characteristics are all applicable to the Cloud service model. This model, according to NIST [8], is composed of three main layers which are: 1)SaaS (Software as A Service) where applications are build on on-demand subscription basics, 2)PaaS (Platform As A Service)where users can deploy their applications using programming languages and tools supported by the Cloud

M. Núñez et al. (Eds.): ICCCI 2015, Part II, LNCS 9330, pp. 34–43, 2015.
DOI: 10.1007/978-3-319-24306-1_4

provider, and 3)IaaS (Infrastructure As A Service) in which providers have control over their data,their middleware,their DBMS and their deployed applications. Indeed, depending on service's type, providers expose and publish their services over the internet in these three layers.

Nowadays, the use of Cloud services is more and more evolving so that the number of Cloud providers is also increasing. Indeed, we distinguish numerous different Cloud providers namely Microsoft Windows Azure, Google AppEngine, Salesforce Force.com, Amazon EC2, etc. Each one of them offers its proper services and the user may require for more than one service which are not necessarily offered by the same Cloud provider. Thus, in this case, we should search for required services in different Cloud providers and furnish it to the final user. Here, we are dealing with the Cloud service discovery mechanism.

To make this search rapid, precise and more performant, and since services are offered by different Clouds, we need to establish relations between these services. This led us to the use of Cloud ontology.

In fact, ontology is a data model which represents a knowledge as a set of concepts and relationships between these concepts within a domain. It's expressed using RDF [2], OWL [1], RDFS [10], etc, which are conceptual languages for representing information. Ontologies data are registered in a knowledge base that exchange informations using SPARQL inference engine [11]. In fact, SPARQL allows to query RDF/OWL data and models and to define globally unambiguous queries.

So, ontology requirements are executed using SPARQL query language, which is a not a trivial task for the final user. Therefore, users should be offered by an interface allowing them to express their requirements without having any knowledge of SPARQL query language. These requirements will then be translated into SPARQL query to search for the appropriate service in Cloud ontology domain.

Let's also note that these requirements are becoming more and more complex so that, sometimes, a single Cloud service cannot meet required tasks. To do this, it needs to be combined and to communicate with other Cloud services that are not necessarily offered by the same Cloud provider. Here, we are dealing with Cloud service composition mechanism. This mechanism provides a new service, named composite service, that consists of various and elementary other services. It is created in order to process no resolved complex tasks.

In this context, we propose a system having as purpose the discovery of different services provided by various Cloud providers based on ontology. To do so, it offers to the user an easy to use and simple interface to express his request without any requirement of SPARQL expertise. Besides, this system allows users to require complex services and tries to resolve it by composing them from different Cloud providers.

This paper is organized as follows. We start in section 2 by defining the general context of this paper. Then, we move to section 3 to explain and detail the proposed system. In the section 4, we cite some related works and compare them to the proposed system. Finally, section 5 concludes the paper and presents some perspectives.

2   General Context

2.1   Cloud Service Discovery Based on Ontology

Cloud computing resolves the problem of time and money that organizations were spending on their servers installation. Indeed, with the Cloud concept, they can just connect to the internet, use different available ressources and pay for what they used.

Cloud architecture consists of three main actors which are:

- Provider: offers three type of services for the three existing layers which are IaaS, PaaS and SaaS.
- Developer: Uses services offered by the IaaS layer to provide platforms to the PaaS layer and further uses the services offered by the PaaS layer to provide softwares to the SaaS layer.
- User: consumes services from the SaaS layer.

Nowadays, different Cloud providers exist. Each one of them offers services for a specific layer. So, in order to select the appropriate one for the final user, it is necessary to use a research mechanism.

Given the increase of Cloud provider number, there is a need to organise their services and express relationships among them to facilitate their access and search. It is in this context that we use Cloud ontology concept.

Indeed, Cloud ontology was exploited in many works to attend different purposes that can be divided in four main points [3]:

- Description of Cloud services and resources: Many Cloud providers used Cloud ontology to define services and to give to their descriptions an understandable semantic by consumers.
- Insurance of Cloud security: It's used to define and even improve Cloud computing security.
- Insurance of Cloud interoperability: Interoperability allows customers to access, using a commun API, to different services which are not necessarily placed in the same Cloud, but can be provided by different ones.
- Discovery and selection of Cloud services: Service selection is not an arbitrary task. It should be processed after performing some study and some comparison between the different services in order to select the right one to the user. To do so, providers used ontology so they can define relations between different services and select the appropriate one.

In this paper, we use ontology to make an intelligent Cloud services discovery and selection and to ensure the interoperability among services offered by more than one Cloud provider.

More precisely, we propose to use a domain ontology in which all services of different Cloud providers are saved and related to each other by some properties to facilitate their access.

2.2   Cloud Service Composition

The composition of business processes is an already used concept during several years in SOA architecture [7] which is implemented using Web services. In fact, user requests are sometimes complex and not obvious. To answer them, we have to compose two or several services to produce a virtual one, called composite service. This issue is very used in SOA architecture and approaches have been proposed to ensure Web service composition namely: Workflow based approach, XML based approach and ontology based approach.

Similarly, user demands for Cloud services are sometimes a bit complex, that only one service can not meet this demand only if it is combined with at least one other service. Note that Cloud services do not necessarily belong to the same Cloud. Each one of them may belong to his own Cloud and should be combined and composed to give birth to an available and composite service for the end user.

Based on lecture of some research works, we classify the use of Cloud service composition into three main types:

- Composition based on QoS parameter: Its general principle is the study of QoS parameter of each Cloud service (cost, availability, reliability, etc.) and their composition according to their parameters to provide a composite one meeting user demand.
- Composition of IT services: We quote here the work of Tran Vu Pham et al. [12] who proposed a framework allowing an on-demand composition and deployment of HPC applications hosted on HPC Clouds.
- Composition of Web applications hosted in the Cloud: its aim is the management of the large number of services provided by organizations as well as the large number of organizations providing these services. In fact, Cloud allows the storage of these services and their access by anyone, anytime and anywhere. But, these services need sometimes to be composed to meet user requirements.

Let's note that some works were oriented to the combination of SOA and Cloud computing architectures. As an example, we cite here our previous work [13] in which we combined these architectures in order to ensure composition of Cloud services. More precisely, we proposed to include SOA architecture between PaaS and SaaS layers so that services can be composed and provided to the final user.

3   Proposed System

This section details the proposed system that ensures the discovery of different Cloud services based on ontology and their composition. This system, explained by figure 1, consists of five main modules which are: User Interface, Translator module,Ontology domain, Research into service ontology component and Composition component. The functionalities of these components are detailed in the following subsections.

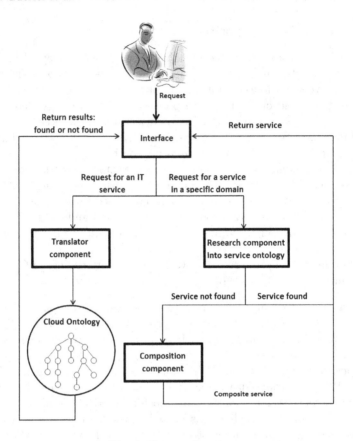

Fig. 1. Proposed system

3.1   User Interface

This interface helps the user to write his request without requiring him to have knowledge of SPARQL. Indeed, the interface is primarily intended for two types of users who are: the user and the developer. The developer can ask for an IT Cloud service (SaaS, PaaS, IaaS) and the user can query for a software service or for services in a certain domain hosted in the Cloud. So, depending on these two types of queries, the interface will react and respond to the user request. More precisely, if:

- the user of this system asks for an IT service. So, the interface will offer him (provider or a simple user)three fields to fulfill which are: concept field, object or data property field and property value field. He can also specify the type of the required service (SaaS, PaaS or IaaS). Indeed, the developer can ask for
  - a concept without requiring any property. So, in this case, he has just to fulfill in the concept field the name of the specific software, platform or infrastructure.

- ○ a concept having a certain property without specifying the value of this property. In this case, he should fulfill the property field.
- ○ a concept having a certain property having a specific value. In this case, he should fulfill the property and value fields.
- • user query a specific service domain. In this case, we offer to the user two fields allowing him to specify the input and the output of service. Note that we are dealing with services hosted in different Clouds.

So, depending on the type of the required service, the system either send it to the "translator component", or to "Research into service ontology component", which are both of them detailed in the next subsections.

3.2 Translator Component

This component intervenes once the user writes his request. It translates his query into SPARQL query to execute it on the defined ontology. Depending on the fields fulfilled by the user, it writes the appropriate SPARQL query. For example, if the user asks for services having as object property: "hasCPU" and as value "2.0", then, the system automatically saves data written in both property and value fields and generates this query:

```
select ?restriction
where{
        ?restriction :hasCPU "2.0"‘xsd:double.
     }
```

3.3 Cloud Domain Ontology

This component is responsible for the classification of services provided by different Clouds in a domain ontology. Doing this, we can use services offered by different Clouds without facing interoperability problems.

Note that the ontology is defined with three main constituants which are: concepts, individuals and properties which can be object or data properties. All services offered by different Cloud providers, are classified in this ontology as individuals, which are the final part of a class. These individuals are related with each other by object and data properties.

To do so, we used Protégé 4.2 ontology editor. Indeed, Protégé is an editor that allows to construct an ontology for a given domain.

So, SPARQL queries of users demand are all processed on this ontology. Once done, we can have two types of results: either the concept is found and the system display it to the final user or the concept is not found and the system displays a message to the final user confirming him the non-existence of the service.

3.4 Research into Service Ontology Component

This component occurs when the user requires a Web service hosted in the Cloud. Indeed, nowadays, numerous Web services are hosted in the Cloud for many reasons namely: its rapid elasticity, its on-demand self services, its resource pooling and its multi-tenancy.

So, many Web services offered by different Cloud providers are hosted in the Cloud and their discovery become a necessary step to be processed.

To make this discovery precise, we also use ontology mechanism. We offer an interface to the final user that allows him to write the input and the output of a service. These data are enrolled in the system and searched then into the ontology that we defined using OWL and given by figure 2.

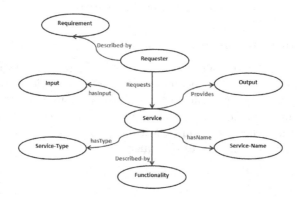

Fig. 2. Service Cloud ontology

Indeed, this component compares the input and the output values written by the user to the input and output parameters of each service. If there is a service which its parameters correspond to the ones written by the user, then, it will be displayed to him, else, the composition component intervenes.

3.5   Composition Component

Requirements are sometimes so complex that they can not be achieved only if we combine two or more services. Indeed, the inter-operation and the communication of different Cloud services will allow the birth of new Cloud service, which we name composed Cloud service. This one can meet user requirements that are becoming more and more complex. More precisely, Cloud services will coexist, support, complement and be combined as a virtual Cloud service.

This composition component will occur if the "Research into service ontology component" did not find the required service. Its function is described as bellows. It compares the input of the user request to the input parameters of different available services, and the output of the user requirement to the output parameters of different available services. Here, three cases can occur:

- The input of the user requirement exists in the ontology as an input parameter to a Cloud service and the output one does not exist. In this case, the required service does not exist and the system does return a result to the final user.

- The output of the user requirement exists in the ontology as an output parameter to a Cloud service and the input one does not exist. In this case, the required service does not exist and the system does return a result to the final user.
- The input of the user requirement exists in the ontology as an input parameter to a Cloud service S1 and the output of the user requirement exists as an output parameter to another Cloud service S2. In this case, we move to compare the output parameter of S1 to the input one of S2. Here, two others cases can occur:
  - These two parameters are the same. Thus, the system composes these two services and provides it as a single service to the final user.
  - These two parameters are not the same. Thus, the system compares these two parameters to the ones of all other available services. Once there is a connection between the input and the output of each service, beginning with the input given by the user coming to his output, then the system composes all these services and provides them as a single virtual service to the end user. Otherwise, if the services could not be connected based on their inputs and outputs, then, no results will be returned to the user.

To better understand the role of this component, we propose to use a simple and short case study that requires the intervention of two services and their composition in order to meet the user request.

Let's take the following scenario. Two Cloud services are enrolled in the ontology which are: S1="Euro Money conversion" and S2="Coffee". S1 has as input parameter "Euro" and as output parameter "dollar". S2 has as input parameter "dollar" and as output parameters: "Capuchin" and "Coffee". A user of the system may want a Coffee and pay 1Euro. So, he writes "1Euro" as an input parameter and "Coffee" as an output one. This demand requires the intervention of the composition component because there is no service which its input and output parameter values correspond to the values entered by the user. However, the values of these two parameters exist in the ontology but belong to two different services: S1 and S2. Then, the system proceeds to the next step which is the comparison between the output parameter of S1 and the input parameter of S2. In this scenario, they have the same value which is:"dollar". So, the system composes these two services, converts the euro to the dollar using S1 and sends the result to the S2 that, one receiving the dollar amount, provides the coffee to the user.

4   Related Works

Cloud service discovery mechanism based on ontology became an interesting topic. Many algorithms , systems and approaches were proposed to ensure this task. We will try to quote the most recent ones in this section.

Miranda Zhang et al. [15] proposed an ontology, named CoCoOn, specified for the first Cloud layer, IaaS. They also proposed a system, CloudRecommender having as a main goal the discovery for different resources offered by this layer. The discovery in this work is limited to this IaaS layer.

We quote also the work of Francesco Moscato et al. [9] who proposed an API for ensuring the interoperability of multiple services offered by different Clouds. This will facilitate and make the discovery,based on ontology, more precise and intelligent.

We can not finish this part without citing works of Sim et al. [5,6]. Indeed, they proposed Cloudle, an engine for cloud service discovery based on ontology. To do so, they defined their proper ontology that they called CO-1 and CO-2. The engine searches, basing on ontology, for the appropriate services and ranks them to the user according to their utilities, pricing and availability constraints to the final user.

We note that these works are limited in the discovery of Cloud services. They didn't propose their composition in case of their non-existence. However, our proposed system allows the user to require for complex services. Moreover, it treats even services hosted in the Cloud in a certain domain and not just IT services.

This does not prevent the existence of some works which focus on this issue.

For example, we quote in this part the work of Jiehan Zhou et al. [14] who proposed a layer, CM4SC, located between PaaS and SaaS layers to ensure the composition process. Indeed, this layer is designed using a platform offered by Windows Azure cloud provider and contains all web services that will be discovered and composed in complex requirement cases.

The composition of services relied on one Cloud provider whereas in our work, we rely on different Cloud providers to offer more choices for the user to select the appropriate service. Besides, our discovery is based on ontology.

Indeed, our proposition does not stop at the discovery process or composition one, but combines them to ensure a complete mechanism that has as main goal the answering of user requirements, regardless of their complexities, based on ontology.

5   Conclusion

The contribution of this paper is the proposition of a system having as main purpose these two main points: 1) discovery of Cloud services based on ontology and 2) their composition in case of complex requirements. This system is characterized by its easy to use interface which can be executed by anyone without requiring any expertise from him. Discovery mechanism is based on ontology already defined and designed with protégé editor ontology. The system resolves even complex requirements by composing different available services to provide them in the final step as a virtual service.

As a future work, we propose to add similarity research based on ontology to enhance discovery mechanism. Besides, we will formally verify services, involved in composition process, before providing them to the final user. We can check for example their compatibility and soundness using our tool D&A4WSC [4]. This highlights the performance of this final provided service.

References

1. Antoniou, G., van Harmelen, F.: Web ontology language: Owl. In: Staab, S., Studer, R. (eds.) Handbook on Ontologies in Information Systems, pp. 76–92. Springer-Verlag (2003)
2. Candan, K.S., Liu, H., Suvarna, R.: Resource description framework: Metadata and its applications. SIGKDD Explor. Newsl. **3**(1), 6–19 (2001)
3. Darko, A., Neven, V., Jurica, S.: Cloud computing ontologies: A systematic review, pp. 9–14, April 2012
4. Guerfel, R., Sbaï, Z.: D&A4WSC as a design and analysis framework of web services composition, pp. 337–338 (2014)
5. Han, T., Sim, K.M.: An ontology-enhanced cloud service discovery system. In: Proceedings of the International MultiConference of Engineers and Computer Scientists, March 2010
6. Kang, J., Sim, K.M.: Cloudle : an agent-based cloud search engine that consults a cloud ontology. In: 2011 International Conference on System Science and Engineering (ICSSE), Macao, June 2011, pp. 276–281. IEEE (2011)
7. Lewis, G.: Getting started with service oriented architecture (soa) terminology. Software Engineering Institute (2010)
8. Mell, P., Grance, T.: The nist definition of cloud computing. Technical Report 800–145, National Institute of Standards and Technology (NIST), Gaithersburg, MD, September 2011
9. Moscato, F., Aversa, R., Di Martino, B., Fortis, T.-F., Munteanu, V.: An analysis of mosaic ontology for cloud resources annotation. In: FedCSIS 2011, pp. 973–980 (2011)
10. Nejdl, W., Wolpers, M., Capelle, C., Wissensverarbeitung, R., Hannover, U.: The rdf schema specification revisited. In: Modellierung 2000, p. 2000 (2000)
11. Pérez, J., Arenas, M., Gutierrez, C.: Semantics and complexity of sparql. ACM Trans. Database Syst. **34**(3), 16:1–16:45 (2009)
12. Pham, T.V., Jamjoom, H., Jordan, K., Shae, Z.-Y.: A service composition framework for market-oriented high performance computing cloud. In: Proceedings of the 19th ACM International Symposium on High Performance Distributed Computing, HPDC 2010, New York, NY, USA, 2010, pp. 284–287. ACM (2010)
13. Rawand, G., Zohra, S., Rahma, B.A.: On service composition in cloud computing: a survey and an ongoing architecture. In: Proceedings 6th Internatinal Conference on Cloud Computing Technology and Science, Singapore (2014)
14. Ylianttila, M., Riekki, J., Zhou, J., Athukorala, K., Gilman, E.: Cloud architecture for dynamic service composition. Int. J. Grid High Perform. Comput. **4**(2), 17–31 (2012)
15. Zhang, M., Ranjan, R., Haller, A., Georgakopoulos, D., Menzel, M., Nepal, S.: An ontology-based system for cloud infrastructure services' discovery. In: CollaborateCom, pp. 524–530. IEEE (2012)

Modeling the Impact of Visual Components on Verbal Communication in Online Advertising

Jarosław Jankowski[1,2(✉)], Jarosław Wątróbski[1], and Paweł Ziemba[3]

[1] West Pomeranian University of Technology, Faculty of Computer Science and Information Technology, Żołnierska 49, 71-210, Szczecin, Poland
{jjankowski,jwatrobski}@wi.zut.edu.pl
[2] Department of Computational Intelligence, Wrocław University of Technology, Wybrzeże Wyspiańskiego 27, 50-370, Wrocław, Poland
[3] The Jacob of Paradyż University of Applied Sciences in Gorzów Wielkopolski, Teatralna 25, 66-400, Gorzów Wielkopolski, Poland
pziemba@pwsz.pl

Abstract. Together with the development of electronic marketing, the key factors that influence the effectiveness of interactive media were identified in the earlier research. This article presents approach which makes possible to determine the relationship between the textual and visual components and their impact on results of the campaign. The structure of interactive advertising unit with adjustable visual and verbal influence is presented followed by results from the field experiment performed with the usage of proposed approach.

Keywords: Online advertising · Human-computer interaction · Web design

1 Introduction

The growing importance of digital marketing influences the need for environmental studies of the interactive phenomena that occur in online marketing. Research in this area is focused mainly on increasing the effectiveness of advertising campaigns, modelling their impact on audience and analyzing the effects [9]. The basis of many solutions in this area is marketing engineering, which develops in conjunction with computer science and data processing algorithms the analytical systems towards adaptive solutions [13]. There is also integration of knowledge from fields such as sociology, psychology and social engineering and connecting them with analysis of user experience and different methods of website evaluation [16]. The areas of research in this field relate to modelling levels of perception and interaction with the impact of the marketing message on the recipient. Phenomenon's such as habituation or sensory adaptation, which reduce the impact of advertising as a result of the conscious and unconscious elimination of advertising content takes place [15]. This contributes to the need to find solutions with the ability to model the structures and content of media and forms of communication, providing results at a certain level. The article presents the assumptions of interactive advertising objects with varying levels of impact on

© Springer International Publishing Switzerland 2015
M. Núñez et al. (Eds.): ICCCI 2015, Part II, LNCS 9330, pp. 44–53, 2015.
DOI: 10.1007/978-3-319-24306-1_5

target users and possibilities of multidimensional measurements. The presented concept allows the selection of design options and analysis of the interaction of verbal and visual elements in terms of their impact on the obtained effects. Results from empirical study showed practical applications of presented solutions and the relations between visual and text elements within online advertisements.

2   Role of Visual Components in the Interactive Communication

The use of electronic media in marketing enables two-way communication using visual components with the customer and allows researchers to perform measurement and analysis in several areas [9]. The concept of interaction and the available definitions were created in different stages of development of electronic systems. S. Rafaeli defines a recursive interaction as an interpersonal communication in which the exchange of data or information that relates to an earlier stage of the communication process is variable and the roles of sender and recipient are identified [10]. C. Heeter lists six dimensions of interaction that identify the information functions and the complexity of the communication process [4]. Different areas of mapping similarities between interpersonal communication and interaction with the electronic system were analyzed by J. Steuer [11] with determined the level of possible changes to the content and form of an interactive environment. These approaches draw a clear distinction between the interaction with the system and the human communication through electronic systems. D. Hoffman and T.P. Novak combine earlier definitions and refer them to the Internet, where interaction can occur both with the system in a form of machine interactivity and personal interactivity [5]. Interaction elements together with visual elements are combined with marketing messages and are affecting the efficiency and consumer behavior. Communication implemented in the online environment and visual components can take different levels of interaction that affect the perception of the media. Research in this area refers to the impact of media implemented in the form of banner ads on the effects and the role they play in the brand recognition or awareness [13]. The results presented by L. Hairong and J.L. Bukovac indicate the cognitive basis of the relationship between ad format, size, graphics, animation elements and their impact on the direct effects [3]. Important role in the communication process play elements in the form of verbal messages and graphics. They generate stimuli and impact is increased when the media components stand out among the previously transmitted messages or parts of website. There is an analogy to the traditional media, for which the characteristics of the impact and increase the effectiveness by changing the intensity of interactions and visual communication were analyzed among others by A.E. Beattie and A.A. Mitchell [1]. Within the websites increase the level of interaction can occur in both the visual and the verbal communication together with cognitive processes [14]. For the visual elements can be analyzed the impact of different communication options, ranging from less invasive forms to a high level of vividness and distinction. Limited perception creates the need for increase the influence of persuasive messages. This includes animation elements which impact on the results analyzed in the first phase of the development of interactive advertising, including in the work of J.H. Ellsworth [2]. S. Sundar and others have studied online animated elements and determined their impact for similar static ads placed in print media [12]. Some of these aspects were addressed in our research based on fuzzy modelling and searching for compromise solutions using static aggregated data [6][8] and measuring the impact on web users with the use of difference of influence between repeated contacts with the website [7]. While complexity of online marketing content grows this requires the most common use of a combination of elements of verbal and visual impact. At the same time, the possibility of reducing the impact of verbal elements should be taken into account due to the introduction of graphic elements that absorb the attention of the recipient and help to reduce the significance of a text message in the decision-making process. The presence of many factors influencing the effectiveness of online advertising and dependencies between components, graphics, and text messages means that it is important to use analytical methods and tools to measure these phenomena and the use of acquired knowledge. The following research presented in this paper shows the concept of advertising object with the structure generated in an automated manner using different levels of impact on the recipient. This gives the ability to determine the design options that enable the integration of both information functions involving elements of text and visual impact in such a way that did not follow their mutual compensation.

3   Interactive Advertising Object with Varying Levels of Influence

A review of the literature reveals the complex factors that impact the effectiveness of interactive media. When designing advertising objects, the main goal of communication, as well as the levels of verbal and visual influence, must be chosen to achieve certain effects. The next part of the paper presents the concept of interactive component, which integrates the various elements of the impact in both verbal and visual for and generates ad unit structure in an automated manner with different levels of persuasion. This approach makes it possible to analyze the impact of verbal and graphic communication in a unified measurement environment and allows a comparison of the results obtained for different variants of transmission. The set of $R_i = \{G_i, T_i\}$ advertising components is designed to provide specific functionality and impact on the recipient and is divided into subset G_i with included graphic elements and T_i with text elements. The subsets G_i and T_i include graphic components $G_i = \{G_1, G_2, ..., G_n\}$ and text components $T_i = \{T_1, T_2, ..., T_m\}$ of advertising object through the generalized structure shown in Fig. 1.

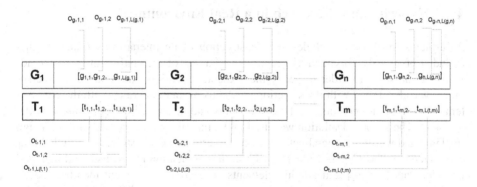

Fig. 1. Structure of the advertising object with varying levels of influence

Each item is determined by the number of options available in the case of graphic elements $G_i = \{g_{i,1}, g_{i,2}, ..., g_{i,Lg,i}\}$ and text elements $T_j = \{t_{j,1}, t_{j,2},..., t_{j,Lt,j}\}$, where $L_{g,i}$ determines the number of variants of a graphic element G_i, $L_{t,j}$ specifies the number of variants of the text element T_j. For each variant $g_{i,j}$ and $t_{i,j}$ levels of influence $o_{g,i,j}$ and $o_{t,i,j}$ of object are defined. The overall measure of the impact I_i of the object R_i consist of summarized L_G graphic elements and L_T text elements, expressed by the formula:

$$I_i = \sum_{j=1}^{L_G}\left(w_{g,i,j} * o_{g,i,j}\right) + \sum_{j=1}^{L_T}\left(w_{t,i,j} * o_{t,i,j}\right) \tag{1}$$

where $o_{g,i,j}$ represents the level of impact of a graphic element, $o_{t,i,j}$ describes the level of impact of a text element, according to $w_{g,i,j}$ and $w_{t,i,j}$ defined as a weights adopted for a given elements, which determines the strength of its impact on the remaining components. Thus constructed, an object can be used to test the system and determine the system model. In the process of exposure of an interactive object in the test user group request from the website is followed by selection of components and recording interactions. When generating a variant followed by selection of elements from the given selection function fs(E_i,R,n), which is responsible for the selection of an element E_i from the set R for call transfer n, the main objective is to replicate the system responses for different variants and to design and build a model of the system. The selection process takes place to generate an advertising object variant design and direct it to the user's browser. The measuring system collects data on individual exposures and outcomes achieved. Data is passed as a feed to the system model, where the provisioning model structures are in the feedback. In the following section, experimental research was conducted in a real environment, which aimed to verify the design and testing in relation to verbal and graphic elements of the media and their interactions.

4   Experimental Research in a Real Environment

A further step of work included the construction of an interactive component that included previously presented assumptions and a test of the configuration of the advertising message. In order to verify the presented solution experimental interactive object was developed with the selection mechanisms and integrated with the real system. Within the defined object, the components had different levels of impact on the recipient. The proposed solution was used in the module associated with the campaign implemented within the social network. The advertising object structure included four components of $T = \{T_1, T_2,\}$ $G= \{G_1, G_2\}$ with versification mechanism and included text elements $\{T_1, T_2\}$ and graphic elements $\{G_1, G_2\}$. T_1 element identifies seven versions of textual variants $\{t_{1,1}, t_{1,2}, t_{1,3}, t_{1,4}, t_{1,5}, t_{1,6}, t_{1,7}\}$ with different levels of interaction and integration of textual expressions with call for action messages like in this case *Click* or *Click now*. The text structure of indexes $t_{1,1}, t_{1,2}, t_{1,3}$ are pointed to the different characteristics of the social platform. G_1 element was carried out in the form of a graphic button in seven variants $\{g_{1,1}, g_{1,2}, g_{1,3}, g_{1,4}, g_{1,5}, g_{1,6}, g_{1,7}\}$. The first two variants $g_{1,1}$ and $g_{1,2}$ contained static objects without animation. The first showed less contrast with the background, while the second variant was more pronounced on the background. $g_{1,4}$ element contained animated text on a static background. $g_{1,5}$ variant contained animated elements with a higher level of impact. The most aggressive forms of animation were combined with elements of $g_{1,6}$ and $g_{1,7}$. Item G_2 included additional graphical information indicating the possibility of setting up a user account without costs. An animated version of the element was used for $g_{2,3}$ and the static version for $g_{2,2}$. For variant $g_{2,1}$ element G_2 was not visible. In the analyzed period component was displayed 249,149 times. Every possible combination of elements was shown 282 times on average. The message was generated for 27,338 unique visitors and 698 were registered during this period of interaction with measuring effectiveness using the click-through ratio. In the first step an ANOVA analysis was performed for all media options including the graphical elements, both animated and static, and verbal components. The analysis (Table 1) indicates the greatest impact on

Table 1. Analysis of data involving animated elements

Element	Effect	E	p	–95 %	+95 %	W
T_1	−0,010438	0,047150	0,824899	−0,103112	0,082235	−0,005219
T_2	−0,025659	0,037410	0,493154	−0,099188	0,047870	−0,012830
G_1	**0,099495**	**0,045404**	**0,028964**	**0,010254**	**0,188736**	**0,049747**
G_2	**0,078476**	**0,037475**	**0,036835**	**0,004820**	**0,152133**	**0,039238**
1L vs 2L	0,088441	0,057518	0,124877	−0,024610	0,201492	0,044220
1L vs 3L	0,054469	0,069385	0,432869	−0,081905	0,190843	0,027234
1L vs 4L	0,072525	0,057371	0,206861	−0,040236	0,185286	0,036262
2L vs 3L	−0,085040	0,055268	0,124615	−0,193668	0,023588	−0,042520
2L vs 4L	0,034075	0,045991	0,459152	−0,056319	0,124470	0,017038
3L vs 4L	0,043490	0,056396	0,441036	−0,067355	0,154335	0,021745

the aggregated results was represented by graphical element G_1 and G_2 element of levels of significance respectively $p(G_1) = 0.0289637$ and $p(G_2) = 0.0368348$. The results indicate the limited importance of the text message t_1 and t_2 when switching on animated graphics.

The next step was analyzed with the usage of response surface modelling used in the analysis of experimental results. The analysis shows the impact of Pareto effect level G_1 at the level of $P(G_1) = 2.19$, and the effect on the level of $P(G_2) = 2.09$. Positive values indicate an increasing change of the effects generated by the element on the level $G_1 = 1$ to $G_1 = 7$ and the G_1 element from $G_2 = 1$ to $G_2 = 3$. For the text element T_1 with value $P(T_1) = -0.22$, and for the element T_2 value $P(T_2) = -0.68$. The negative values reflect the direction of the impact element, which in this case occurs at the change from the seventh to the first variant of the element T_1, and the third to the first element of G_1. A small value indicates a minimal impact on the effects of text elements using different graphical components. In some situations, it may reduce the information function of the interface, because all the attention is focused on the recipient's graphic elements. The next step was to analyses the effects at different values of input parameters. The surface response, which represents the effect of changing the version of graphical element G_1 and G_2 for the effects of the variants of the text elements $T_2 = 7$ and $T_1 = 1$, is shown in Fig. 2. The analysis of the effects of changes in the value of $T_2 = 7$ and $T_1 = 2$ is shown in Fig. 3.

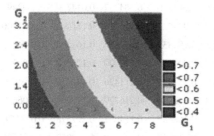 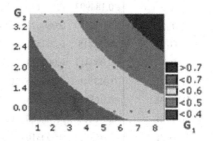

Fig. 2. Dependence of the effects of G_1 and G_2 for a constant value of $T_2 = 7$ and $T_1 = 1$

Fig. 3. Dependence of the effects of the values of G_1 and G_2 for the constant $T_2 = 7$ and $T_1 = 2$

These graphs show that when changing graphics variations, text elements do not substantially affect the results change. In this case, verbal communication has little effect on the call interaction, and its importance to the process of communication is negligible. This is confirmed by the analysis carried out for different levels of volatility text elements, the results of which are shown in Fig. 4 and 5. Changes to T_1 textual variants of the first embodiment to the seventh have virtually no influence on the obtained results. The analysis shows that the occurrence of animated graphical elements within media reduces the importance of the impact of verbal communication. Despite the difference in potency and text elements of persuasion used, the obtained results were not affected.

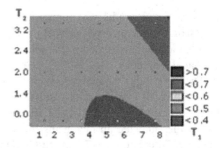

Fig. 4. Dependence of the effects of the value of T_1 and T_2 for constant G_1=3 and G_2 = 3

Fig. 5. Dependence of the effects of values of T_1 and T_2 for constant G_1 = 2 and G_2 = 3

To assess their impact on the course of action, a set of variants containing graphic elements of the static low level of invasiveness was analyzed. This approach was designed to determine the effect of text elements on the obtained results and to determine their importance in the excluded animated elements. Table 2 presents the results of analyses carried out on media exposure to the exclusion of animated elements.

Table 2. Analysis of the effects of the absence of animated elements

Element	Effect	E	t(70)	p	−95,%	+95,%	W
T_1	−0,184993	0,100150	−1,84716	0,068950	−0,384737	0,014750	−0,092497
T_2	−0,009400	0,080628	−0,11658	0,907523	−0,170208	0,151408	−0,004700
G_1	0,033818	0,064521	0,52414	0,601839	−0,094866	0,162502	0,016909
G_2	−0,000417	0,063814	−0,00653	0,994810	−0,127690	0,126857	−0,000208
1L vs 2L	0,017946	0,064866	0,27667	0,782852	−0,111426	0,147318	0,008973
1L vs 3L	−0,013340	0,102003	−0,13078	0,896328	−0,216779	0,190099	−0,006670
1L vs 4L	0,046308	0,080860	0,57269	0,568687	−0,114962	0,207578	0,023154
2L vs 3L	−0,045112	0,099310	−0,45425	0,651054	−0,243180	0,152957	−0,022556
2L vs 4L	0,068797	0,079953	0,86046	0,392470	−0,090665	0,228258	0,034398
3L vs 4L	0,003517	0,132919	0,02646	0,978966	−0,261581	0,268615	0,001758

All variants were considered for the transmission of text elements T_1 and T_2, and a first and a second embodiment of both the element T_1 and the T_2 component. The analysis indicated the greatest effect of the text element T_1 (p = 0.068950), as shown in Table 2. Pareto analysis results for this set of data indicate the impact of the element T_1 at $P(T_1)$ = -1.84. Orientation effects on growth effect are the same as in the previous analysis, the change values as $P(T_1)$ = -1.84 obtained when incorporated with animated elements level $P(T_1)$ = -0.22 indicates that text was increasing the impact of graphic elements. For the text element T_2, the effect is much smaller and takes the level of $P(T_2)$ = -0.11. This is consistent with the concept of the project and the choice of font sizes with the larger format for an element T_1. Fig. 6 shows the influence of the effects of changes in textual variants with static media versions G_1 and G_2.

With the limited impact of graphics, changes of T_1 and textual variants are reflected in the obtained results. Fig. 7 shows dependences of design variant $G_1 = 2$ and $G_2 = 2$, which represents the static graphic elements contrasting with the background site better than with variants $G_1 = 1$ and $G_2 = 1$. Even with such a selection of textual variants, the influence is clearly visible.

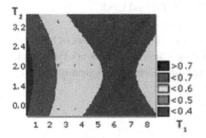 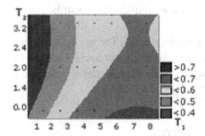

Fig. 6. Effects depending on the value of T_1 and T_2 parameters for static graphic elements $G_1 = 2$ and $G_2 = 1$

Fig. 7. Dependence of the effects of the values of T_1 and T_2 for static graphics $G_1 = 2$ and $G_2 = 2$

The analyses indicate the possibility of reducing the importance of the impact of verbal communication represented by textual messages in an interactive environment with the exposure of associated graphics with a high impact on the recipient. This effect should be taken into account when the purpose of advertising is to direct text messages of a certain informational value. In this case, it is important to reduce the visual impact of other elements. The results also point to the need for analysis of the issue of advertising within websites with a large number of graphic elements that absorb the attention of the recipient and hinder the perception of verbal communication. The selection of components may be implemented in an automated manner by eliminating variations towards best results. The conducted experimental studies show one of the areas of application of ad units with varying versions of the components that make it possible to carry out detailed analyses of the effectiveness of advertising campaigns.

5   Summary

An analysis of elements influencing the effectiveness of interactive media indicates the presence of a number of factors that determine the effectiveness of advertising campaigns in the online environment. The advertising message integrates many elements, both verbal and graphic, which in combination can act on the recipient with varying intensity. The proposed approach to advertising design, i.e. breaking it down into its components, makes it possible to test different design options and determine the relationships that exist between the components. This approach uses the electronic properties of the environment and the ability to integrate with adaptive websites. The conducted experimental studies indicated that the impact of individual components and reduced the share of verbal communication in parallel using the graphical elements of the greater intensity of impact. Simultaneous use of these elements makes verbal communication have a large effect on the interactions generated by the user. The introduction of dynamically generated elements with different levels of interaction allows the selection of the structure of the message, which provides an acceptable level of results. Future work assumes testing different scales of influence and adding more testing dimensions based on vividness effect, animations with the different intensity and frequencies towards more generalized results and model.

Acknowledgments. The work was partially supported by Fellowship co-Financed by European Union within European Social Fund, by European Union's Seventh Framework Programme for research, technological development and demonstration under grant agreement no 316097 [ENGINE].

References

1. Beattie, A.E., Mitchell, A.A.: The relationship between advertising recall and persuasion: an experimental investigation. In: Alwitt, L.F., Mitchell, A.A. (eds.) Psychological Processes and Advertising Effects: Theory, Research, and Application, pp. 129–155. Lawrence Erlbaum Associates, Hillsdale (1985)
2. Ellsworth, J.H., Ellsworth, M.V.: Marketing on the Internet: Multimedia Strategies for the World Wide Web. John Wiley & Sons Inc., New York (1995)
3. Hairong, L., Bukovac, J.L.: Cognitive Impact of Banner Ad Characteristics: An Experimental Study. Journalism and Mass Communication Quarterly **76**(2), 341–355 (1999)
4. Heeter, C.: Implications of new interactive technologies for conceptualizing communication. In: Savaggio, J., Bryant, J. (eds.) Media Use in the Information Age: Emerging Patterns of Adoption and Consumer Use, pp. 217–235. Lawrence Erlbaum Associates, Hillsdale (1989)
5. Hoffman, D.L., Novak, T.P.: Marketing in Hypermedia Computer-Mediated Environments: Conceptual Foundations. Journal of Marketing **60**(3), 50–68 (1996)
6. Jankowski, J.: Balanced approach to the design of conversion oriented websites with limited negative impact on the users. In: Bădică, C., Nguyen, N.T., Brezovan, M. (eds.) ICCCI 2013. LNCS, vol. 8083, pp. 527–536. Springer, Heidelberg (2013)
7. Jankowski, J.: Increasing website conversions using content repetitions with different levels of persuasion. In: Selamat, A., Nguyen, N.T., Haron, H. (eds.) ACIIDS 2013, Part II. LNCS, vol. 7803, pp. 439–448. Springer, Heidelberg (2013)
8. Jankowski, J.: Modeling the structure of recommending interfaces with adjustable influence on users. In: Selamat, A., Nguyen, N.T., Haron, H. (eds.) ACIIDS 2013, Part II. LNCS, vol. 7803, pp. 429–438. Springer, Heidelberg (2013)
9. McCoy, S., Everard, A., Polak, P., Galletta, D.F.: The Effects of Online Advertising. Communications of the ACM **50**(3), 84–88 (2007)
10. Rafaeli, S.: Interactivity: from new media to communication. In: Hawkins, R.P., Wieman, J.M., Pingree, S. (eds.) Advanced Communication Science: Merging Mass and Interpersonal Processes, pp. 110–134. Sage, Newbury Park (1988)
11. Stauer, J.: Defining Virtual Reality: Dimensions Determining Telepresence. Journal of Communication **42**(4), 73–93 (1992)

12. Sundar, S., Sunetra, N., Obregon, R., Uppal, C.: Does Web Advertising Work? Memory for Print vs. Online Media. Journalism and Mass. Communication Quarterly **75**(4), 822–835 (1999)
13. Urban, G.L., Liberali, G., MacDonald, E., Bordley, R., Hauser, J.R.: Morphing Banner Advertising. Marketing Science **33**(1), 27–46 (2014)
14. Yoo, C.Y., Kim, K.: Processing of Animation in Online Banner Advertising: The Role of Cognitive and Emotional Responses. Journal of Interactive Marketing **19**(4), 18–34 (2005)
15. Zha, W., Wu, H.D.: The Impact of Online Disruptive Ads on Users' Comprehension, Evaluation of Site Credibility and Sentiment of Intrusiveness. American Communication Journal **16**(2), 15–28 (2014)
16. Ziemba, P., Piwowarski, M., Jankowski, J., Wątróbski, J.: Method of criteria selection and weights calculation in the process of web projects evaluation. In: Hwang, D., Jung, J.J., Nguyen, N.-T. (eds.) ICCCI 2014. LNCS, vol. 8733, pp. 684–693. Springer, Heidelberg (2014)

Countering Social Engineering Through Social Media: An Enterprise Security Perspective

Heidi Wilcox[✉] and Maumita Bhattacharya

School of Computing & Mathematics, Charles Sturt University, Albury, Australia
{hwilcox,mbhattacharya}@csu.edu.au

Abstract. The increasing threat of social engineers targeting social media channels to advance their attack effectiveness on company data has seen many organizations introducing initiatives to better understand these vulnerabilities. This paper examines concerns of social engineering through social media within the enterprise and explores countermeasures undertaken to stem ensuing risk. Also included is an analysis of existing social media security policies and guidelines within the public and private sectors.

1   Introduction and Background

Social media sites such as Facebook, Myspace, LinkedIn, and Twitter are a data mining goldmine for readily available personal and sensitive information made publicly for the web, especially when the majority of participants are using default privacy settings. (King, 2008; Furnell, 2008; Slonka, 2014; Nayak Prince & Robinson 2014; Wong et al 2014). The increased adoption of social media technologies and failing to protect company information may result in data leakage, business continuity failures and compliance breaches, reputational risks through loss of valuable intellectual property, consumer confidence and competitive advantage (Colwill, 2009; Almeida, 2012). Traditional security countermeasures are not keeping up with these changes in the workplace as more businesses are encountering breaches targeting the human elements, such as social engineering. (Colwill, 2009; Rudman, 2010; He, 2012). Social engineers exploit human behaviour idiosyncrasies to form an attack from the outside that leads them to gain inconspicuous entry into protected areas of the company for their own illicit use. (Mitnick & Simon, 2001).

As the line between business use and personal use is often blurred, social engineers can gather sensitive data from any number of social media accounts to form a personal resume on a targeted employee. (Meister & Willyerd, 2010). Traditionally, email was the primary vector for spam and phishing exploits, however, the popularity and scope for large volume targets in social media has seen these threats moving away from email and on to social platforms. Web based attacks such as phishing are consistently found to be the leading transport vectors for cyber-attacks; social media and social gaming provides the perfect vehicle or attack surface for delivering lures and payloads. (Arachchilage & Love, 2014; Ikhalia, 2014). The top three social media issues negatively experienced by organizations include: employees sharing too much

© Springer International Publishing Switzerland 2015
M. Núñez et al. (Eds.): ICCCI 2015, Part II, LNCS 9330, pp. 54–64, 2015.
DOI: 10.1007/978-3-319-24306-1_6

information, the loss of confidential information, and increased exposure to litigation (Symantec, 2011). Other equally important results include losses concerning employee productivity and increased risk of exposure to virus and malware (Almeida, 2012). These platforms enable social engineers to operate freely, efficiently and cost effectively with low margins for getting caught. (Franchi, Poggi & Tomaiuolo, 2014).

Boudreaux (2010a), and Foreshew (2012), propose that organizations protecting information assets through effective security policies and governance will more effectively manage the business risks of the future. Social media policies and guidelines provide advice on how social media participation will be applied to all of the members of an organization (Bell, 2010). It is also reported that the most effective security countermeasure against social engineering is to increase employee awareness of the many tricks employed by social engineers against them in the workplace. (Bada & Sasse, 2014).

The remainder of the paper is organized as follows. In Section 2 we explore those countermeasures currently offered by enterprise to address the challenges faced by social engineering through social media concerning people, process and technology. We also review existing social media policies from opposing sectors and compare areas of coverage in Section 3 with some concluding remarks presented in Section 4.

2   Countering Social Engineering through Social Media: Current Perspectives

The following global perspectives underline information security practices concerning people, process and technology that are currently used by enterprise in an attempt to decrease loss attributed to their social media usage.

2.1   People

Global organizations are embracing new technologies that explore huge business benefits but also bring catastrophic organizational risk. (Almeida, 2012). Countermeasures for online social engineering concerning people and employees have had various levels of success. Information security taskforces aligned to assess such threats are now heading in a positive direction towards understanding the motivations behind these attacks. These paths include establishing types of threat vectors and introducing awareness initiatives that effectively reduce business risk. (VMIA, 2010). Current practices focus on creating individual employee awareness and training in both the public and private sectors; whether they are online at home or work. There is a general consensus from both sectors that there needs to be collaboration from government and business in all industries to increase cyber security effectiveness.

The UK and US governments are guiding business and consumers with awareness initiatives in response to a trend in developed nations to adopt a web 2.0 and social business model framework for all government departments and processes. Recent awareness campaigns include the introduction of a national computer emergency response team (CERT-UK) in the UK to focus on new cybersecurity policies (Luxford, 2014) and a '10 steps to cyber security' awareness program; The Cyber Security Awareness Campaign organized by the National Institute of Electronics and Information Technology in India; Go Safe Online (including annual Awareness Day) in Singapore and national Cybersecurity Policy Review in 2013 from the US government initiated a 'Stop.Think.Connect' Awareness campaign with a national Awareness month.

Australian culture embraces mateship and social interaction as an integral part of everyday life. This intrinsic character trait therefore carries over to the workplace to form primarily trusting and open social norms between the organization's management, employees and customer base. By adopting social media as a technological tool to unite business processes, Australians – as employers or employees – are prime targets for social engineers to exploit this trusting 'weakness'. (Mitnick & Simon, 2001).

In response to the increasing threats to online security the Australian government under Prime Minister Julia Gillard introduced an amalgamation of cyber security and information security specialist organizations to form the Australian Cyber Security Centre (ACSC). The Centre will host experts in their fields representing Defense Signals Directorate, Defense Intelligence Organization, the Australian Security Intelligence Organization, the Attorney General's Department's CERT-Australia, Australian Federal Police and Australian Crime Commission. The initiative will analyze and assess incoming cyber threats, and work closely with various industries and private sector parties to formulate effective countermeasures and create public awareness. Other major Australian awareness initiatives already in place include Stay Smart Online including an annual Awareness Week, the annual Cyber Security Challenge Australia, and Cybersmart aimed at educating schools and school age children.

2.2   Policy

According to PriceWaterhouseCoopers (2013), fewer than 30% of global organizations have existing policies that include countermeasures covering social media usage. Results from this paper's policy review also found the vast majority of reviewed policies from global organizations did not include countering measures for dealing with online social engineering or phishing attempts while participating in social media platforms. General inconsistency of current policy design suggests that organizations are disparate in their views to countermeasure social media risk, and are lacking clarity in ways to precede legally and ethically.

In 2010, governments within the US and Australia declared the Government 2.0 initiative which would allow communications between Government and the people to be 'open' and transparent. (AG, 2010). In response a major restructure of current security practices ensued. These governments are still trying to figure out how to put boundaries around an employee's personal, professional, and official agency use. Each use has different security, legal, and managerial implications and government agencies are tasked with striking a balance between using social media for official agency interests only, and allowing all employees access for personal and professional interests.

The majority of Government agencies are managing online and social media access in two ways:

1. Controlling the number or types of employees who are allowed access to social media sites or
2. Limiting the types of sites that are approved for employee access.

A third approach sees agencies providing customized social media channels, such as GovLoop (2014), 'in-house' for employees, behind the organization's firewalls.

Effective examples of Australian guidelines can be viewed from The Australian Taxation Office (ATO), Department of Immigration and Citizenship (DIAC) and Department of Human Services (DHS). Additionally, The Australian Signals Directorate (ASD) in conjunction with The Department of Defense maintains a wealth of all access publications for all industries including the comprehensive Australian Government Information Security Manual which is produced by ASD to guide Australian Government ICT professionals on security of national systems.

Despite this proactive approach by government, private sectors seem to be looking cautiously inwards for security research and strategy. The Ponemon Institute (2011) observed that only 33% of private organizations in Australia had adequate policy relating to social media usage. Some of the better examples of private sector policy development include Telstra, Dell, IBM and Kodak. US Tech giant CISCO has also released a comprehensive policy duo with the CISCO Social Media Playbook: Best Practice Sharing; and the Social Media Policy, Guidelines & FAQs.

The authors of this paper also undertook an analysis of social media policies currently existing in social media management for organizations. Tables 2 and 3 have been drawn to compare current standards for 24 organizational social media policies and guidelines. These policies were selected randomly from the online publically available database of social media governance documents compiled by Chris Boudreaux (2014). A balance was sought between public and private sectors in policy selection (19 public and 19 private sector) for comparison. The samples were limited to those obtainable on publically based internet search and were further limited to those policies written in English (purposed for the understanding of the author). This small cross section acted as a guide to contribute to the author's understanding of the research topic. Delivering the information in table format provides a clarity to observing a pattern of social media issue coverage areas included as a general rule within these documents. The aim for these tables is to provide review of as much documentation of current, publicly available policy covering organizational use of social media to formulate resources for a best practice framework for future research direction. Primary to our purpose of analysis, these policies were gathered to investigate coverage of security risks to employees and advice on mitigating social engineering through these technologies. To aid in achievement of this goal the documents were coded with the use of Atlas.ti software. The qualitative process involved coding of key points from all 24 documents with Atlas.ti (2015) which revealed groupings of the data set for comparison. The comparison through Atlas.ti coding and extraction ascertained a trend for six primary coverage areas (Table 1) most likely to appear in a policy or guideline to act as educational advice to employees, or as a means of legal protection for the organization. As the documents varied widely in their approach on an individual level, this representation will serve as a general overview of issues addressed and listed for employee guidance. This is certainly the case when dealing with new technologies such as social media adoption. Social engineers focus on people as their targets primarily, using the human element to cause corruption to their own organizational technical tools and assets to infiltrate deeper within the organization.

Table 1. Description of the six social media coverage areas used for policy review.

Coverage Area & Description	
1 Social Media Account Management	*Policy Coverage Percentage*
Applies to the creation, maintenance and destruction of accounts designed for organizational purposes. Official accounts are predominantly operated by designated employees trained or certified in social media communications security.	N=11 46%
2 Acceptable Use	*Policy Coverage Percentage*
This area of policy is relevant to how a company's employees will use social media technologies either at work in a professional capacity, or for personal use while at work. Policies may state if certain sites are restricted from use, and also if personal use is condoned while on working time. Consequences of any violations would be listed here.	N=23 96%
3 Social Media Content Management	*Policy Coverage Percentage*
In conjunction with account access and creation, policies need to cover what content is to be published online. Official accounts usually have responses drafted from CEO's or Department Managers, carefully sanctioned by organizational leaders. Professional accounts serve the purpose for regular business processes such as customer support or marketing (similar to email functionality).	N=15 62%
4 Employee Conduct	*Policy Coverage Percentage*
Guides and policies include terms of ethical use and conduct for those participating in work-related online engagement. Most policies will either refer to existing Codes of Conduct or elaborate further into the "do's and don'ts" of social media use.	N=21 87%
5 Legal	*Policy Coverage Percentage*
Policy coverage includes any reference to laws and regulations that may have an impact on the company or the individual while using online communications. This can be a general point where employees are expected to abide by applicable laws but with no mention of any one in particular. Some policies elaborate further by explaining impact to action with each specific law, especially privacy and confidentiality and; copyright and intellectual property.	N=22 92%
6 Security	*Policy Coverage Percentage*
Policies cover technical and behavioural security issues. Of particular interest to this research is the inclusion of advice or awareness to the risks attributed to social engineering. It is imperative to include awareness on social engineering attacks (spearfishing, click-jacking) directed at employees using social media information. Technical security measures such as password protection, identification authentication (PKI), and virus scans can be included here.	N=10 42%

Table 2. Social media policies reviewed from the public sector.

Social Media Policies- Public Sector					
Organisation Department of Information Technology Government of India					
Policy Type Social Media Guide					
Objectives Provides a basic framework and guidelines for government agencies and e-projects					
Social Media Account Mgt[1]	**Acceptable Use[2]**	**Social Media Content Mgt[3]**	**Employee Conduct[4]**	**Legal[5]**	**Security[6]**
✓	✓	✓		✓	✓
Organisation US Federal CIO Council					
Policy Type Social Media Guide					
Objectives Provide best practices and recommendations for federal use of social media and cloud computing resources					
	✓			✓	✓
Organisation US Environmental Protection Agency					
Policy Type Social Media Policy					
Objectives This policy applies to EPA employees, contractors, and other personnel acting in an official capacity on behalf of EPA.					
	✓	✓	✓	✓	
Organisation Harvard University					
Policy Type Social Media Guide					
Objectives For those required to speak on behalf of Harvard at an individual level.					
	✓		✓	✓	
Organisation US Department of Defence					
Policy Type Social Media Policy					
Objectives For all members of all departments in US defence force while using internet services including social networking and twitter					
	✓	✓		✓	
Organisation UK Ministry of Defence					
Policy Type Social Media Policy					
Objectives Enables Service and MOD personnel to make full use of online presences while protecting their own, Service, and Departmental interests					
✓	✓	✓	✓	✓	✓
Organisation National Library of Australia					
Policy Type Social Media Policy					
Objectives Applies to all Library employees using or having a need to participate in online social media activity for official Library communications and through personal accounts which they have created and administer themselves.					
	✓	✓	✓	✓	✓
Organisation Department of Justice Victoria, Australia					
Policy Type Social Media Policy					
Objectives Recommended for members and contractors of the department for online use					
✓	✓	✓	✓	✓	
Organisation Australian Public Service Commission					
Policy Type Social Media Guide					
Objectives Guidance for APS employees and Agency Heads to help APS employees understand the issues to take into account when considering making public comment, including online					
	✓		✓		

Table 2 (*Continued*)

Organisation Australian Department of Finance **Policy Type** Social Media Guide **Objectives** Guide for employees of the Australian Government Department when participating in online engagement					
✓	✓	✓	✓		✓
Organisation The NSW Police Department **Policy Type** Social Media Policy + Social Media Guide **Objectives** To be used by all NSW Police Force employees when using social media					
✓	✓	✓	✓	✓	✓
Organisation Queensland Government **Policy Type** Social Media Guide **Objectives** The guidelines apply to all departments and covers officially established, publicly available and departmentally-managed social media accounts, but does not require the establishment of the accounts. It does not apply to use of social media on a personal or professional basis or cover use of social media for political or internal government purposes.					
✓	✓	✓	✓	✓	✓

Table 3. Social media policies reviewed from the private sector.

Social Media Policies- Private Sector					
Organisation Intel **Policy Type** Social Media Guide **Objectives** Applies to Intel employees and contractors participating in Social Media for Intel					
Social Media Account Mgt$_1$	**Acceptable Use**$_2$	**Social Media Content Mgt**$_3$	**Employee Conduct**$_4$	**Legal**$_5$	**Security**$_6$
			✓		✓
Organisation Kodak **Policy Type** Social Media Guide **Objectives** Aims at Policy Developers, guiding principles					
	✓		✓		✓
Organisation IBM **Policy Type** Social Media Guide **Objectives** Employees of IBM, however does not state specifically					
	✓		✓		
Organisation Cisco **Policy Type** Social Media Policy and Social Media Guide **Objectives** Applies to all Cisco employees, vendors and contractors who are contributing or creating on social media in appropriate and effective engagement					
✓	✓	✓	✓		
Organisation Coca-Cola **Policy Type** Social Media Guide **Objectives** Applies to company employees and agency associates when using social media to help market the company and promote the brand					
	✓		✓		
Organisation BBC - News **Policy Type** Social Media Guide **Objectives** Applies to BBC officials and employees such as reporters, presenters and correspondents relating to news coverage and online activity					
✓	✓	✓	✓		

Table 3 (*Continued*)

Organisation				
Organisation Chartered Institute of Public Relations **Policy Type** Social Media Guide **Objectives** Designed to help UK members of CIPR navigate a rapidly evolving communications landscape				
✓	✓	✓		✓
Organisation Ford **Policy Type** Social Media Guide **Objectives** Designed for personnel of Ford Motor Company when participating in online social media				
✓		✓		
Organisation Walmart **Policy Type** Social Media Guide **Objectives** Applies to customers using social media to contact Walmart, a small section applies to associates of Walmart responding on behalf of the company				
✓	✓		✓	
Organisation Telstra **Policy Type** Social Media Policy **Objectives** Representing, Responsibility and Respect. Applies to all Telstra employees and contractors				
✓	✓	✓	✓	
Organisation Lifeline **Policy Type** Social Media Policy **Objectives** This policy is intended to provide employees, volunteers and supporters of Lifeline with clarity on the use of social media platforms				
✓	✓	✓	✓	
Organisation Football NSW **Policy Type** Social Media Policy **Objectives** This social media policy aims to provide some guiding principles to follow when using social media. It applies to the entire membership including players, coaches and referees.				
✓	✓	✓		

2.3   Technology

Traditional network security techniques such as anti-virus, firewalls and access control now combine focus with the vigilance of monitoring operations. This includes, but is not limited to, data loss prevention tools; active monitoring and analysis of security intelligence/internal auditing via penetration testers (seeking social engineering entry points). These tools will aid security professionals in becoming more accustomed to 'reading between the lines' of mined data. (Oxley, 2011). Table 4 lists the ten most applied technical controls used in Australian business, as surveyed by CERT-Australia in 2012.

Table 3. Popular technical controls used in Information Security within the enterprise.

Technical Control	Popularity	Technical Control	Popularity
Anti-virus software	93%	Digital Certificate	72%
Firewalls	93%	Encrypted login sessions	68%
Anti-spam filters	90%	Intrusion detection	60%
Access control	83%	Encrypted media	50%
VPNs	81%	Two factor authentication	48%

3   Key Trends Observed

Trends indicate that corporations are being forced to recognize emerging online threats as an overall business or strategic risk as opposed to only a technology risk. (Rudman, 2010). Overwhelmingly, all industries are discovering the need to establish trust and enhance business process transparency, due to the pervasive and instant nature of social networks. Reputation can be lost or gained rapidly through the voice of the online communities. (Almeida, 2012). The increased adoption of employees using social networking for both business benefit and personal socializing has created an immediate need for management to reassess their current security culture, making employees a primary focus. This is happening at a slower pace than these technologies are being introduced. It is evident that traditional policies and training are not effectively covering all aspects of information security. The result being, there are more cyber-attacks related to online social engineering than ever before.

Figure 1 illustrates patterns ascertained from our review of 24 enterprise social media policies and guidelines. Coverage for areas concerning ethical employee conduct and terms of acceptable use for social media usage shows prevalence. There are also high levels of inclusion relating to legal issues where coverage of applicable laws and regulations aids in online litigation protection. It is very concerning, however, that organizations are embracing social media usage without offering judicious advice relating to the security issues encountered within these technologies.

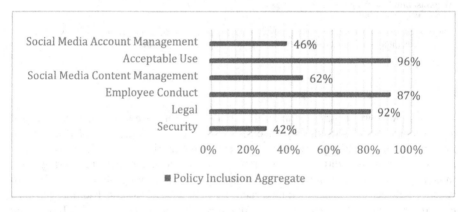

Fig. 1. Comparing Social Media Policies. **Based on our tablature style review of 24 public and private sector social media policies.

From the selection of policies available, very few mentioned guidance for securing social media technologies specifically, and even fewer associated social engineering with prominent security threats. This alludes to users lacking awareness of social engineers operating over social media data, and the types of information that could be used against the employee in further attacks. They also seem to be ill-informed as to how these attacks (phishing) can be implemented and what real examples look like. What is clear from aspects of this investigation is that the problems associated with social engineering through social media are not well understood. The security obstacles continue to deepen because of the lack of clarity from management on why, when and where to apply effective countermeasures.

4   Conclusion

Social engineering defies traditional security efforts due to the method of attack relying on human naiveté or error. The vast amount of information now made publicly available to social engineers through online social networks is facilitating methods of attack which rely on some form of human error to enable infiltration into company networks. This investigation confirms social engineering through social media channels targeting organizational employees as one of the most challenging information security threats. There is a worrying trend to rush these technologies into the workplace without initiating effective security strategies involving social media use. We have contributed to research by addressing the gaps concerning social media policy development and the lack of advice given to employees regarding social engineering. Social engineering through social media confirms the crucial need for employees to be made aware of attack methods through a combination of policy development and employee education, alongside traditional technical countermeasures.

References

Almeida, F.: Web 2.0 technologies and social networking security fears in enterprises (2012). arXiv preprint arXiv:1204.1824

Amigorena, F.: The threat from within: how to start taking internal security more seriously. Computer Fraud & Security **2014**(7), 5–7 (2014)

Arachchilage, N.A.G., Love, S.: Security awareness of computer users: A phishing threat avoidance perspective. Computers in Human Behavior **38**, 304–312 (2014)

Bada, M., Sasse, A.: Cyber Security Awareness Campaigns Why do they fail to change behaviour? (2014)

Bell, J.: Clearing the AIR. Communication World **27**(1), 27–30 (2010)

Boudreaux, C.: Social media policies. In: The Social Media Management Handbook, pp. 274–285

CERT-Australia: 2012 Cybercrime and Security Survey. Report from CERT-Australia. Australian Government, Attorney General's Department (2012)

Colwill, C.: Human factors in information security: The insider threat–Who can you trust these days? Information Security Technical Report **14**(4), 186–196 (2009)

Foreshew, J.: Companies should develop their own social media policies, The Australian online (2012)

Franchi, E., Poggi, A., Tomaiuolo, M.: Information Attacks on Online Social Networks. Journal of Information Technology Research (JITR) **7**(3), 54–71 (2014)

Furnell, S.: End User Security Culture – A Lesson That Will Never Be Learnt? Computer Fraud & Security, 6–9, April 2008

GovLoop: Knowledge Network for Government. Website (2014)

He, W.: A review of social media security risks and mitigation techniques. Journal of Systems and Information Technology **14**(2), 171–180 (2012)

Ikhalia, E.J.: A New Social Media Security Model (SMSM)

King, P.: Cyber Crooks Target Social Networking Sites. Point for Credit Research & Advice, 9, January 1, 2008

Luxford, H.: UK Lauches New Cybersecurity Initiative. Plans to work with infrastructure providers to create a safer business environment. Article from Datacenter Dynamics (2014)

Meister, J.C., Willyerd, K.: The 2020 workplace. HarperCollins, New York (2010)

Mitnick, K.D., Simon, W.L.: The art of deception: Controlling the human element of security. John Wiley & Sons (2001)

Nayak, D., Prince, S., Robinson, R.: Information privacy risk assessment of facebook graph search. In: Science and Information Conference (SAI), pp. 1005–1006. IEEE, August 2014

Oxley, A.: A best practices guide for mitigating risk in the use of social media. IBM Center for The Business of Government (2011)

PriceWaterhouseCoopers: The Global State of Information Security Survey. Price Waterhouse Coopers (2014)

Rudman, R.J.: Framework to identify and manage risks in Web 2.0 applications. African Journal of Business Management **4**(13), 3251–3264 (2010)

Slonka, K.J.: Awareness of malicious social engineering among facebook users (Doctoral dissertation, Robert Morris University) (2014)

Symantec: Social Media Protection Flash Poll Global Results. Powerpoint Presentation (2011)

VMIA: Risk Insight. Social Media: What's the risk? Report from VMIA Risk Roundtable (2010)

Wilcox, H., Bhattacharya, M., Islam, R: Social engineering through social media: an investigation on enterprise security. In: ATIS 2014, CCIS 490, pp. 243–255 (2014) (in press)

Wong, K., Wong, A., Yeung, A., Fan, W., Tang, S.K.: Trust and Privacy Exploitation in Online Social Networks. IT Professional **16**(5), 28–33 (2014)

Loading Speed of Modern Websites and Reliability of Online Speed Test Services

Aneta Bartuskova$^{(\boxtimes)}$ and Ondrej Krejcar

Faculty of Informatics and Management, Center for Basic and Applied Research,
University of Hradec Kralove, Rokitanskeho 62, 500 03, Hradec Kralove, Czech Republic
aneta.bartuskova@uhk.cz, ondrej@krejcar.cz

Abstract. Loading speed of websites reflects the website's performance and has a significant influence on user experience and satisfaction. In this paper we analyze causes and consequences of slow loading of websites, followed by recommendations and comments on usual practice and feasibility. In the next part of the article we discuss a choice of the right online speed test in order to achieve an optimized performance of websites. Selected free online services are analyzed, with a focus on available functions and settings. Finally the issue of reliability and default settings of these services is discussed and tested by performing a series of repeated speed tests with a combination of settings.

Keywords: Loading speed · Website speed testing · Web performance · Usability

1 Introduction

This paper is focused on loading speed of modern websites and reliability of online speed test services. With the abundance of available online resources to choose from, web visitors are getting less tolerant about slow-loading websites. It appears that users are much more impatient than previously thought [1]. Slow-loading sites are therefore a major frustration and turnoff for web surfers [2]. The role and importance of loading speed is discussed in more detail in the State of the Art.

In this paper we analyze causes and consequences of slow loading of websites. The main causes include size of the images, responsive web design and misuse of JavaScript scripting language. A set of recommendations follows, along with comments on usual practice and feasibility. These general rules are a good starting point, however we should aspire for an optimized performance for the particular website. We suggest using an online speed test for this purpose. The next part of this paper is therefore dedicated to a choice of the right online speed test, reliability of these tests and how can we overcome discovered limitations.

An overview of free online services for speed testing is presented. Then available functions and settings are analyzed, with a focus on choice of testing locations and browser. Basic settings of speed tests are shown on the selected speed test service. Finally the issue with reliability and default settings of these services is discussed and tested by performing a series of repeated speed test with a combination of settings.

© Springer International Publishing Switzerland 2015
M. Núñez et al. (Eds.): ICCCI 2015, Part II, LNCS 9330, pp. 65–74, 2015.
DOI: 10.1007/978-3-319-24306-1_7

2   State of the Art

There are many terms specifying the time needed for loading a website in the browser window, e.g. loading speed, site speed, page load speed, response time, speed of data display or download delay. One of these alternative attributes is usually included in various sets of usability or design attributes researched in literature. Lee and Kozar included some of these variations in their thorough analysis of website usability constructs and placed them all in one category called "Simplicity", along with such terms as efficiency, minimal action or simple layout [3]. Similarly Morkes and Nielsen claimed that users want to get their information quickly, they want fast response times for hypertext links and at the same time they like well-organized sites that make important information easy to find [4]. Rosen and Purinton also connected loading speed with website's simplicity. According to these authors simplicity of design makes the site more appealing and also faster to load [2].

Constantinides classified site speed as one of usability factors among the main building blocks of web experience, in other words one of the elements which enhance web site usability [5]. Gehrke and Turban suggested among other usability factors also page loading and download time [6]. Aladwani classified speed of page loading as one of technical characteristics of the website and important attribute of the website quality [7]. Loiacono et. al. introduced response time as one of the dimensions for Web site evaluation in WebQual, an instrument for consumer evaluation of web sites [8]. Cebi included speed as one of website design parameters under technical adequacy with following description: "The site should provide quick loading, accessing, and using" [9].

Green and Pearson presented web site usability dimensions, among them also download delay. They characterized this dimension by e.g. these wordings: "The rate at which the information was displayed was fast enough" or "The speed in which the computer provided information was fast enough" [10]. Download delay is also among five factors included in the Palmer instrument for measuring of Web site usability [11]. Download delay is defined as the initial request for access to the page and then each subsequent request for changing pages with the site [12].

The issue of loading speed is very important not only on presentation websites but also in e-commerce. Online customers expect fast loading Web pages [13]. Loading speed is also influencing a user's preference for a particular website [14]. Rosen and Purinton pointed out that web surfers are not very patient and some web design experts have estimated that they have exactly 10 seconds to lure people into a site. Lindgaard et. al. suggests that time needed for assessing a visual appeal of a website is actually about 50 ms [15]. Loading speed is then very important for user experience, because slow loading speed means that user is forced to watch blank white screen or only partially displayed content of a website for a certain period of time. However first impression of a web site is very important if the user continues to use the web site [14]. Therefore it is not surprising that slow loading sites are a major frustration and turnoff for web surfers [2]. Loading time is also a major contributing factor to page abandonment - the average user has no patience for a webpage that takes too long to load [16].

3   Cause of Slow Loading of Websites

It is well researched that loading speed of the website has a significant influence on a user. The topic of this section is about what causes slower loading speed and how can we simply avoid it. Sometimes a website loads slowly uselessly simply because it is not optimized, sometimes lower speed is an intentional payment for better user experience. As Work stated, page loading time is an important part of any website's user experience but many times we'll let it slide to accommodate better aesthetic design or new functionality [16].

3.1   Size of the Images

Of course, web site designers can choose not to include slow loading elements [green]. These are especially photos and other images, which have big file size in order to keep high quality. The cost of pictures is in download speed, frequently mentioned as a concern to on-line users [17]. Including many images and videos on a web page results in a slow loading time, which can be very frustrating [18]. Morkes and Nielsen also argued that users want fast-loading graphics and they want to choose whether to download large (slow) graphics [4]. The situation is much more difficult for mobile phones - networks are slower, hardware is less capable, and you have to deal with the messy world of data limitations and transcoding methods [18, 28, 29].

Despite of these conclusions, modern web design trends tend to prioritize large high-quality photos and graphics. When we look at modern websites or at web design templates available on the internet today, either free or for a price, very often there is a large picture over the whole screen. If it is a responsive website, the image is usually wide enough to fit on the widescreen monitor. These background or header images are often not even content-related and sometimes quite ambiguous, while they consume significant amount of downloading time and negatively affect loading speed. As for user experience point of view, it was proven by eye-tracking methodology that people ignore more images than they look at on the web, and they look at images for just a fraction of a second [19]. Also on web pages with multiple superfluous images, people treat the entire page as an obstacle course they must navigate, as a result they look at text around images but not at images [19]. In the study of Morkes and Nielsen, some users said that graphics that add nothing to the text are a distraction and waste of time [4]. Even though their role is not well-founded, large images play a significant role in today's web design, probably because it is "modern" and it looks "professional".

3.2   Responsive Web Design

In 2011 Marcotte wrote that we're designing for more devices, more input types, more resolutions than ever before [20]. He answered this challenge with "responsive design". Responsive website can display web content differently according to the particular device - its screen size and capabilities - on which it is being viewed. Resizing a desktop image to fit a mobile device's screen however implies downloading an image that's been suited for a desktop environment, which is an unnecessary large file [21].

Many web designers tend to use the simplest method available - common way of dealing with this issue is hiding too large images (and other content) for smaller displays. This however does not solve the problem with downloading too much data when it is not needed. The images will not be displayed, but they will be still downloaded, causing delay in page loading speed [18]. And if you use CSS media query to replace the background image with a mobile version, in some cases it would actually download both images [21]. Another common way of dealing with images is "shrinking" - as responsive web design uses fluid images to match the different screen sizes, desktop-grade image is downloaded every time, even when loaded on a much smaller screen [18].

3.3   JavaScript

According to Barker, JavaScript is potentially the largest area for improvement when trying to address the total performance of the website [22]. Websites relying heavily on JavaScript, especially those with AJAX (asynchronous JavaScript and XML) can experience performance issues. Souders tested the Alexa Top 100 URLs with and without JavaScript and demonstrated improvement of an average performance of 31% when removing JavaScript from a website [23]. Of course solution to faster loading speed is not removing JavaScript completely, web designers should rather learn to use it more efficiently. A motivation to learn programming properly is however a big problem these days. Because of the limitless resources available on the internet, one does not have to learn a lot to create a plausibly functional website. This trend can be simply confirmed e.g. by looking at "a question and answer site for professional and enthusiast programmers" called Stack Overflow (available at http://stackoverflow.com/). Often the questions are so basic with respect to the relevant field of knowledge, that it is apparent that people who ask them are lacking of elemental knowledge, yet they are trying to create a website, some of them are even trying to do it for business.

Another issue of JavaScript, apart from bad programming techniques, is its overuse. Inexperienced developers often download whole JavaScript libraries in order to add functionality, which could be accomplished in cleaner and more efficient way. Sometimes they do not even need the libraries, they just incorporate them into the webpage as a part of "useful" package of functions because it is simple. The same goes for front-end frameworks such as Bootstrap (available at http://getbootstrap.com/) or Foundation (available at http://foundation.zurb.com/), which offer a collection of code, styles and functions.

4   Recommendations

Many of the discussed performance issues can be solved. However we need to distinguish between an ignorance or inactivity regarding online performance and design intent with calculated balance between website attractiveness and performance on the other hand. In this section we summarize and propose some of the general available solutions, which can be applied without further investigation by speed tests.

4.1 Size of the Images

The best recommendation concerning images would be of course not using many images or at least not many large images. If we really want to keep large images on the websites, we should at least optimize their compression rate and avoid multiple images alternating in carousel which are hidden but still being downloaded with the first display of a website. Issue with many images on a website can be solved e.g. by technique called lazy loading, which delays loading of images in long web pages. Images outside of viewport will not be loaded before user scrolls to them [24].

Considering graphics, in the past many effects could be made only by Photoshop and similar editors and they resulted in extra files which had to be downloaded with the website. Nowadays web designers can use CSS3 and achieve similarly attractive results. Also maintaining a CSS3-based design is easier than making changes to background images through a graphics program [25].

4.2 Responsive Web Design

Handling of images is still an open issue in responsive web design, but there are several ways of serving different sizes of images to different devices (and thus saving loading time for smaller, slower, low bandwidth mobile devices). One of the solution is "Adaptive Images", which detects user's screen size and automatically creates, caches, and delivers device appropriate re-scaled versions of webpage's embedded HTML images [26].

Another improvement in loading speed would be to separate styles for individual resolutions and load them conditionally. It is also better to use mobile-first approach, which uses many default values, instead of desktop-first, which would overwrite all styles and then overwrite again, often to default values.

4.3 JavaScript

As was already mentioned, web designers should learn to use JavaScript more efficiently. One of the simplest techniques how to make our code more efficient is minifying the code. JavaScript minification is quite simple to do and saves on total file size of .js files (and also loading speed) [22]. Then there are of course best practices of how to optimize the code regarding performance, lying in the basics like use of variables, functions or loops.

We can improve loading speed by not including libraries and functions which the particular website does not need for its functioning. Also we can improve loading speed significantly by coding in pure JavaScript instead of coding with the help of a library, which however requires better knowledge of JavaScript. Barker showed a difference in performance (and also loading speed) between use of pure JavaScript and jQuery library (available at http://jquery.com/). He demonstrated e.g. 93% improvement of average benchmark time by using pure JavaScript to access the DOM in Firefox browser instead of using JQuery library [22].

5   Speed Test and Settings

General recommendations are sufficient for websites with rather small audience which function as an additional presentation medium. When the website plays a key presentation role and expects many new or returning users, its performance should be analyzed more thoroughly. Speed tests are suitable for this purpose. They measure loading speed of the website, show what elements cause the biggest delay in website´s response time and also offer comparison and recommendations.

5.1   Choosing the Right Speed Test

We used the following approach to choose a suitable speed test for purposes of this study. Firstly we analyzed functionality of several online services, which were picked out according to their rank in Google search with keywords "website speed test". First ten results (Google PageSpeed appeared two times) are presented in [Table 1].

Table 1. Online services available for website speed testing

Service name	URL address	Status
pingdom	http://tools.pingdom.com/	functional
WebPagetest	http://www.webpagetest.org/	functional
GTmetrix	http://gtmetrix.com/	functional
dotcom-monitor	https://www.dotcom-tools.com/website-speed-test.aspx	functional
Google PageSpeed	https://developers.google.com/speed/pages peed/insights/	functional but only verbal evaluation
Web Page Analyzer	http://www.websiteoptimization.com/servic es/analyze/	unmaintained
Website Speed Test	http://www.pagescoring.com/website-speed-test/	non-functional
WEBSITETEST	http://www.websitetest.com/	functional
Website speed check	http://rapid.searchmetrics.com/en/seo-tools/site-analysis/website-speed-test,46.html	non-functional

Five functional and suitable services remained from the original set, which were analysed more thoroughly in the next step. The analysis focused on the available settings of speed test, including registration if the service had that possibility. The basic settings possibilities are presented in the following table [Table 2].

On the basis of this analysis of online services for website speed testing, WebPagetest service was finally chosen as the most complex service, offering many choices of connection type, location and browser of the test server and many advanced choices of test processing. This service was also recommended in Barker's Pro JavaScript Performance book. The number and range of parameters, which we can configure for WebPagetest, is extraordinarily robust [22].

Table 2. Further analysis of services for website speed testing

Service name	settings without login	added settings after login
pingdom	location (3)	N/A (long-term monitoring of one site, paid for more sites)
WebPagetest	location (47), connection (8) browser (8 desktop, 3 mobile) *)	N/A (registration is available only for a discussion forum)
GTmetrix	x	location (7), connection (6) browser (2 desktop, 1 mobile)
dotcom-monitor	location (22) browser (5 desktop, 1 mobile on several devices)	N/A (paid service for long-term monitoring of websites)
WEBSITETEST	location (14), connection (4) browser (4 desktop, 0 mobile)	N/A (new registrations are not available)

*) browsers and devices depend on selected location.

5.2   Basic Settings of the Speed Test

In the previous section WebPagetest was selected as an online service for website speed testing in this study. Because we will test websites with czech domains ".cz", we will use location of the web server Prague, Czech Republic. This location offers testing on browsers IE11, Chrome, Canary, Firefox and Safari for Windows. We can find the most used web browsers in the Czech Republic on gemiusRanking website (available at http://www.rankings.cz/). In January 2015, the share of Google Chrome group (all versions) in the Czech Republic was 30%, Firefox group reached 24,6%, Internet Explorer version 11 reached 15,8% and Safari 0,9% [27]. Testing with combination of browsers Chrome, Firefox and IE11 will reach over 70% coverage on internet usage in the Czech Republic, which is acceptable for our purposes.

This configuration will however not test performance on mobile devices. That is especially WebKit / Chrome Mobile with 12,8% coverage in the Czech Republic. WebPagetest at the time of writing this paper offered only one location for testing on mobile devices (Dulles, VA USA). Only one other speed test offered testing on mobile devices - dotcom-monitor - however it offered only Safari browser. Type of internet connection was kept at native connection with no traffic shaping.

5.3   Reliability of Speed Tests

If we want to measure a website's speed more than once (we are interested in the development of loading speed) or we want to compare several websites regarding their performance, we should be interested in the reliability of repeated speed tests. Otherwise the difference in measured values may be not the real difference in performance but rather the difference caused by inconsistent testing.

This issue is connected with a problem of default settings. Usually we expect that default settings are recommended settings and if do not understand it very well; we should keep it at default values. However the majority of analyzed services at default settings arbitrarily assigned a location and browser (used for the particular speed testing) to our request. The choice of browser is connected to the choice of location and some services don't offer more browsers in one place. Information about the browser is sometimes even not included in the report.

6 Performed Speed Test

We have conducted a series of speed tests with WebPagetest to analyze the impact of inconsistent settings on speed testing reliability. For testing, we have selected two websites, which ranked the highest in Google search on "webove stranky" in czech, which translates to "websites" in English. Both of these websites are representative presentations of web agencies, which offer web services. At the same time these two websites are different enough to show significant differences in loading speed. The second tested website (www.inpage.cz) uses responsive design and more images and JavaScript in contrast to the first tested website (www.webnode.cz). Three tests were performed for every browser-location combination to eliminate the majority of random effects. Average values from these three tests are shown in [Table 3] together with standard deviations, which represent variability between individual settings.

Table 3. Speed test by WebPagetest performed on the website Webnode (www.webnode.cz)

Location	Average loading speed [s]			Standard deviation
	Chrome	Firefox	IE11	
Prague	1,89	0,97	0,72	0,62
London	3,15	1,71	1,78	0,81
Vienna	4,82	3,01	2,8	1,11
Reykjavik	3,07	2,06	2	0,60
Amsterdam	4,69	2,91	4,74	1,04
Standard deviation	1,23	0,85	1,5	

Table 4. Speed test by WebPagetest performed on the website InPage (www.inpage.cz)

Location	Average loading speed [s]			Standard deviation
	Chrome	Firefox	IE11	
Prague	3,85	3,54	1,91	1,04
London	7,7	5,33	7,82	1,41
Vienna	10,92	6,79	54,22	26,27
Reykjavik	6,56	7,83	7,20	0,63
Amsterdam	10,11	5,39	11,9	3,37
Standard deviation	2,84	1,63	21,3	

It is apparent from our results, that choice of both location and browser have quite a significant effect on measured values. The values are more variant in the case of larger websites, which takes longer to load. In our case, the size of Webnode was 556kb while the size of InPage was 2060kb, which is about four times larger. The biggest difference was registered on InPage website with IE11 browser in Vienna. It is probable that some error occurred in this particular testing session, corrective speed test performed for this combination resulted in value 15,33s. However regular user cannot distinguish between a correct result and faulty one, so this occurrence also belong to the evaluation of reliability of speed tests.

7   Conclusions

Our results indicate that the choice of browser and location affects significantly testing of website loading speed. This turns out to be a very serious issue in the case of comparing the test results, which we however need in most cases. Either we compare between our website and competitive websites, or we compare previous versions of our site with the current one. The lengthy and costly process of redesigning a website is often based on these results, so it is very important that they are reliable. If we rely on default settings, we get significant hidden inaccuracies in the measured values, which are caused by inconsistent settings.

First, we recommend taking use of known issues and general recommendations regarding website performance. Second, we recommend incorporating speed testing into designing and redesigning processes, under these conditions: 1) choosing a speed test service which offers choice between various browsers and testing locations, 2) use consistent settings in every test, 3) perform repeated tests and use average values in order to reduce inconsistencies caused by issues, which we cannot influence.

Acknowledgment. This work and the contribution were also supported by project "Smart Solutions for Ubiquitous Computing Environments" FIM, University of Hradec Kralove, Czech Republic (under ID: UHK-FIM-SP-2015-2103).

References

1. Galletta, D., Henry, R., McCoy, S., Polak, P.: Website Delays: How tolerant are users? Journal of the AIS **5**(1), Article 1 (2004)
2. Rosen, D.E., Purinton, E.: Website design: viewing the web as a cognitive landscape. Journal of Business Research **57**, 787–794 (2004)
3. Lee, Y., Kozar, K.A.: Understanding of website usability: Specifying and measuring constructs and their relationships. Decision Support Systems **52**(2), 450–463 (2012)
4. Morkes, J., Nielsen, J.: Concise, Scannable, and Objective: How to Write for the Web (1997). http://www.useit.com/papers/webwriting/writing.html
5. Constantinides, E.: Influencing the online consumer's behavior: the Web experience. Internet Research **14**(2), 111–126 (2004)
6. Gehrke, D., Turban, E.: Determinants of successful web site design: relative importance and recommendations for effectiveness. In: Proceedings of the 31st Hawaii International Conference on Information Systems (1999)

7. Aladwani, A.: An empirical test of the link between web site quality and forward enterprise integration with web customers. Business Process Management Journal **12**(2), 178–190 (2006)
8. Loiacono, E., Watson, R., Goodhue, D.: WebQual: an instrument for consumer evaluation of web sites. International Journal of Electronic Commerce **11**(3), 51–87 (2007)
9. Cebi, S.: Determining importance degrees of website design parameters based on interactions and types of websites. Decision Support Systems **54**(2), 1030–1043 (2013)
10. Green, D.T., Pearson, J.M.: The Examination of Two Web Site Usability Instruments for Use in B2C eCommerce Organizations. Journal of Information Systems **49**(4), 19–32 (2009)
11. Behan, M., Krejcar, O.: Adaptive graphical user interface solution for modern user devices. In: Pan, J.-S., Chen, S.-M., Nguyen, N.T. (eds.) ACIIDS 2012, Part II. LNCS, vol. 7197, pp. 411–420. Springer, Heidelberg (2012)
12. Rose, G., Straub, D.W.: Predicting General IT Use: Applying TAM to the Arabic World. Journal of Global Information Management **6**(3), 39–46 (1998)
13. Cockburn, A., McKenzie, B.: What do Web users do? An empirical analysis of Web use. International Journal of Human-Computer Studies **54**, 903–922 (2001)
14. Schenkman, B., Jönsson, F.: Aesthetics and preferences of web pages. Behaviour and Information Technology **19**(5), 367–377 (2000)
15. Bartuskova, A., Krejcar, O.: Evaluation framework for user preference research implemented as web application. In: Bădică, C., Nguyen, N.T., Brezovan, M. (eds.) ICCCI 2013. LNCS, vol. 8083, pp. 537–548. Springer, Heidelberg (2013)
16. Work, S.: How Loading Time Affects Your Bottom Line (2015). https://blog.kissmetrics.com/loading-time/
17. Lightner, N.J., Eastman, C.: User preference for product information in remote purchase environments. Journal of Electronic Commerce Research **3**(3), 174–186 (2002)
18. Kadlec, T.: Implementing responsive design: Building sites for an anywhere, everywhere web. New Riders, Berkeley (2013)
19. Nielsen, J., Pernice, K.: Eyetracking Web Usability. New Riders Publishing, Thousand Oaks (2009)
20. Marcotte, E.: Responsive Web Design. A Book Apart (2011)
21. Sarmiento, J.P.: Web Design Shock. 11 reasons why Responsive Design isn't that cool! (2015). http://www.webdesignshock.com/responsive-design-problems/
22. Barker, T.: Pro JavaScript Performance: Monitoring and Visualisation. Apress, New York (2012)
23. Souders, S.: JavaScript Performance. Impact of JavaScript (2012). http://www.stevesouders.com/blog/2012/01/13/javascript-performance/
24. Tuupola, M.: Lazy load (2014). http://plugins.jquery.com/lazyload/
25. Weyl, E., Lazaris, L., Goldstein, A.: HTML5 & CSS3 For The Real World. SitePoint (2011). ISBN:978-0-9808469-0-4
26. Wilcox, M.: Adaptive Images. Deliver small images to small devices (2014). http://adaptive-images.com/
27. gemiusRanking: Web browsers (2015). http://rankings.cz/en/rankings/web-browsers.html
28. Mashinchi, R., Selamat, A., Ibrahim, S., Krejcar, O.: Granular-rule extraction to simplify data. In: Nguyen, N.T., Trawiński, B., Kosala, R. (eds.) ACIIDS 2015. LNCS, vol. 9012, pp. 421–429. Springer, Heidelberg (2015)
29. Penhaker, M., Krejcar, O., Kasik, V., Snášel, V.: Cloud computing environments for biomedical data services. In: Yin, H., Costa, J.A., Barreto, G. (eds.) IDEAL 2012. LNCS, vol. 7435, pp. 336–343. Springer, Heidelberg (2012)

Knowledge Management in Website Quality Evaluation Domain

Paweł Ziemba[1(✉)], Jarosław Jankowski[2,3], Jarosław Wątróbski[2], and Jarosław Becker[1]

[1] The Jacob of Paradyż University of Applied Sciences in Gorzów Wielkopolski,
Teatralna 25, 66-400, Gorzów Wielkopolski, Poland
{pziemba,jbecker}@pwsz.pl
[2] West Pomeranian University of Technology, Żołnierska 49, 71-210, Szczecin, Poland
{jjankowski,jwatrobski}@wi.zut.edu.pl
[3] Department of Computational Intelligence, Wrocław University of Technology,
Wybrzeże Wyspiańskiego 27, 50-370, Wrocław, Poland

Abstract. This paper deals with the problem of building a repository of knowledge about the methods and models of assessing the quality of Internet services. The repository has been constructed in the form of ontologies representing the various methods of quality assessment. For this purpose, the algorithm was developed based on ontologies with conceptualization of knowledge contained in the individual methods, and each completed ontology was evaluated. As a result of the research, domain ontologies were implemented reflecting the website quality assessment methods.

Keywords: Ontology evaluation · Knowledge conceptualization · Website quality

1 Introduction

In the sectors related to e-commerce and online advertising, revenue is directly determined by the number of users visiting corporate websites, blogs, portals and social platforms. More users increase the potential of advertising and this has a direct impact on the number of transactions and the amount of revenue, as well as attraction of new customers [3]. In the United States, recent revenues online advertising amounted to 42.78 billion dollars [1], while in Europe this figure was 27.3 billion euros [2]. It is worth noting that for businesses using a website to generate transactions, the website's quality can have a major impact on sales [4]. Poor quality of service and user experience can cause existing Internet customers [4], potential sales and repeat visits to be lost [5]. Therefore, in order to maximize profits from electronic commerce or online advertising, website owners should take care to offer only the highest quality.

The quality of a website can be understood as the attribute that specifies how well it meets the needs of users [6] [35]. It should be noted that quality is defined by a model composed of characteristics and features/criteria describing its various components [14]. In the literature, there are many methods used to assess the quality of Internet services, with the most formalized including: eQual [7], Ahn [8] SiteQual [9],

© Springer International Publishing Switzerland 2015
M. Núñez et al. (Eds.): ICCCI 2015, Part II, LNCS 9330, pp. 75–85, 2015.
DOI: 10.1007/978-3-319-24306-1_8

Web Portal Site Quality [10] and Website Evaluation Questionnaire [11]. They have been widely used in both academic work [12] and business practice [13]. Analysis of the literature and areas of practical use of methods and models for the assessment of website quality indicates a gap in the area of a research repository of knowledge. The possible construction of such a repository in the form of ontologies allows formal specification and analysis of the various methods of assessment and specific influencing factors on the quality of a website [18], and consequently the sharing and reuse of the resulting area of domain knowledge [15]. The ontological form provides access to the knowledge and important is the ability to use the built ontology, e.g. in service quality assessment systems and their integration into the larger domain ontologies.

This paper presents an algorithm for constructing ontologies for different methods of assessing the quality of Internet services. In accordance with the developed algorithm, ontologies were constructed and evaluated based on the inference models and questions of competence. The article concludes with a presentation of research findings and possible future directions of work.

2   Literature Review

The term "ontology" in computer science is defined as "the specification of conceptualization" [16] and it allows concepts and domain knowledge to be captured. A similar definition says that an ontology is treated as a data structure and a tool for data representation, allowing knowledge to be shared and reused in artificial intelligence systems that use a common vocabulary [17]. Therefore ontology seems to be a natural form of representation of the repository of knowledge about the methods of quality assessment. This is due to the fact that the use of ontologies will create conceptual models explaining the structure of the different methods of evaluation criteria. The use of ontologies will also be shared and repeated use of such structures is possible to facilitate management. The possibility of using ontologies as a repository of knowledge is confirmed by work [26] where a biomedical knowledge base was created using ontologies. Ontologies are also implemented in the knowledge bases of systems, e.g. an expert system for the study of company financial ratios [19] or a decision support system for the construction of railway infrastructure [20]. In [21] ontologies, in addition to rule-based expert systems, other modules are present within the agent system. The role of ontologies in this system is to enter user queries, return results to users, and provide a system of knowledge from experts and knowledge engineers. Analysis of the literature shows that ontologies are also used in the systems and methods of quality assessment. For example, the work [18] presents an ontology quality that formalizes the knowledge necessary to evaluate the quality of e-government. This ontology is then used in a self-adaptive quality monitoring system, used to test the quality of services provided within e-government platforms [22]. In this system, in addition to ontology quality, ontologies also use characteristics of portals, user behavior and the problems encountered by users while using the portal [23]. In contrast, [24] consider the use of ontologies in the assessment of the quality of tourism services websites. In this case, the ontology would be part of the provision of a description of issues related to the field of tourism and it could assist in identifying the requirements for this type of website. Its application is presented by the authors in comparison to other possible methods to determine the requirements for tourism services [25]. However, analysis of the literature showed that there is a repository of knowledge covering a few important methods for assessing the quality of services. Meanwhile, the reasons discussed above and the use of ontologies show that the construction of a repository of knowledge about the methods of assessing the quality of web services based on ontologies is justified.

3 Ontology Building Framework

The ontology construction methodology is frequently used, differing in the degree of formalization, destination and detail [29]. The most formal and detailed methodology includes Methontology [27] and NeOn [28]. Methontology defines in detail the process of conceptualization, while NeOn largely formalizes the problem of ontology specification. Therefore, the author's quality assessment methods are based on these two methodologies. This algorithm is shown in Fig. 1.

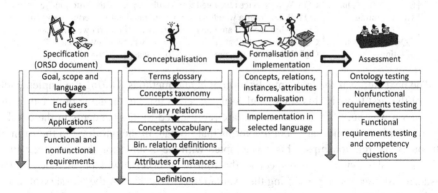

Fig. 1. Algorithm for ontology implementation

The first stage of the algorithm is a specification involving the preparation of the document ORSD (Ontology Requirements Specification Document) [31], which has the goal of building an ontology, the language of its implementation, and questions of competence used to verify the correctness of the ontology after its construction. The second phase includes the tasks leading to the conceptualization of knowledge, which is to include the ontology. The next step is to formalize the ontology based on a previously made conceptualization and implement it in the selected language and use the appropriate ontology editor. The last step in the construction of the ontology is its assessment, checking the consistency of ontology reasoners by using and verifying answers to the questions of competence defined in the specification phase. Fig. 1 shows the algorithm used to build an ontology of conceptual methods for assessing the quality of websites, i.e.: eQual, Ahn, SiteQual, Website Evaluation Questionnaire and Web Portal Site Quality.

4   Building Ontologies of Websites' Quality Evaluation Methods

The first built ontology was eQual. According to the developed algorithm implemented at the beginning stage of the specification, the effect was the document ORSD which characterized the requirements of the ontology'. Part of this document is presented in Table 1.

Table 1. Partial requirements of the ORSD eQual ontology

1.	Purpose - Reflected in the ontology quality assessment method eQual
2.	Scope - The ontology includes characteristics, criteria and the evaluation scale model eQual
3.	Implementation language - OWL 2 DL
4.	Users - Experts evaluating the quality of websites
5.	Applications - Evaluation of quality websites
6a	Non-functional requirements - NFR1. The names of classes and instances begin with a capital letter (excluding proper names and names of criteria). NFR2. The names of the attributes and relationships consist of a verb written with an initial lower case letter and a noun written with an initial capital letter, e.g. "hasValue".
6b	Functional requirements
	CQG1. Belonging and weight	CQ1. What criteria do the quality characteristics of "Empathy" have?
		CQ2. What criteria have a weight factor of $> = 5$?
	CQG2. The value of services ratings	CQ3. What services are rated $> = 5$ with respect to the criterion "reputation"?
		CQ4. What criteria weighing $> = 5.0$ are rated < 4 for service "website2"?
		CQ5. What services are rated $> = 6$ in terms of criteria weighing $> = 5.5$?
	CQG3. Completeness of ratings	CQ6. In terms of what criteria is service "website2" evaluated?
		CQ7. Which services are assessed for at least 8 criteria?

The next step was the conceptualization of eQual methods. During the implementation of the first tasks in this stage, a glossary of terms used in the table of eQual method criteria was built [7]. These criteria are represented by the concepts in the ontology. In addition, the glossary contains terms that operate in the ontology in the form of attributes and relationships. This was adopted with the assumption that the assessment of individual services with respect to the following criteria will be included in the instances of concepts representing the evaluation criteria. Part of the glossary of terms in the eQual ontology is presented in Table 2, where C is the concept, R is the relationship, A_I is the attribute instance, and A_C is the attribute class/concept.

Table 2. Partial glossary of terms for the eQual ontology

Term	Description	Type
Personalization	Feature describing the degree of personalization service	C
Information quality	Characteristics of grouping attributes relating to the quality of information	C
eQual	The name of the quality assessment method	C
isCriterion(C,Ch)	Feature C is the criterion associated with characteristics Ch	R
hasCriterion(Ch,C)	Characteristics Ch criterion is assigned to C	R
hasEvaluation(I_w,I_c)	I_w has a rating service for the instance and features of C	R
isEvaluation(I_c,I_w)	An instance of I features of C contains an assessment of the I_w service	R
hasEvaluationValue (I_c,Integer)	An instance of I features of C is an integer value assessment	A_I
hasWeightValue (C,Double)	Feature C is the criterion of having a floating-point value of the weight	A_C

Another task carried out in the conceptualization phase was the construction of a taxonomy of concepts. Fig. 2 shows the hierarchy of the separate attributes (criteria) of the quality of the individual characteristics of the grouping criteria.

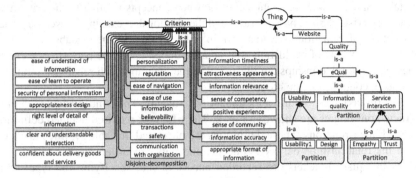

Fig. 2. Taxonomy of concepts in the eQual ontology

This approach allowed more transparent applications of the ontology. This is also consistent with the representation of open and closed worlds in the ontology [32]. Namely, the quality of each model contained in the various assessment methods is a closed model. This means that it is complete and a new one cannot be added to it. The quality evaluation criteria are the open portion of the world, which means that there may be additional criteria not included in the ontology to date.

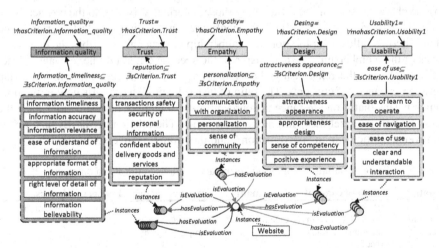

Fig. 3. Diagram of the ad hoc binary relations of the eQual ontology

For these reasons, the part reflecting the ontology of the eQual quality model was used to partition taxonomic relationships between concepts occurring at the same level of the hierarchy. In contrast, separability of the taxonomic relationship was among the criteria used, since these criteria describe the different parts' quality, but there may be other criteria describing the quality.

The next task was to build an ad hoc diagram of binary relations. It was decided to provide all the relationships here on one diagram, because it allows us to more clearly demonstrate the concept articulated in the ontology. The diagram is shown in Figure 3, and it takes into account the relationships between concepts relating to the criteria and characteristics (isCriterion, hasCriterion) and the relationship existing between instances of concepts (hasEvaluation, isEvaluation). In order to improve the readability of Fig. 3, the different concepts of evaluation criteria are grouped under the performance characteristics. Specific relationships are illustrated in the diagram; each relationship refers to each concept contained in the group.

The fourth task was to build a dictionary of concepts. The concepts included in the constructed ontologies are shown in Fig. 2. The fifth task required detailed definitions of ad hoc binary relations. These definitions are presented in Table 3, which takes into account the relationships between concepts and between instances of concepts. The concepts in square brackets are recorded ancestors (i.e. specifying the type) or instances involved in the relationship. The relations "isCriterion" and "hasCriterion" exist between concepts. When the relationship is "isCriterion", the source concepts are different quality criteria and the concepts are the specific characteristics of the target, which, according to the model eQual, include criteria. In the other work relationship, "hasCriterion", the concept of both the source and the destination is the same relating to the characteristics of the quality of the model eQual. This solution, in a situation where C1 represents the characteristics of the concept and the concept C2 means the criteria, can be understood as follows: (a) there are certain criteria in the C2 group that belong to the characteristics of the C1 group ("C2 isCriterion some C1"), (b) the characteristics of the C1 are only those criteria that belong to the characteristic ("C1 hasCriterion only C1"). This configuration relationship "isCriterion" and "hasCriterion" allows exploration of membership criteria and describes the characteristics of the membership as a conclusion to the specific criteria in the relevant characteristics. As for the relationships "isEvaluation" and "hasEvaluation", they overlap between instances of concepts. The ratio of "isEvaluation" has a jurisdiction function, which means that if one instance of source refers to a number of destinations, then the reasoner will interpret these instances as the same. The relation of "hasEvaluation" is an inverse relationship to "isEvaluation", and therefore it is inverse functional.

Table 3. Definitions of the ad hoc binary relations of the eQual ontology

Relation	Source concept	Cardinality	Target concept	Property	Inverse
isCriterion (condition necessary)	[Criterion] *undefined area* in example personalization	N - existential quantification	[eQual] *undefined range* in example Empathy	-	has Criterion
hasCriterion (condition necessary and sufficient)	[eQual] *undefined area* in example Empathy	1 - universal quantification	[eQual] *undefined range* in example Empathy	-	is Criterion
isEvaluation	[criterion] Website1_reputation	-	[Website] Website1	Functional	has Evaluation
hasEvaluation	[Website] Website1	-	[Criterion] Website1_reputation	Inverse functional	is Evaluation

The sixth task was to define attributes of specific instances and classes. For instance, in the ontology the eQual attribute ("hasEvaluationValue") determines the value of the evaluation of a given service award criterion. Specific values for this attribute are assigned to individual instances of concepts corresponding to the evaluation criteria. According to the eQual method, evaluation value is an integer in the range 1-7. This step also defines two classes of attributes, i.e. "hasEvaluationValue" and "hasWeightValue", the task of which is to limit the range of possible values for the attributes and classes, for instance from 1 to 7. They provide a mechanism to control the accuracy of the applicant weights and grades. In addition, the attribute "hasWeightValue" is also an attribute of the concepts contained in the class "Criterion", and it is to be stored in the weights of the criteria. The seventh task in the conceptualization stage was to create detailed definitions of constants that will be used in the ontology. There are two constants that relate to the scope of the assessment (assessment scale) and weights. The constant "Evaluation" is an integer and has a value point in the range of 1-7. The second constant, "Weight", is a double, and its possible values are well within the range of <1.0, 7.0>. In this step, the applied ontology editor Protege [33] was formalized and implemented. With its help, we formalized concepts, relationships, instances and attributes and implemented the eQual ontology language OWL 2 DL [34] on the basis of the conceptualization presented earlier.

The last step was the construction of an ontology evaluation involving, among others, the classification of concepts using a reasoner and checking the answers to the questions of competence ontology [28]. The eQual ontology is available in [36] and the effect of the reasoner is given in [37]. By analyzing the form of the ontology reasoner, it can be seen that different criteria are correctly assigned to the relevant performance characteristics. The hierarchy of concepts is generated by the reasoner, so it is consistent with the model of quality contained in the eQual method. The individual quality criteria are also included in the concept of "Criterion" so that the ontology is clear. As for the question of competence contained in Table 1, it checks the response of the ontology, and it was necessary to introduce the sample data presented in Table 4. The criteria that are not included in Table 4 were assigned a weight of 1.0. In addition, the ontology did not include their instances, and therefore none of the sites had assessments against the criteria listed in Table 4.

Table 4. Sample weighting of the criteria and evaluation in the eQual ontology

Characteristics	Criterion	Weight	Rate1	Rate2
Empathy	Communication with organization	4.1	5	1
	Personalization	2.5	4	6
	Sense of community	3.7	7	7
Trust	Transaction safety	1.4	5	3
	Security of personal information	5.1	2	4
	Confident about delivery of goods and services	4.3	3	-
	Reputation	6.2	5	2
Information quality	Information timeliness	6.7	6	2

The answers to the questions of competence ontology included in Table 1 are shown in Fig. 4. A comparison of Table 4 and Fig. 4 shows that all the answers are correct ontologies. Consequently, it is clear that the ontology is built correctly and fully reflects the quality assessment method of the eQual websites. According to the presented algorithm, there is another ontology structure that also reflects the evaluation of the quality of websites, which in turn presents the ontology in the basic form and is deduced by the reasoner: Ahn [38] SiteQual [39] Website Evaluation Questionnaire [40] and Web Portal Site Quality [41].

Fig. 4. Answers to the competence questions of the eQual ontology

5   Summary

This article discusses the problem of using conceptualization methods to evaluate the quality of websites. Five methods were used and the conceptualization of domain ontologies was then evaluated. In the evaluation process, the mechanism requesting competency questions was used, addressed to the built ontology. Recognition of the different methods in the form of an ontology allows the reuse and sharing of knowledge contained in the individual methods. Therefore, the natural direction of further research seems to be to build a combination of ontologies to enable integration of data from a variety of assessment methods. This would also allow for the evaluation of services through a variety of methods contained in the integrated ontology and for a comparison of the results of the assessment in a terminology and a reference plane. Such an ontology, along with the assessment criteria in the selection process presented in [30] could serve as the core of the expert system of assessment of website quality. An additional direction of future work should include the development of an ontology about the possibility of environmental data records and to build and maintain a repository of detailed use cases of a developed ontology.

Acknowledgments. The work was partially supported by Fellowship co-Financed by European Union within European Social Fund, by European Union's Seventh Framework Programme for research, technological development and demonstration under grant agreement no 316097 [ENGINE].

References

1. IAB: IAB internet advertising revenue report. 2013 full year results. Pricewaterhouse-Coopers LLP (2014)
2. IAB Europe: Adex Benchmark 2013. European online advertising expenditure. IAB Europe (2014)
3. Chiou, W.C., Lin, C.C., Perng, C.: A strategic framework for website evaluation based on a review of the literature from 1995–2006. Information & Management **47**, 282–290 (2010)
4. Kim, S., Stoel, L.: Dimensional hierarchy of retail website quality. Information & Management **41**, 619–633 (2004)
5. Hwang, J., Yoon, Y.S., Park, N.H.: Structural effects of cognitive and affective reponses to web advertisements, website and brand attitudes, and purchase intentions: The case of casual-dining restaurants. International Journal of Hospitality Management **30**, 897–907 (2011)
6. Chou, W.C., Cheng, Y.: A hybrid fuzzy MCDM approach for evaluating website quality of professional accounting firms. Expert Systems with Applications **39**, 2783–2793 (2012)
7. Barnes, S.J., Vidgen, R.: Data triangulation and web quality metrics: A case study in e-government. Information & Management **43**, 767–777 (2006)
8. Ahn, T., Ryu, S., Han, I.: The impact of Web quality and playfulness on user acceptance of online retailing. Information & Management **44**, 263–275 (2007)
9. Webb, H.W., Webb, L.A.: SiteQual: an integrated measure of Web site quality. Journal of Enterprise Information Management **17**, 430–440 (2004)
10. Yang, Z., Cai, S., Zhou, Z., Zhou, N.: Development and validation of an instrument to measure user perceived service quality of information presenting Web Portals. Information & Management **42**, 575–589 (2005)
11. Elling, S., Lentz, L., de Jong, M., van den Bergh, H.: Measuring the quality of governmental websites in a controlled versus an online setting with the 'Website Evaluation Questionnaire'. Government Information Quarterly **29**, 383–393 (2012)
12. Sorum, H., Andersen, K.N., Clemmensen, T.: Website quality in government: Exploring the webmaster's perception and explanation of website quality. Transforming Government: People, Process and Policy **7**, 322–341 (2013)
13. Kaya, T.: Multi-attribute Evaluation of Website Quality in E-business Using an Integrated Fuzzy AHPTOPSIS Methodology. International Journal of Computational Intelligence Systems **3**, 301–314 (2010)
14. ISO/IEC 25010: 2010(E): Systems and software engineering — Systems and software Quality Requirements and Evaluation (SQuaRE) — System and software quality models.
15. Hepp, M.: Ontologies: state of the art, business potential, and grand challenges. In: Hepp, M., de Leenheer, P., de Moor, A., Sure, Y. (eds.) Ontology Management. Semantic Web, Semantic Web Services, and Business Applications. Springer, Heidelberg (2008)
16. Gruber, T.R.: A translation approach to portable ontology specifications. Knowledge Acquisition **5**, 199–220 (1993)

17. Guzman-Arenas, A., Cuevas, A.D.: Knowledge accumulation through automatic merging of ontologies. Expert Systems with Applications **37**, 1991–2005 (2010)
18. Magoutas, B., Halaris, C., Mentzas, G.: An ontology for the multi-perspective evaluation of quality in e-government services. In: Wimmer, M.A., Scholl, J., Grönlund, Å. (eds.) EGOV. LNCS, vol. 4656, pp. 318–329. Springer, Heidelberg (2007)
19. Shue, L.Y., Chen, C.W., Shiue, W.: The development of an ontology-based expert system for corporate financial rating. Expert Systems with Applications **36**, 2130–2142 (2009)
20. Saa, R., Garcia, A., Gomez, C., Carretero, J., Garcia-Carballeira, F.: An ontology-driven decision support system for high-performance and cost-optimized design of complex railway portal frames. Expert Systems with Applications **39**, 8784–8792 (2012)
21. Orłowski, C., Ziółkowski, A., Czarnecki, A.: Validation of an Agent and Ontology-Based Information Technology Assessment System. Cybernetics and Systems **41**, 62–74 (2010)
22. Magoutas, B., Mentzas, G.: SALT: A semantic adaptive framework for monitoring citizen satisfaction from e-government services. Expert Systems with Applications **37**, 4292–4300 (2010)
23. Magoutas, B., Schmidt, K.U., Mentzas, G., Stojanovic, L.: An adaptive e-questionnaire for measuring user perceived portal quality. International Journal of Human-Computer Studies **68**, 729–745 (2010)
24. Mich, L., Franch, M.: Instantiating web sites quality models: an ontologies driven approach. In: Proc. of the CAISE 2005 Workshop on Web Oriented Software Technologies (2005)
25. Mich, L., Franch, M., Inverardi, P.N.: Choosing the "Rightweight" model for web site quality evaluation. In: Cueva Lovelle, J.M., González Rodríguez, B.M., Gayo, J.E.L., del Puerto Paule Ruiz, M., Aguilar, L.J. (eds.) ICWE 2003. LNCS, vol. 2722. Springer, Heidelberg (2003)
26. Villanueva-Rosales, N., Dumontier, M.: yOWL: An ontology-driven knowledge base for yeast biologists. Journal of Biomedical Informatics **41**, 779–789 (2008)
27. Corcho, Ó., Fernández-López, M., Gómez-Pérez, A., López-Cima, A.: Building legal ontologies with METHONTOLOGY and WebODE. In: Benjamins, V., Casanovas, P., Breuker, J., Gangemi, A. (eds.) Law and the Semantic Web. LNCS (LNAI), vol. 3369, pp. 142–157. Springer, Heidelberg (2005)
28. Villazon-Terrazas, B., Ramirez, J., Suarez-Figueroa, M.C., Gomez-Perez, A.: A network of ontology networks for building e-employment advanced systems. Expert Systems with Applications **38**, 13612–13624 (2011)
29. Casellas, N.: Legal Ontology Engineering. Methodologies, Modelling Trends, and the Ontology of Professional Judicial Knowledge. Springer, Dordrecht (2011)
30. Ziemba, P., Piwowarski, M., Jankowski, J., Wątróbski, J.: Method of criteria selection and weights calculation in the process of web projects evaluation. In: Hwang, D., Jung, J.J., Nguyen, N.-T. (eds.) ICCCI 2014. LNCS, vol. 8733, pp. 684–693. Springer, Heidelberg (2014)
31. Suárez-Figueroa, M.C., Gómez-Pérez, A., Villazón-Terrazas, B.: How to write and use the ontology requirements specification document. In: Meersman, R., Dillon, T., Herrero, P. (eds.) OTM 2009, Part II. LNCS, vol. 5871, pp. 966–982. Springer, Heidelberg (2009)
32. Knorr, M., Alferes, J.J., Hitzler, P.: Local closed world reasoning with description logics under the well-founded semantics. Artificial Intelligence **175**, 1528–1554 (2011)
33. Stanford Center for Biomedical Informatics Research. http://protege.stanford.edu/
34. Grau, B.C., Horrocks, I., Motik, B., Parsia, B., Patel-Schneider, P., Sattler, U.: OWL 2: The Next Step for OWL. Web Semantics: Science, Services and Agents on the World Wide Web **6**, 309–322 (2008)

35. Jankowski, J.: Integration of collective knowledge in fuzzy models supporting web design process. In: Jędrzejowicz, P., Nguyen, N.T., Hoang, K. (eds.) ICCCI 2011, Part II. LNCS, vol. 6923, pp. 395–404. Springer, Heidelberg (2011)
36. http://tinyurl.com/WQM-eQual-example-inst
37. http://tinyurl.com/WQM-eQual-inferred-example-ins
38. http://tinyurl.com/WQM-Ahn; http://tinyurl.com/WQM-Ahn-inferred
39. http://tinyurl.com/WQM-SiteQual; http://tinyurl.com/WQM-SiteQual-inferred
40. http://tinyurl.com/WQM-WEQ; http://tinyurl.com/WQM-WEQ-inferred
41. http://tinyurl.com/WQM-WPSQ; http://tinyurl.com/WQM-WPSQ-inferred

CSI: Special Session on Computational Swarm Intelligence

A New Algorithm to Categorize E-mail Messages to Folders with Social Networks Analysis

Urszula Boryczka, Barbara Probierz[✉], and Jan Kozak

Institute of Computer Science, University of Silesia,
Będzińska 39, 41–200 Sosnowiec, Poland
{urszula.boryczka,barbara.probierz,jan.kozak}@us.edu.pl

Abstract. This paper presents a new approach to an automatic categorization of email messages into folders. The aim of this paper is to create a new algorithm that will allow one to improve the accuracy with which emails are assigned to folders (Email Foldering Problem) by using solutions that have been applied in Ant Colony Optimization algorithms (ACO) and Social Networks Analysis (SNA). The new algorithm that is proposed here has been tested on the publicly available Enron email data set. The obtained results confirm that this approach allows one to better organize new emails into folders based on an analysis of previous correspondence.

Keywords: E-mail Foldering Problem · Ant Colony Optimization · Enron E-mail · Social Network Analysis

1 Introduction

In today's world, the media and social communication are among the most important factors in and methods of shaping contemporary social and economic reality. Communication via the Internet is a very simple yet convenient way of transmitting information. However, the biggest problem for the users is how to properly organize electronic mail and assign email messages to particular folders, especially when these are to be categorized automatically.

Email foldering is a complex problem because an automatic classification method can work for one user while for another it can lead to errors, which is often because certain users create new catalogs, but they also stop using some of the folders that were created earlier. At the same time folders do not always correspond to email subjects and some of them can refer to the tasks to do, project groups and certain recipients, while others only make sense in relation to previous email messages. Moreover, information items come at different times, which creates additional difficulties. A similar problem was dealt with in a paper [3] that presented a case study of benchmark email foldering based on the example of the Enron email data set. R. Bekkerman et al. classified emails from seven email boxes into thematic folders based on four classifiers: Maximum Entropy, Naive Bayes, Support Vector Machine and Wide-Margin Winnow.

Based on an analysis of this problem, the aim of the study was established, i.e. to create a new algorithm which would make it possible to better match new

© Springer International Publishing Switzerland 2015
M. Núñez et al. (Eds.): ICCCI 2015, Part II, LNCS 9330, pp. 89–98, 2015.
DOI: 10.1007/978-3-319-24306-1_9

emails to particular folders. The proposed method is based on the Ant Colony Optimization algorithm methodology and social networks. Social network analysis concerns the contacts between the sender and the recipients of email messages and it deals with studying the structure of folders that have been created in particular inboxes, which here was additionally marked with the pheromone trail.

This algorithm was used in a previously prepared data set of emails which had been obtained from the public Enron email data set especially for this purpose. The experiments that were conducted have confirmed that the proposed algorithm makes it possible to improve the accuracy with which email messages are assigned to folders.

This article is organized as follows. Section 1 comprises an introduction to the subject of this article. Section 2 describes Enron E-mail Dataset. Section 3 Social Network Analysis is presented. In section 4, describes Ant Colony Optimization. Section 5 focuses on the presented, new version approach based on ACO algorithm and SNA. Section 6 presents the experimental study that has been conducted to evaluate the performance of the proposed algorithm, taking into consideration Enron E-mail Dataset. Finally, we conclude with general remarks on this work and a few directions for future research are pointed out.

2   Enron E-mail Dataset

The Enron email data set constitutes a set of data which were collected and prepared as part of the CALO Project (a Cognitive Assistant that Learns and Organizes). It contains more than 600,000 email messages which were sent or received by 158 senior employees of the Enron Corporation. A copy of the database was purchased by Leslie Kaelbling with the Massachusetts Institute of Technology (MIT), and then it turned out that there were serious problems associated with data integrity. As a result of work carried out by a team from SRI International, especially by Melinda Gervasio, the data were corrected and made available to other scientists for research purposes.

This database consists of real email messages that are available to the public, which is usually problematic as far as other data sets are concerned due to data privacy. These emails are assigned to personal email accounts and divided into folders. There are no email attachments in the data set and certain messages were deleted because their duplicate copies could be found in other folders. The missing information was reconstructed, as far as possible, based on other information items; if it was not possible to identify the recipient the phrase no_address@enron.com was used. The Enron data set is commonly used for the purpose of studies that deal with social network analysis, natural language processing and machine learning.

The Enron email data set is often used to create a network of connections that represent the informal structure of an organization [13] and the flow of information within a company [7]. J. Shetty and J. Adibi constructed a database scheme on the basis of the Enron data set [15] and they also carried out research by using the entropy model to identify the most interesting and the most important nodes in the graph [16]. Studies on the dynamic surveillance of suspicious individuals were conducted by Heesung Do, Peter Choi and Heejo Lee and presented in a paper [9] by using a method based on the Enron data set.

3   Social Network Analysis

A Social Network is a multidimensional structure that consists of a set of social entities and the connections between them. Social entities are individuals who function within a given network, whereas connections reflect various social relations between particular individuals. The first studies of social networks were conducted in 1923 by Jacob L. Moreno, who is regarded as one of the founders of social network analysis. SNA is a branch of sociology which deals with the quantitative assessment of the individual's role in a group or community by analyzing the network of connections between individuals. Moreno's 1934 book that is titled Who Shall Survive presents the first graphical representations of social networks as well as definitions of key terms that are used in an analysis of social networks and sociometric networks [14].

A social network is usually represented as a graph. According to the mathematical definition, a graph is an ordered pair $G = (V, E)$, where V denotes a finite set of a graph's vertices, and E denotes a finite set of all two-element subsets of set V that are called edges, which link particular vertices such that:

$$E \subseteq \{\{u, v\} : u, v \in V, u \neq v\}. \tag{1}$$

vertices represent objects in a graph whereas edges represent the relations between these objects. Depending on whether this relation is symmetrical, a graph which is used to describe a network can be directed or undirected.

Edges in a social network represent interactions, the flow of information and goods, similarity, affiliation or social relationships. Therefore, the strength of a connection is measured based on the frequency, reciprocity and type of interactions or information flow, but this strength also depends on the attributes of the nodes that are connected to each other and the structure of their neighborhood.

The degrees of vertices and the degree centrality of vertices are additional indicators that characterize a given social network. The degree of a vertex (indegree and outdegree) denotes the number of head endpoints or tail endpoints adjacent to a given node. Degree centrality is useful in determining which nodes are critical as far as the dissemination of information or the influence exerted on immediate neighbors is concerned. Centrality is often a measure of these nodes' popularity or influence. The probability that the immediate neighbors of vertex v are also each other's immediate neighbors is described by the clustering coefficient gc_v of vertex v such that:

$$gc_v = \frac{2E_v}{k_v(k_v - 1)}, \quad k_v > 1, \tag{2}$$

where E_v is the number of edges k_v between the neighbors of vertex v.

Many studies that were carried out as part of SNA were aimed at finding correlation between a network's social structure and efficiency [12]. At the beginning, social network analysis was conducted based on questionnaires that were filled out by hand by the participants [8]. However, research carried out by using email messages has become popular over time [1]. Some of the studies found that research teams were more creative if they had more social capital [11]. Social networks are also associated with discovering communication networks. The database which was used in the experiments that are presented in this article can be used to analyze this problem. G. C. Wilson and W. Banzhaf, among others, discussed such an approach, which they described in their article [19].

4   Ant Colony Optimization – Ant Colony Decision Trees

The inspiration for creating Ant Colony Optimization algorithms came from the desire to learn how animals that are almost blind, such as ants, are able to find the shortest path from the nest to the food source. The studies and experiments that were conducted by S. Goss and J.L. Deneubourg as well as those that were presented in papers [2,17,18] and aimed to explain how nature accomplishes this task constituted the first step toward implementing this solution in algorithms. However, only the attempts made by M. Dorigo [10], to create an artificial ant colony system and use it to find the shortest path between the vertices of a given graph were the key step toward creating ACO algorithms.

While searching for food, agent-ants create paths, on which they lay a pheromone trail. This allows them to quickly return to the nest and communicate information about the location of food to other ants. Based on feedback, the shortest paths, i.e. paths that are assigned large pheromone trail values, between the nest and the food source are created.

An agent-ant makes every decision concerning another step in accordance with the formula:

$$j = \begin{cases} \arg\max_{r \in J_i^k}\{[\tau_{ir}(t)] \cdot [\eta_{ir}]^\beta\}, & \text{if } q \le q_0 \\ p_{ij}^k(t), & \text{otherwise,} \end{cases} \tag{3}$$

where τ_{ij} is the value of the reward, i.e. the degree of usefulness of the decision option that is being considered; η_{ir} is the value of the quality of a transition from state i to state r which was estimated heuristically; β is the parameter describing the importance of values η_{ir}; $p_{ij}^k(t)$ is the next step (decision) which was randomly selected by using the probabilities:

$$p_{ij}^k(t) = \begin{cases} \frac{\tau_{ij}(t) \cdot [\eta_{ij}]^\beta}{\sum_{r \in J_i^k} \tau_{ir}(t) \cdot [\eta_{ir}]^\beta}, & \text{if } j \in J_i^k \\ 0, & \text{otherwise,} \end{cases} \tag{4}$$

where J_i^k denotes the set of decisions that ant k can make while being in state i.

After a single ant-agent has covered the whole distance, pheromone trail is laid out on each edge it has visited (i, j). Let us assume that τ_{ij} denotes the pheromone trail value on the edge (i, j):

$$\tau_{ij}(t+1) = (1 - \rho) \cdot \tau_{ij}(t) + \rho \cdot \tau_0, \tag{5}$$

where ρ denotes a coefficient such that $0 \le \rho \le 1$ represents the residue of the pheromone trail, whereas τ_0 value equal to the initial pheromone value.

Ant Colony Decision Trees (ACDT) algorithm [5] employs Ant Colony Optimization techniques for constructing decision trees. As mentioned before, the value of the heuristic function is determined according to the splitting rule employed in CART approach [6]. These algorithms were described in [4].

The evaluation function for decision trees will be calculated according to the following equation:

$$Q(T) = \phi \cdot w(T) + \psi \cdot a(T, P), \tag{6}$$

where $w(T)$ is the size (number of nodes) of the decision tree T; $a(T, P)$ is the accuracy of the classification object from a training set P by the tree T; ϕ and ψ are constants determining the relative importance of $w(T)$ and $a(T, P)$.

5   Proposed Algorithm

The previously conducted studies that are described in paper [4] confirmed the validity of using decision tables and ACO algorithms, whereas the creation of the social network of electronic mail users as well as an analysis of this network provided inspiration for creating a new algorithm that would make it possible to properly assign email messages to folders. The way in which this algorithm works is presented in Fig. 1.

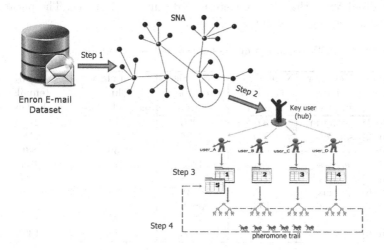

Fig. 1. Diagram of the proposed algorithm

The first step (Fig. 1 - step 1, Alg. 1 - line 1) that was taken to use this method for improving the accuracy with which email messages are assigned to folders was to create a social network based on the contacts between the sender and the recipients of emails obtained from the Enron data set. All users of the mailboxes obtained from the Enron email data set constitute the vertices, whereas the connections for which the frequency of interactions, i.e. the number of emails sent between particular individuals, was more than 10 constitute the edges. Then, during the analysis of this network, key users (hubs) and their immediate neighbors were selected (Fig. 1 - step 2, Alg. 1 - line 2), thus creating groups of users within the network, which are presented in Tab. 1.

Table 1. Selected groups of users

Name of groups	Key user (hub)	Hub's closest neighbors
Group 1	lokay-m	hyatt-k, mcconnell-m, schoolcraft-d, scott-s, watson-k
Group 2	sanders-r	cash-m, dasovich-j, haedicke-m, kean-s, sager-e, steffes-j
Group 3	shackleton-s	jones-t, mann-k, stclair-c, taylor-m, ward-k, williams-j
Group 4	steffes-j	dasovich-j, gilbertsmith-d, presto-k, sanders-r, shapiro-r
Group 5	symes-k	scholtes-d, semperger-c, williams-w3
Group 6	williams-w3	mann-k, semperger-c, solberg-g, symes-k
Group 7	farmer-d	bass-e, beck-s, griffith-j, nemec-g, perlingiere-d, smith-m
Group 8	beck-s	buy-r, delainey-d, hayslett-r, kaminski-v, kitchen-l, may-l, mcconnell-m, shankman-j, white-s
Group 9	rogers-b	baughman-d, davis-d, griffith-j, kitchen-l, lay-k

Key users constitute the network's nodes with the highest vertex degree. The users who have no connections within the network are those individuals who received fewer than the predetermined number of emails. The parameters of selected groups are presented in Tab. 2.

Table 2. Parameters of selected groups of users.

Name of groups	Key user (hub) in group	N. of classes (folders) in group	N. of objects (emails) in group	N. of classes (folders) for the key user	N. of objects (emails) for the key user
Group 1	lokay-m	87	4963	11	2493
Group 2	sanders-r	230	8024	29	1181
Group 3	shackleton-s	126	4131	39	886
Group 4	steffes-j	154	4247	21	617
Group 5	symes-k	35	3598	11	767
Group 6	williams-w3	58	5138	18	2767
Group 7	farmer-d	156	5998	24	3538
Group 8	beck-s	185	6522	84	1703
Group 9	rogers-b	81	2963	14	1395

The next step that is to be taken in order to implement the proposed algorithm is to transform the set of data from the Enron email data set into a decision table, separately for each mailbox within a given group (Fig. 1 - step 3, Alg. 1 - line 6). The prepared decision table consists of six conditional attributes and one decision attribute *category*, which defines the folder to which the email is assigned. Conditional attributes were selected to define the most important information about each message. They consist of the information from the sender field, the first three words from the email subject (with the exception of basic words and copulas) where additionally, words which belong to the set of decision classes are supported. The length of the message, and the information about the person who received the message was added to a courtesy copy (CC) (as the Boolean value) is also checked by the conditional attributes. If the person is not in a courtesy copy - the person is the recipient.

After the ACO algorithm has been run, a classifier is repeatedly constructed based on training data (Alg. 1 - line 14). Each classifier is tested on test data (Alg. 1 - line 15) and a pheromone trail is laid depending on the obtained results (Alg. 1 - line 20). While a classifier is being developed for each user, the communication network of the group that is assigned to this user is being analyzed. A classifier is built for a selected user via the algorithm, and then a classifier is constructed for each subsequent individual in the group by using the same pheromone trail matrix. After the classifiers have been built and the pheromone trail matrices have stabilized, the final classifier for a given user is developed (Alg. 1 - lines 21-24), in accordance with the diagram in Fig. 1 - step 4, which makes it possible to retain information related to the decisions made by the other members of the group (via the pheromone trail).

Algorithm 1. Pseudo code of the proposed algorithm

```
 1  set_of_hubs = social_network_analysis(enron_e-mail_dataset);
 2  user's_group = SNA_group(set_of_hubs); //For the chosen hub //eq(2)
 3  //Hub is the first
 4  for person=1 to number_of_users_in_the_group do
 5    dataset[person]=prepare_decision_tables(person)
 6  endFor
 7  pheromone = initialization_pheromone_trail(); //Common to all users
 8  //The first and last iteration for hub
 9  for person=1 to (number_of_users_in_the_group+1) do
10    for i=1 to (number_of_iterations / (number_of_users+1)) do
11      best_classifier = NULL;
12      for j=1 to number_of_ants do
13        new_classifier = build_classifier_aACDT(pheromone, dataset[person]);//eq(3)
14        assessment_of_the_usefulness_classifier(new_classifier);
15        if new_classifier is_higher_usefulness_than best_classifier then//eq(6)
16          best_classifier = new_classifier;
17        endIf
18      endFor
19      update_pheromone_trail(best_classifier, pheromone); //eq(5)
20      //Only in last iteration - for hub
21      if person == (number_of_users_in_the_group+1) then
22        if best_classifier is_higher_usefulness_than best_built_class then//eq(6)
23          best_built_class = best_classifier;
24        endIf
25      endIf
26    endFor
27  endFor
28  result = best_built_class;
```

Information about the pheromone trail that has been laid is then passed on as feedback to the classifier that is being built for a given user, which has an influence on which folder will be chosen by this classifier. After the algorithm has finished running, the best classifier is obtained, which was developed based on test and training data (Alg. 1 - line 29). Validation data are regarded as future data, for which the accuracy with which email messages are assigned to folders is computed. The way in which this algorithm works is presented in Alg. 1.

6 Experiments

In order to check the usefulness of the proposed solution, experiments were conducted, the results of which are presented in Tab. 3. The proposed algorithm was implemented in C++. All computations were carried out on a computer with an Intel Core i5 2.27 GHz processor, 2.9 GB RAM, running on the Debian GNU/Linux operating system.

The construction of the social network and the selection of the group of users based on an analysis of this social network were carried out in a deterministic manner, as described in Section 5. As for the ACO algorithm, experiments were conducted and repeated 30 times for each of the data sets. Each experiment was carried out for 250 generations consisting of 25 agent-ants. The algorithm parameters were as follows: $q_0 = 0.3$; $\beta = 3$; $\rho = 0.1$. The results shown in Tab. 3 represent the results from the average of 30 executions.

Table 3. Comparison of all results with regard to the usefulness of the decision with which emails are assigned to folders

dataset	Classical algorithms presented in [3]				Previous algorithm presented in[4]	Proposed algorithm
	MaxEnt	Naive Bayes	SVM	WMW		
beck-s	0.558	0.320	0.564	0.499	0.583	**0.600**
farmer-d	0.766	0.648	0.775	0.746	0.811	**0.834**
lokay-m	0.836	0.750	0.827	0.818	0.888	**0.891**
sanders-r	0.716	0.568	0.730	0.721	0.829	**0.871**
williams-w3	0.944	0.922	0.946	0.945	**0.962**	0.960
rogers-b	–	0.772	–	–	**0.911**	0.900
shackleton-s	–	0.667	–	–	0.709	**0.751**
steffes-j	–	0.755	–	–	0.841	**0.863**
symes-k	–	0.789	–	–	0.930	**0.937**

The obtained results, which are presented in Tab. 3 and Fig. 2, indicate that there was a significant improvement in the classification of emails when the proposed algorithm was used. The results for the other algorithms are cited based on papers [3,4].

The proposed algorithm each time generated better results, as compared to classical algorithms, for nine data sets that had been created based on nine groups of users. For some of the sets (sanders-r, symes-k), the accuracy with which email messages were assigned to folders increased by as much as 15%.

For two of the sets (rogers-b, williams-w3), the accuracy with which email messages are assigned to folders is very high (93-96%), but the use of social network analysis did not lead to the improvement of the results.

In addition, the comparison between the results yielded by the proposed algorithm and those produced by the algorithm that is described in paper [4] shows that the accuracy with which folders are assigned to emails improved by 1-3% (beck-s, farmer-d, steffes-j, symes-k) , whereas for sets sanders-r and shackleton-s this accuracy is even 5% higher, which confirms the validity of using social networks and analyzing mailboxes for a group of users, rather than individuals.

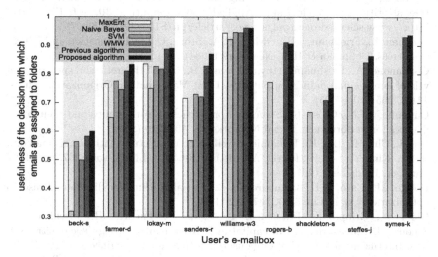

Fig. 2. The correctness of the proposed categorization method

7   Conclusions

Based on the conducted experiments, it has been confirmed that the accuracy with which email messages are automatically classified into folders improves when Ant Colony Optimization algorithms and social network analysis are used. The results improved significantly after creating a social network, based on which communication between the users had been analyzed.

The aim of this study has been achieved and the results that have been obtained are highly satisfactory, given that the tests were carried out on uncleaned data sets. However, the observations that were made during the experiments allow one to assume that this algorithm can be successfully used to suggest that new folders be created based on an analysis of the members of a group. If a rigid structure of folders is imposed on the users in a company, this should even further improve the accuracy with which email messages are assigned to proper folders, which we intend to study in the future.

References

1. Aral, S., Van Alstyne, M.: Network structure & information advantage (2007)
2. Beckers, R., Goss, S., Deneubourg, J., Pasteels, J.M.: Colony size, communication and ant foraging strategy. Psyche **96**, 239–256 (1989)
3. Bekkerman, R., McCallum, A., Huang, G.: Automatic categorization of email into folders: Benchmark experiments on enron and sri corpora. Center for Intelligent Information Retrieval, Technical Report IR (2004)
4. Boryczka, U., Probierz, B., Kozak, J.: An ant colony optimization algorithm for an automatic categorization of emails. In: Hwang, D., Jung, J.J., Nguyen, N.-T. (eds.) ICCCI 2014. LNCS, vol. 8733, pp. 583–592. Springer, Heidelberg (2014)
5. Boryczka, U., Kozak, J.: Enhancing the effectiveness of ant colony decision tree algorithms by co-learning. Applied Soft Computing **30**, 166–178 (2015). http://www.sciencedirect.com/science/article/pii/S1568494615000575
6. Breiman, L., Friedman, J.H., Olshen, R.A., Stone, C.J.: Classification and Regression Trees. Chapman & Hall, New York (1984)
7. Chapanond, A., Krishnamoorthy, M., Yener, B.: Graph theoretic and spectral analysis of enron email data. Computational and Mathematical Organization Theory **11**(3), 265–281 (2005)
8. Cummings, J.N., Cross, R.: Structural properties of work groups and their consequences for performance. Social Networks **25**, 197–210 (2003)
9. Do, H., Choi, P., Lee, H.: Dynamic surveillance: a case study with enron email data set. In: Kim, Y., Lee, H., Perrig, A. (eds.) WISA 2013. LNCS, vol. 8267, pp. 81–99. Springer, Heidelberg (2014)
10. Dorigo, M., Caro, G.D., Gambardella, L.: Ant algorithms for distributed discrete optimization. Artif. Life **5**(2), 137–172 (1999)
11. Gloor, P., Grippa, F., Putzke, J., Lassenius, C., Fuehres, H., Fischbach, K., Schoder, D.: Measuring social capital in creative teams through sociometric sensors. International Journal of Organisational Design and Engineering (2012)
12. Gloor, P.A.: Swarm Creativity: Competitive Advantage through Collaborative Innovation Networks. Oxford University Press, USA (2006)
13. Keila, P., Skillicorn, D.: Structure in the enron email dataset. Computational and Mathematical Organization Theory **11**(3), 183–199 (2005)
14. Moreno, J.L.: Who Shall Survive? Foundations of Sociometry, Group Psychotherapy and Sociodrama. Beacon House, Beacon (1953, 1978)
15. Shetty, J., Adibi, J.: The enron email dataset database schema and brief statistical report. Tech. rep. (2004)
16. Shetty, J., Adibi, J.: Discovering important nodes through graph entropy the case of enron email database. In: Proceedings of the 3rd International Workshop on Link Discovery, LinkKDD 2005, pp. 74–81. ACM, New York (2005)
17. Theraulaz, G., Goss, S., Gervet, J., Deneubourg, J.: Swarm intelligence in wasps colonies: an example of task assignment in multiagents systems. In: Proceedings of the 1990 IEEE International Symposium on Intelligent Control, pp. 135–143 (1990)
18. Verhaeghe, J., Deneubourg, J.: Experimental study and modelling of food recruitment in the ant tetramorium impurum. Insectes Sociaux **30**, 347–360 (1983)
19. Wilson, G.C., Banzhaf, W.: Discovery of email communication networks from the enron corpus with a genetic algorithm using social network analysis (2009)

Population-Based Ant Colony Optimization
for Sequential Ordering Problem

Rafał Skinderowicz[(✉)]

Institute of Computer Science, Silesia University, Sosnowiec, Poland
rafal.skinderowicz@us.edu.pl

Abstract. The population-based ant colony optimization (PACO) algorithm uses a pheromone memory model based on a population of solutions stored in a solution archive. Pheromone updates in the PACO are performed only when a solution enters or leaves the archive. Absence of the local pheromone update rule makes the pheromone memory less flexible compared to other ACO algorithms but saves computational time. In this work, we present a novel application of the PACO for solving the sequential ordering problem (SOP). In particular, we investigate how different values of the PACO parameters affect its performance and identify some problems regarding the diversity of solutions stored in the solution archive. A comparison with the state-of-the-art algorithm for the SOP shows that the PACO can be a very competitive tool.

Keywords: Population based Ant Colony Optimization (PACO) · Ant Colony System · Sequential Ordering Problem

1 Introduction

Ant Colony Optimization (ACO) is a nature-inspired metaheuristic that is mainly used to find good quality solutions to combinatorial optimization problems [1]. In nature, some species of ants use chemical substances called *pheromones* as a method of an *indirect* communication with other members of the colony. Ant which found a food source deposits small amounts of the pheromone on the path leading from the nest to the food source. The pheromone trail attracts other ants and guides them to the food source. In the ACO a number of artificial ants iteratively construct solutions to the problem tackled based on the heuristic information and artificial pheromone trails deposited on the available solution components. The pheromone trails play a role of a collective memory usually referred to as a *pheromone memory*.

The pheromone memory allows the artificial ants to *learn* and hence it is essential for the good performance of the algorithm. Many ACO modifications were proposed in the literature to improve its convergence speed. Most of them change the pheromone memory operation to increase the exploitation of the solution search space. A detailed overview of the most successful approaches can be found in [1]. One of the most interesting modifications of the pheromone

© Springer International Publishing Switzerland 2015
M. Núñez et al. (Eds.): ICCCI 2015, Part II, LNCS 9330, pp. 99–109, 2015.
DOI: 10.1007/978-3-319-24306-1_10

memory usage can be found in the population-based ACO (PACO) proposed by
Guntch et al. [6]. In the PACO the values of pheromone trails directly depend
on a *solution archive*. The archive contains a population of solutions selected
from the solutions generated by the ants in preceding iterations. Each time a
new solution enters the archive a fixed amount of pheromone is deposited on the
corresponding pheromone trails. When the solution is removed from the archive
the pheromone is decreased (evaporated) reversing the previous change.

The pheromone memory used in the PACO has a few advantages. Firstly, the
update concerns only a small subset of available pheromone trails while in the
ACO all the trails are changed in every iteration of the algorithm. For example,
in the case of the TSP the pheromone update can be performed in $O(n)$ time,
where n denotes the number of nodes (cities), while in the ACO the update
requires $O(n^2)$ time. This also applies to $\mathcal{MAX}$–$\mathcal{MIN}$ Ant System (MMAS)
which is one of the most efficient ACO algorithms [1]. Secondly, by controlling
the size of the archive one can easily affect the exploitation to exploration ratio of
the algorithm, which may result in a faster convergence to good quality solutions.

Oliveira et al. conducted a detailed analysis of the PACO for the travelling
salesman problem (TSP) and the quadratic assignment problem (QAP) [7]. The
authors concluded that the PACO is competitive to the state-of-the-art ACO
algorithms with the advantage of finding good quality solutions in a shorter
time, especially if a *restart* procedure is periodically applied to the solution
archive. Our main objective in this paper is to present a computational study
of the PACO for the SOP, what, to the best of our knowledge, was not done
before. We investigate the PACO convergence with various parameters settings,
both without and with a local search (LS) applied. The results are also compared
with the current state-of-the-art algorithm for the SOP which is the combination
of the ACS and an efficient local search heuristic [3]. Additionally, we identify
problems with the solution archive diversity and suggest some possible solutions.

This paper is organized as follows. Firstly, we describe the SOP and the dedicated local search heuristic. Next, we characterise the PACO pheromone memory model and its difference from the ACS's. Fourth section presents results of
the computational experiments on SOP instances from the well known TSPLIB
repository [8]. The last section presents conclusions.

2   Sequential Ordering Problem

The sequential ordering problem is usually formulated as a generalization of the
asymmetric TSP (ATSP), but it can also be stated in terms of a production
planning system [2]. A complete directed graph $G = (V, A)$ is given, where V is
a set of nodes (cities) and $A = \{(i, j) \mid i, j \in V, i \neq j\}$ is a set of directed arcs
corresponding to the roads between the cities. Each arc $(i, j) \in A$ has a positive
weight d_{ij} denoting the length of the arc. Without loss of generality, it can be
assumed that every solution starts at the node 0 and ends at the node $n - 1$,
where $n = |V|$.

Also a precedence digraph $H = (V, R)$ is given, where V is the same set of
nodes as before. R is a set of edges denoting precedence relations between pairs

of nodes. Specifically, an arc $(i, j) \in R$ reflects the precedence constraint between the nodes i and j, such that i has to precede j in every *feasible* solution. By definition, the node 0 precedes every other node, i.e. $(0, i) \in R \ \forall i \in V \setminus \{0\}$. Analogically, the node $n-1$ has to be preceded by every other node, i.e. $(i, n-1) \in R \ \forall i \in \{n - 1\}$. Obviously, the precedence graph must be acyclic in order for a set of feasible solutions to be non-empty.

A solution to the SOP is a Hamiltonian path in which the sum of the edges weights is minimal and all the precedence constraints are satisfied. The precedence constraints make the SOP more difficult to solve than the ATSP, because the dedicated solution construction and improvement algorithms have to include solution feasibility verification [3].

Ant Colony Optimization algorithms are usually paired with a local search method applied after the ants have constructed their solutions. In our work we used the local search algorithm proposed by Gambardella et al. [3]. The algorithm is based on the 3-Opt heuristic for the TSP but uses clever *labelling procedure* to deal with the precedence constraints what allows to reduce the computational complexity to $O(n^3)$ compared with $O(n^5)$ for a naive implementation.

3   Ant Colony Algorithms

Many ACO algorithms were proposed in the literature, but we focus on the two relevant to our work. Firstly, we briefly describe the Ant Colony System which paired with the efficient local search is the current state-of-the-art algorithm for the SOP [4]. Next, we describe the population-based ACO algorithm in reference to the ACS.

3.1   Standard Ant Colony System

The ACS algorithm is based on the ACO but includes a few modifications which increase its convergence speed. In every iteration of the algorithm each ant constructs a complete solution to the problem. An ant starts its construction process at the node 0 (in the case of the SOP) and in every subsequent step appends one of the unvisited nodes to its partial solution. The nodes are selected based on the heuristic information associated with each edge $(i, j) \in A$, denoted by $\eta_{ij} = 1/d_{ij}$, and the pheromone trail, denoted by τ_{ij}. With probability q_0, where $0 \leq q_0 < 1$ is a parameter, an ant k at node i selects a next node j from the set of unvisited nodes, that maximizes $\tau_{ij} \cdot \eta_{ij}^{\beta}$, where β is a parameter. With probability $1 - q_0$ the next city is chosen according to probability distribution determined by:

$$p_{ij} = \frac{\tau_{ij} \cdot \eta_{ij}^{\beta}}{\sum_{h \in S_k} \tau_{ih} \cdot \eta_{ih}^{\beta}}, \tag{1}$$

where S_k is a set of unvisited nodes of the ant k. For high values (close to 1) of the parameter q the search process becomes mostly deterministic and *exploitative*– driven by the heuristic information and the knowledge accumulated in the pheromone memory.

Two kinds of pheromone update are applied in the ACS. The first, called *local pheromone update*, is applied after an ant has selected a next node and consists in evaporating a small amount of pheromone on the corresponding edge according to: $\tau_{ij} = (1 - \rho) \cdot \tau_{ij} + \rho \cdot \tau_0$, where ρ is a parameter and τ_0 is the initial pheromone trail value computed at the beginning of the algorithm. Local pheromone update was introduced to decrease the probability of selecting the same node by other ants. The second pheromone update rule is called *global* and consists in deposition of the pheromone on the edges belonging to the best solution found to date.

Often so called *candidate lists* are used to speed up the ACO algorithms. A candidate list for a node contains a subset of its cl nearest neighbours [1]. When selecting a next node, the decision is restricted to unvisited nodes which belong to the candidate list. If there are no such nodes the remaining ones are considered.

3.2 Population-Based ACO

PACO differs in several aspects from the ACS and ACO [6]. The main difference is the introduction of a new pheromone memory model in which pheromone values are calculated based on the solutions stored in the *solution archive* (the contents of the archive is called population), denoted by P. The algorithm starts with the empty archive and all the pheromone trails set to τ_0. After each iteration the *iteration best* solution may enter the archive depending on the *update strategy* used. The size of the archive, denoted by K, is limited, hence when it reaches the limit adding a new solution requires removal of one of the solutions contained in the archive. Each time a solution enters or leaves the archive the corresponding pheromone trails are increased or decreased, respectively.

The general form of the PACO pheromone matrix is given by:

$$\tau_{ij} = \tau_0 + \Delta \cdot |\{\pi \in P \,|\, (i, j) \in \pi\}| \,, \tag{2}$$

where $\Delta = (\tau_{\max} - \tau_0)/K$. As can be seen, the maximum value of a pheromone trail equals $K \cdot \Delta$ and is obtained if all solutions in the archive contain the corresponding edge. It is worth noting that in the PACO there is no local pheromone update what slightly reduces the computation time compared with the ACS and significantly compared with the ACO. In the ACS the local pheromone update has complexity $O(mn)$, where n is the size of the problem instance and m is the number of ants. In the ACO the pheromone is evaporated from all trails in every iteration what requires $O(n^2)$ time.

The archive update strategy plays essential role for the performance of the PACO. Guntch proposed [6] several update strategies: the age-based strategy, the quality-based strategy, the probabilistic quality-based strategy, age & probability combination and elitist-based strategy. In our work we focused on the *age-based*, *quality-based* and *elitist-based* strategies also selected in [7].

Age-Based Strategy. In this strategy a new solution is always appended to the population, but when the population size reaches the limit, K, the eldest solution is removed. It is easy to notice, that after K iterations the solution construction process is influenced only by the last K iteration best solutions found.

Quality-Based Strategy. Instead of keeping the last K iteration best solutions, the solution archive contains K best solutions found in all the previous iterations. A new solution enters the population (archive) only if it is of better quality than the worst solution in the population, which in such case is removed, otherwise the population remains unchanged. Obviously, this strategy is more exploitation oriented than the age-based strategy.

Elitist-Based Strategy. In the age-based strategy it is possible that the best solution found to date is lost after K iterations. This strategy prevents this situation by keeping the *elitist solution*, denoted by e, which is always equal to the globally best solution found so far. The elitist solution can be replaced only be a better one. The pheromone trails corresponding to the elements of the elitist solution receive $\omega_e \cdot \tau_{\max}$, where $\omega_e \in [0,1]$ is a parameter, the remaining solutions of the population are updated with a weight $\Delta \cdot (1 - \omega_e)/(K - 1)$.

4   Experiments

In order to make a fair comparison between the ACS and the PACO several experiments were conducted on the SOP instances selected from the TSPLIB repository [8]: *prob.42*, *ft70.1*, *ft70.2*, *ft70.3*, *ft70.4*, *prob.100*, *kro124p.1*, *kro124p.2*, *kro124p.3*, *kro124p.4*, *rbg150a*, *rbg174a*, *rbg253a*, *rbg323a*, *rbg341a*, *rbg358a*, *rbg378a*. For the instances from *prob.42* to *kro124p.4* the time limit was set to 50 seconds and for the rest to 100 seconds.

Our main point of focus was the PACO algorithm, hence all the ACS parameters remained constant in our experiments and were chosen based on our preliminary computations and the recommendations found in the literature [1]. The common parameters values for both algorithms were as follows: $m = 10$ ants, $\beta = 4$, $q_0 = 0.7$ and size of the candidate lists $cl = 30$. The values of local and global pheromone evaporation factors in the ACS were set to $\alpha = 0.1$ and $\rho = 0.2$, respectively. The initial pheromone value in the ACS was set to $(n \cdot L_{\mathrm{nn}})^{-1}$, where L_{nn} is the solution length produced by the nearest neighbour heuristic [1].

The ACO family of algorithms is usually used in combination with an efficient local search (LS) algorithm. In order to allow for a more accurate comparison each of the algorithms was run without and with the local search heuristic applied. The LS heuristic used was the same as proposed by Gambardella in [3] and was applied to all the solutions produced by the ants in each iteration. It is worth noting, that the LS is much more computationally expensive than the solution construction process, $O(n^3)$ vs $O(n^2)$, thus the number of solutions constructed in the same time is significantly smaller if the LS is used.

For each combination of the parameters values the algorithms were run 30 times. To minimize the influence of the implementation efficiency on the algorithms performance both were implemented using the same framework and shared most of the code. The algorithms were implemented in C++ and compiled with the GCC v4.8.1 compiler with -O2 switch. The computations were carried on a computer with Intel Xeon X5650 six-core 2.660GHz processor with 12MB cache running under the control of Scientific Linux. A single processor core was utilised by the algorithm.

4.1 Analysis of PACO Parameter Settings

The PACO algorithm performance depends on the three main settings: the size of the solution archive K, the archive update strategy and $\tau_{\max}$. The other parameters include the initial (and also minimum) pheromone value, τ_0, which was always set to $1/(n-1)$ as proposed in [6]. In the case of the elitist-based update strategy one also has to set the ω_e parameter used by the elitist update. In our experiments we used $omega_e = 0.5$ as suggested in [6].

Both the archive solution size limit, K, and the update strategy have significant influence on the pheromone memory in the PACO, hence to find the best configuration we run the algorithm using various combinations of values of the two. Specifically, we run the algorithm with the aged-based, quality-based and elitist-based update strategies. The values of K were set following Guntch [6], that is, $K = 1, 5, 25$, and we also added $K = 2$. The algorithms were run on two SOP instances, namely *kro124p.1* and *prob.100*, of which the first one can be considered relatively easy and the latter difficult [5]. Figure 1 shows the PACO

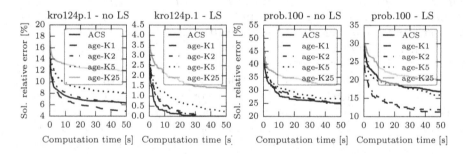

Fig. 1. Solution quality over time of the ACS and PACO algorithms using different values of K for the *age-based* solution archive update strategy, without and with the local search (LS) on instances *kro124p.1* and *prob.100*.

and the ACS convergence for the *age-based* update strategy on the two problem instances. As can be seen, the solution archive size K has strong influence on the PACO performance regardless of the instance. The best solution quality was obtained for the smallest sizes of the solution archive, namely $K = 1$ and $K = 2$. For those values the PACO search process is strongly focused on the areas of the solution search space containing the most recent solutions and ignoring the longer search history. For the small values of K the quality of solutions found

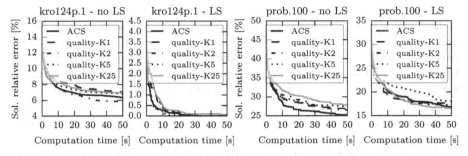

Fig. 2. Solution quality over time of the ACS and PACO algorithms using different values of K for the *quality-based* solution archive update strategy, without and with the local search (LS) on instances *kro124p.1* and *prob.100*.

by the both algorithms was similar, except for the *prob.100* with LS for which the PACO obtained much better results.

The PACO with the *quality-based* strategy proved to be much less sensitive to changes in K as shown in Fig. 2. For the both instances and regardless of the local search use the PACO convergence remained close to the ACS, especially for *kro124p.1* instance. For *prob.100* the differences between the performance for the various values of K were more visible, but varied depending on the use of local search. On the one hand, for $K = 25$ the PACO obtained the worst quality solutions, on the other hand for the same value but with the LS the algorithm obtained the best results.

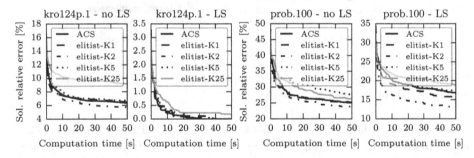

Fig. 3. Solution quality over time of the ACS and PACO algorithms using different values of K for the *elitist-based* solution archive update strategy, without and with the local search (LS) on instances *kro124p.1* and *prob.100*.

The elitist-based strategy is a mix of the two previous, because it favours the best solution found to date but also allows recent iteration-best solutions to enter the population of solutions. With this strategy the results for different values of K were more consistent than for the age-based strategy but less than for the quality-based strategy, as can be seen in Fig. 3. The worst results for the instance *kro124p.1* were obtained for the largest archive size, i.e. $K = 25$. Similar results were obtained for the age-based strategy, however when comparing the two strategies it is clear that the elitist-based strategy wins because the best solution found so far is always present in the archive and strongly influences the search process increasing the exploitation to exploration ratio. The best convergence was observed for $K = 2$ in which case the archive contains only two elements: the best solution so far (elitist solution) and the most recent iteration best solution. In such a case ants try to improve the elitist solution by mixing it with the edges from the most recent iteration best solution. Figure 3 shows that this behaviour proved to be especially successful for *prob.100* instance with the LS applied.

Summarizing, both the solution archive size, K, and the update strategy have strong influence on the PACO convergence. The best results were obtained for the small values of K, i.e. $K = 1$ or $K = 2$, and the worst for the largest size: $K = 25$. These observations are in opposition with the results obtained for the TSP and the QAP problems as reported by Oliveira et al. in [7]. To explain the differences we need to consider that the problems (TSP, QAP, SOP) have different structure of the solution search space. The precedence constraints in

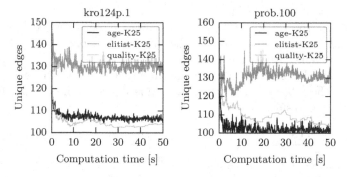

Fig. 4. Mean number of unique (different) edges in the solutions contained in the PACO solution archive over time for the *age-based, quality-based, elitist-based* update strategies and the archive size: $K = 25$.

the SOP make it difficult to find feasible solutions and even a small change can entail more changes resulting from the necessity to satisfy all the precedence constraints. Better solutions are found rarely, hence the iteration best solutions are mostly the same and the solution archive is likely to contain multiple copies of a single solution. The confirmation can be found in Fig. 4 which shows a mean number of different (unique) edges which are parts of the solutions in the archive for $K = 25$. As we can see, even that the 25 solutions contain a total of over 2500 edges (for the instances *kro124p.1* and *prob.100*), the mean number of different edges was below 140 for the *elitist-based* strategy. For the *age-based* and *quality-based* the number was even smaller, often below 110. The more copies of a solution are contained in the archive the larger the values of pheromone trails become (see Eq. (2)). In other words, the larger archive results in a stronger emphasis on the exploitation than on the exploration of the solution search space.

4.2   Comparison with ACS

In order to compare the PACO with the ACS we run both algorithms on several instances from the TSPLIB repository [8]. Based on the findings presented above, the PACO without the LS was run with the *elitist-based* archive update strategy, $K = 2$ and $\tau_{\max} = 0.5$. The version with the LS differed by using the *age-based* update strategy. The results are presented in Tab. 1. Generally, the differences between the algorithms seem small, but the *non-parametric Wilcoxon rank-sum test* was used to check if the differences were statistically significant at the significance level $\alpha = 0.01$. The cases in which one of the algorithms obtained significantly better results than the other are shown in bold. When no LS was used, the ACS obtained significantly better results in two cases (*rgb358a*, *rbg378a*) and the PACO in one case (*ft70.1*) out of total seventeen. When the LS was used, the ACS obtained better results in four cases and the PACO in two. Both algorithms were able to find the best known solutions in the most cases. In four other cases the solutions were within 1% from the best known, and in one case (*prob.100*) the PACO found a solution about 5% worse than the best known upper bound. As we can see, the PACO is definitely competitive with the ACS, regardless of the local search usage.

Table 1. Percentage deviation of the average solution quality obtained by the PACO and the ACS from the best known solutions. Instances with no publicly available best known solutions were compared with the best know upper bounds [5]. Values in bold refer to a significant difference to the other algorithm.

Instance	Best known	ACS best	PACO best	ACS	PACO	ACS+LS	PACO+LS
prob.42	243	243	243	3.95	2.66	0.25	**0**
ft70.1	39313	39313	39313	3.79	**2.93**	0	0
ft70.2	40419	40419	40419	5.13	4.26	**0.02**	0.08
ft70.3	42535	42535	42535	5.15	4.98	0	0
ft70.4	53530	53530	53530	2.38	2.57	0.01	0
prob.100	1190	1303	1251	25.23	25.77	16.86	**10.23**
kro124p.1	39420	39420	39420	6.33	5.47	0	0.02
kro124p.2	41336	41336	41336	6.37	6.37	0	0
kro124p.3	49499	49499	49534	8.05	6.55	0.35	0.14
kro124p.4	76103	76103	76103	4.33	4.34	**0.05**	0.25
rbg150a	1750	1750	1750	0.63	0.65	0	0
rbg174a	2033	2033	2033	0.65	0.68	**0**	0.02
rbg253a	2950	2950	2950	0.87	0.92	0	0
rbg323a	3141	3141	3151	4.61	5.23	**0.18**	0.44
rbg341a	2570	2588	2615	5.98	5.64	2.24	2.68
rbg358a	2545	2598	2594	**10.83**	12.4	2.93	2.8
rbg378a	2816	2840	2860	**12.26**	14.01	**1.61**	2.14

5   Conclusions

In the present paper we discussed and analysed the PACO for the SOP. Specifically, we have considered three solution archive (population) update strategies and a few archive sizes. The results were compared with the state-of-the-art algorithm, namely the ACS with an effective LS heuristic proposed in [3]. We have found that the best results were obtained for the *elitist-based* strategy when no LS was used, and for the *age-based* strategy when the LS was applied. Surprisingly, the best convergence was observed when the solution archive size K was equal to 1 or 2, and the worst for $K = 25$. It can be explained by the relative difficulty of finding good quality *feasible* solutions to the SOP, hence the diversity of solutions in the population remains low. In other words, the solution archive contains multiple copies of a single solution, what unnecessary increases the emphasis on exploitation, and this effect becomes stronger for the larger values of K. One possible solution to this problem could be to introduce a threshold for the number of copies of a single solution that can be put in the archive. Another solution could be to include a pheromone restart procedure similar to the one proposed by Oliveira et al. [7].

Generally, with the right parameters values the PACO is competitive with the ACS. Although the analysis showed no clear winner, the PACO has a few advantages. Firstly, there is no need to store the pheromone matrix explicitly as the pheromone values can be computed from the solutions stored in the archive [9]. This allows to reduce a memory usage from $O(n^2)$ to $O(Kn)$, where K is the size of the solution archive. If the K is kept constant this reduces to $O(n)$. Secondly, the lack of the local pheromone update in the PACO makes it easier to create an efficient parallel implementation of the algorithm as the number of modifications to the shared memory (pheromone memory) is reduced. It would also be interesting to compare the PACO pheromone memory with the selective pheromone memory proposed for the ACS in [10].

Acknowledgments. This research was supported in part by PL-Grid Infrastructure.

References

1. Dorigo, M., Stützle, T.: Ant colony optimization. MIT Press (2004)
2. Escudero, L.F.: An inexact algorithm for the sequential ordering problem. European Journal of Operational Research **37**(2), 236–249 (1988)
3. Gambardella, L.M., Dorigo, M.: An ant colony system hybridized with a new local search for the sequential ordering problem. INFORMS Journal on Computing **12**(3), 237–255 (2000)
4. Gambardella, L.M., Montemanni, R., Weyland, D.: An enhanced ant colony system for the sequential ordering problem. In: Operations Research Proceedings 2011, pp. 355–360. Springer (2012)
5. Guerriero, F., Mancini, M.: A cooperative parallel rollout algorithm for the sequential ordering problem. Parallel Computing **29**(5), 663–677 (2003)
6. Guntsch, M., Middendorf, M.: A population based approach for ACO. In: Cagnoni, S., Gottlieb, J., Hart, E., Middendorf, M., Raidl, G.R. (eds.) EvoWorkshops 2002. LNCS, vol. 2279, pp. 72–81. Springer, Heidelberg (2002)
7. Oliveira, S.M., Hussin, M.S., Stützle, T., Roli, A., Dorigo, M.: A detailed analysis of the population-based ant colony optimization algorithm for the tsp and the qap. In: Proceedings of the 13th Annual Conference Companion on Genetic and Evolutionary Computation, pp. 13–14. ACM (2011)

8. Reinelt,   G.:   Tsplib95.   http://www.iwr.uni-heidelberg.de/groups/comopt/-software/tsplib95/index.html

9. Scheuermann, B., So, K., Guntsch, M., Middendorf, M., Diessel, O., ElGindy, H., Schmeck, H.: Fpga implementation of population-based ant colony optimization. Applied Soft Computing **4**(3), 303–322 (2004)

10. Skinderowicz, R.: Ant colony system with selective pheromone memory for TSP. In: Nguyen, N.-T., Hoang, K., Jędrzejowicz, P. (eds.) ICCCI 2012, Part II. LNCS, vol. 7654, pp. 483–492. Springer, Heidelberg (2012)

Designing a Universal Suboptimal Tree Integration Algorithm

Marcin Maleszka[(✉)]

Wroclaw University of Technology, St. Wyspianskiego 27, 50-370 Wroclaw, Poland
marcin.maleszka@pwr.edu.pl

Abstract. Processing hierarchical data structures, due to their increasing popularity, becomes more and more important. In our previous works we have developed many exact integration algorithms for trees, but designing a new algorithm for each new integration aim is resource consuming. In this paper we propose a universal integration method based on genetic programming heuristic. Experimental evaluation shows that this approach is slightly slower than exact algorithms and sometimes provides worse solutions. One important advantage however is that it may be applied with no additional effort to any problem, without the need for a long design phase.

Keywords: Tree integration · Integration algorithms · Genetic programming · Suboptimal integration

1 Introduction

Hierarchical data structures are commonly used in multiple theoretical and practical applications. Nowadays, even documents are stored in the hierarchical XML format. In consequence, there is now a need for developing new methods for processing hierarchical structures. In this paper we focus on one such method: a heuristic approach to tree integration.

Hierarchical data structures integration is a large research area, consisting of many different problems to be solved. Our focus in this and previous research ([9,10]) is the merging aspect of integration. The problem is to determine a single tree that is the best representative of the given set of input trees. A common approach to this problem is the consensus theory, including determining a median tree. There are multiple possible aims (applications) of tree integration and each requires a specialized mathematical description and integration algorithm. In our research we have developed a common definition of the hierarchical data structure and of the integration task, that we may apply to multiple problems and integration aims. This way we only use one mathematical description for every problem. This still requires one to design an entirely new algorithm for each new integration problem (each new aim).

In this paper we try to mitigate this problem by using a heuristic approach (genetic programming) operating directly on the common mathematical definition of the problem. The approach may be used to a new problem by simply

© Springer International Publishing Switzerland 2015
M. Núñez et al. (Eds.): ICCCI 2015, Part II, LNCS 9330, pp. 110–118, 2015.
DOI: 10.1007/978-3-319-24306-1_11

swapping the equation describing the problem. This is not a perfect solution, as the heuristic approach does not guarantee finding the best result of integration and it is slower than developed exact algorithms. The advantage of this approach is that the heuristic may start providing solutions long before the design phase of developing a new algorithm is finished, thus offering a simple stop-gap measure for multiple basic applications.

This paper is organized as follows: in Section 2 we provide a brief overview of existing approaches to tree integration and other relevant works; in Section 3 we define the complex tree and the integration task for trees as the basis on which our research is conducted; Section 4 contains brief overview of our previous tree integration algorithms, the motivation for using heuristics for integration and the design and evaluation of the genetic programming we use for tree integration; Section 5 closes this paper with final conclusions on the overall process of integrating complex trees.

2   Related Works

A comprehensive description of inconsistency solving by determining the consensus may be found in [13]. The underlying issue of the consensus theory is defined as follows: There is a given set X (of alternatives or objects) that is a subset of some universe U. The selection is to determine some subset of X based on some criteria. If the selection is deterministic, then repeating it for the same X should always give the same result. In that case it may be called some selection function. This is extended by the consensus theory, which further states that the selection may not be an element of X.

At the beginning of the consensus theory research, the authors were concerned with simple structures of the universe U, like the linear order, or partial order. With more advanced techniques, more complex structures of U were researched (i.e., divisions, coverage, complex relations, etc.). Tree structures were first integrated by means of consensus in [1,2,5]. In those works a problem of finding a median tree was defined for n-trees. The median tree minimizes the sum of distances to all other structures. This is the basis of optimality criterion used in our research.

A survey by Rahm and Bernstein [14] presents an interesting classification of matching approaches. Schema matching is in general a much wider area than just tree integration, but the current widespread use of XML (and it's schemas) makes the practical applications of schema matching a case of tree integration. The research done by Do [6] goes further, describing some criteria for XML schema integration (some of them are general enough to use also to relational and other schemas), divided into four general areas: input criteria, output criteria, quality measures, and effort criteria. The most relevant criteria to hierarchical structures integration are named by the authors as: schema information (a criterion based on the size of input schemas), schema similarity (the closer the inputs are, the smaller the space to find the output in will be), element representation (if the elements are found in the output), cardinality (cardinality of relationships in the output), precision and recall (as defined in information retrieval). Our previous research (see Section 3) is based on those works.

In this paper we combine tree integration with genetic algorithms and programming [4]. Since their conception genetic algorithms have been used to solve a variety of problems and were a basis for many more advanced techniques. The classical problem solved by GA is the travelling salesman problem [7]. Genetic Algorithms and related heuristics, like genetic progamming or differential evolution, have been also used to generate robot trajectory, design aircraft [4], solve cryptanalysis problems [3] or find strategies in coordination games [8]. We also used a specific genetic programming heuristic to integrate a specific type of trees in [11].

3   Complex Tree Integration

In our research we use a modified definition of a tree data structure, as first proposed in [12]. We define the tree on several levels. The most basic one is the simple data structure, that we call dendrite (the definition is presented only to estabilish the notation):

Definition 1. **Dendrite** *is a directed rooted tree* $D = (W, E, r)$, *where:*

- W *is a finite set of nodes,*
- E *is a set of edges* $E \subseteq W \times W$.
- $r \in W$ *is the root.*

We use the set of attributes A. Let V_a be the domain of $a \in A$. $V = \bigcup_{a \in A} V_a$ is the set of all attribute values. Let also T be the set of node types. A type determines the attributes of a node.

We define a *Complex Tree* as follows:

Definition 2. **Complex Tree** *is a five* $CT = (W, E, r, T_{CT}, A_{CT}, V_{CT})$, *where:*

- (W, E, r) *is a dendrite of the tree,*
- $T_{CT} : W \rightarrow T$ *is a function that assigns each node a type,*
- $A_{CT} : W \rightarrow 2^A$ *is a function that assigns each node its attributes,*
- $V_{CT} : W \times A \rightarrow V$ *is a partial function that assigns each node and attribute a value.*

This definition of a tree is used as it allows processing (especially integration) at different levels. In Section 4 of this paper we will use only the most basic level of *dendrite*.

There are multiple definitions of complex tree integration task. Here, we use the following one:

Definition 3. **Complex Tree Integration Function** *is every function* I:

$$I : \Pi(\overline{CT}) \rightarrow 2^{\overline{CT}},$$

where $\overline{CT}$ *is the set of all complex trees and* $\Pi(A)$ *is a partition of* A; *such that for each* $CT \in I(\Pi(\overline{CT}))$ *one or more criterion* K_j *is satisfied.*

The criteria are the most important part in this definition, as they are used to determine the aim of integration. In our previous research [12] we had defined several criteria that may be used to describe different aims:

Criteria for consolidating the knowledge and keeping all data – a group of criteria defined as completeness. Several variants of completeness are possible, including:

- Schema completeness – types, labels and attributes remain unchanged;
- Structure completeness – all nodes in the tree remain, but may change position;
- Data completeness – attribute values remain unchanged;
- Relationship completeness – all relations in tree remain;
- Path completeness – all paths from input trees remain;
- Semantic completeness – all knowledge in the tree remains.

Criteria for keeping the same sub-trees – this a specific variant of completeness called sub-tree agreement. Variants of this include:

- Input Sub-Tree Agreement – the sub-trees in the integrated tree are identical as in the input trees;
- Output Sub-Tree Agreement – as above, but the mathematical calculation of the measure is reversed;
- Final Sub-Tree Agreement – the number of unique sub-trees remains the same.

Criteria for minimizing the size of the output tree and avoiding duplicated knowledge – this is possible with the criteria we call minimality and precision. Several variants exist including:

- Precision – no duplicate data entry (node) exists in the tree;
- Depth minimality – the integrated tree has a depth equal or smaller than the input tree with the highest depth;
- Cardinal minimality – the integrated tree has equal or smaller number or nodes than the largest input tree.

Criteria for finding a median/average solution (the consensus) – this is done with the various optimality criteria, including:

- 1-optimality (O1 consensus), which is often used to find the best existing representative of a set of elements;
- 2-optimality (O2 consensus), which is used to create a new representative for a set of elements.

As an example for criteria use we will describe schema completeness.

Let as assume two complex trees describing traffic. Leafs in the tree represent road fragments where measurements are done and other nodes in the tree are larger parts of roads (e.g. entire road through town with tens of crossroads). The trees descibe the same area and have only differences in types or roads or attributes used, as they where created based on different data (i.e. different traffic measuring equipment). The final aim of integration is to create a single system detailing traffic. The key in integrating trees is to keep all data on types of monitored roads (or crossings) and the types of measurements. As the traffic changes often, keeping the values of attributes is unimportant. Once the integrated system is online these values will be calculated anew. Thus the schema completeness, dealing with types and attributes of complex tree, is important to use in integration. Some other criterion should be however considered for the structure of the tree (with similar reasoning leading to its use).

4   Tree Integration Algorithm

4.1   Optimal and Suboptimal Integration Algorithms

In our previous works [9,10] we have developed several algorithms for integrating complex trees using specific criteria. These include both exact and approximate algorithms with very good approximation (over 99% of criterion value achieved). The criteria that may be satisfied due to these algorithms range from basic one like structure completeness to complex ones like 1-optimality and sub-tree agreement.

One problem we encountered is that each new criterion or set of criteria we want satisfied require a new designed algorithm. While an exact algorithm with low computational complexity may usually be found, the design process usually takes some time. In most cases this is a one-time expenditure, after which we have good and fast solutions. In rare cases we need to integrate the trees using a specific set of criteria only once and we don't need exact results. In that case designing a new algorithm is comparatively very costly. Due to this reason we have considered using a heuristic algorithm to integration.

There are several heuristic algorithms that may potentially be used to integrate complex trees. Unfortunately, most of them limit the number of available solutions. The most egregious example is the simplest implementation of Tabu Search with predefined set of possible results. It is possible to search all variants of complex trees using a more advanced version of Tabu Search, but this leads to another disadvantage: the number of possible neighboring trees to check in subsequent iterations rises exponentially.

We have determined that genetic programming are the meta-heuristic allowing best balance between the limit of the set of possible solutions and the time to search it. Additionally, there is one important advantage of using genetic programming for complex tree integration. Due to the definition of the task (see Section 3), the integration criteria we want satisfied may be used directly as the feasibility (fitness) function of the genetic programming.

4.2   Genetic Programming for Tree Integration

A genetic programming uses a number of genetic operators and a specific representation of the chromosome (solution). The solution in tree integration is a single complex tree (in our case: dendrite). The chromosome consists of two parts – nodes and structure (relations) – as in Definition 1. The chromosome does not have a predefined size, as any size of output tree is allowed – the limit is enforced by the feasibility function.

As previously noted, we use a criterion (or a linear combination of criteria) as the feasibility function. The following genetic operators are used:

- **S**. Selection. A standard roulette selection method is used to determine pairs for crossover. The chance of using a tree is proportional to the value of its feasibility.
- **E**. Elitist Selection. The best result of each generation is always preserved as several copies in the next generation. This operator is parametrized by the percentage of copies preserved P_n.
- **C1**. Crossover. A random sub-tree in tree one is substituted with a random sub-tree in the second tree. A sub-tree located at any level may be used, including substituting a single leaf node (smallest sub-tree) with a child of the root with its whole sub-tree (largest sub-tree). This operator is parametrized by crossover chance C_p. For each pair of trees selected for crossover the chance of it occurring is C_p. The sub-trees are selected with uniform distribution.
- **M1**. Mutation, first variant. A node is substituted by a random one. The change is only applied to the node, the relations to nodes above and below in the structure remain unchanged. New node is generated using the same procedure as when creating the starting population – it may have any type or attributes. This operator is parametrized by mutation chance $M1_p$. A chance of changing a node is equal to $M1_p$ for each node in each tree.
- **M2**. Mutation, second variant. A new child is added to the node. The new child is generated using the same procedure as when creating the starting population – it may have any type or attributes. This operator is parametrized by mutation chance $M2_p$. A chance of adding a child to a node is equal to $M2_p$ for each node in each tree.
- **M3**. Mutation, third variant. A child of the node is removed. The whole sub-tree connected with the child is also removed. This operator is parametrized by mutation chance $M3_p$. A chance of removing a child from a node is equal to $M3_p$ for each node in each tree.

We followed the classic procedure for generating consecutive generations of chromosomes: Selection (including Elitist Selection), followed by Crossover, then each tree in resulting population was subject to each type of mutation.

4.3 Evaluation

We experimentally determined best parameters for each integration criterion (each feasibility function) and also a set of parameters that gave best results on average if applied to any criterion. These parameters are as follows:

- For Structure Completeness Criterion: population size – 100, max number of generations – 500, $P_n = 0.02$, $C_p = 0.8$, $M1_p = 0.1$, $M2_p = 0.01$, $M3_p = 0.005$.
- For Relation Completeness Criterion: population size – 200, max number of generations – 1000, $P_n = 0.02$, $C_p = 0.9$, $M1_p = 0.05$, $M2_p = 0.01$, $M3_p = 0.005$.
- For 1-Optimality Criterion: population size – 500, max number of generations – 10000, $P_n = 0.01$, $C_p = 0.9$, $M1_p = 0.1$, $M2_p = 0.03$, $M3_p = 0.01$.
- *Average* parameters: population size – 200, max number of generations – 10000, $P_n = 0.01$, $C_p = 0.9$, $M1_p = 0.1$, $M2_p = 0.02$, $M3_p = 0.01$.

With these parameters we experimentally compared previous algorithms with the genetic programming (both best parameters for each criterion and average parameters for each criterion). We compared criterion value and time needed to calculate it, averaged over 10 runs of each algorithm. Each algorithm was run on the same physical machine, with identical background load. Partial results of the evaluation are presented in tables 1 and 2.

Table 1. Algorithm comparison for 20 trees, 10 nodes per tree, 1-100 range of nodes

Algorithm	Criterion Value	Computation Time
Structure Completeness		
Exact Algorithm	1	< 1s
Genetic Programming (best)	≥ 0.7	< 1s
Genetic Programming (best)	1	2s
Genetic Programming (average)	≥ 0.7	5s
Genetic Programming (average)	1	11s
Relation Completeness		
Exact Algorithm	1	< 1s
Genetic Programming (best)	≥ 0.4	< 1s
Genetic Programming (best)	≥ 0.9	10s
Genetic Programming (best)	1	over ten minutes
Genetic Programming (average)	≥ 0.4	< 1s
Genetic Programming (average)	≥ 0.9	14s
Genetic Programming (average)	1	over ten minutes
1-Optimality		
Exact Algorithm	1	several hours
Approximate Algorithm	≥ 0.99	54s
Genetic Programming (best)	≥ 0.9	2m 22s
Genetic Programming (best)	≥ 0.99	25m 13s
Genetic Programming (average)	≥ 0.9	2m 46s
Genetic Programming (average)	≥ 0.99	31m 24s

The results show the expected result, that the genetic programming requires more time to obtain the same result as previous algorithms. Additionally, genetic programming require more time to obtain even a worse result. Nevertheless

Table 2. Algorithm comparison for 100 trees, 100 nodes per tree, 1-1000 range of nodes

Algorithm	Criterion Value	Computation Time
Structure Completeness		
Exact Algorithm	1	3s
Genetic Programming (best)	≥ 0.7	2s
Genetic Programming (best)	1	5s
Genetic Programming (average)	≥ 0.7	8s
Genetic Programming (average)	1	18s
Relation Completeness		
Exact Algorithm	1	8s
Genetic Programming (best)	≥ 0.4	< 1s
Genetic Programming (best)	≥ 0.9	16s
Genetic Programming (best)	1	over ten minutes
Genetic Programming (average)	≥ 0.4	< 1s
Genetic Programming (average)	≥ 0.9	28s
Genetic Programming (average)	1	over ten minutes
1-Optimality		
Exact Algorithm	1	several hours
Approximate Algorithm	≥ 0.99	5m 56s
Genetic Programming (best)	≥ 0.9	5m 37s
Genetic Programming (best)	≥ 0.99	32m 59s
Genetic Programming (average)	≥ 0.9	10m 19s
Genetic Programming (average)	≥ 0.99	47m 02s

genetic programming can obtain good (or best) results in a fraction of the time required to *develop and run* an exact algorithm.

We have also run the genetic programming for criteria where no exact or approximate algorithms were defined yet and obtained satisfactory results in several minutes, including time required for programming the criterion (feasibility function). The same may be done for many other criteria or linear combinations of criteria, due to the very definition of the integration task.

5   Conclusion

In this paper we describe a universal genetic programming solution for complex tree integration. This method gives suboptimal results in time comparable to exact and approximate algorithms designed in our previous research, and optimal results in longer time. These results were expected when applying the heuristic approach to complex tree integration problems. An important advantage given by heuristic approach is that it allows immediate start, without time needed for designing a criterion. This is especially important when applying the method to integration criteria which do not have an exact algorithm designed yet. We postulate that the following is the best procedure to integrate trees using new criteria: a) start the universal suboptimal integration algorithm described in this paper for the given criterion b) simultaneously work on designing an exact (or approximate) algorithm for the problem; in short time a suboptimal solution will be given by the genetic algorithm, and if necessary, after designing and implementing the other algorithm an optimal solution will be found. This may be especially important if the integration is a once-only occurrence and the time factor is more important that the precision of the result. If integration is repeated often, the time required to design an exact algorithm becomes short in comparison. Thus the method should find a limited number of applications.

Acknowledgments. This research was funded by a Polish Ministry of Science and Higher Education grant.

References

1. Adams, E.N.: N-Trees as Nestings: Complexity, Similarity, and Consensus. Journal of Classification **3**, 299–317 (1986)
2. Barthelemy, J.P., McMorris, F.R.: The Median Procedure for n-Trees. Journal of Classification **3**, 329–334 (1986)
3. Boryczka, U., Dworak, K.: Cryptanalysis of transposition cipher using evolutionary algorithms. In: Hwang, D., Jung, J.J., Nguyen, N.-T. (eds.) ICCCI 2014. LNCS, vol. 8733, pp. 623–632. Springer, Heidelberg (2014)
4. Davis, L.: Handbook of genetic algorithms, vol. 115. Van Nostrand Reinhold, New York (1991)
5. Day, W.H.E.: Optimal Algorithms for Comparing Trees with Labeled Leaves. Journal of Classification **2**, 7–28 (1985)
6. Do, H.-H., Melnik, S., Rahm, E.: Comparison of schema matching evaluations. In: Chaudhri, A.B., Jeckle, M., Rahm, E., Unland, R. (eds.) NODe-WS 2002. LNCS, vol. 2593, pp. 221–237. Springer, Heidelberg (2003)
7. Grefenstette, J., Gopal, R., Rosmaita, B., Van Gucht, D.: Genetic algorithms for the traveling salesman problem. In: Proceedings of the First International Conference on Genetic Algorithms and their Applications, pp. 160–168. Lawrence Erlbaum, New Jersey (1985)
8. Juszczuk, P.: Finding optimal strategies in the coordination games. In: Hwang, D., Jung, J.J., Nguyen, N.-T. (eds.) ICCCI 2014. LNCS, vol. 8733, pp. 613–622. Springer, Heidelberg (2014)
9. Maleszka, M., Nguyen, N.T.: Approximate Algorithms for Solving O1 Consensus Problems Using Complex Tree Structure. T. Computational Collective Intelligence **8**, 214–227 (2012)
10. Maleszka, M., Nguyen, N.T.: Path-oriented integration method for complex trees. In: Jezic, G., Kusek, M., Nguyen, N.-T., Howlett, R.J., Jain, L.C. (eds.) KES-AMSTA 2012. LNCS, vol. 7327, pp. 84–93. Springer, Heidelberg (2012)
11. Maleszka, M., Mianowska, B., Nguyen, N.-T.: A heuristic method for collaborative recommendation using hierarchical user profiles. In: Nguyen, N.-T., Hoang, K., Jędrzejowicz, P. (eds.) ICCCI 2012, Part I. LNCS, vol. 7653, pp. 11–20. Springer, Heidelberg (2012)
12. Marcin, M., Nguyen, N.T.: A Method for Complex Hierarchical Data Integration. Cybernetics and Systems **42**(5), 358–378 (2011)
13. Nguyen, N.T.: Advanced Methods for Inconsistent Knowledge Management. Springer, London (2008)
14. Rahm, E., Bernstein, P.A.: A survey of approaches to automatic schema matching. The VLDB Journal **10**, 334–350 (2001)

Heterogeneous DPSO Algorithm for DTSP

Urszula Boryczka and Łukasz Strąk[(✉)]

Institute of Computer Science, University of Silesia,
Będzińska 39, 41-205 Sosnowiec, Poland
urszula.boryczka@us.edu.pl, lukasz.strak@gmail.com

Abstract. A new heterogeneous algorithm of Discrete Particle Swarm Optimization has been proposed in this paper to solve the Dynamic Traveling Salesman problem. The test environment is random and variable in time, what requires a rapid adaptation of the algorithm to changes. An inappropriate selection of the algorithm parameters leads to stagnation and quality deterioration of obtained results. The modification, proposed in this paper, enables to reduce the number of the algorithm parameters, regarding the swarm size, the number of iterations and the size of neighborhood. A higher diversity of particles vs. the homogeneous version positively influences the quality of obtained results, what was demonstrated in various experiments.

Keywords: Dynamic traveling salesman problem · Pheromone · Discrete Particle Swarm Optimization · Heterogeneous

1 Introduction

There has been a growing interest in studying evolutionary algorithms in dynamic environments in recent years due to its importance in real word applications [1]. A problem where input data are changeable depending on time is called a Dynamic Optimization Problem (DOP). The purpose of optimization for DOPs is to continuously track and adapt to changes through time and to quickly find the currently best solution [2]. An example of DOP is a dynamic traveling salesman problem (TSP). It consists of sequences of static TSPs. Each subsequent subproblem forms on the basis of the previous one. A fragment of data is transferred to the next subproblem (the unchanged part), while the rest is modified.

Particle Swarm Optimization (PSO) is a technique based on a swarm population created by Russell Eberhart and James Kennedy in 1995 [3]. This technique is inspired by the social behavior of a bird flocking or fish schooling. The algorithm was created primarily to optimize the functions of continuous space. Each particle position is one problem solution. This technique employs the mutual interactions of swarm building particles. This influence is determined by various parameters, such as the social factor - the significance of the optimal position, found by the particle itself, and the cognitive factor - the significance of the optimal position, founded by the entire swarm. The PSO algorithm has quickly

© Springer International Publishing Switzerland 2015
M. Núñez et al. (Eds.): ICCCI 2015, Part II, LNCS 9330, pp. 119–128, 2015.
DOI: 10.1007/978-3-319-24306-1_12

become popular, due to the fact that is easy to implement [3]. The Discrete Particle Swarm Optimisation (DPSO) algorithm is an attempt to transfer the herding behavior to discrete space.

The original PSO algorithm and its discrete version are homogeneous [4] - all particles reveal the same values of parameters and the same pattern of searching through the space of solutions. This is good regarding the simplicity of the algorithm. However, it is that heterogeneous populations are common in the natural environment [5]. A homogeneous PSO population consists of subjects having the same parameters. The update of velocity and position is carried out in the same way for each particle (all the indices are of the same value). Hereby, each particle employs the same pattern of searching through the space of solutions. One of the major problems in population algorithms is the balance between exploration and exploitation. In heterogeneous population, particles may reveal different behaviors and different patterns of searching through the space of solutions. In addition, certain parameters may turn out useful at the onset, while other may be expedient in later iterations. In this way, it is easier to control the balance between exploration and exploitation [4,6].

The structure of this paper is as follows: section 2 includes a review of the literature, concerning the dynamic traveling salesman problem and the heterogeneous version of particle swarm optimization. Section three provides a description of the proposed solution. Section four presents the performed studies. The last section includes summary and conclusions.

2   Background

Dynamic TSP is expressed through changes in both the number of vertices and a cost matrix. Any change in the input data may imply the change of local and global optima. Formula (1) describes the distance matrix, applied in the problem:

$$D(t) = \{d_{ij}(t)\}_{n(t) \times n(t)},$$ (1)

where: t denotes the time parameter or iteration, i and j are endpoints and n denotes vertices count. This problem can be understood as a series of static TSP problems. Each element strongly resembles its predecessor, as only some data are modified. In this paper, the change applies to the distance between the vertices only. The number of vertices is constant. In order to trace the algorithm results (for testing purposes), the optimum should be estimated after the data are changed.

Heterogeneity is defined as lack of uniformity. In population algorithms, it may be indicated in many ways. The taxonomy for the PSO algorithm has been presented in [4]. The authors have divided heterogeneity into four categories, taking into account neighborhood heterogeneity, the model-of-influence heterogeneity, the update-rule heterogeneity and the parameter heterogeneity.

The neighborhood heterogeneity applies to a case, in which the neighborhood size varies for each particle. The topology of particles is then irregular (the vertices are different degree). Some particles impose then a stronger influence on other particles (this feature is used in calculation of a subsequent position).

The model-of-influence heterogeneity means a different approach of each subject to the selection of the best particle which controls each of its subsequent positions. For example, one particle can update its position, following the best particle in its entire neighborhood, while other particles are fully informed.

The variety of position update strategies are based on grouping of particles with regards to the mode of searching through the space of solutions - assigning them definite specialization. For example, one group may explore the space of solutions, while another may be involved in local searching. This type of heterogeneity diversifies the population in an efficient way, as each particle may differently search through the space of solutions.

The diversity of parameters in a swarm is that either each individual particle or a group of certain particles may possess parameters values, which are different from those in other particles or in groups in the population. In this way, some particles may have high inertia (ω) and explore the space of solutions, while other may reveal low inertia and search for a global optimum (around the best identified position).

Heterogeneity may also reveal adaptive character. In case, when certain behaviors enable to obtain better results, the algorithm may increase the number of subjects with these features. In a reverse variant, i.e. when no good results are at hand, the algorithm may accordingly reduce the number of subjects with undesired features. In this way, the algorithm may adapt to a given problem. Such a strategy is used by the heterogeneous DPSO algorithm, which is described in a more detail below.

3 DPSO with Pheromone

In order to adapt the PSO algorithm to operation in discrete space, it was necessary to redefine certain concepts. The discrete version of the algorithm for DTSP is based on edges and sets of edges. Edge is a tuple of the following three factors (a, x, y), where a denotes the probability of choosing the edge to the next position, x and y are the endpoints. Variable a is constrained to a number between 0 and 1 and $(a, x, y) = (a, y, x)$. A feasible TSP solution $\{(1, 2), (2, 3), (3, 4), (4, 1)\}$ in the discrete version is equal to $\{(1,1,2), (1,2,3), (1,3,4), (1,4,1)\}$ [7,8].

At the beginning of the algorithm, the population is randomly distributed in the space of solutions. Out of randomized positions, the best particle is selected ($gBest$). Then, its subsequent velocity is calculated for each particle and then another subsequent position. See Fig. 3 for the graphic presentation of the subsequent position creation process for each particle in the swarm. In the first step ($k = 0$), subtraction of two sets is carried out: $gBest$ and the actual position (X, one of the problem solutions). The result of this operation are edges, by which the solution, marked as $gBest$ is better from the actual solution (X). The same action is performed for $pBest$ sets (the best solution, founded by the modified particle). The $pBest$ set in the first step is equivalent to the actual position. Then the velocity (V^k) is added in both procedures; the velocity in the first

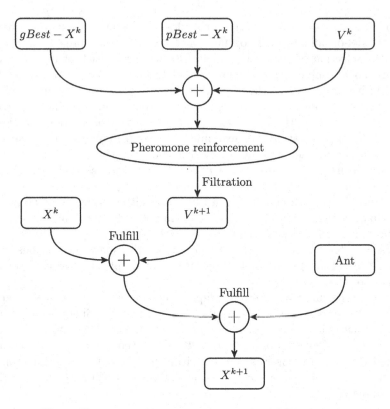

Fig. 1. Formation of a new position in the DPSO algorithm

iteration is an empty set. The resulting set is reinforced by the pheromone - for each edge coefficient a the pheromone value is added. The reinforced set is then filtered, based on the a parameter of each edge, in which it is contained. This operation has been described in a more detail below. After this stage, formed collection becomes particle velocity (V^{k+1}). Usually, this set does not form the Hamiltonian's cycle (TSP solution). The task for subsequent steps is the completion of the velocity set. A created set will be a subsequent position of (X^{k+1}) particle.

The above diagram is expressed by equations (2) and (3), which describe particle movement in the discrete space of solutions. These formulas is based on the classical PSO algorithm equation:

$$V_i^{k+1} = c_2 rand() \cdot (gBest - X_i^k)$$
$$+ c_1 rand() \cdot (pBest_i - X_i^k) \tag{2}$$
$$+ \omega \cdot V_i^k ,$$
$$X_i^{k+1} = \Delta \tau^k(V_i^{k+1}) \oplus c_3 rand() \cdot X_i^k , \tag{3}$$

where: i denotes the number of particles, k - iteration number and $rand()$ is a random value with uniform distribution from $[0,1]$. The operations of addition and subtraction are the sum and the difference of sets and variable w is called an inertia weight. Parameters c_1 and c_2 are the cognitive and social parameters for scaling $pBest$ and $gBest$ respectively. The $\oplus$ operator completes the next position with the edges from the previous position if the added edge does not make an infeasible tour. Function $\Delta\tau^k(V_i^{k+1})$ is a reinforcement probability of edge selection to the next position given by formula (4) below.

$$\Delta\tau^k(V_i^{k+1}) = (\tau_e - 0.5) \cdot \frac{k}{k_c} \quad \forall e \in V_i^{k+1} \subseteq E, \tag{4}$$

where e denotes edge, τ_e is the edge pheromone value extracted from the pheromone matrix, k denotes iteration number and k_c is iterations count (assumed before algorithm run). Factor $\frac{k}{k_c}$ scales edge reinforcement depending on the progress of the algorithm. The result of equation (4) is the edge set with increased or decreased the probability component a for each edge. The probability of the selection is reinforced by the amount of pheromone associated with this edge. The pheromone evaporation process will be responsible for the function of penalty, if the edge does not improve the tour. To apply the pheromone, we created formula (4), which is responsible for the allocation of the pheromone for the edges.

Algorithm 1 presents a pseudo code for the DPSO algorithm, the filtration and completion operations.

After creating the velocity set (equation (3)), the complement to the full Hamilton cycle begins. The operation is done by adding missing edges from the nearest neighborhood. The neighborhood is created based on the nearest neighbor heuristic using α-measure [9]. Pheromone update (evaporation) and pheromone initialization are identical to that in the MMAS algorithm [10].

The DPSO algorithm, presented in this study, equipped with the virtual pheromone, has self-adaptation features which enable the adaptation to the input data changes. In each subsequent subproblem, the distance matrix is changed only partially. The proposed solution utilizes the information about the solutions from before the changes, in order to minimize the objective function after the changes. Therefore, the algorithm searches more effectively through the space of solutions. In the study, measure-based neighborhood α-measure has also been applied, which uses the theory of tolerance [11].

A heterogeneous swarm of particles has been applied in the study. Each particle belongs to a certain group. The population within the group is homogeneous (all the particles have the same parameters values).

In this paper we propose two variants of the algorithm. In the first version, each group has a constant number of particles. In the second version, the number of the particles in the group is variable (adaptation based on feedback). After calculating the next position of the particle, its contribution to improving $gBest$ is calculated according to the formula:

$$p_{gain} = 0.9 \cdot \sup\{gBest - X_i, 0\} + 0.1 \cdot \sup\{pBest - X_i, 0\}, \tag{5}$$

Algorithm 1. DPSO algorithm outline

Create random initial swarm
Create neighborhood for each vertex
for $k = 0 \rightarrow Iterations$ **do**
   for $i = 0 \rightarrow SwarmSize$ **do**
      Calculate velocity V_i^{k+1} (equation (2))
      Filtration stage
      $X_i^{k+1} = \emptyset$
      for all $(a, x, y) \in V_i^{k+1}$ **do**   ▷ eq. (3)
         Increase a by pheromone at edge (x,y)
         Random value $r \in [0, 1]$
         if $r \leq a$ **then**
            $X_i^{k+1} \cup (1, x, y)$
         end if
      end for
      for all $(a, x, y) \in X_i^k$ **do**   ▷ eq. (3)
         Random value $r \in [0, 1]$
         $\bar{a} = r \cdot c_3$
         Random value $r \in [0, 1]$
         if $r \leq \bar{a}$ **then**
            $X_i^{k+1} \cup (1, x, y)$
         end if
      end for
      Fulfill stage
      Complete X_i^{k+1} using neighborhood
      Set $V_i^k = V_i^{k+1}$
      Set $X_i^k = X_i^{k+1}$
      Update $pBest_i$ and $gBest$
   end for
   Evaporate pheromone
end for

---

where: p_{gain} denotes particle gain, $\sup\{gBest - X_i, 0\}$ improvement of the best identified solution by the swarm, $\sup\{pBest - X_i, 0\}$ denotes the improvement of its own best solution. Group gain is the sum of the gains of all particles in the group. Changing the size of the group is triggered at a certain number of iterations. In this way, out of the pool of initial groups only those remain in the last iterations of the algorithm which enabled to obtain the best results. The number of particles in a group changes according to formula:

$$f(gr) = \frac{gr_{gain}}{\sum\limits_{g=0}^{groups} g_{gain}} \cdot SwarmSize, \tag{6}$$

where: gr denotes the group, for which size is calculated, whereas *groups* denotes the number of all groups. The gr_{gain} variable means a cumulative contribution of all the particles in this group gr in optimum identification. This value is calculated for each particle from the formula (5).

A change of the group size is associated with a particle transfer to another group. In this way, it inherits the parameters of the target group (without position change). The size of the entire swarm is constant in each algorithm iteration. A transfer of one particle enlarges the target group by one particle, while the source group is reduced by one element. The algorithm reviews the leaves of the group, disregarding its structure. It is considered only in case of particle transfer.

During the start of the algorithm for each subproblem, the values of group parameters c_1, c_2, c_3, ω are random. The number of groups is equal to the size of the swarm. The group size can only be changed in case of a second variant of the algorithm. Parameter c_1 has the following probability of selection values: $P(0.1)$ = 0,4, $P(0.75)=0.15$, $P(1.5)=0.3$, $P(1.75)=0.15$, parameter c_2, c_3: $P(0.1)=0.4$ $P(1)=0.15$, $P(1.5)=0.15$, $P(2)=0.3$, parameter ω: $P(0.1)=0.4$, $P(0.25)=0.2$, $P(0.5)=0.4$.

The probability of selecting parameter values is based on the behavior of particles. Small values of c_1, c_2, c_3, ω cause the particle to rapidly change position, what may improve the convergence. Over time, by increasing pheromone reinforcement, small values of parameters give more controls to pheromone. For large values, the pheromone influence is smaller.

4   Experimental Results

The studies were carried out on a Benchmark Generator. It was dictated by the lack of a unified and a consistent approach to testing of the algorithms, solving the dynamic traveling salesman problem, as in case of the static version of the problem (TSPLIB repository). Pseudo code 2 shows the process of the algorithm that takes into account the dynamism of DTSP.

Algorithm 2. DTSP Benchmark

Read static TSP problem
while Stop criterion is not met **do**
    Create swarm and neighborhood
    Optimize problem using DPSO   ▷ Solve problem
    Update $gBest$
    Change input data by $n\%$
    Compare obtained value with exact value
end while

The algorithms compared in the experiment section are based on the same input data (the same changes). For this reason, they can be directly compared. For the purpose of this paper, eleven subproblems were assumed (the original one with TSPLIB and ten modifications). An exception is the *gr666*, problem which consists of one subproblem only (the original TSPLIB problem).

The tests were repeated 30 times and the results have been averaged. The calculations, associated with the size of particular groups, are activated every

Table 1. DPSO algorithm settings in homogenous and heterogeneous version

Homogeneous DPSO					Heterogeneous DPSO					Common parameters		
Problem	c_1	c_2	c_3	ω	Problem	c_1	c_2	c_3	ω	Iterations	SwarmSize	Neighborhood
berlin52	0.5	0.5	0.5	0.2	berlin52,					312	32	7
kroA100	0.5	0.5	0.5	0.5	kroA100,					700	64	7
kroA200	0.5	0.5	0.5	0.5	kroA200,	Random settings				1800	80	7
gr202	0.5	0.5	0.5	0.5	gr202,					2424	101	10
gr666	0.5	1.0	1.5	0.6	gr666					6144	112	30

25th iteration of the algorithm. The parameters, presented in Tab. 1, were used for all the studies. The number of iterations is given to one subproblem. The total number of iterations is the value read from the table and multiplied by eleven (the number of subproblems). The exception is the problem of *gr666* for which the number of subproblems is equal to one.

4.1   The Group Analysis

Tab. 2 shows the group gain in the algorithm with a constant number of particles in the group (each group consists of single particle), in the last algorithm iterations. Although the best set of parameters values for a homogeneous version of the algorithm is not in the list, other parameters replace it. This is due to the fact that the pheromone is growing up in compared to the first iteration of the algorithm. Without increasing the pheromone value, parameters with very small values (eg., item 6 in the ranking) would cause the particle to change position in a chaotic manner. Small values of the the gain (Tab. 2) means that most of them comes from the improvement of the own best solution (*pBest*). However, at this stage of the calculation, the solution is close to the best solution found by the swarm, which may (in future iterations) causes the improvement of the *gBest*.

Table 2. The gain of group calculated using the formula (5) in the last iteration of the algorithm with a constant number particles in the group. The problem is *gr666*

Rank	c_1	c_2	c_3	ω	Group gain	Rank	c_1	c_2	c_3	ω	Group gain
1	0.1	2	1	0.5	441.1	6	0.1	0.1	0.1	0.5	221.5
2	1.75	2	1.5	0.25	350.4	7	1.75	2	2	0.1	219
3	0.1	0.1	2	0.5	343.7	8	0.1	1	2	0.5	218.4
4	0.1	0.1	2	0.1	262.8	9	0.1	2	2	0.1	203.5
5	0.75	1.5	1	0.5	261	10	0.75	2	2	0.25	201.5

4.2   Comparison of the Algorithms

Successive studies compared the three variants of the DPSO algorithm for Dynamic Traveling Salesman Problem. Tab. 3 summarizes the results obtained for different problems.

Table 3. Comparison of different variants of the DPSO algorithm - Heterogeneous (b) means constant group size version while (b) the resize group version. T. denotes the algorithm running time, G. denotes the distance from the optimum, D. denotes the average standard deviation

Problem	DPSO algorithms								
	Homogeneous			Heterogeneous (a)			Heterogeneous (b)		
	T. [s]	G. [%]	D.	T. [s]	G. [%]	D.	T. [s]	G. [%]	D.
berlin52	0.2	**0.01**	3.8	0.24	0.02	5.2	0.2	0.02	4.3
kroA100	1.55	0.95	197	1.63	**0.88**	147.1	1.54	0.89	139.5
kroA200	8.66	2.14	306.8	7.3	2.31	279.3	7.28	**1.78**	314.4
gr202	20.4	1.41	237.3	19.02	**1.39**	225	18.68	1.54	369.1
pcb442	78.33	3.45	988.9	74.39	**1.99**	415.1	99.13	9.02	1017.7
gr666	223.83	4.09	1784.3	334.59	**3.94**	1819.2	454.43	11.24	5125.6

Only for *berlin52* instance the heterogeneous version did not achieve a better result. In other cases, at least one version of heterogeneous outperform homogeneous result. The largest gain was reported for the problem *pcb442*. The most important fact is that the heterogeneous version with a constant number particles in a group can obtain similar results for both small and large problem instances. The version with adaptive number of particles in the group achieved very poor results for larger instances of DTSP problems. It was caused by the fact that too many groups, which could improve $gBest$ in the last iterations of the algorithm, were removed in early stages of the calculations. Analysis of the average standard deviations (11 subproblems) showed that, although the parameters are random, the returned results were similar. Average optimal tour for every analyzed DTSP problem are as follows (order like in Tab. 3) 7635, 22357.3, 31073, 40582.1, 52497.5, 294358. The results can be improved through the use of local search (not used).

5   Conclusions

In literature many reports can be traced addressing the issue of the broadly understood heterogeneity. It can be observed from the fact that it often improves the quality of obtained results. It applies both to the classic PSO and to the hybrid version, described in the paper. The main objective of this study was to reduce the number of parameters. Additionally, presented algorithms improved the results obtained from the standard homogeneous version (DPSO). The improvement is not significant, but the achievement of the indirect goal allows to further explore the heterogeneity in the context of Dynamic Traveling Salesman Problem.

In future, we are going to concentrate on various models of the particle gain measure. Further studies will also address the impact of heterogeneity on the solution quality in the context of the number of the algorithm iterations.

References

1. Branke, J.: Evolutionary approaches to dynamic environments. In: GECCO Workshop on Evolutionary Algorithms for Dynamics Optimization Problems (2001)
2. Li, W.: A parallel multi-start search algorithm for dynamic traveling salesman problem. In: Pardalos, P.M., Rebennack, S. (eds.) SEA 2011. LNCS, vol. 6630, pp. 65–75. Springer, Heidelberg (2011)
3. Kennedy, J., Eberhart, R.: Particle swarm optimization. In: Proceedings of the IEEE International Conference on Neural Networks, pp. 1942–1948 (1995)
4. de Oca, M.A.M., Peña, J., Stützle, T., Pinciroli, C., Dorigo, M.: Heterogeneous particle swarm optimizers. In: Haddow, P., et al. (eds.) IEEE Congress on Evolutionary Computation, CEC 2009, pp. 698–705. IEEE Press, Piscatway (2009)
5. Hansell, M.: Built by Animals. Oxford University Press, New York (2007)
6. Nepomuceno, F.V., Engelbrecht, A.P.: A self-adaptive heterogeneous PSO inspired by ants. In: Dorigo, M., Birattari, M., Blum, C., Christensen, A.L., Engelbrecht, A.P., Groß, R., Stützle, T. (eds.) ANTS 2012. LNCS, vol. 7461, pp. 188–195. Springer, Heidelberg (2012)
7. Boryczka, U., Strąk, Ł.: Efficient DPSO neighbourhood for dynamic traveling salesman problem. In: Bădică, C., Nguyen, N.T., Brezovan, M. (eds.) ICCCI 2013. LNCS, vol. 8083, pp. 721–730. Springer, Heidelberg (2013)
8. Zhong, W.L., Zhang, J., Chen, W.N.: A novel set-based particle swarm optimization method for discrete optimization problems. In: Evolutionary Computation, CEC 2007, vol. 14, pp. 3283–3287. IEEE (1997)
9. Helsgaun, K.: An effective implementation of k-opt moves for the Lin-Kernighan TSP heuristic. Technical report, Roskilde University (2006)
10. Stützle, T., Hoos, H.H.: Max-min ant system. Future Generation Computer Systems **16**(8), 889–914 (2000)
11. Helsgaun, K.: An effective implementation of the lin-kernighan traveling salesman heuristic. European Journal of Operational Research **126**, 106–130 (2000)

Decomposition Based Multiobjective Hyper Heuristic with Differential Evolution

Richard A. Gonçalves$^{(\boxtimes)}$, Josiel N. Kuk,
Carolina P. Almeida, and Sandra M. Venske

Department of Computer Science, UNICENTRO, Guarapuava, Brazil
{richard,josiel,carol,ssvenske}@unicentro.br

Abstract. Hyper-Heuristics is a high-level methodology for selection or generation of heuristics for solving complex problems. Despite their success, there is a lack of multi-objective hyper-heuristics. Our approach, named MOEA/D-HH$_{SW}$, is a multi-objective selection hyper-heuristic that expands the MOEA/D framework. MOEA/D decomposes a multi-objective optimization problem into a number of subproblems, where each subproblem is handled by an agent in a collaborative manner. MOEA/D-HH$_{SW}$ uses an adaptive choice function with sliding window proposed in this work to determine the low level heuristic (Differential Evolution mutation strategy) that should be applied by each agent during a MOEA/D execution. MOEA/D-HH$_{SW}$ was tested in a well established set of 10 instances from the CEC 2009 MOEA Competition. MOEA/D-HH$_{SW}$ was favourably compared with state-of-the-art multi-objective optimization algorithms.

Keywords: Hyper-heuristic · MOEA/D · Choice function

1 Introduction

Some real problems are multiobjective consisting of several conflicting objectives to optimize and multiobjective evolutionary algorithm based on decomposition (MOEA/D) has become an increasingly popular framework for solving them. In MOEA/D, N agents collaborate to solve N subproblems and the selection is a process of choosing solutions by agents [9]. At each step, nearby agents collaborate to solve related subproblems by applying operator to modify its solutions and selecting among the nearby agents' solutions or its own.

Obtaining the best performance of MOEA/D and other MOEAs in a specific domain needs setting a number of parameters and selecting operators, making hard their application for ordinary users. Hyper-Heuristics emerged to alleviate this problem. A Hyper-heuristic is a high-level methodology for automatic selection or generation of heuristics for solving complex problems [1]. The use of hyper heuristics to solve multiobjective optimization problems (MOPs) is still underexplored. So, the main objective of this work is to develop a multi-objective hyper-heuristic, called MOEA/D-HH$_{SW}$ (MOEA/D Hyper-Heuristic

© Springer International Publishing Switzerland 2015
M. Núñez et al. (Eds.): ICCCI 2015, Part II, LNCS 9330, pp. 129–138, 2015.
DOI: 10.1007/978-3-319-24306-1_13

with Sliding Window), which chooses the best Differential Evolution operator to be applied by each MOEA/D agent. MOEA/D-HH$_{SW}$ was tested in a set of 10 unconstrained MOPs from the CEC 2009 MOEA Competition.

The main contributions of this paper are: (i) the proposal of a new multi-objective hyper-heuristic and (ii) a new adaptive choice function that uses a sliding window to deal with the dynamic nature of the operator selection.

The remainder of this paper is organized as follows. Section 2 gives a brief overview of Multiobjective Optimization and MOEA/D-DRA. An overview of Hyper-Heuristics is presented in 3 while the fundamental concepts related to Differential Evolution are shown in Section 4. MOEA/D-HH$_{SW}$ and its new adaptive choice function are detailed in Section 5. Experiments and results are presented and discussed in Section 6 while Section 7 concludes the paper.

2   Multiobjective Optimization

A Multiobjective Optimization Problem (MOP) can be stated as Min (or Max) $\mathbf{f}(\mathbf{x}) = (f_1(\mathbf{x}), ..., f_M(\mathbf{x}))$ subject to $g_i(\mathbf{x}) \leq 0$, i $= \{1, ..., G\}$, and $h_j(\mathbf{x}) = 0$, j $= \{1, ..., H\}$ $\mathbf{x} \in \Omega$, and where the integer M ≥ 2 is the number of objectives. A solution minimizes (or maximizes) the components of the objective vector $\mathbf{f}(\mathbf{x})$ where $\mathbf{x}$ is a n-dimensional decision variable vector $\mathbf{x} = (x_1, ..., x_n) \in \Omega$.

This work combines the MOEA/D framework, Differential Evolution and Hyper-Heuristics to solve multiobjective optimization problems. MOEA/D applies a decomposition approach to transform the MOP into a number of single objective optimization subproblems, each subproblem is solved by an agent. The objective of each subproblem is a weighted aggregation of all individual objectives in the MOP. Neighborhood relations among these subproblems depend on distances among their aggregation weight vectors. Each subproblem is simultaneously optimized using mainly information from its neighboring subproblems. Dynamical Resource Allocation [15] is used to allocate different computational efforts to each agent. An utility value is computed for the solution of each agent and the computational effort allocated to the agents is based on these values.

3   Hyper-Heuristics

Hyper-Heuristics permit the design of flexible solvers by means of the automatic selection and/or generation of heuristic components (low level heuristics). One important distinction between heuristic and hyper-heuristics is that the first searches the space of solutions while the latter search the space of heuristics [1].

Two main hyper-heuristic categories can be considered: heuristic selection (methodologies for choosing or selecting existing heuristics) and heuristic generation (methodologies for generating new heuristics from the components of existing ones). Decisions taken by the hyper-heuristics are based on feedback from the search process. This learning process can be classified as on-line or off-line. In on-line hyper-heuristics (used in this work) information is obtained during the execution of the algorithm. While in off-line hyper-heuristics the information is obtained before the search process starts (*a priori*)[1].

Our proposed algorithm is an on-line selection hyper-heuristic. A general on-line selection hyper-heuristic is briefly described as follows [1]. An initial set of solutions is generated and iteratively improved using low level heuristics until a termination criterion is satisfied. During each iteration, the heuristic selection decides which low level heuristic will be used next based on feedback obtained in previous iterations. After the selected heuristic is applied, a decision is made whether to accept the new solution or not using an acceptance method.

The Choice Function [3,10] is one of the most successful selection heuristics described in the hyper-heuristic literature. The Choice Function chooses a heuristic by the combination of three terms: performance of each heuristic, the heuristic performance when it is applied following the last heuristic applied (performance of heuristic pairs) and information about the last time it was applied. The first two terms favour intensification while the latter favours diversification.

The heuristic selection mechanism used in our work is a new adaptive choice function (Subsection 5.1) and the low level heuristics used correspond to Differential Evolution mutation and crossover strategies (Section 4). The acceptance method used corresponds to the MOEA/D population update step (Section 5). Multi-objective hyper-heuristics can be found in [2,6,10,13].

4   Differential Evolution

DE is a population-based stochastic search technique developed by Storn and Price [12]. DE works with three control parameters: N, F and CR. N is the population size. The scaling factor parameter (F) is used to scale the difference between vectors, which is further added to a target vector $\hat{\mathbf{x}}$. This process is called *mutation* in DE. The vector resulting from mutation, named *trial vector*, is combined with a parent vector $\mathbf{x}^p$ in the crossover operation, according to the Crossover Rate parameter (CR). Finally, the offspring is compared with its parent vector to decide who will "survive" to the next generation. There are some variations to the basic DE. In this paper we used the following variations: DE/*rand*/1/*bin*, DE/*rand*/2/*bin*, DE/*current-to-rand*/1/*bin*, DE/ *current-to-rand*/2/*bin*, and DE/*nonlinear*.

For DE/*rand*/1/*bin* and DE/*rand*/2/*bin*, a random individual (*rand*) is selected for target vector $\hat{\mathbf{x}}$, there is one (or there are two) pair(s) of solutions which is (are) randomly chosen to calculate the differential mutation and the binomial crossover is used (the probability of choosing a component from the parent vector or the trial vector is given by a binomial distribution). For DE/*current-to-rand*/1/*bin* and DE/*current-to-rand*/2/*bin*, the current individual is selected for the target vector $\hat{\mathbf{x}}$ and there is one (or there are two) pair(s) of solutions randomly chosen to calculate the differential mutation and the binomial crossover [12]. The last mutation strategy, called in this work DE/*nonlinear*, was proposed in [11] for MOEA/D framework, and disregards the values of CR and F. It is a hybrid operator based on polynomials, where each polynomial represents the offspring which takes the form $\boldsymbol{\rho}(w) = w^2\mathbf{c}_a + w\mathbf{c}_b + \mathbf{c}_c$, where w

is generated based on an interpolation probability, P_{inter} [11]. Assuming that $rand \in U[0,1]$ (i.e., $rand$ is a value between 0 and 1 randomly generated by a uniform distribution) we have $w \in U[0,2]$, if $rand \leq P_{inter}$ and $w \in U[2,3]$ otherwise. Individuals $\mathbf{c}_a$, $\mathbf{c}_b$ e $\mathbf{c}_c$ are defined as $\mathbf{c}_a = (\mathbf{x}_c - 2\mathbf{x}_b + \mathbf{x}_a)/2$, $\mathbf{c}_b = (4\mathbf{x}_b - 3\mathbf{x}_c - \mathbf{x}_a)/2$, $\mathbf{c}_c = \mathbf{x}_c$, where $\mathbf{x}_c$, $\mathbf{x}_b$ and $\mathbf{x}_a$ are individuals randomly chosen from the current population. In MOEA/D-HH$_{SW}$, the mutation strategies are used as low level heuristics and chose by the proposed choice function with sliding window.

5   Proposed Approach

MOEA/D-HH$_{SW}$ is a selection hyper-heuristic that uses an adaptive choice function to determine the DE mutation strategy applied by each agent (individual) during a MOEA/D-DRA [15] execution. Algorithm 1 presents the pseudocode of the proposed approach.

The first steps of MOEA/D-HH$_{SW}$ initialize various data structures (steps 1 to 6). The weight vectors $\boldsymbol{\lambda}^i$, $i = 1, ..., N$ are generated using a uniform distribution. The neighborhood ($B^i = \{i_1, \cdots, i_C\}$) of weight vector $\boldsymbol{\lambda}^i$ is comprised by the indices of the C weight vectors closest to $\boldsymbol{\lambda}^i$. The initial population (pool of agents) is randomly generated and evaluated. Each individual ($\mathbf{x}^i$) is associated with the ith weight vector. The empirical ideal point ($\mathbf{z}^*$) is initialized as the minimum value of each objective found in the initial population and the generation (g) is set to 1.

Afterwards, the algorithm enters its main loop (steps 7 to 33). Firstly, it is determined which individuals (agents) from the population will be processed. A 10-tournament selection based on the utility value of each subproblem is used to determine these individuals [5].

Step 10 selects the low level heuristic s used to generate a new individual. The selection is performed based on the Choice Function value of each strategy i ($CF(i)$), which is calculated and updated in step 27. The heuristic chosen is the one with higher $CF(i)$ value. In this work, the low level heuristics used correspond to Differential Evolution mutation and crossover strategies.

The low level heuristics are applied considering individuals randomly selected from *scope*. In this work, *scope* can swap from the neighborhood to the entire population (and vice-versa) along the evolutionary process of MOEA/D-HH$_{SW}$. It is composed by the indices of chromosomes from either the neighborhood B^i (with probability δ) or from the entire population (with probability $1 - \delta$). Applying the chosen low level heuristic, a modified chromosome $\mathbf{y}$ is generated in step 16. The polynomial mutation in step 17 generates $\mathbf{y}' = (y_1', \cdots, y_n')$ from $\mathbf{y}$. In step 18, if the new chromosome $\mathbf{y}'$ has an objective value better than the value stored in the empirical ideal point, $\mathbf{z}^*$ is updated with this value.

The next steps involve the population update process (steps 19 to 25) which is based on the comparison of the fitness of individuals. In the MOEA/D framework, the fitness of an individual is measured accordingly to a decomposition function. In this work the *Tchebycheff function* (Min $g^{te}(\mathbf{x} \mid \boldsymbol{\lambda}, \mathbf{z}^*) = \max_{1 \leq j \leq M} \{\lambda_j \mid f_j(\mathbf{x}) - z_j^* \mid\}$) is used.

Algorithm 1. MOEA/D-HH$_{SW}$.

1: Generate N weight vectors $\boldsymbol{\lambda}^i = (\lambda_1^i, \lambda_2^i, ...\lambda_M^i), i = 1, ..., N$
2: For $i = 1, \cdots, N$, define the set of indices $B^i = \{i_1, \cdots, i_C\}$ where $\{\boldsymbol{\lambda}^{i_1}, .., \boldsymbol{\lambda}^{i_C}\}$
  are the C closest weight vectors to $\boldsymbol{\lambda}^i$ (by the Euclidean distance)
3: Generate an initial population $P^0 = \{\mathbf{x}^1, \cdots, \mathbf{x}^N\}$, $\mathbf{x}^i = (x_1^i, x_2^i, ...x_n^i)$
4: Evaluate each individual in the initial population P^0 and associate $\mathbf{x}^i$ with $\boldsymbol{\lambda}^i$
5: Initialize $\mathbf{z}^* = (z_1^*, \cdots, z_M^*)$ by setting $z_j^* = min_{1 \leq i \leq N} f_j(\mathbf{x}^i)$
6: $g = 1$
7: **repeat**
8:     Let all the indices of the subproblems whose objectives are MOP individual
  objectives f_i compose the initial I. By using 10-tournament selection based on π^i,
  select other $N/5 - M$ indices and add them to I.
9:     **for** each individual $\mathbf{x}^i$ in I **do**
10:         s = argmax$_{i=1...LLH}$ CF(i) where LLH is the number of low level heuristics

11:         **if** $rand < \delta$ **then** //Determining the scope ($rand$ in U[0,1])
12:             $scope = B^i$
13:         **else**
14:             $scope = \{1, \cdots, N\}$
15:         **end if**
16:         Generate a new solution $\mathbf{y}$ by strategy s (repair it if necessary)
17:         Apply polynomial mutation to produce $\mathbf{y}'$ (repair it if necessary)
18:         Update $\mathbf{z}^*$, $z_j^* = min(z_j^*, f_j(\mathbf{y}'))$
19:         **for** each subproblem k (k randomly selected from $scope$) **do**
20:             **if** $g^{te}(\mathbf{y}' \mid \boldsymbol{\lambda}^k, \mathbf{z}^*) < g^{te}(\mathbf{x}^k \mid \boldsymbol{\lambda}^k, \mathbf{z}^*)$ **then**
21:                 **if** a new replacement may occur **then**
22:                     Replace $\mathbf{x}^k$ by $\mathbf{y}'$ and increment n_r
23:                 **end if**
24:             **end if**
25:         **end for**
26:         reward = $(g^{te}(\mathbf{x}^i \mid \boldsymbol{\lambda}^k, \mathbf{z}^*) - g^{te}(\mathbf{y}' \mid \boldsymbol{\lambda}^k, \mathbf{z}^*))/(g^{te}(\mathbf{x}^i \mid \boldsymbol{\lambda}^k, \mathbf{z}^*))$
27:         Update CF(s) accordingly to Algorithm 2
28:     **end for**
29:     **if** g modulo 50 == 0 **then**
30:         Update the utility π^i of each subproblem i
31:     **end if**
32:     $g = g + 1$;
33: **until** $g >$MAX-EV

Accordingly to what is selected for the *scope* (steps 12 or 14), the neighborhood or the entire population is updated. The population update is as follows: if a new replacement may occur, (i.e. while $n_r < NR$ and there are unselected indices in *scope*), a random index (k) from *scope* is chosen. If $\mathbf{y}'$ has a better Tchebycheff value than $\mathbf{x}^k$ (both using the kth weight vector - $\boldsymbol{\lambda}^k$) then $\mathbf{y}'$ replaces $\mathbf{x}^k$ and the number of updated chromosomes (n_r) is incremented. To avoid the overproliferation of $\mathbf{y}'$, a maximum number of updates (NR) is used.

Then, the reward obtained by the application of the selected low level heuristic is calculated as the difference between the Tchebycheff value of the parent and child divided by the value of the parent. This reward is used in step 27 to update the Choice Function values. If the current generation is a multiple of 50, then the utility value of each subproblem is updated. The evolutionary process stops when the maximum number of evaluations (MAX-EV) is reached and MOEA/D-HH$_{SW}$ outputs the Pareto set and Pareto front approximations.

5.1 Proposed Choice Function with Sliding Window

One of the most important parts of a hyper-heuristic is the selection method, which is responsible for choosing the best heuristic used by each agent (individual) during each stage of the optimization process. The Choice Function scores heuristics based on a combination of different measures and the heuristic to apply at each stage is chosen based on these scores. The first measure (f_1) records the previous performance of each individual heuristic. The second measure (f_2) considers the pair-wise relationship between heuristics. The last measure (f_3) corresponds to the time elapsed since each heuristic was last selected by the Choice Function. The Classical Choice Functions is $CF(i) = \alpha \cdot f_1(i) + \beta \cdot f_2(i) + \gamma \cdot f_3(i)$ [7]. It has some limitations and, therefore, some improved versions have been proposed in the literature [4,5].

Algorithm 2. New Choice Function with Sliding Window

1: Reset all choice functions for all operators
2: **for** i = 1 ... SlidingWindowSize **do**
3:     op = SlidingWindowOperator[i]
4:     reward = SlidingWindowReward[i]
5:     $f_1(s) = f_1(s)$ + reward
6:     **if** i $\neq$ 1 **then**
7:         f_2(lastOperator, op) = f_2(lastOperator, op) + lastReward + reward
8:     **end if**
9:     **for** i = 1 ... LLH **do**
10:         $f_3(i) = f_3(i)$ + SlidingWindowTimeSpentOn(i)
11:     **end for**
12:     f_3(op) = 0
13:     lastReward = SlidingWindowReward[i]
14:     lastOperator = op
15: **end for**
16: **for** i = 1 ... LLH **do**
17:     CF(i) = SF $\cdot$ ($\phi \cdot f_1(i) + \phi \cdot f_2(s, i)$) + $\delta \cdot f_3(i)$
18: **end for**

In this paper further improvements to the choice function are proposed by the use of relative rewards (step 27 of Algorithm 1) and a sliding window. The relative rewards normalize the raw rewards in the interval (0,1) and has little

Algorithm 3. Update of the New Choice Function with Sliding Window

1: Insert the data related to the low level heuristic application (operator identifier, reward and time spent on it) into the Sliding Window
2: **if** SlidingWindowSize $\geq$ MaxSliding **then**
3:     Remove the oldest data from the SlidingWindow
4: **end if**
5: **if** reward > 0 **then**
6:     $\phi = 0.99$
7:     $\delta = 0.01$
8: **else**
9:     $\phi = \phi - 0.01$
10:     $\delta = 1.0 - \phi$
11: **end if**

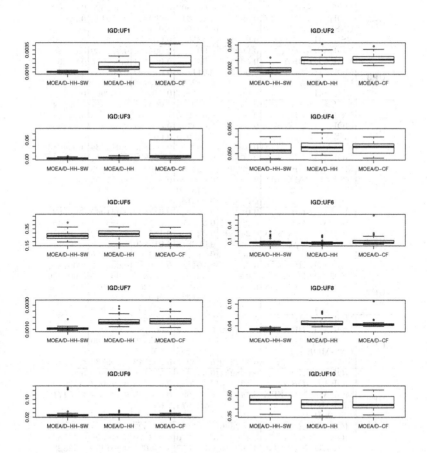

Fig. 1. Boxplots for IGD values.

fluctuation throughout the optimization process, therefore, being less sensitive to scale differences and the current evolutionary stage. The sliding window stores just a recent subset of the low level heuristics application and, so, discards old event (data) that has little or no relevance in the current stage of the dynamic process of optimization. The combination of these two features provide a more robust choice function. It is important to note that the proposed approach uses

Table 1. IGD statistics. Dark gray cells emphasize the best results while light gray cells emphasize the second best results.

CEC09	Algorithm	Median	Mean	Std	Min	Max
UF1	MOEA/D-HH$_{SW}$	0.001005	0.001005	0.000074	0.000884	0.001183
	MOEA/D-DRA	0.001503	0.001526	0.000090	0.001417	0.001757
	MOEAD-CMX-SPX	0.004171	0.004292	0.000263	0.003985	0.005129
	ENS-MOEA/D	-	0.001642	0.000125	-	-
	NSGA-II	0.095186	0.094731	0.003249	0.088511	0.103222
UF2	MOEA/D-HH$_{SW}$	0.001715	0.001834	0.000475	0.001335	0.003362
	MOEA/D-DRA	0.003375	0.003502	0.001039	0.002326	0.006638
	MOEAD-CMX-SPX	0.005472	0.005615	0.000412	0.005149	0.006778
	ENS-MOEA/D	-	0.004048	0.001005	-	-
	NSGA-II	0.035151	0.035071	0.001479	0.032968	0.039164
UF3	MOEA/D-HH$_{SW}$	0.002507	0.003165	0.009852	0.001039	0.055887
	MOEA/D-DRA	0.001488	0.003948	0.004131	0.001086	0.014019
	MOEAD-CMX-SPX	0.005313	0.011165	0.013093	0.004155	0.068412
	ENS-MOEA/D	-	0.002591	0.000456	-	-
	NSGA-II	0.089894	0.090817	0.016815	0.062901	0.126556
UF4	MOEA/D-HH$_{SW}$	0.051989	0.052916	0.003478	0.046915	0.060247
	MOEA/D-DRA	0.060765	0.060285	0.004757	0.051492	0.070912
	MOEAD-CMX-SPX	0.063524	0.064145	0.004241	0.055457	0.075361
	ENS-MOEA/D	-	0.042070	0.001325	-	-
	NSGA-II	0.080935	0.080737	0.002809	0.074034	0.084683
UF5	MOEA/D-HH$_{SW}$	0.266449	0.273792	0.055874	0.193141	0.424148
	MOEA/D-DRA	0.220083	0.254930	0.089484	0.146933	0.511464
	MOEAD-CMX-SPX	0.379241	0.418508	0.135554	0.211058	0.707093
	ENS-MOEA/D	-	0.248110	0.042555	-	-
	NSGA-II	0.214958	0.220145	0.051622	0.154673	0.331853
UF6	MOEA/D-HH$_{SW}$	0.073173	0.096334	0.062852	0.041048	0.286627
	MOEA/D-DRA	0.207831	0.326176	0.287571	0.053371	0.823381
	MOEAD-CMX-SPX	0.248898	0.327356	0.185717	0.056972	0.792910
	ENS-MOEA/D	-	0.060847	0.019840	-	-
	NSGA-II	0.080177	0.080728	0.006460	0.067996	0.090331
UF7	MOEA/D-HH$_{SW}$	0.001051	0.001106	0.000166	0.000920	0.001845
	MOEA/D-DRA	0.001569	0.001945	0.001364	0.001336	0.008796
	MOEAD-CMX-SPX	0.004745	0.006262	0.003307	0.003971	0.014662
	ENS-MOEA/D	-	0.001728	0.000852	-	-
	NSGA-II	0.048873	0.048977	0.001959	0.044634	0.051968
UF8	MOEA/D-HH$_{SW}$	0.030973	0.031072	0.002449	0.027362	0.037827
	MOEA/D-DRA	0.040352	0.040667	0.003788	0.033777	0.050412
	MOEAD-CMX-SPX	0.056872	0.057443	0.003366	0.051800	0.065620
	ENS-MOEA/D	-	0.031006	0.003005	-	-
	NSGA-II	0.112219	0.113226	0.002742	0.109836	0.121190
UF9	MOEA/D-HH$_{SW}$	0.028231	0.044430	0.041128	0.024260	0.152107
	MOEA/D-DRA	0.137856	0.123078	0.039018	0.025008	0.139986
	MOEAD-CMX-SPX	0.144673	0.097693	0.054285	0.033314	0.151719
	ENS-MOEA/D	-	0.027874	0.009573	-	-
	NSGA-II	0.106841	0.106981	0.000681	0.105806	0.108729
UF10	MOEA/D-HH$_{SW}$	0.467566	0.467229	0.049327	0.362297	0.560866
	MOEA/D-DRA	0.406094	0.408770	0.066770	0.210500	0.553572
	MOEAD-CMX-SPX	0.467715	0.462653	0.038698	0.391496	0.533234
	ENS-MOEA/D	-	0.21173	0.019866	-	-
	NSGA-II	0.257846	0.259851	0.012541	0.234047	0.288036

the Scaling Factor proposed in [5] and the adaptive mechanism proposed in [4]. Algorithm 2 presents the pseudocode of the selection of the low level heuristic by the proposed choice function while Algorithm 3 describes its update procedure.

6 Experiments and Results

In order to test MOEA/D-HH$_{SW}$, we perform some experiments considering the 10 unconstrained instances from the CEC 2009 multi-objective benchmark [14]. The search space dimension used is n=30 for all the instances. We consider the IGD-metric of the final approximation over 30 independent executions.

MOEA/D-HH$_{SW}$ is first confronted with two multiobjective hyper-heuristics based on choice function: MOEA/D-HH[5] and MOEA/D+Modified Choice Function [4]. Figure 1 shows that MOEA/D-HH$_{SW}$ outperforms the other approaches considered in six of the ten instances (UF1, UF2, UF3, UF4, UF7 and UF8) and is the second best algorithm in two other instances (UF5 and UF6). So, it is an improved version of the other multiobjective hyper-heuristics. The proposed algorithm is also compared with some well known evolutionary multiobjective algorithms: MOEA/D-DRA, MOEA/D-DRA-CMX+SPX and NSGA-II. Table 1 presents median, standard deviation (std), minimum (min) and maximum (max) of IGD-metric values. IGD-metric values of MOEA/D-DRA-CMX+SPX and ENS-MOEA/D are extracted from [8,16], respectively. MOEA/D-HH$_{SW}$ is the best method in UF1, UF2, UF4, UF6, UF7, UF8 and UF9 and the second best method in UF3. Therefore, it can be considered a state-of-the-art multiobjective algorithm.

7 Conclusions

This paper proposed a new multi-objective hyper-heuristic, called MOEA/D-HH$_{SW}$, based on a new adaptive choice function with sliding window to determine the low level heuristics (operators) in the MOEA/D framework. Our pool of low level heuristics was constituted by five Differential Evolution operators.

The proposed choice function uses relative rewards and a sliding window. The advantages of using relative rewards is that the scale is indifferent from the problem (instance) being solved and the current stage of evolution while the sliding window maintains a set of relevant events for current state of the optimization, discarding old events that may not be important due to the dynamic nature of the optimzation process.

We applied our proposed approach on test instances from the CEC 2009 MOEA Competition. The experiments conducted showed that the proposed adaptive choice function performs better than a modified choice function recently proposed in [4] and an improved version of the modified choice function used in MOEA/D-HH [5]. The proposed algorithm (MOEA/D-HH$_{SW}$) was favourably compared with state-of-the-art methods and constitutes a promising approach.

Acknowledgments. The authors acknowledge Fundação Araucária for the partial financial support with project number 23.116/2012.

References

1. Burke, E.K., Gendreau, M., Hyde, M., Kendall, G., Ochoa, G., Ozcan, E., Qu, R.: Hyper-heuristics. J. Oper. Res. Soc. **64**(12), 1695–1724 (2013)
2. Burke, E.K., Silva, J.L., Silva, A., Soubeiga, E.: Multi-objective hyper-heuristic approaches for space allocation and timetabling. In: Meta-heuristics: Progress as Real Problem Solvers, p. 129. Springer (2003)
3. Cowling, P.I., Kendall, G., Soubeiga, E.: A hyperheuristic approach to scheduling a sales summit. In: Burke, E., Erben, W. (eds.) PATAT 2000. LNCS, vol. 2079, pp. 176–190. Springer, Heidelberg (2001)
4. Drake, J.H., Özcan, E., Burke, E.K.: An improved choice function heuristic selection for cross domain heuristic search. In: Coello, C.A.C., Cutello, V., Deb, K., Forrest, S., Nicosia, G., Pavone, M. (eds.) PPSN 2012, Part II. LNCS, vol. 7492, pp. 307–316. Springer, Heidelberg (2012)
5. Gonçalves, R.A., Kuk, J.N., Almeida, C.P., Venske, S.M.: MOEA/D-HH: a hyper-heuristic for multi-objective problems. In: Gaspar-Cunha, A., Henggeler Antunes, C., Coello, C.C. (eds.) EMO 2015. LNCS, vol. 9018, pp. 94–108. Springer, Heidelberg (2015)
6. Kateb, D.E., Fouquet, F., Bourcier, J., Traon, Y.L.: Artificial mutation inspired hyper-heuristic for runtime usage of multi-objective algorithms. CoRR abs/1402.4442 (2014). http://arxiv.org/abs/1402.4442
7. Kendall, G., Soubeiga, E., Cowling, P.: Choice function and random hyperheuristics. In: Proceedings of the Fourth Asia-Pacific Conference on Simulated Evolution and Learning, SEAL, pp. 667–671. Springer (2002)
8. Mashwani, W.K., Salhi, A.: A decomposition-based hybrid multiobjective evolutionary algorithm with dynamic resource allocation. Appl. Soft Comput. **12**(9), 2765–2780 (2012)
9. Li, K., Kwong, S., Zhang, Q., Deb, K.: Interrelationship-based selection for decomposition multiobjective optimization. IEEE Transactions on Cybernetics **PP**(99), 1 (2014)
10. Maashi, M., Özcan, E., Kendall, G.: A multi-objective hyper-heuristic based on choice function. Expert Systems with Applications **41**(9), 4475–4493 (2014)
11. Sindhya, K., Ruuska, S., Haanpää, T., Miettinen, K.: A new hybrid mutation operator for multiobjective optimization with differential evolution. Soft Comput. **15**(10), 2041–2055 (2011)
12. Storn, R., Price, K.: Differential evolution - a simple and efficient heuristic for global optimization over continuous spaces. J. Global Optim. **11**(4), 341–359 (1997)
13. Vazquez-Rodriguez, J.A., Petrovic, S.: A mixture experiments multi-objective hyper-heuristic. J. Oper. Res. Soc. **64**(11), 1664–1675 (2013)
14. Zhang, Q., Zhou, A., Zhao, S., Suganthan, P.N., Liu, W., Tiwari, S.: Multiobjective optimization test instances for the CEC 2009 special session and competition. Tech. rep., University of Essex and Nanyang Technological University, CES-487 (2008)
15. Zhang, Q., Liu, W., Li, H.: The performance of a new version of MOEA/D on CEC09 unconstrained MOP test instances. In: Congress on Evolutionary Computation, pp. 203–208 (2009)
16. Zhao, S.Z., Suganthan, P.N., Zhang, Q.: Decomposition-based multiobjective evolutionary algorithm with an ensemble of neighborhood sizes. IEEE Trans. Evol. Comput. **16**(3), 442–446 (2012)

Multi-start Differential Evolution Approach in the Process of Finding Equilibrium Points in the Covariant Games

Przemyslaw Juszczuk[(✉)]

Institute of Computer Science, University of Silesia,
ul.Bedzinska 39, Sosnowiec, Poland
przemyslaw.juszczuk@us.edu.pl

Abstract. In this article we present the application of the Differential Evolution algorithm to the problem of finding optimal strategies in the covariant games. Covariant game is the class of games in the normal form, in which there is at least one Nash equilibrium, and some payoffs of players may be in some way correlated. We used the concept, that there is possibility, that the approximate solution of the game exists, when players use only the small subset of strategies. The Differential Evolution algorithm is presented as the multi-start method, in which sampling of the solution is used. If preliminary estimation of the solution is not satisfactory, the new subset of strategies for all players is selected and the new population of individuals is created.

Keywords: Differential evolution · Game theory · Equlibrium point · Covariant game

1 Introduction

Evolutionary computation technique is one of the most important optimization techniques. Although many modifications and new algorithms were introduced, there is no universal technique, which could give the good results for all optimization problems. One of the most popular and efficient techniques is the Differential Evolution. It is simple, population-based algorithm mainly used in the field of numerical optimization. Its aim is to find solution:

$$x^* = \arg \min_{x \in X} F(x) \tag{1}$$

for the quality function $F(x)$, where X denotes a set of feasible solutions and $x \in X$ is a vector $x = [x^1, x^2, \ldots, x^{n_F}]$. n_F is a number of function dimensions.

A major impediment for numerical optimization is multimodality of function $F(x)$, since the found solution may not be the global optimum. Such assumption rejects many groups of algorithms in which exploitation of the search space is dominant over the exploration and leads to locally optimal solutions. As a solution to above problems, many algorithms use some random factors to increase the

© Springer International Publishing Switzerland 2015
M. Núñez et al. (Eds.): ICCCI 2015, Part II, LNCS 9330, pp. 139–148, 2015.
DOI: 10.1007/978-3-319-24306-1_14

impact of the exploration of the search space. One of the mostly used approaches are genetic operators similar to the mutation in the classical evolutionary algorithms. Such mutation allows to randomly modify the single genotype and eventually release the object from the attraction area of the local optimum. Some less popular techniques involves replacing similar objects in the population with the new, randomly generated elements. One of such methods are so-called multi-start algorithms, in which the best solution in the single algorithm run is remembered, and the whole population start the optimization process from the scratch.

In this article we present the multi-start Differential Evolution algorithm adapted to the one of the most famous game theory problems, which is generating Nash equilibria (also called optimal strategies). We will focus on the covariant games in the normal form for n players generated with the GAMUT program [13]. It is the standard test suite used in the game theory and it allows to generate different classes of games. Proposed method allows to find multiple solutions in one single algorithm run. Every considered game has n players and m number of strategies. We assume, that the single solution involves using the subset of strategies - and the remaining strategies have zero probability of being chosen. After a few hundreds of iterations it is possible to estimate quality of candidate solution. The candidate solution that not meet some initial prerequisites is rejected. The new subset of strategies is selected and the whole Differential Evolution population is once again generated. The result of the single algorithm run is the subset of approximate Nash equilibria.

Article is organized as follows: first, we give some examples of multi-start algorithms. Some examples of algorithms for calculating the Nash equilibria are also given. Next section brings us some basic definitions related to the game theory and covariant games. In the section four we describe the multi-start Differential Evolution algorithm and explain details of the solution. We end article with experiments and conclusions.

2   Related Works

Multi-start techniques in evolutionary computation are not very popular. The most of the algorithms focus mainly to introduce some randomness inside the population instead of restarting the whole algorithm. Besides the classic modifications including for example mutation in the swarm algorithms [16], there are methods in which life span of the single individual is introduced. One of such earlier algorithms is the particle swarm optimization with the life span [20]. Other group of algorithms connects multi-start with the local search method [10]. One of the earlier methods of this type was described as a multi-start technique controlled by an annealing schedule [14]. Approaches with multi-start method used for combinatorial problems may be found in [1,5].

The main algorithm for computing Nash equilibria is the Lemke–Howson (LH) algorithm [4], which is a pivoting algorithm similar to the simplex algorithm for linear programming. The Lemke–Howson is often considered as the state of art algorithm. Unfortunately, the classical method developed by Lemke and

Howson (1964) not always return the solution in the game. The most popular algorithms used for calculating the exact Nash equilibrium may be found in Gambit program [11]. It contains algorithms like Lemke Howson or support enumeration [3]. Methods based on the acquiring approximate solution (ϵ-Nash equilibrium) mainly used in the class of random games are less popular than classical approaches. These methods are mainly based on the computation of the pessimistic approximation of the Nash equilibrium. Such algorithms give $\epsilon = \frac{3}{4}$ [6], $\epsilon = \frac{1}{2}$ [2] and $\epsilon = 0.3393$ [19], where ϵ means the maximum deviation from the optimal strategy for the worst player.

3   Game Theory and Covariant Games

The simple game in the strategic form may be described as:

$$\Gamma = \langle N, \{A_i\}, M \rangle, i = 1, 2, ..., n \tag{2}$$

where:

- $N = \{1, 2, ..., n\}$ is the set of players;
- $\{A_i\}$ is the finite set of strategies for the i-th player with m-strategies;
- $M = \{\mu_1, \mu_2, ..., \mu_n\}$ is the finite set of the payoff functions.

The strategy profile will be defined as follows: $a = (a_1, ..., a_n)$ for all players, where every element of the a is the mixed strategy for the player.

$$a_i = (P(a_{i_1}), P(a_{i_2}), ..., P(a_{i_m})), \tag{3}$$

where $P(a_{i_1})$ will be the probability of choosing the strategy 1 by the player i and m is the number of strategies for single player. It is possible to calculate the strategy profile of all players excluding single element.

$$a_{-i} = (a_1, ..., a_{i-1}, a_{i+1}, ..., a_n), \tag{4}$$

where i is the excluded player index. Excluded player is not allowed to play any of the mixed strategy, and his only option is to play the pure strategy.

One of the most important problems in all game theories is finding the Nash equilibrium. In general, the Nash equilibrium is a strategy profile such that no deviating player could achieve a payoff higher than the one that the specific profile gives him:

$$\forall_i, \forall_j \ \mu_i(a) \geq \mu_i(a_{i_j}, a_{-i}), \tag{5}$$

where i is the i-th player, j is the number of the strategies for given player, $\mu_i(a)$ is the payoff for the i-th player for the strategy profile a and $\mu_i(a_{i_j}, a_{-i})$ is the payoff for the i-th player using strategy j against the profile a_{-i}.

In this article we focus only on the covariant games, which are basically the random games with some random covariance between strategies. The highest payoffs are for all outcomes in which every player chooses the same action, although it is not always the case that all of these outcomes yield the same payoffs. There is at least one Nash equilibrium in every game (according to the famous John Nash theorem [12]). There is lack of information about the number of ϵ equilibria in coordination games.

4 Multi-start Differential Evolution

Differential evolution (DE) is a stochastic, population-based search strategy developed by R. Storn and K. Price [17], and deeply studied by Jouni Lampinen, Ivan Zelinka [7,8], and others. Like other evolutionary techniques, it starts with randomly generated set of potential solutions, which are modified over time. One of the main differences between other similar methods is number of populations. The Differential Evolution has three equal sized populations: the parent population, the trial population and the child population. The algorithm creates trial individual by by perturbing the solutions with a scaled difference of two randomly selected parent vectors. The crossover process is very simple and allows to create offspring. The replacement of the parent takes place only if the newly generated offspring is better adapted to the search space. The simple DE schema is presented on the fig. 1.

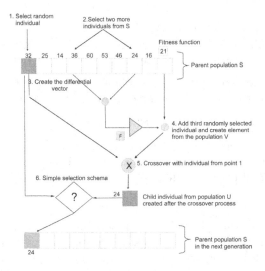

Fig. 1. The Differential Evolution schema

A non-cooperative game in strategic form consists of a set of players, and, for each player, a set of strategies available to him as well as a payoff function mapping each strategy profile (i.e. each combination of strategies, one for each player) to a real number. That number captures the preferences of the player over the possible outcomes of the game. In this situation, the basic problem is to transform the normal form game into the single genotype. Every element of the genotype may be described as the probability of choosing single strategy of the player. In this article we assume, that the mixed strategy for every player is based only on the subset of strategies. In other words, for the single game, there is possibility to find the approximate Nash equilibrium on the basis of different subset of strategies.

4.1 Genetic Operators and Multi-start

In games, where there is over a dozen strategies - the most of them has usually zero probability of being chosen. This leads to the conclusion, that number of strategies, that have an impact on the solution is usually very small. This assumption was confirmed in [9]. Of course selecting all strategies for every player is also possible and it could give satisfactory results. It is worth noting that in [15] author suggest that using less active strategies gives better results for the player. A subset containing only a few active strategies in this situation seems to be better choice.

Proposed multi-start method is based on sampling of the solution space. The convergence of the population is the fastest in the first few hundred iterations. So with high probability it is possible to estimate quality of solution after the first g iterations, where $g \ll Max_{iterations}$. Every subset of active strategies, in which the ϵ value after the generation g is higher than the expected value is rejected. The new subset of strategies is randomly selected, and the whole sampling process is started from scratch. When the potential solution passes the sampling phase, it is solved by the classical Differential Evolution algorithm, in which mutation, crossover and selection allows to direct the search towards the global optimum.

The mutation is the main operator, and its role is dominant over the crossover. The whole process is based on so called differential vector, which is calculated on the basis of two randomly selected individuals. Third random individual is added to the difference vector:

$$\forall_i \forall_j U_{i,j} = P_{r_1,j} + F \cdot (P_{r_2,j} - P_{r_3,j}) \tag{6}$$

where P is the population of parents. Numbers r_1, r_2 and r_3 determine index of the individual. The crossover schema is based on the combination of the genes from the parent and the trial individual. The lower crossover factor CR allows to put more emphasis on the elements from the parent individual, which leads to the less aggressive search. On the other hand, the higher crossover factor promotes the trial individuals which impact on the greater variety in the population:

$$\forall_i, \forall_j, U_{i,j} = \begin{cases} V_{i,j} \text{ when } RAND[0,1) < CR, \\ P_{i,j} \text{ in other case,} \end{cases}$$

A selection schema used in this algorithm guarantees that individuals promoted to the next generation are not worse than individuals in actual generation.

$$\forall_i, X^*_i = \begin{cases} U_i \text{ when } f(U_i) \leq f(X_i), \\ X_i \text{ in other case,} \end{cases}$$

where X^* is the individual in the next generation, $f()$ is the fitness function.

4.2   Fitness Function

Solution of the problem is finding the Nash equilibrium. As we said earlier in the article, equilibrium is the situation, in which none of the players cannot change his strategy to get more payoff. So every deviation from this situation is the approximation of the global optimum. We completely omit the value of the payoff - our goal is only to find equilibrium, and not maximizing it. After transforming the game into the continuous function optimization problem it is possible to calculate the minimal deviation from the optimal strategy for every player. This may be achieved as follows:

$$f_1 = \sum_{i=1}^{n} \sum_{j=1}^{n} c \cdot |u_i - u_j|; i \neq j \tag{7}$$

where u_i is the payoff for the i-th player, and c is constant value. Every part of the genotype is the probability of choosing the single strategy for the player, so the sum of probabilities for single player must be equal to 1. Above statement must be met for every player:

$$f_2 = \sum_{i=1}^{n} |1 - \sum_{j=1}^{m} P(a_{ij})| \tag{8}$$

f_1 means the maximum deviation from the optimal strategy - in other words, the worst case scenario. It should be equal to 0 - which helds for the global optimum. Every deviation is denoted as the ϵ.

5   Experiments

Recently it was shown, that the Differential Evolution may be very efficient tool in the continuous optimization problems. Process of computing Nash equilibria was transformed into such problem, and the main goal for this article is to show, that the proposed solution is capable to solve covariant games. As it was mentioned before, we generated a coordination games on the basis of the GAMUT program. All generated games had solutions in the pure and mixed strategy profiles. The number of players was equal to 3 and 4, and the number of strategies for the single player was in range $\langle 5, 25 \rangle$. Size of the problems is more difficult, that often considered standard examples with only 2 players. Number of strategies for the single player was chosen arbitrarily. All payoffs in the games were normalized to the range $\langle 0, 1 \rangle$, and for every problem size experiment was repeated 30 times. The set of parameters in the Differential Evolution was as follows:

- we used classical mutation and crossover schema with mutation factor F equal to 0.7 and the crossover factor CR equal to 0.5 (as it was proposed in the [18]);

- population size was set to number of strategies multiplied by the square of the number of players;
- sampling length - 300 iterations and number of iterations - 1000 for 3 players games and 2000 for 4 players games.

All games in the experiments were described with two elements: number of players and number of strategies. For example 3p10s means 3 player's game with 10 strategies for every player.

5.1   Quality of Solutions

ϵ value means the maximal deviation from the optimal strategy, so $\epsilon = 0$ is considered as the global optimum. In the table 1 quality of solution is given. The Differential Evolution was able to solve every game, but in the other hand, none exact solution was found. We considered solutions, in which every player used 2 and 3 strategies from the set of all available strategies. Results are very promising - it is worth noting, that the best polynomial approximation algorithms for two players give the result with ϵ around 0.33.

Moreover, the Differential Evolution is capable to give multiple solutions in the single algorithm run. In the Table 1 we can see: minimum, first quartile, average, third quartile, maximum and standard deviation of the ϵ values. There

Table 1. Basic statistics for the ϵ values - different type of covariant games with 2 and 3 active strategies.

	2 strategies per player					
	Minimum	1 Quartile	Average	3 Quartile	Maximum	Standard deviation
3p10s	0.0216	0.0519	0.0694	0.0942	0.1050	0.0320
3p12s	0.0382	0.0475	0.0652	0.0768	0.1173	0.0306
3p14s	0.0297	0.0561	0.0838	0.0834	0.1321	0.0346
3p15s	0.1179	0.1233	0.1353	0.1369	0.1669	0.0176
4p5s	0.0143	0.0256	0.0301	0.0366	0.0461	0.0093
4p7s	0.0330	0.0535	0.0712	0.0867	0.1014	0.0249
4p9s	0.0661	0.0921	0.1165	0.1398	0.1650	0.0323
4p10s	0.0629	0.0831	0.1158	0.1467	0.1642	0.0416
	3 strategies per player					
	Minimum	1 Quartile	Average	3 Quartile	Maximum	Standard deviation
3p10s	0.0594	0.0691	0.0845	0.1020	0.1134	0.0219
3p12s	0.0613	0.0737	0.0979	0.1160	0.1299	0.0288
3p14s	0.1082	0.1216	0.1312	0.1410	0.1495	0.0154
3p15s	0.0949	0.1086	0.1284	0.1506	0.1583	0.0263
4p5s	0.0371	0.0498	0.0597	0.0688	0.0961	0.0173
4p7s	0.0752	0.0962	0.1089	0.1218	0.1465	0.0249
4p9s	0.0849	0.1209	0.1331	0.1574	0.1663	0.0304
4p10s	0.0737	0.1216	0.1445	0.1793	0.1923	0.0449

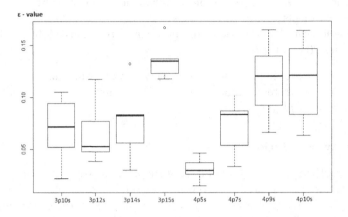

Fig. 2. ε values for different type of covariant games - 2 active strategies per player

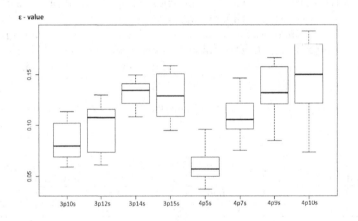

Fig. 3. ε values for different type of covariant games - 3 active strategies per player

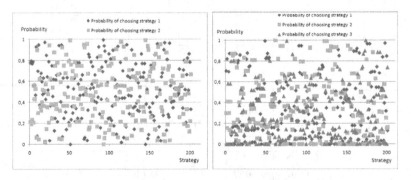

Fig. 4. Dispersion charts with probabilities of chosing single strategy. a) - games with 2 active strategies per player; b) - games with 3 active strategies per player.

was no game, in which global optimum was found (ϵ equal 0), on the other hand, standard deviation for every problem was very small, so we can assume that results obtained by the algorithm were repeatable.

On the fig. 2 and fig. 3 we can see boxplots for the same problems, in which players used respectively two active strategies, and three active strategies. The most important thing, is that regardless to the number of active strategies, obtained results were quite similar. It indirectly confirms inference from the article [15] - covariant games are similar to the classical random games problem, and the low number of active strategies also leads to the very good results.

5.2 Probability of Choosing Single Strategy

The ϵ-well supported Nash equilibrium is concept of solution, in which every strategy for the player has non-zero probability of being chosen. According to the above assumption, the large number of active strategies should lead to solutions in which active strategies have lower probability of being chosen - and this probabilities are distributed in some way uniformly over the set of strategies. In general, when number of strategies for the single player increases, the probability of chosing strategy should decreasing. On the fig. 4 we compared 200 different solutions (chosen randomly from all generated solution set). Every point in the chart is probability for the first or the second strategy. For the chart a), we can see, that the points are scattered around the chart more or less equally. Situation is different in the chart b) (points represent probability of choosing first, second or third strategy), where there is accumulation of the points in the lower bound of the chart. In other words, one from set of three strategies very often has a very low probability of being chosen - opposite to that, what we would expect according to the ϵ-well supported equilibrium concept.

6 Conclusions

In this article we presented the game theory problem, which can be easily transformed into the continuous function optimization problem. The best obtained results are very promising, however it was not possible to find exact Nash equilibrium in any game. It is also interesting, that preliminary studies on the probability of choosing single strategies was very promising. We have shown, that even for the small set of active strategies it was still possible to indicate, that more often exist approximate equilibrium with only 2 active strategies. Increasing the number of active strategies available for single player not necessarily gives better results even for large instance of games. Of course this problem still requires further experiments and will be examined in further works.

References

1. Bräysy, O., Haslea, G., Dullaertb, W.: A multi-start local search algorithm for the vehicle routing problem with time windows. European Journal of Operational Research **159**(3), 586–605 (2004)

2. Daskalakis, C., Mehta, A., Papadimitriou, C.: A note on approximate Nash equilibria. Journal Theoretical Computer Science **410**, 1581–1588 (2009)
3. Dickhaut, J., Kaplan, T.: A program for finding Nash equilibria, Working papers, University of Minnesota, Department of Economics (1991)
4. Lemke, C.E., Howson, J.T.: Equilibrium Points of Bimatrix Games. Society for Industrial and Applied Mathematics **12**, 413–423 (1964)
5. Boese, K.D., Kahng, A.B., Sudhakar, M.: A new adaptive multi-start technique for combinatorial global optimizations. Operations Research Letters **16**(2), 101–113 (1994)
6. Kontogiannis, S.C., Panagopoulou, P.N., Spirakis, P.G.: Polynomial algorithms for approximating Nash equilibria of bimatrix games. Journal Theoretical Computer Science **410**, 1599–1606 (2009)
7. Lampinen, J., Zelinka, I.: Mixed variable non-linear optimization by differential evolution. Proceedings of Nostradamus (1999)
8. Lampinen, J., Zelinka, I.: On stagnation of the differential evolution algorithm. In: Proceedings of Mendel, 6th International Mendel Conference on Soft Computing (2000)
9. Lipton, R., Markakis, E., Mehta, A.: Playing large games using simple strategies, pp. 36–41 (2003)
10. Marchiori, E.: Genetic, iterated and multistart local search for the maximum clique problem. In: Cagnoni, S., Gottlieb, J., Hart, E., Middendorf, M., Raidl, G.R. (eds.) EvoWorkshops 2002. LNCS, vol. 2279, pp. 112–121. Springer, Heidelberg (2002)
11. McKelvey, R.D., McLennan, A.M., Turocy, T.L.: Gambit: Software Tools for Game Theory, Version 0.2010.09.01 (2010). http://www.gambit-project.org
12. Nash, J.F.: Non-cooperative games. Annals of Mathematics **54**(2), 286–295 (1951)
13. Nudelman, E., Wortman, J., Shoham, Y., Leyton-Brown, K.: Run the GAMUT: a comprehensive approach to evaluating game-theoretic algorithms. In: Proceedings of the Third International Joint Conference on Autonomous Agents and Multiagent Systems, vol. 2 (2004)
14. Piccioni, M.: A Combined Multistart-Annealing Algorithm for Continuous Global Optimization, Institute for Systems Research Technical Reports (1987)
15. Rubinstein, A.: Modelling Bounded Rationality. MIT Press (1998)
16. Stacey, A.: Particle swarm optimization with mutation. In: The 2003 Congress on Evolutionary Computation, vol. 2, pp. 1425–1430 (2003)
17. Storn, R., Price, K.: Differential evolution - a simple and efficient heuristic for global optimization over continuous spaces. Journal of Global Optimization **11**(4), 341–359 (1997)
18. Storn, R.: On the usage of differential evolution for function optimization. In: NAFIPS 1996, pp. 519–523. IEEE (1996)
19. Tsaknakis, H., Spirakis, P.G.: An optimization approach for approximate nash equilibria. In: Deng, X., Graham, F.C. (eds.) WINE 2007. LNCS, vol. 4858, pp. 42–56. Springer, Heidelberg (2007)
20. Zhang, Y.-W., Wang, L., Wu, Q.-D.: Mortal particles: particle swarm optimization with life span. In: Tan, Y., Shi, Y., Chai, Y., Wang, G. (eds.) ICSI 2011, Part I. LNCS, vol. 6728, pp. 138–146. Springer, Heidelberg (2011)

Genetic Algorithm in Stream Cipher Cryptanalysis

Iwona Polak$^{(\boxtimes)}$ and Mariusz Boryczka

Institute of Computer Science, University of Silesia,
Będzińska 39, 41-200 Sosnowiec, Poland
{iwona.polak,mariusz.boryczka}@us.edu.pl

Abstract. Cryptography nowadays is a very important field of protecting information from falling into wrong hands. One of modern cryptography branch is stream cipher cryptography. This paper focuses on cryptanalysis of such ciphers using genetic algorithm. Genetic algorithm as one of optimisation methods isn't quite obvious to use in the field of cryptography, nevertheless it can give interesting results. In this article authors look for the shortest equivalent linear system which approximate given keystream with linear shift feedback register.

Keywords: Genetic algorithm · Cryptanalysis · RC4 · Stream cipher · LFSR

1 Introduction

Modern cryptology is divided into two main blocks: symmetric-key cryptography and public-key cryptography. The first group is next divided into subgroups: block ciphers and stream ciphers (fig. 1). Stream cipher operating principle is such that for every bit of plain text, there is corresponding bit of keystream – these two are added together with XOR operation and the result gives one bit of cipher text. The main issue here is that every bit of plain text is coded independently of all others.

In modern cryptography some metaheuristic techniques were used many times for cryptanalysis of such block ciphers as: XTEA [6], DES [8,11,12] or AES [5]. In all these papers there are weakened versions used – with shorter keys or smaller amount of rounds.

Attacking stream ciphers with genetic algorithms were already tested in [7,14]. Authors of these papers look for shortest equivalent linear system which approximate given keystream with linear shift feedback register. Authors study registers of length 5 to 8, whereas this paper focuses on finding the shortest equivalent linear system which approximate given keystream with linear shift feedback register. Our previous research were conducted on LFSR itself [9] and A5/1 and A5/2 stream ciphers [10].

The paper is organized as follows. In section 2 there is short description of Linear Feedback Shift Register. Section 3 gives overview of genetic algorithm.

© Springer International Publishing Switzerland 2015
M. Núñez et al. (Eds.): ICCCI 2015, Part II, LNCS 9330, pp. 149–158, 2015.
DOI: 10.1007/978-3-319-24306-1_15

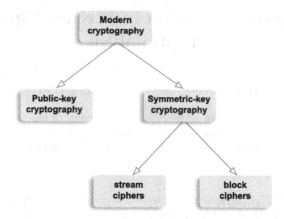

Fig. 1. Modern cryptography classification

Stream cipher RC4 is introduced in 4. Section 5 reports experiments' details and results, and section 6 draws conclusions.

2   Linear Feedback Shift Register

Binary series produced by shift registers are used in many cryptographic structures and pseudorandom number generators. Linear Feedback Shift Registers (LFSR) are used as part of many stream ciphers, e.g. in mobile GSM encryption by A5/1 and A5/2 or in Bluetooth by E_0. LFSRs are common because they are very easy for hardware implementation. Breaking component LFSR could be the first step in breaking the whole stream cipher.

LFSR is such a structure in which in every step:

- first bit is some function of all other bits from previous state (usually XOR of chosen bits, called taps),
- all other bits are just shifted one position to the right,
- output stream is appended with last bit from the previous state.

LFSRs are described by equations of the form $x^a + x^b + ... + 1$, where every power of x shows places where taps occur. The highest power is the length of the LFSR. Register shown in fig. 2 is based on the following equation:

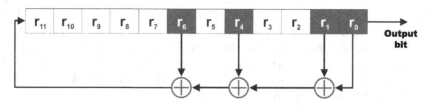

Fig. 2. LFSR scheme from equation $T(x) = x^{12} + x^6 + x^4 + x^1 + 1$

$$T(x) = x^{12} + x^6 + x^4 + x^1 + 1 \qquad (1)$$

3 Genetic Algorithm

A method called genetic algorithm (GA) was first introduced by J. H. Holland [3,4]. Its functioning is based on evolution of living organisms. It uses phenomena such as natural selection and evolutionary operators like crossover and mutation. Genetic algorithm works on specified population which consists of finite number of individuals. Each individual is represented in a specific way suited best for considered task. Proper selection of GA's functions and individual's representation is essential to obtain satisfying results.

Genetic algorithm belongs to the set of probabilistic methods, which means that for every run of the algorithm different results could be obtained. It also gives approximate solution, very good one, but not necessarily the best one.

In this work basic version of GA was used, which acts as follows:

Algorithm 1. Genetic Algorithm

---
1   randomly generate initial population
2   evaluate the fitness function for every individual
3   **while** *termination condition has not been reached* **do**
4 | apply for chosen individuals:
5 | − crossover
6 | − mutation
7 | replace old population with new one using selection and reproduction
8 | evaluate the fitness function for every individual
9   **return** *the best solution found*

Individual, that is LFSR register, is represented as binary array of length n. Taps are marked as 1s (fig. 3). Tap at least significant bit is obligate. Otherwise it will not be a proper LFSR.

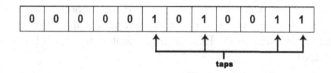

Fig. 3. Representation of an individual from equation 1 and fig. 2

In this work there are two models of crossover considered and they both are one-point crossover. One crossover point is selected on both parents. Two children are formed by swaping bits beyond this crossover point. In this case individuals may vary in length, which is not a classical situation. Crossover operator had to be adapted to this situation.

First model is crossover to the right (fig. 4) in order to guarantee producing only correct individuals – least significant bit will always have value 1, so in parents and children. Crossover to the left is the second model used (fig. 5). It is not so simple as the first model because during the crossover process there may arise individuals which are not a valid LFSR registers. This will happen when crossover point is selected at the end of shorter individual, and there exists no tap on corresponding place in the longer individual. In such situation the repair process were implemented: right-most bit is set as 1 (fig. 6).

Mutation introduces random changes to genotype. It can transfer population to the region, which by crossover wouldn't be available. For example if all individuals begin with 0 (0010, 0010, 0101, ...) whereas the best solution begins with 1 (1010) there is no possible way to generate this solution relying solely on crossover. The only chance is if the first bit of some individual mutates from 0 to 1. On the other hand mutation can damage good solutions pulling the population from region of promising search. So the probability of mutation isn't very high. This also has grounds in nature – as in living organisms evolution mutation also occurs rarely. Designing mutation properly and giving a right probability to it, have very significant affect on success and speed of genetic algorithm.

In this paper there are two models of mutation considered. They were chosen among several mutation models based on best results yielded in preliminary studies. Random swap mutation draws one tap and moves it to some other place (fig. 7). Stretch mutation does not interfere with taps. This mutation stretches individual's length for one bit (fig. 8).

As a selection method Roulette Wheel Selection (RWS) was used. In RWS the chance of an individual being selected is proportional to its fitness function value (section 5.1). The higher the value, the higher chance of being selected to the reproduction process. It is noteworthy that RWS does not guarantee the best individual to be selected.

4   RC4

RC4 is a stream cipher invented in RSA Securites. At the beginning it was a trade secret, but few years later it was posted anonymously on Cypherpunks mailing list [1]. The keystream is the same length as plain text and they are also independent of each other. RC4 is implemented in SSL/TLS protocol and in WEP standard.

0 0 1 0 0 0 1 0 1 0 0 0 0 0 1
0 1 0 0 0 1 1 0 0 0 1
0 1 0 0 1 0 0 0 0 0 1
0 0 1 0 0 0 1 0 0 1 1 0 0 0 1

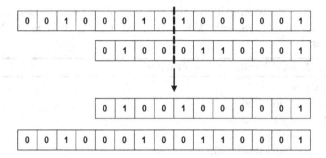

Fig. 4. Crossover to the right

0 0 1 0 0 0 1 0 1 0 0 0 0 0 1
0 1 0 0 0 1 1 0 0 0 1
0 1 0 0 0 1 1 0 1 0 0 0 0 0 1
0 0 1 0 0 0 1 0 0 0 1

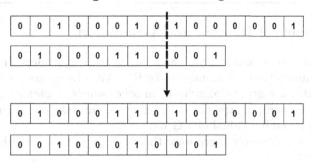

Fig. 5. Crossover to the left

0 0 1 0 0 0 1 0 1 0 0 0 0 0 1
0 1 0 0 0 1 1 1
0 1 0 0 0 1 1 1 0 1 1 0 0 0 1
0 0 1 0 0 0 1 1

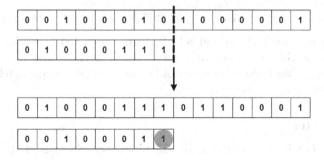

Fig. 6. Crossover to the left with the repair process

0 0 0 1 0 0 0 1 0 1 0 0 0 0 0 1
0 1 0 1 0 0 1 0 0 0 0 0 0 0 0 1

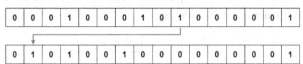

Fig. 7. Random swap mutation

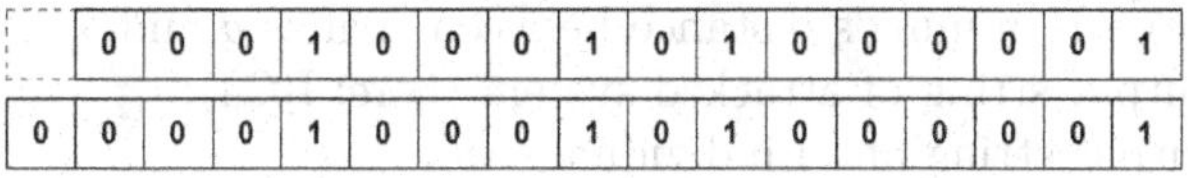

0 0 0 1 0 0 0 1 0 1 0 0 0 0 0 1
0 0 0 0 1 0 0 0 1 0 1 0 0 0 0 0 1

Fig. 8. Stretch mutation

RC4 has variable key length and the key initializes permutation in the array S. After initialization keystream k is generated as long as it is needed by the following rules:

Algorithm 2. RC4 keystream generation

---
1 **while** *true* **do**
2    $i = (i + 1)$ mod 256;
3    $j = (j + S[i])$ mod 256;
4   Swap $S[i]$ and $S[j]$;
5    $t = (S[i] + S[j])$ mod 256;
6    $k = S[t]$;

---

5   Experimental Results and Discussion

The aim of the experiments was to find the shortest equivalent linear system which approximates the keystream given by RC4 with stream given by LFSR and to achieve that using genetic algorithm. In other words cryptanalysis consisted in comparing both output streams – one generated by attacked cipher, the other one generetad by an individual of a GA.

During the experiments there were three test keys used taken from official specification [13]:

- k = 0x641910833222772a (key length: 64 bits)
- k = 0x8b37641910833222772a (key length: 80 bits)
- k = 0xebb46227c6cc8b37641910833222772a (key length: 128 bits)

For every key listed above and set of parameters listed in section 5.1 there were 50 runs of GA performed and subsequently average and median values were calculated. This is done because GA is a probabilistic method, which could generate different results for every run.

5.1   Parameters

Individual in GA is represented as one LFSR register (details of the representation are given in section 3). Every individual produces some output stream. This stream is then compared with keystream from attacked RC4 cipher. The fitness function is calculated as follows:

$$f_{fit} = \frac{H(a[n + 1..l], b[n + 1..l])}{l - n} \tag{2}$$

where:

        $H(a, b)$ – Hamming's distance between strings a and b,
        a – output string of attacked system (here: RC4),
        b – output string of an individual,
        l – output length,
        n – length of an individual.

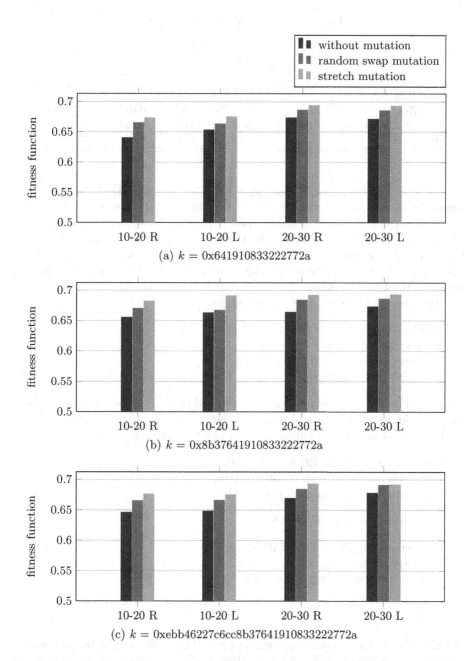

Fig. 9. Average values for keys k grouped by parameteres used. Descriptions on the OX axis mean: *10-20* – registers of length 10-20 bits; *20-30* – registers of length 20-30 bits; R – crossover to the right; L – crossover to the left. Results without mutation, with random swap mutation and with stretch mutation are presented in different colours. On the OY axis is the value of fitness function.

Table 1. Results for RC4 cryptanalysis

k	best	worst without mutation	worst with mutation	avg	median
0x641910833222772a	0.750	0.591	0.625	0.673	0.675
0x8b37641910833222772a	0.750	0.625	0.627	0.678	0.675
0xebb46227c6cc8b37641910833222772a	0.803	0.607	0.621	0.674	0.671
altogether	0.803	0.591	0.621	0.675	0.675

Initial bits are skipped in this calculation, as they are always correct. The number of bits equal to the length of generated individual is removed. The fitness function is also normalized to the interval $[0,1]$ in order to compare results given by the individuals of different length. This also means that for $f_{fit} = 0$ none of the output bits match (0%), for $f_{fit} = 0.5$ half of the output bits match (50%) and for $f_{fit} = 1$ all of the output bits match (100%). The higher the fitness function, the better results.

Parameters of the GA were set as follows:

- number of individuals: 40,
- number of generations: 100,
- crossover probability: 0.5,
- mutation probability: 0.05.

Number of individuals and generations give good results. Crossover and mutation probabilities were chosen on the basis of the preliminary experiments.

For LFSRs also additional parameters had to be specified:

- output length (set to 100 for all tests),
- minimum and maximum register length,
- minimum and maximum taps amount.

Experiments were performed for:

- registers of length 10-20 and 20-30 bits,
- crossover to the right and to the left,
- random swap and stretch mutation.

5.2   Results

Results of the experiments are shown in fig. 9. They are grouped by the parameters used: the length of the registers, the crossover and the mutation types.

Cumulative score for all keys considered are presented in table 1, where: *best* is the best result returned in all experiments for a given key k; *worst without/with mutation* is the worst result returned in all experiments for a given key taking not or taking into account mutation; *avg* is the average value for all results returned for a given key and *median* is median value for all results returned for a given key. The last line in the table presents summary results for all considered keys and all experiments performed.

The worst results were achieved without any mutation ($0.591 \leq f_{fit} \leq 0.625$). So as it can be seen that mutation has improved the results ($0.621 \leq f_{fit} \leq 0.627$). The stretch mutation gives sligthly better results than random swap mutation.

There is no significant difference between results given with the crossover to the right compared to the crossover to the left. The individuals' length had more impact – the results were better for longer registers.

Results for different keys' length are quite similar. Although they differ a bit for the best and the worst course of the algorithm, but average and median values are almost the same.

The value of fitness function f_{fit} indicates that such amount of the stream produced by the best individual matches the stream produced by RC4 cipher. So the best results found by the GA are the same in 59-80% of bits as the stream produced by RC4. It could be enough information to decode the message, at least partially. Average and median values are 67-68% with low standard deviation, so the presented solution is stable.

6 Conclusion

In this paper the cryptanalysis of RC4 stream cipher was presented. For this purpose the genetic algorithm was used. This method is effective for not only A5/1 and A5/2 as was shown in previous research [10], but also for RC4 as was demonstrated by experiments described in this work. Probably the attack could be expanded to other stream ciphers.

In future applying nature-inspired techniques in modern cryptography will be further developed. It is planned to compare presented method with tabu search [2].

References

1. RC4 Source Code. Cypherpunks, September 1994. http://cypherpunks.venona.com/archive/1994/09/msg00304.html
2. Glover, F.: Future Paths for Integer Programming and Links to Artificial Intelligence. Comput. Oper. Res. **13**(5), 533–549 (1986)
3. Goldberg, D.E.: Genetic algorithms in search, optimization, and machine learning, 3rd edn. Wydawnictwa Naukowo-Techniczne, Warszawa (2003). (in Polish)
4. Holland, J.H.: Adaptation in Natural and Artificial Systems. University of Michigan Press, Ann Arbor (1975)
5. Hospodar, G., Gierlichs, B., De Mulder, E., Verbauwhede, I., Vandewalle, J.: Machine learning in side-channel analysis: a first study. Journal of Cryptographic Engineering **1**(4), 293–302 (2011)
6. Itaim, P., Riff, M.C.: Applying Differential Cryptanalysis for XTEA using a Genetic Algorithm (2008)

7. Karearn, H.A.: Attacking stream cipher systems using genetic algorithm. Master's thesis, College of Science, University of Basrah, August 2000
8. Laskari, E.C., Meletiou, G.C., Stamatiou, Y.C., Vrahatis, M.N.: Applying evolutionary computation methods for the cryptanalysis of Feistel ciphers. Applied Mathematics and Computation **184**(1), 63–72 (2007). Special issue of the International Conference on Computational Methods in Sciences and Engineering 2004 (ICCMSE-2004)
9. Polak, I., Boryczka, M.: Breaking LFSR using genetic algorithm. In: Bădică, C., Nguyen, N.T., Brezovan, M. (eds.) ICCCI 2013. LNCS, vol. 8083, pp. 731–738. Springer, Heidelberg (2013)
10. Polak, I., Boryczka, M.: Cryptoanalysis of A5/1 and A5/2 using genetic algorithm. In: Systemy Inteligencji Obliczeniowej, pp. 145–153. Instytut Informatyki Uniwersytetu Śląskiego, Katowice (2014). (in Polish)
11. Selvi, G., Purusothaman, T.: Cryptanalysis of Simple Block Ciphers using Extensive Heuristic Attacks. European Journal of Scientific Research **78**(2), 198–221 (2012)
12. Song, J., Zhang, H., Meng, Q., Wang, Z.: Cryptanalysis of four-round DES based on genetic algorithm. In: International Conference on Wireless Communications, Networking and Mobile Computing, WiCom 2007, pp. 2326–2329, September 2007
13. Strombergson, J., Josefsson, S.: Test Vectors for the Stream Cipher RC4, May 2011
14. Younis, H.A., Awad, W.S., Abd, A.A.: Attacking of Stream Cipher Systems Using a Genetic Algorithm. Journal of University of Thi-Qar **8**(3), 78–84 (2013)

Cryptanalysis of SDES Using Modified Version of Binary Particle Swarm Optimization

Kamil Dworak[1,2]([✉]) and Urszula Boryczka[2]

[1] Future Processing, Bojkowska 37A, 44-100 Gliwice, Poland
[2] Institute of Computer Science, University of Silesia,
Bedzinska 39, 41-200 Sosnowiec, Poland
kamil.dworak@us.edu.pl, urszula.boryczka@us.edu.pl
http://www.ii.us.edu.pl/en
http://www.future-processing.com

Abstract. Nowadays, information security is based on ciphers and cryptographic systems. What evaluates the quality of such security measures is cryptanalysis. This paper presents a new cryptanalysis attack aimed at a ciphertext generated with the use of the *SDES* (Simplified Data Encryption Standard). The attack was carried out with a modified version of the *BPSO* (Binary Particle Swarm Optimization) algorithm. A well-adjusted version of this method can have a positive effect on the quality of the results obtained in a given period of time.

Keywords: Simplified Data Encryption Standard · Binary Particle Swarm Optimization · Particle Swarm Optimization · Cryptanalysis · Cryptography

1 Introduction

Security based on authentication is not sufficient these days. Intercepting any sensitive data may happen even during the process of developing computer systems. In these cases, cryptography is becoming more and more popular. In particular, this field deals with ciphering and deciphering information, but it also designs complex cryptographic systems assuring safe data access [6]. Its main objective is not to obscure the existence of the message, but to transform it in such a way that it can only be read by its sender and intended recipients [12]. Cryptography is closely related to cryptanalysis, a field of study which is about deciphering ciphertexts without knowing the right deciphering key [5]. It is also used in the process of finding bugs in existing cryptographic systems to increase and upgrade their level of security [1]. Cryptanalysis processes are not the fastest ones, their operation time is very long. In order to optimize the time of these operations, one of the optimization techniques called *BPSO* was used.

Techniques based on artificial intelligence are more commonly used in the field of computer security. Over recent years, many publications on the application of various evolutionary techniques such as Tabu Search, evolutionary algorithms or Simulated Annealing in cryptanalysis, have been presented. In 1998

© Springer International Publishing Switzerland 2015
M. Núñez et al. (Eds.): ICCCI 2015, Part II, LNCS 9330, pp. 159–168, 2015.
DOI: 10.1007/978-3-319-24306-1_16

Andrew John Clark [2] presented the results of his research on the cryptanalysis of classic ciphers with the use of evolutionary algorithms. The results were not accurate every time, though in most cases it was possible to read the decrypted message. Poonan Garg in [4] presented an interesting attack carried out using a memetic algorithm on a ciphertext generated with the *SDES*, which had very good results. Nalini and Raghavendra presented in their article [10] a cryptanalysis attack using *PSO* with grouping against the modification of *DES*, developed by the authors themselves. Techniques based on evolutionary algorithms are gaining more interest in the field of computer security. It leads to many successes, yet there are still numerous problems to be discussed [1]. The next chapter of this paper presents the specification of the *SDES* used to generate a ciphertext that will be subjected to the attack. The third chapter contains the basic information on *PSO* and *BPSO* algorithms and their modifications for the purpose of cryptanalysis. The next chapter concerns the experiments, observations and the results of decrypting the ciphertext using the developed attack and comparison between proposed attack and the simple random walk algorithm. The last chapter contains conclusions and further plans.

2 Simplified Data Encryption Standard

The *SDES* is a simplified version of a well-known Data Encryption Standard (*DES*). It was designed in 1996 by Edward Schaefer for general academic purposes [11]. *SDES* is a symmetric cipher, that is, the same key is used to encrypt and decrypt information [12]. It is a block cipher. The length of a single encrypted message block equals 8 bits and the data is encrypted by a 10-bit key [11]. The *SDES* operates on strings of bits instead of normal characters. The decryption process uses the same encrypting algorithm but the subkeys are provided in reverse order. *SDES* is a two-round algorithm, based on two basic operations: combination done with permutations and dispersion [9]. The block diagram of the encryption is shown in Fig. 1.

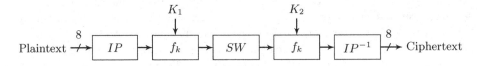

Fig. 1. Encryption with the use of *SDES*

The first round is preceeded by the initial permutation *IP* [11]. The bits of a message are relocated with the use of an appropriate, determined shift table. The second bit is put to the first position, the sixth one to the second etc. A moved string is then divided into two equal 4-bit parts (*L* - left and *R* - right), and the f_k, function, known as the round function, is executed:

$$f_{K_i}(L_i, R_i) = (L_{i-1} \oplus f(R_{i-1}, K_i), R_{i-1}), \tag{1}$$

where:

f_{K_i} - round function,
L/R - left/right part of the binary (input) string,
K_i - round key,
f - deciphering function,
i - round's ID.

The process of swapping the right part to the left is executed by the SW function (swap). Then the second round begins. This time the SW function is omitted. At the end, the result is put through the final IP^{-1} permutation, which is the opposite of the initial permutation IP [9]. In this way, an 8-bit block of ciphertext is generated. The operation is repeated until the plaintext is encrypted. The register of all essential permutations looks as follows:

$$
\begin{aligned}
IP &= [\; 2 \;\; 6 \;\; 3 \;\; 1 \;\; 4 \;\; 8 \;\; 5 \;\; 7 \;] \\
P10 &= [\; 3 \;\; 5 \;\; 2 \;\; 7 \;\; 4 \;\; 10 \;\; 1 \;\; 9 \;\; 8 \;\; 6 \;] \\
P8 &= [\; 6 \;\; 3 \;\; 7 \;\; 4 \;\; 8 \;\; 5 \;\; 10 \;\; 9 \;] \\
E &= [\; 4 \;\; 1 \;\; 2 \;\; 3 \;\; 2 \;\; 3 \;\; 4 \;\; 1 \;] \\
P4 &= [\; 2 \;\; 4 \;\; 3 \;\; 1 \;] \\
IP^{-1} &= [\; 4 \;\; 1 \;\; 3 \;\; 5 \;\; 7 \;\; 2 \;\; 8 \;\; 6 \;]
\end{aligned}
$$

2.1   Generating Subkeys

In order to properly calculate the value of the f_k function, it is necessary to generate the subkeys. For every round, the key is modified with such operations as reduction, shifting and permutation [11], the aim of which is to generate an auxiliary key. At the beginning, the encryption key is subjected to the given permutation $P10$. A received string is divided into two 5-bits parts. A single cyclical shift to the left is performed on each of them. The shifted parts are concatenated in a binary string. The final string is subjected to the $P8$, permutation, thereby generating the K_1 subkey. The K_2 subkey is the result of an auxiliary, double, cyclical shift at the stage of dividing the encryption key. A received string is also compressed with the $P8$ permutation.

2.2 f_{K_i} and f Functions

The f_{K_i} function is executed by the *Feistal network* (Fig. 2). The right part of the string is automatically moved to the left side. The left part is subjected to the addition modulo 2 with the result of the ciphering function f, the arguments of which are the K_i subkey and the right part of the string of R_{i-1}.

The function f is the right ciphering algorithm. A flow chart illustrating its performance is shown in (Fig. 3). At the beginning, the expanding permutation E is executed. To increase the coefficient of the avalanche effect, two new settings

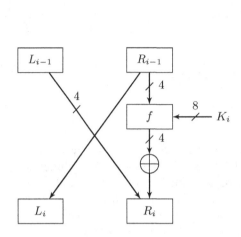 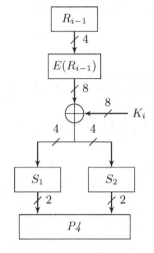

Fig. 2. Feistal function f_{K_i} **Fig. 3.** Deciphering function f

from the determined permutation E [12] are assigned to every bit. The right part of the string is expanded from 4 to 8 bits. In this way, it is possible to perform the additional modulo 2 with the given K_i subkey. The received string is divided into two equal parts. The left one is moved to the *S-Block* S_1, whereas the right one to S_2. Each block is a specially designed 4x4 matrix of the elements from the set 0, 1, 2, 3 written in the binary system [11]. The *S-Blocks* are the only non-linear element of majority of block ciphers [12,14]. The 4-bit strings of data are the input. They are converted into 2-bit output. The first and last bit of each substring represents the number of a row, whereas the second and third bit represents the number of a column. The *S-Blocks* are presented below. The two chosen digits are concatenated with each other. The result is finally subjected to the $P4$ permutation.

$$S_1 = \begin{bmatrix} 01 & 00 & 11 & 10 \\ 11 & 10 & 01 & 00 \\ 00 & 10 & 01 & 11 \\ 11 & 01 & 11 & 10 \end{bmatrix}, \quad S_2 = \begin{bmatrix} 00 & 01 & 10 & 11 \\ 10 & 00 & 01 & 11 \\ 11 & 00 & 01 & 00 \\ 10 & 01 & 00 & 11 \end{bmatrix}.$$

3   Particle Swarm Optimization

The *PSO* was first presented in 1995 by Kennedy and Eberhart [7]. The authors were inspired by the behavior of animals in troops. They attempted to simulate the behavior of a school of fish moving in a precise and specific manner. In case of danger, the behavior of an individual has a dramatic impact on the behavior of the whole population. Another example of their inspiration is a bevy of birds searching for food. When one of the birds finds a good source of food, the whole bevy moves in this direction. Similarly to evolutionary algorithms, a single individual is represented by a *particle*, whereas the population is a swarm [3]. The particles are dispersed within a multidimensional space of solutions in order to find the global extreme. They move according to their own experience or the one of other, neighboring particles [3]. Every individual is characterized by the speed vector $\vec{V}$ and the position vector $\vec{X}$ [7]. Initially, the particles are randomly placed in the space. In each iteration, the value of the adjustment function is calculated for every particle. During this operation, so far the best adjustment of a given P_{BEST} particle and so far the best adjustment among all of the particles of the swarm G_{BEST} are determined. Then, for every individual, a new value of the speed vector (the below formula) is calculated and the particle is moved to a new position:

$$\vec{V}(t+1) = \vec{V} + c_1 r_1 \cdot (\vec{X}_{P_{BEST}} - \vec{X}) + c_2 r_2 \cdot (\vec{X}_{G_{BEST}} - \vec{X}), \qquad (2)$$

where:
 $\vec{V}$ - current speed vector,
 c_1, c_2 - positive acceleration constants,
 r_1, r_2 - random numbers from $[0, 1]$,
 $\vec{X}$ - current particle position vector,
 $\vec{X}_{P_{BEST}}$ - best position of the particle so far,
 $\vec{X}_{G_{BEST}}$ - best position among all the particles so far,
 t - iteration of the algorithm.

In the case where the speed value $\vec{V}(t+1)$ is greater than the established acceptable particle speed V_{MAX}, the following formula applies:

$$\vec{V}(t+1) = V_{MAX}. \qquad (3)$$

3.1 Binary Particle Swarm Optimization

In 1997 Kennedy and Eberhart developed a modified version of the *PSO*, dedicated to problems characterized by a binary space of solutions, i.e. *BPSO* [8]. Every element of the vector is represented by binary values, namely $X_{i,j} \in \{0, 1\}$. Changing the position of a particle is reduced to negation of bits of particular indexes [3]. One of the most important changes concerning *BPSO* is an update of the $\vec{X}$ position vector, which is determined with the formula [3]:

$$\vec{X}_{i,j}(t+1) = \begin{cases} 1 & if \quad r_{3_j}(t) < sig(\vec{V}_{i,j}(t+1)), \\ 0 & otherwise \end{cases} \qquad (4)$$

where:
 r_3 - random numbers from $[0, 1]$,
 sig - sigmoidal function,
 i - particle index,
 j - position in the speed / position vector,
 t - iteration of the algorithm.

The sigmoidal function value can be calculated using the following formula [3]:

$$sig(\overrightarrow{V}_{i,j}(t)) = \frac{1}{1 + e^{-\overrightarrow{V}_{i,j}(t)}} \quad , \quad sig(\overrightarrow{V}) \in [0,1]. \tag{5}$$

3.2   Proposed Attack

The suggested algorithm is based on the chosen-plaintext attack, which is one of the most popular cryptanalytical methods. This attack is referred to as "Particle Swarm Attack". The attacker is in the possession of the ciphertext enciphered with the *SDES* and the corresponding plaintext [5,6]. Using these data, the cryptanalyst tries to guess what the deciphering key is, the aim of which is to decipher the rest of the intercepted ciphertexts. The algorithm begins with generating a defined number of particles with the speed equal to 0. The $\overrightarrow{X}$ position vector is equated with the example of a decryption key. In the initial iteration the keys are generated randomly. Next, based on the Hamming distance and according to the formula below, the value of the adjustment function is calculated:

$$F_f = H(P,D) = \sum_{i=1}^{n} P_i \oplus D_i, \tag{6}$$

where:
   D_i - the character of the deciphered text,
   P_i - the character of the plaintext,
   H - Hamming distance,
   n - message length.

Then, the P_{BEST} and G_{BEST} values are determined. To do so, the modification to the classic version of *PSO* [13] will be applied. Having calculated the adjustment value for every particle, an auxiliary steering parameter called statistical probability α is introduced. In most cases, it is represented by a random number from [13]. In the proposed attack it will have one constant value. Then, based on the α and speed vector $\overrightarrow{V}$, the particle can update the chosen fragments of the $\overrightarrow{X}$ position vector in relation to its best so far position P_{BEST} or the best position in the G_{BEST} swarm. It is done according to the following formula:

$$if \quad (\alpha < \overrightarrow{V}_i \le (\frac{1}{2}(1+\alpha)), \quad then \overrightarrow{X}_i = P_{BEST_i}, \tag{7}$$

$$if \quad (\frac{1}{2}(1+\alpha) < \overrightarrow{V}_i \le 1), \quad then \overrightarrow{X}_i = G_{BESTi}. \tag{8}$$

Otherwise, when $(0 < \overrightarrow{V}_i \le \alpha)$, the original version of the modification presented in [13], considers the application of the previous position vector.

In the developed attack, it is suggested to calculate new value of the vector according to the classic *BPSO* algorithm:

$$\vec{V}_{i+1}(t+1) = \vec{V}_i + c_1 r_1 \cdot (\vec{X}_{P_{BEST_i}} - \vec{X}_i) + c_2 r_2 \cdot (\vec{X}_{G_{BEST_i}} - \vec{X}_i), \qquad (9)$$

where:

i - particle's index,

j - position in the speed/position vector.

A new value of the position vector is calculated using the eq. 4.

4   Experiments

The aim of the research was to prove that the algorithm presented in this article is able to find the correct ciphering key for a given ciphertext. The attack was implemented in C++ and tested on a computer with an Intel i7 processor clocked at 2.1 GHz. All the tests were performed on one core. The tested ciphertexts were of various length and were created with randomly generated cipher keys. The plain texts were written in English. The algorithm is based on the chosen-plaintext attack. The suggested approach assumes that the ciphering key will be found without searching every possible combination.

All of the control parameters were extracted in an empirical way. The α coefficient is a floating-point number selected randomly. The c_1 cognitive coefficient was set at 0.5, the social factor c_2 at 0.8. The V_{MAX} particle speed was adjusted to the sigmoidal function, to the value 4. The swarm consisted of 20 particles. For every case, the algorithm run 30 times. The experiments were divided according to the length of the ciphertexts into 2 rounds (250 and 1000 characters). Maximum number of iterations was set to 30. *SDES* has only 1024 possible keys so larger amount of iterations is not required.

Table 1 presents the results: the minimum, maximum, median, average and standard deviation of the values of the adjustment functions of the consecutive tests, carried out for Particle Swarm and Single Walk algorithms, for 250 character ciphertexts. In addition, information about the sought after and the best decryption key found are included.

In the first round, 18 out of 30 ciphertexts were cracked by Particle Swarm attack. The Random Walk algorithm cracked only 8 ciphertexts. We decided to compare the guessed keys with the correct ones. In most cases the difference concerned two bits. Fig. 4 shows a boxplot for a few 250-character ciphertexts for Particle Swarm attack. We can see the minimum, maximum and median values for given iterations of the algorithm. The solution is shown in the chart as a unique value.

Table 2 juxtaposes the results for the 1000 character ciphertexts. In this case 24 ciphertexts were properly deciphered by Particle Swarm oppose to the 10 by Random Walk algorithm. Similarly to the previous tests, incorrect decryptions differed by two bits. Fig. 5 shows a box chart for a few 1000-character ciphertexts for Particle Swarm attack.

Table 1. F_f values for each consecutive 250 character ciphertexts for both attacking algorithms

ID	Minimum	Median	Maximum	Average	Standard Deviation	Correct Key	Obtained Key	Iteration
				Particle Swarm Attack				
1	0	984	1140	922	251	1000011000	1000011000	17
2	0	1021	1325	972	270	0110101101	0110101101	3
3	0	957	1388	931	290	1010000100	1010000100	12
4	526	940	1202	931	191	000⓪01①010	000①01⓪010	-
5	0	952	1258	916	258	0101001100	0101001100	10
6	494	1054	1311	1006	200	00⓪100⓪①00	00①100①⓪00	-
7	666	904	1266	945	175	110①00⓪100	110⓪00①100	-
8	0	1032	1831	983	320	1000001101	1000001101	7
9	526	738	1134	785	186	110①00⓪011	110⓪00①011	-
10	0	1032	1831	983	320	1011010101	1011010101	10
				Random Walk Algorithm				
1	638	939	1321	965	31525	10⓪⓪⓪11000	10①⓪①11000	-
2	731	995	1278	1010	21908	011⓪101101	011①101101	-
3	769	1010	1313	1013	22889	101⓪00⓪100	101①00①100	-
4	517	1039	1149	987	22151	000⓪01①010	000①01⓪010	-
5	0	918	1161	895	50530	0101001100	0101001100	27
6	590	999	1280	979	30384	000①00⓪100	000⓪00①100	-
7	727	987	1240	978	19897	1⓪0000110①	1①0000110⓪	-
8	0	984	1324	933	86186	1000001101	1000001101	12
9	646	982	1210	982	20067	110①①01011	110⓪⓪10011	-
10	0	983	1172	938	62002	1011010101	1011010101	23

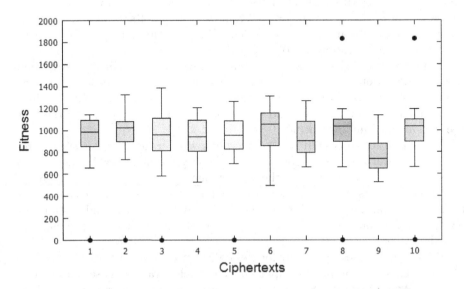

Fig. 4. The boxplot of a randomly picked 250 character ciphertexts for Particle Swarm Attack

Table 2. F_f values for each consecutive 1000 character ciphertexts for both attacking algorithms

ID	Minimum	Median	Maximum	Average	Standard Deviation	Correct Key	Obtained Key	Iteration
Particle Swarm Attack								
11	0	4416	5521	3982	1436	0001101101	0001101101	5
12	0	4063	5336	3993	1077	0000111001	0000111001	7
13	347	3240	4541	2772	1456	001[0]10[0]111	001[1]10[1]111	-
14	0	3811	5340	3614	1453	1000111100	1000111100	17
15	642	4188	5540	3875	1170	100[0]10[0]010	100[1]10[1]010	-
16	0	4554	5522	4062	1404	1010001011	1010001011	12
17	0	4390	5535	4061	1256	1110101000	1110101000	10
18	0	4285	5745	3966	1252	0000110101	0000110101	9
19	0	4103	5086	3777	1106	0001111111	0001111111	5
20	0	4129	4992	3842	1078	1111010000	1111010000	11
Random Walk Algorithm								
11	3007	3845	4816	3874	332190	000[1]101101	000[0]101101	-
12	2744	3558	4414	3566	239546	000[0]111001	000[1]111001	-
13	0	3982	4780	3470	1532160	0010100111	0010100111	23
14	2908	4000	5162	4006	372737	100011[1]100	100011[0]100	-
15	2942	4021	4899	4057	203871	[1000100001]0	[010110100]0	-
16	0	4105	4971	3769	1245140	1010001011	1010001011	21
17	3184	4193	5073	4179	288314	111[0]10[1]000	111[1]10[0]000	-
18	0	3994	5531	3591	1974630	0000110101	0000110101	16
19	1804	3851	5616	4004	685290	000111[1]111	000111[0]111	-
20	2382	4067	4841	3858	419308	111[1]010000	111[0]010000	-

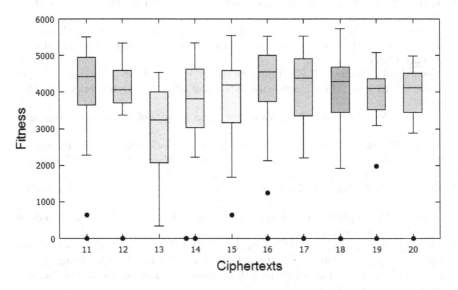

Fig. 5. The boxplot of a randomly picked 1000 character ciphertexts for Particle Swarm Attack

5 Summary

The suggested Particle Swarm cryptanalysis attack allows to obtain satisfying results in the *SDES* cipher cryptanalysis. Correct deciphering can be achieved on the level of up to 80%, depending on the length of the ciphertext.

In the situation when the cryptanalysis attack failed, an interesting correlation between a correct key and the best one found was observed. Mostly, the non-matching bits were symmetrical to each other, as in the test 11. Undoubtedly, it sets out the course for further research in terms of applying modifications in order to improve the quality of obtained results. Moreover, this method should be tested for other ciphering algorithms like *DES* or *Blowfish*. It should be remembered that *BPSO* is one of the possibilities. Another interesting alternative is the Discrete *PSO*, an algorithm, which can perform manage quite well with such problems.

References

1. Boryczka, U., Dworak, K.: Genetic transformation techniques in cryptanalysis. In: Nguyen, N.T., Attachoo, B., Trawiński, B., Somboonviwat, K. (eds.) ACIIDS 2014, Part II. LNCS, vol. 8398, pp. 147–156. Springer, Heidelberg (2014)
2. Clark, A.J.: Optimisation Heuristics for Cryptology. PhD thesis. Information Research Centre Faculty of Information Technology Queensland (1998)
3. Engelbrecht, A.P.: Fundamentals of computional swarm intelligence. Wiley (2005)
4. Garg, P.: Cryptanalysis of SDES via Evolutionary Computation Techniques. International Journal of Computer Science and Information Security **1**(1) (2009)
5. Kahn, D.: The Code-breakers. Scribner (1996)
6. Kenan, K.: Cryptography in the databases. The Last Line of Defense. Addison Wesley Publishing Company (2005)
7. Kennedy, J., Eberhart, R.C.: New optimizer using particle swarm theory. In: Proceedings of the Sixth International Symposium on Micromachine and Human Science (1995)
8. Kennedy, J., Eberhart, R.C.: A discrete binary version of the particle swarm optimization. In: Proc. of the Conference on Systems, Man, and Cybernetics SMC 1997 (1997)
9. Menezes, A.J., Oorschot, P.C., Vanstone, S.A.: Handbook of Applied Cryptography. CRC Press (1997)
10. Nalini, N., Raghavendra, R.: Cryptanalysis of Block Ciphers via Improvised Particle Swarm Optimization and Extended Simulated Annealing Techniques. International Journal of Network Security **6**(3) (2008)
11. Schaefer, E.: A simplified data encryption standard algorithm. Cryptologia **20**(1) (1996)
12. Schneier, B.: Applied Cryptography: Protocols, Algorithms, and Source Code in C, 2nd edn. Wiley (1996)
13. Shen, Q., Jiang, J.H., Jiao, C.X., Shen, G.L., Yu, R.Q.: Modified particle swarm optimization algorithm for variable selection in MLR and PLS modeling: QSAR studies of antagonism of angiotensin II antagonists. European Journal of Pharmaceutical Sciences **22** (2004)
14. Stinson, D.S.: Cryptography. Theory and Practice. Chapman & Hall/CRC Taylor & Francois Group, Boca Raton (2006)

CSDMO: Special Session on Cooperative Strategies for Decision Making and Optimization

InterCriteria Analysis of Parameters Relations in Fermentation Processes Models

Olympia Roeva, Peter Vassilev, Maria Angelova$^{(\boxtimes)}$, and Tania Pencheva

Institute of Biophysics and Biomedical Engineering,
Bulgarian Academy of Science, Sofia, Bulgaria
{olympia,maria.angelova,tania.pencheva}@biomed.bas.bg,
peter.vassilev@gmail.com

Abstract. In this paper the application of InterCriteria Analysis (ICA) is presented. The approach is based on the apparatuses of index matrices and intuitionistic fuzzy sets. ICA is applied to establish the relations and dependencies of defined parameters in non-linear models of *Escherichia coli* MC4110 and *Saccharomyces cerevisiae* fermentation processes. Parameter identification of both fed-batch process models has been done using three kinds of genetic algorithms (GA) – standard single population GA (SGA) and two SGA modifications. The obtained results are discussed in the lights of ICA and some conclusions about existing relations and dependencies between model parameters are derived.

Keywords: InterCriteria Analysis · Genetic algorithms · Binary-coded · Real-coded · Fermentation process · *E. coli* · *S. cerevisiae*

1 Introduction

InterCriteria Analysis (ICA) has been recently developed aiming to gain insight into the nature of the criteria involved and discover on this basis existing correlations between the criteria themselves [3]. A pioneer application of ICA in the field of parameter identification of fermentation processes (FP) models has been presented in [18]. The authors apply ICA for establishing relations and dependencies between two of the main genetic algorithms (GA) parameters, namely number of individuals and number of performed generations, on one hand, and convergence time, model accuracy and model parameters on the other hand. The current investigation is another attempt for successful application of ICA to establish the correlations between FP model parameters.

In general, there are typical FP model structures based on systems of non-linear differential equations with several specific growth rates [8]. One of the most challenging problem in FP modelling is the choice of appropriate model identification procedure. To solve this problem various techniques and approaches, namely exact algorithms, heuristics, and metaheuristics, etc., could be applied. Nowadays, the use of metaheuristics has received more and more attention. These methods offer good solutions, even global optima, within reasonable computing time [9,12].

© Springer International Publishing Switzerland 2015
M. Núñez et al. (Eds.): ICCCI 2015, Part II, LNCS 9330, pp. 171–181, 2015.
DOI: 10.1007/978-3-319-24306-1_17

GA [12], as a representatives of biologically inspired optimization techniques, have proven to be one of the most widely used optimization techniques for global search in different research fields. The effectiveness of different GA have already been demonstrated also for model parameter identification considering FP, among them of Simple GA (SGA) [1,15,16], Modified GA and a SGA modification without mutation operator (going to be referred in the current investigation, respectively as SGA_Mod1 and SGA_Mod2) [1,19,20].

In this investigation SGA and the two SGA modifications are applied to model parameter identification. The three GA are applied using binary and real valued alphabet. The binary representation have dominated most of the works using GA, since the efficiency of the binary coded GA is theoretically proven [12]. The real-coded GA can perform similar to the binary-coded GA [10,13,14] and can be used efficiently in solving continuous research space problems.

Two fed-batch FP, of bacteria *E. coli* and yeast *S. cerevisiae*, are considered as case studies. These microorganisms are among the most important ones with various applications in food and pharmaceutical industry. Based on the results of parameters identification procedures the ICA is applied to identify the model parameters relations.

The paper is organized as follows: the problem formulation is given in Section 2, while Section 3 presents the background of ICA. Numerical results and discussion are presented in Section 4 and conclusion remarks are given in Section 5.

2   Problem Formulation

2.1   Mathematical Models of Fermentation Processes

Case study 1. *E. coli* Fed-batch Fermentation Model. The mathematical model of the considered here *E. coli* fed-batch process is presented by the following non-linear differential equations system [8,17]:

$$\frac{dX}{dt} = \mu X - \frac{F_{in}}{V} X \tag{1}$$

$$\frac{dS}{dt} = -q_S X + \frac{F_{in}}{V}(S_{in} - S) \tag{2}$$

$$\frac{dV}{dt} = F_{in} \tag{3}$$

where

$$\mu = \mu_{max}\frac{S}{k_S + S}, \quad q_S = \frac{1}{Y_{S/X}}\mu \tag{4}$$

and X is the biomass concentration, [g/l]; S is the substrate concentration, [g/l]; F_{in} is the feeding rate, [l/h]; V is the bioreactor volume, [l]; S_{in} is the substrate concentration in the feeding solution, [g/l]; μ and q_S are the specific rate functions, [1/h]; μ_{max} is the maximum value of the μ, [1/h]; k_S is the saturation constant, [g/l]; $Y_{S/X}$ is the yield coefficient, [-].

Case study 2. *S. cerevisiae* Fed-batch Fermentation Model. Fed-batch FP of *S. cerevisiae* is also described by the non-linear differential equations system (Eqs. (1-3)). Due to the specifics of considered process, better behaviour is obtained using following specific rates [17]:

$$\mu = \mu_{2S}\frac{S}{S + k_S} + \mu_{2E}\frac{E}{E + k_E}, \quad q_S = \frac{\mu_{2S}}{Y_{S/X}}\frac{S}{S + k_S} \tag{5}$$

where S, k_S, $Y_{S/X}$ keep their meaning as described above. Additionally, E is the concentration of ethanol, [g/l], μ_{2S} and μ_{2E} are the maximum values of μ, [1/h], and k_E is the saturation constant, [g/l].

For the considered here models (Eqs. (1-4), Eqs. (1-3), (5)), the following parameter vectors should be identified: $p_1 = [\mu_{max} \; k_S \; Y_{S/X}]$ and $p_2 = [\mu_{2S} \; \mu_{2E} \; k_S \; k_E \; Y_{S/X}]$. Model parameters identification is performed based on real experimental data for biomass and glucose concentration. The detailed description of process conditions and experimental data sets are presented in [17].

2.2 Genetic Algorithms for Model Parameter Identification

Simple Genetic Algorithm (SGA). GA are a heuristic search technique that belongs to the class of evolutionary algorithms [12]. GA work with a population of coded parameters set called "chromosomes". As a population-based technique GA search the suitable solutions using genetic operators, such as selection, crossover and mutation.

The implementation of SGA starts with a creation of an initial population, generated randomly. Each solution is then evaluated and assigned a fitness value. According to the fitness function, the most suitable solutions are selected and recombined to form a new offspring. The execution of SGA is terminated when some condition (i.e. reached number of populations, or found solution with a specified tolerance, etc.) is fulfilled [12].

Modified SGA (SGA_Mod1). A simple modification of SGA [20], namely the reproduction being processed after application of both crossover and mutation, is considered here. In this way the loss of reached good solution by either crossover or mutation or both operators is prevented.

Modified SGA without Mutation (SGA_Mod2). Another modification of SGA presented in [19] is considered here as well. This is a modification of SGA without the performance of the mutation operator. This SGA modification reduces the convergence time and increases the identification procedure effectiveness [1,19].

The presented above three kinds of SGA are applied in this investigation in both binary and real-valued coding realizations.

Optimization Criterion. The aim is the adjustment of model parameters of the considered non-linear mathematical models functions in order the best fit to a data set to be obtained. The objective function is defined as:

$$J = \sum_{i=1}^{m} (X_{\exp}(i) - X_{\text{mod}}(i))^2 + \sum_{i=1}^{n} (S_{\exp}(i) - S_{\text{mod}}(i))^2 \to \min \qquad (6)$$

where m and n are the experimental data dimensions; $X_{\exp}$ and $S_{\exp}$ are the available experimental data for biomass and substrate; X_{mod} and S_{mod} are the model predictions for biomass and substrate with a given parameter vector.

3   InterCriteria Analysis

Here the idea proposed in [3] is expanded. Following [2,3] an Intuitionistic Fuzzy Pair (IFP) ([4]) as an estimation of the degrees of "agreement" and "disagreement" between two criteria applied to different objects is obtained. Recall that an IFP is an ordered pair of real non-negative numbers $\langle a, b \rangle$ such that: $a + b \leq 1$.

Consider an Index Matrix (IM) ([5,6]) whose index sets consist of the criteria (for rows) and objects (for columns). The elements of this IM are further assumed to be real numbers. An IM with index sets consisting of the criteria (for rows and for columns) with elements IFPs corresponding to the "agreement" and "disagreement" between the respective criteria is then constructed.

Let O denotes the set of all objects being evaluated, and $C(O)$ be the set of values assigned by a given criteria C to the objects, i.e.

$$O \stackrel{\text{def}}{=} \{O_1, O_2, \ldots, O_n\}, \quad C(O) \stackrel{\text{def}}{=} \{C(O_1), C(O_2), \ldots, C(O_n)\},$$

$$C^*(O) \stackrel{\text{def}}{=} \{\langle x, y \rangle \mid x \neq y \ \& \ \langle x, y \rangle \in C(O) \times C(O)\}$$

In order to find the "agreement" of two criteria the vector of all internal comparisons of each criteria, which fulfill exactly one of three relations R, $\overline{R}$ and $\tilde{R}$, is constructed. In other words, it is required that for a fixed criterion C and any ordered pair $\langle x, y \rangle \in C^*(O)$ it is true:

$$\langle x, y \rangle \in R \Leftrightarrow \langle y, x \rangle \in \overline{R} \qquad (7)$$

$$\langle x, y \rangle \in \tilde{R} \Leftrightarrow \langle x, y \rangle \notin (R \cup \overline{R}) \qquad (8)$$

$$R \cup \overline{R} \cup \tilde{R} = C^*(O) \qquad (9)$$

From the above it is seen that only a subset of $C(O) \times C(O)$ needs to be considered for the effective calculation of the vector of internal comparisons (further denoted by $V(C)$) since from Eqs. (7-9) it follows that if the relation between x and y is known, then so is the relation between y and x. Thus of interest are only the lexicographically ordered pairs $\langle x, y \rangle$. Denote for brevity:

$$C_{i,j} = \langle C(O_i), C(O_j) \rangle$$

Then for a fixed criterion C, the vector with $\frac{n(n-1)}{2}$ elements is obtained:

$$V(C) = \{C_{1,2}, C_{1,3}, \ldots, C_{1,n}, C_{2,3}, C_{2,4}, \ldots, C_{2,n}, C_{3,4}, \ldots, C_{3,n}, \ldots, C_{n-1,n}\}$$

Let $V(C)$ be replaced by $\hat{V}(C)$, where for the k-th component ($1 \le k \le \frac{n(n-1)}{2}$):

$$\hat{V}_k(C) = \begin{cases} 1, & \text{iff } V_k(C) \in R \\ -1, & \text{iff } V_k(C) \in \overline{R} \\ 0, & \text{otherwise} \end{cases}$$

When comparing two criteria the degree of "agreement" is determined as the number of matching components of the respective vectors (divided by the length of the vector for normalization purposes). This can be done in several ways, e.g. by counting the matches or by taking the complement of the Hamming distance. The degree of "disagreement" is the number of components of opposing signs in the two vectors (again normalized by the length). An example pseudocode for two criteria C and C' is presented below.

Algorithm 1. Calculating "agreement" and "disagreement" between two criteria

Require: Vectors $\hat{V}(C)$ and $\hat{V}(C')$

```
 1: function DEGREES OF AGREEMENT AND DISAGREEMENT(V̂(C), V̂(C'))
 2:     V ← V̂(C) − V̂(C')
 3:     μ ← 0
 4:     ν ← 0
 5:     for i ← 1 to n(n−1)/2 do
 6:         if Vᵢ = 0 then
 7:             μ ← μ + 1
 8:         else if abs(Vᵢ) = 2 then          ▷ abs(Vᵢ): the absolute value of Vᵢ
 9:             ν ← ν + 1
10:         end if
11:     end for
12:     μ ← 2/(n(n−1)) μ
13:     ν ← 2/(n(n−1)) ν
14:     return μ, ν
15: end function
```

If the respective degrees of "agreement" and "disagreement" are denoted by $\mu_{C,C'}$ and $\nu_{C,C'}$, it is obvious (from the way of computation) that $\mu_{C,C'} = \mu_{C',C}$ and $\nu_{C,C'} = \nu_{C',C}$. Also it is true that $\langle \mu_{C,C'}, \nu_{C,C'} \rangle$ is an IFP.

4 Numerical Results and Discussion

In this investigation three kinds of SGA, in both binary- and real-coded realizations, are applied for the two considered FP model parameters identification.

In both case studies crossover operators are double point (binary) and extended intermediate recombination (real); mutation operators – bit inversion (binary) and real-value mutation (real); selection operators – roulette wheel selection (for both binary and real); number of generations – $maxgen = 100$; generation gap – $ggap = 0.80$; crossover rate – $xovr = 0.85$; mutation rate – $mutr$ $= 0.01$ and number of individuals – $nind = 100$ (Case study 1) and $nind = 20$ (Case study 2).

Thirty independent runs of altogether six realizations of SGA are performed to obtain the parameters estimations for models Eqs. (1-4) and Eqs. (1-3),(5). Based on the averaged results from the identification procedures for computational time (T), objective function (J) and model parameters (parameter vectors p_1 and p_2), two IMs are constructed – IM A_1 for Case study 1 and IM A_2 for Case study 2.

Case Study 1

$$A_1 = \begin{array}{c|cccccc} & O_1 & O_2 & O_3 & O_4 & O_5 & O_6 \\ \hline C_1 & 65.2632 & 55.0379 & 37.1563 & 34.0625 & 40.1240 & 35.3312 \\ C_2 & 4.6841 & 4.4700 & 4.5669 & 4.5045 & 4.6720 & 4.4958 \\ C_3 & 0.4824 & 0.4876 & 0.4634 & 0.4870 & 0.4933 & 0.4962 \\ C_4 & 0.0109 & 0.0118 & 0.0086 & 0.0119 & 0.0129 & 0.0132 \\ C_5 & 0.4944 & 0.4948 & 0.4962 & 0.4952 & 0.4937 & 0.4947 \end{array} \tag{10}$$

where O_1 is the binary-coded SGA, O_2 – real-coded SGA, O_3 – binary-coded SGA_Mod1, O_4 – real-coded SGA_Mod1, O_5 – binary-coded SGA_Mod2, O_6 – real-coded SGA_Mod1, C_1 – computational time (T), C_2 – objective function value (J), C_3 – value of maximum specific growth rate (μ_{max}), C_4 – value of saturation constant (k_S), and C_5 – value of yield coefficient $(Y_{S/X})$.

Considering Case study 1 the obtained results show that the objective function value of binary-coded SGA (O_1) could be improved applying real-coded SGA (O_2). Using binary-coded SGA_Mod1 (O_3) the J-value is improved compared to binary-coded SGA (O_1). Moreover, the result is obtained for the computational time that is almost twice less than the T for O_1. The same results are obtained for real-coded SGA_Mod1 (O_4) and SGA_Mod2 $(O_5$ and $O_6)$ – the J value is improved or at least preserved using less computational resources. The results prove that the real-coded SGA have significant improvements on the computation speed preserving the decision precision or even improving it [11,15,16].

IM A_1 is used to establish the relations of defined parameters in non-linear model of *E. coli* MC4110 (Eqs. (1-4)) and optimization algorithms outcome. Based on the Algorithm 1 (see Section 3), ICA is implemented in the Matlab environment. The resulting IMs that determine the degrees of "agreement" (μ) and "disagreement" (ν) between criteria are as follows:

$$\mu_1 = \begin{array}{c|ccccc} & C_1 & C_2 & C_3 & C_4 & C_5 \\ \hline C_1 & 1 & 0.667 & 0.400 & 0.333 & 0.333 \\ C_2 & 0.667 & 1 & 0.333 & 0.400 & 0.400 \\ C_3 & 0.400 & 0.333 & 1 & 0.933 & 0.267 \\ C_4 & 0.333 & 0.400 & 0.933 & 1 & 0.333 \\ C_5 & 0.333 & 0.400 & 0.267 & 0.333 & 1 \end{array} \tag{11}$$

$$\nu_1 = \begin{array}{c|ccccc} & C_1 & C_2 & C_3 & C_4 & C_5 \\ \hline C_1 & \mathbf{0} & 0.333 & 0.600 & 0.667 & 0.667 \\ C_2 & 0.333 & \mathbf{0} & 0.667 & 0.600 & 0.600 \\ C_3 & 0.600 & 0.667 & \mathbf{0} & 0.067 & 0.733 \\ C_4 & 0.667 & 0.600 & 0.067 & \mathbf{0} & 0.667 \\ C_5 & 0.667 & 0.600 & 0.733 & 0.667 & \mathbf{0} \end{array} \qquad (12)$$

Case Study 2

The notations for the objects (from O_1 to O_6), as well as for the criteria C_1 and C_2 keep their meaning as mentioned above. The rest of the criteria, because of the different specific rates used (Eq. (5)) are as follows: C_3 and C_4 – value of maximum specific growth rates (respectively μ_{2S} and μ_{2E}), C_5 and C_6 – value of saturation constants (k_S and k_E), and C_7 – value of yield coefficient ($Y_{S/X}$).

$$A_2 = \begin{array}{c|cccccc} & O_1 & O_2 & O_3 & O_4 & O_5 & O_6 \\ \hline C_1 & 0.02207 & 0.02215 & 0.02231 & 0.02209 & 0.02361 & 0.02251 \\ C_2 & 355.37808 & 231.94385 & 275.93613 & 458.07666 & 225.10216 & 251.12209 \\ C_3 & 0.97053 & 0.99911 & 0.96310 & 0.98277 & 0.91228 & 0.90669 \\ C_4 & 0.14984 & 0.11822 & 0.09649 & 0.14257 & 0.14161 & 0.09574 \\ C_5 & 0.13710 & 0.13055 & 0.11464 & 0.13862 & 0.14630 & 0.11130 \\ C_6 & 0.79989 & 0.79999 & 0.79988 & 0.80000 & 0.75601 & 0.71148 \\ C_7 & 0.39578 & 0.41188 & 0.43964 & 0.39994 & 0.47393 & 0.40979 \end{array} \qquad (13)$$

In this case the real-coded GA are not always faster as demonstrated in the case of *E. coli* fed-batch FP. In this Case study only the real-coded SGA (O_1) is faster than the binary-coded SGA (O_2). The fastest one among all considered here six kinds of GA is the binary-coded SGA_Mod2 (O_5), but with lower value of model accuracy. Very close to this GA is the real-valued SGA_Mod2 (O_6), obtaining for lamost the same time model accuracy close to the highest one.

Implementing ICA through Algorithm 1, the following resulting IMs that determine the μ- and ν-values between criteria are obtained:

$$\mu_2 = \begin{array}{c|ccccccc} & C_1 & C_2 & C_3 & C_4 & C_5 & C_6 & C_7 \\ \hline C_1 & \mathbf{1} & 0.200 & 0.267 & 0.200 & 0.400 & 0.200 & 0.867 \\ C_2 & 0.200 & \mathbf{1} & 0.667 & 0.600 & 0.533 & 0.733 & 0.200 \\ C_3 & 0.267 & 0.667 & \mathbf{1} & 0.667 & 0.600 & 0.933 & 0.400 \\ C_4 & 0.200 & 0.600 & 0.667 & \mathbf{1} & 0.800 & 0.733 & 0.333 \\ C_5 & 0.400 & 0.533 & 0.600 & 0.800 & \mathbf{1} & 0.667 & 0.533 \\ C_6 & 0.200 & 0.733 & 0.933 & 0.733 & 0.667 & \mathbf{1} & 0.333 \\ C_7 & 0.867 & 0.200 & 0.400 & 0.333 & 0.533 & 0.333 & \mathbf{1} \end{array} \qquad (14)$$

$$\nu_2 = \begin{array}{c|ccccccc} & C_1 & C_2 & C_3 & C_4 & C_5 & C_6 & C_7 \\ \hline C_1 & \mathbf{0} & 0.800 & 0.733 & 0.800 & 0.600 & 0.800 & 0.133 \\ C_2 & 0.800 & \mathbf{0} & 0.333 & 0.400 & 0.467 & 0.267 & 0.800 \\ C_3 & 0.733 & 0.333 & \mathbf{0} & 0.333 & 0.400 & 0.067 & 0.600 \\ C_4 & 0.800 & 0.400 & 0.333 & \mathbf{0} & 0.200 & 0.267 & 0.667 \\ C_5 & 0.600 & 0.467 & 0.400 & 0.200 & \mathbf{0} & 0.333 & 0.467 \\ C_6 & 0.800 & 0.267 & 0.067 & 0.267 & 0.333 & \mathbf{0} & 0.667 \\ C_7 & 0.133 & 0.800 & 0.600 & 0.667 & 0.467 & 0.667 & \mathbf{0} \end{array} \qquad (15)$$

Observing the obtained values for μ and ν, the criteria pairs are classified, based on the proposed in [7] scale (see Table 1), in the following groups: Group I (positive consonance) with criteria pairs that have the degree of "agreement" $\mu \in (0.75 - 1.00]$; Group II (strong dissonance) – $\mu \in (0.43 - 0.57]$; Group III (weak dissonance) – $\mu \in (0.25 - 0.33] \cup (0.67 - 0.75]$; Group IV (dissonance) – $\mu \in (0.33 - 0.43] \cup (0.57 - 0.67]$, and Group V (negative consonance) – $\mu \in (0.00 - 0.25]$. The crietia pairs falling in the defined 5 groups are listed in Table 2.

Table 1. Consonance and dissonance scale [7]

Interval of $\mu_{C,C'}$	Meaning
[0-0.5]	strong negative consonance
(0.5-0.15]	negative consonance
(0.15-0.25]	weak negative consonance
(0.25-0.33]	weak dissonance
(0.33-0.43]	dissonance
(0.43-0.57]	strong dissonance
(0.57-0.67]	dissonance
(0.67-0.75]	weak dissonance
(0.75-0.85]	weak positive consonance
(0.85-0.95]	positive consonance
(0.95-1.00]	strong positive consonance

Table 2. Results from ICA application

Value	Fed-batch fermentation model	
	E. coli	**S. cerevisiae**
Group I	$\mu_{max} \leftrightarrow k_S$	$\mu_{2S} \leftrightarrow k_E, \mu_{2E} \leftrightarrow k_S, Y_{S/X} \leftrightarrow J$
Group II		$k_S \leftrightarrow T, k_S \leftrightarrow Y_{S/X}$
Group III	$\mu_{max} \leftrightarrow Y_{S/X}, \mu_{max} \leftrightarrow J, k_S \leftrightarrow T$ $k_S \leftrightarrow Y_{S/X}, Y_{S/X} \leftrightarrow T$	$\mu_{2S} \leftrightarrow J, T \leftrightarrow k_E, \mu_{2E} \leftrightarrow k_E$
Group IV	$T \leftrightarrow J, \mu_{max} \leftrightarrow T, k_S \leftrightarrow J$ $Y_{S/X} \leftrightarrow J$	$\mu_{2E} \leftrightarrow \mu_{2S}, k_S \leftrightarrow \mu_{2S}, \mu_{2E} \leftrightarrow T$ $\mu_{2S} \leftrightarrow T, k_S \leftrightarrow J, \mu_{2E} \leftrightarrow Y_{S/X}$ $\mu_{2S} \leftrightarrow Y_{S/X}, k_E \leftrightarrow Y_{S/X}, k_S \leftrightarrow k_E$
Group V		$T \leftrightarrow J, \mu_{2E} \leftrightarrow J, k_E \leftrightarrow J, Y_{S/X} \leftrightarrow T$

It is expected that the results from ICA for the two Case studies should refer to some identical or similar relations and dependences between the considered models parameters. This is due to the fact that non-structural models, represented by systems of non-linear differential equations, are considered. On the other hand, the models describing FP of E. coli and S. cerevisiae are based on different specific growth rates, so there should be some differences in the parameters relations. As it can be seen in Table 2, the expected strong connection in pair $k_S \leftrightarrow \mu_{max}$, explained by the physical meaning of these model parameters [8], has been confirmed. The degree of "agreement" between these two parameters falls in the Group I – positive consonance – in both Case studies. Considering

Case study 2, the same results are observed also for the pairs $\mu_{2S} \leftrightarrow k_E$ and $Y_{S/X} \leftrightarrow J$. For both cases the degrees of "agreement" between $k_S \leftrightarrow T$ and $k_S \leftrightarrow Y_{S/X}$, as well as between specific growth rate and J fall in the groups of dissonances (Groups II and III). Absolute coincidence is observed in the Group III between the pairs of specific growth rate (μ_{max} in Case study 1 and μ_{2S} in Case study 2), and the value of the optimization criterion J, while the pairs $k_S \leftrightarrow T$ and $k_S \leftrightarrow Y_{S/X}$ in Case study 2 are separated in the group of strong dissonance (Group II).

In Case study 2 more complicated specific growth rates lead to weak correlations of the rest of the pairs, falling in the Group III. Another coincidence for both Case studies is in Groups IV – between specific growth rates (μ_{max} in Case study 2 and μ_{2S} and μ_{2E} in Case study 2) and T, as well as between k_S and J. In spite of the above discussed similarities, different degrees of "agreement" in pairs with the yield coefficient $Y_{S/X}$ are observed. Such distinction has been expected due to different specific growth rates in both FP (Eqs. (4) and (5)).

Due to the stochastic nature of the considered here six kinds of GA, it is quite difficult for some very strong relations between the optimization criterion J and the convergence time T on the one hand, and model parameters on the other hand, to be established. If the randomly chosen initial conditions are sufficiently good, the relation between J and T will be of higher agreement (Group IV in Case study 1). Or, conversely, GA starting from more "distanced" initial conditions may need more time to find a satisfactory solution, thus leading to a lower degree of "agreement" between J and T (Group V in Case study 2). Although, there are three coincidences for both Case Studies: between specific growth rates and J in Group III, between specific growth rates and T in Group IV, and between k_S and J again in Group IV.

The discussed above results from ICA application show some relations and dependencies caused by the physical meaning of the model parameters – on the one hand, and stochastic nature of the considered metaheuristics – on the other hand.

5   Conclusion

In this paper a new approach, namely InterCriteria Analysis, is applied, to establish relations and dependencies between different model parameters. Moreover, an attempt to discover some relations between model parameters and identification algorithms accuracy and convergence time is conducted. Fermentation processes of bacteria *E. coli* and yeast *S. cerevisiae* are examined and six different GA are applied for their parameter identification.

The application of ICA confirms some known relations and dependences originating from the physical meaning of the considered model parameters. Due to the biggest number of obtained pairs of criteria falling into the groups of dissonance, some more investigations is going to be performed in order to obtain more reliable relations between FP model parameters. Then, having much more correlations and dependencies, some advices might be given to the researchers/operators for the structure/parameter identification of the FP model parameters, and/or tuning of optimization techniques, and/or even for optimal carrying out of the processes in order to support the decision making process.

Acknowledgments. The work is supported by the Bulgarian National Scientific Fund under the grant DFNI-I-02-5 "InterCriteria Analysis – A New Approach to Decision Making".

References

1. Angelova, M.: Modified Genetic Algorithms and Intuitionistic Fuzzy Logic for Parameter Identification of a Fed-batch Cultivation Model. PhD Thesis, Sofia (2014). (in Bulgarian)
2. Atanassov, K.: On Intuitionistic Fuzzy Sets Theory. Springer, Berlin (2012)
3. Atanassov, K., Mavrov, D., Atanassova, V.: Intercriteria Decision Making: A New Approach for Multicriteria Decision Making, Based on Index Matrices and Intuitionistic Fuzzy Sets. Issues in IFSs and GNs **11**, 1–8 (2014)
4. Atanassov, K., Szmidt, E., Kacprzyk, J.: On Intuitionistic Fuzzy Pairs. Notes on IFS **19**(3), 1–13 (2013)
5. Atanassov, K.: On Index Matrices, Part 1: Standard Cases. Advanced Studies in Contemporary Mathematics **20**(2), 291–302 (2010)
6. Atanassov, K.: On Index Matrices, Part 2: Intuitionistic Fuzzy Case. Proceedings of the Jangjeon Mathematical Society **13**(2), 121–126 (2010)
7. Atanassov, K., Atanassova, V., Gluhchev, G.: InterCriteria Analysis: Ideas and Problems. Notes on Intuitionistic Fuzzy Sets **21**(1), 81–88 (2015)
8. Bastin, G., Dochain, D.: On-line Estimation and Adaptive Control of Bioreactors. Elsevier Scientific Publications (1991)
9. Boussaid, I., Lepagnot, J., Siarry, P.: A Survey on Optimization Metaheuristics. Information Sciences **237**, 82–117 (2013)
10. Chen, Z.-Q., Wang, R.-L.: Two Efficient Real-coded Genetic Algorithms for Real Parameter Optimization. International Journal of Innovative Computing, Information and Control **7**(8), 4871–4883 (2011)
11. Gen, M., Cheng, R.: Genetic Algorithms and Engineering Optimization. Wiley, New York (2000)
12. Goldberg, D.E.: Genetic Algorithms in Search, Optimization and Machine Learning. Addison Wesley Longman, London (1989)
13. Goldberg, D.E.: Real-coded Genetic Algorithms, Virtual Alphabets, and Blocking. Complex Systems **5**, 139–167 (1991)
14. Lin, C.-L., Huang, C.-H., Tsai, C.-W.: Structure-specified real coded genetic algorithms with applications. In: Sajja, P.S., Akerkar, R. (eds.) Advanced Knowledge Based Systems: Model, Applications & Research, chap. 8, vol. 1, pp. 160–187 (2010)
15. Mohideen, A.K., Saravanakumar, G., Valarmathi, K., Devaraj, D., Radhakrishnan, T.K.: Real-coded Genetic Algorithm for System Identification and Tuning of a Modified Model Reference Adaptive Controller for a Hybrid Tank System. Applied Mathematical Modelling **37**, 3829–3847 (2013)
16. Peltokangas, R., Sorsa, A.: Real-coded genetic algorithms and nonlinear parameter identification. In: 4th International IEEE Conference on Intelligent Systems - IS 2008, vol. 2, pp. 10–42 - 10–47 (2008)

17. Pencheva, T., Roeva, O., Hristozov, I.: Functional State Approach to Fermentation Processes Modelling. Prof. Marin Drinov Academic Publishing House, Sofia (2006)
18. Pencheva, T., Angelova, M., Atanassova, V., Roeva, O.: InterCriteria Analysis of Genetic Algorithm Parameters in Parameter Identification. Notes on Intuitionistic Fuzzy Sets **21**(2), 99–110 (2015)
19. Roeva, O., Kosev, K., Trenkova, T.: A modified multi-population genetic algorithm for parameter identification of cultivation process models. In: Proc. of the ICEC 2010, Valencia, pp. 348–351 (2010)
20. Roeva, O.: A Modified Genetic Algorithm for a Parameter Identification of Fermentation Processes. Biotechnology and Biotechnological Equipment **20**(1), 202–209 (2006)

Optimization Procedure of the Multi-parameter Assessment and Bidding of Decision-Making Variants in the Computerized Decision Support System

Jarosław Becker[1(✉)] and Ryszard Budziński[2]

[1] The Department of Technology, The Jacob of Paradyż University of Applied Sciences in Gorzów Wielkopolski, Gorzów Wielkopolski, Poland
jbecker@pwsz.pl
[2] University of Szczecin, Faculty of Economics and Management, Szczecin, Poland
ryszard.budzinski@wneiz.pl

Abstract. The aim of the paper is to present the architecture and to use the platform of data organisation based on information notations of the multi-criteria linear programming (MLP). It constitutes the foundation of the construction of the computerized decision support system (DSS 2.0 – by Budziński R., Becker J., 2008-2015). The system was developed for the needs of complex, multi-methodical decision-making analyses. The MLP method is considered in two aspects. The first one includes the development of the platform for the formulation of decision-making tasks in the form of standardized information structures (MLP model templates), where each one is a meta-description for the set of analysed objects. While the second aspect relates to the use of a multi-model (connected partial MLP models representing objects) and the optimization algorithm for searching the best solutions (efficient) with regard of the process of their internal adjustment.

Keywords: Multi-parameter bidding based on optimization · Combining the MLP models · MLP multi-model · Computerized decision support system (DSS)

1 Introduction

An original solution, presented in the article, is the data organization platform based on information notations of the multi-criteria linear programming (MLP) method. It is the foundation of the construction of the computerized decision support system (DSS 2.0, [6]), which has been developed for the needs of the complex, multi-faceted decision-making analysis. Its subject is any category of objects – understood as decision-making variants. This system distinguishes multi-methodical research approach, which involves sharing algorithm of different, complementary methods – AHP (ranking), Electre Tri (grouping), econometric analysis (valuation) and induction of decision rules (the use of rough set theory) – in a simple and useful form.

© Springer International Publishing Switzerland 2015
M. Núñez et al. (Eds.): ICCCI 2015, Part II, LNCS 9330, pp. 182–192, 2015.
DOI: 10.1007/978-3-319-24306-1_18

The multi-criteria decision-making analysis was extended to simulate the competitiveness of the rejected decision-making variants, with respect to the set of variants, which are the optimal solution. This competition can be iterative (like in multi-parameter biddings). The system supports participants representing the objects by answering the question: *What should be done to be on certain ranking lists at the lowest cost?* Functionality of the obtained prototyping solutions was illustrated on the practical example of the multi-criteria assessment of the employees of a company. Data identifying the employees were coded due to the confidential nature of the study.

2   Determinants of the Knowledge Sources Integration in the Decision Support System

Complexity of decision-making analyses is a major problem of the information system engineering. For example, information which is not clearly expressed in the form of decision criteria, resulting only from the context of the given decision-making situation is very difficult for modelling and including the multi-criteria methods in the calculations. Studies relevant for this type of scientific inquiries are presented in the papers: [9, 10], [20]. The complexity of the decision-making process is also influenced by the dispersion of data sources and decision variant assessments. This feature can be also identified with the subject of attributes and may occur in a situation, when their values come from many specialised external sources, like e.g. expert opinions and statistical data. A way popular in the literature [9], [12], [19, 20, 21] to represent the verbal descriptions and qualitative data is the use of linguistic sets, which allow to take into considerations the probabilistic nature of the human ratings.

Literature [8], [15], [17] contains a variety of procedures and methods of multiple criteria decision making (MCDM). According to [8] they can be divided into methods based on the functional model (American school) and relational model (European school). The fact that the theory of decisions creates methodological foundations for the analysis and generating best solutions is not about the utility of the information system in practice. In fact, the needs of management translate into the essential factors that should be taken into account in the design of computerized DSS (Fig. 1), namely:

- multi-stage nature of the decision-making process,
- structure of criteria – simple (criteria vector) or complex (hierarchical or network),
- number of decision-makers and experts,
- scale of the decision problem (few or mass problems),
- flexibility of decision variants (customising the parameter values),
- linguistics of data (statements of experts or respondents).

The complexity of the description of the decisive situation causes that it is difficult to emerge the method that would be universal, to which we could attribute the possibility to obtain the best solution of many different decision-making problems.

In the subject of method integration, the creation of hybrid approaches, which combine the application of different methods for solving the decision-making problem, a few interesting approaches have been proposed. For example, the use of the

AHP procedure and PROMETHEE as the hybrid method to select the outsourcing strategy of the computer system was presented in the paper [18]. A similar connection of the TOPSIS method with the AHP method is proposed in [13]. The paper [7] presents the practical combination of the relational model (the ELECTRE method) with the functional model (MAUT). Another interesting approach is presented in the paper [4], the authors used the PROMETHEE II method and linear programming.

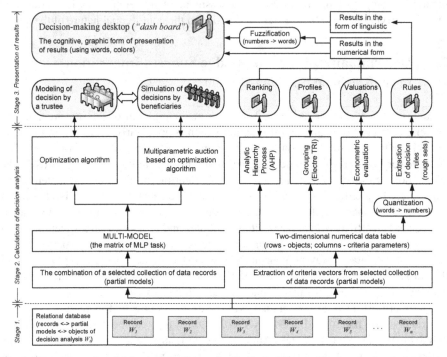

Fig. 1. The scope of multi-faceted (multi-methodical) decision-making analysis of objects in the system supporting decision-making (source: [6])

In the context of the listed conditions of the decision-making process and experiences in creation of the hybrid solutions, a novel solution has been proposed. It is the integration of knowledge sources – measurement data, expert opinions and competences, unified structures of mathematical models and collections of selected methods – in the information system (DSS 2.0), in an important moment for the information and decision-making process, which is the decision-making game [2]. The goal of each game is the selection of the solutions from the best available ones. Integration of methods in the DSS consists of the use of their functionality on a common set of data (Fig. 1) within a coherent information-decisive process consisting of:

a) *optimization of decisions* from the point of view of interests of the trustee's resources and from the perspective of beneficiaries competing for the resources,

b) *multi-criteria analysis,* in which there were used the approaches connected with the achievements of: the American school (AHP method [16]), European (Electre Tri [14]) and Polish school (Rough Set Theory [11]),

c) *identification* in terms of quantitative methods of the econometric analysis [2].

The base of the methods integration is the coherent and flexible information structure of the system (MLP platform). It allows you to define the template for the decision-making task (pattern of mathematical model, Fig. 2). According to the template to the system there are introduced data of objects (decision variants: W_1, W_2, ..., W_n; Fig. 1). Technical and economic parameters of each variant can be expressed in the form of numerical values and linguistic assessments (fuzzy values) from the ordinal scale defined by experts or respondents.

The system allows the presentation of detailed results for each method separately and together, in the form of the decision-making desktop (Fig. 1, *"dash board"*), within which the applied methods function on the basis of a consultation of experts diagnosing the state of the tested objects. The desktop has the cognitive form of presentation of results [2].

3   Model of MLP Platform – Data Organization in the Decision Support System

Defining decision problems (tasks) in the system is inseparable with the determination of the structure of the mathematical model template in the specially developed for this purpose module of the MLP model generator [1]. Its service was divided into thematic groups (described in Fig. 2, blocks: A, B, C and D) concerning the variables, balances and equations of partial goals and changes of labels (names, units of measurement and character relationships).

In the system there was used the approach proposed by Budzinski [5], not requiring the determination a priori of the values of goals for implementation, based on the axiom of a "goal game", difference of non-negative quality indicators ($q_k = x_{(z+k)}$) beneficial features and undesirable features ($q_k = -x_{(z+k)}$) for $k = 1,2, ..., r$ and $z = n \times s$. In this method the partial goals (from block D, fig. 2)

$$\forall W_t \exists x_j^{(t)} \{ f_k(\mathbf{x}_j) = \sum_{t=1}^n d_{kj}^{(t)} x_j^{(t)} \to (\min \vee \max) \}, \tag{1}$$

where $\mathbf{x}_j = \{ x_j^{(1)}, x_j^{(2)}, ..., x_j^{(n)} \}$, are recorded in the form of balances

$$\forall W_t \exists x_j^{(t)} \{ (\sum_{t=1}^n d_{kj}^{(t)} x_j^{(t)}) - u_k x_{(z+k)} = 0 \}, \tag{2}$$

then their synthesis is performed to the form of the goal function

$$F(q_1, ..., q_r) = \sum_{k=1}^r f(q_k) = \sum_{k=1}^r w_k q_k \to \max, \tag{3}$$

where
$$q_k = \begin{cases} x_{(z+k)} & \text{for } f_k(\mathbf{x}_j) \to \max \\ -x_{(z+k)} & \text{for } f_k(\mathbf{x}_j) \to \min \end{cases}. \tag{4}$$

While w_1, w_2, ..., w_r are the ranks of validity, preferences of reaching different goals. While u_k are the technical parameters of normalization bringing k partial goals to their equal rank in optimization calculations

$$u_k = \frac{100 l_k}{\sum_{t=1}^{n} \left| d_{kj}^{(t)} \right|},$$

(5)

where: $\left| d_{kj}^{(t)} \right|$ – the absolute values of parameters, l_k – the accepted for calculations number of non-zero elements in the k row of partial goals [5].

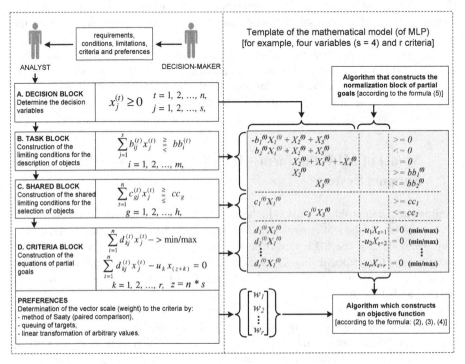

Fig. 2. The idea of constructing a pattern (template) of the mathematical model (source: [2])

There was accepted the principle that every object is the partial model and at the same time the data record (with variable lengths from the point of view of various decision tasks), and the whole task formally fulfils the condition of the relational database with its all attributes. In the engineering approach there was proposed an original method based on a special structure of meta data, called the *converter*, which is used to connect homogenous and partial mathematical models in the given task[2].

4   Optimization Procedure of Object Selection with the Iterative Process of Their Multi-parameter Adjustment

The system examined two cases of multi-criteria optimization, generating decision-making solutions for the decision-maker (resources trustee) and supporting beneficiaries (representing the objects in the task) in appeal processes in the context of improving the basic solution of the decision-maker. In the first case we seek objects (decision variants), which meet the decision-maker's expectations. In the second one, for all objects, which were not in the area of solutions accepted by the decision-maker, we seek the "cheapest" way to improve the value of criteria, guaranteeing their competitiveness in the next step of the decision-making game (optimization). This process is usually of the iterative nature. We seek better solutions by the gradual reduction of the *cost type* of criteria values (the less is on the criterion, the better for the trustee) and vice versa, by a mild increase of the *profit type* of criteria values.

The object selection procedure consists of 6 steps, described in Fig. 3. Generating decision-making solutions for the decision-maker runs from step 1 to 4. As a result of optimization of a multi-model there is obtained the division of the considered set of objects into the *accepted and rejected*. In the winning group there are variants for which the value of utility functions reaches maximum and at the same time satisfies all limiting conditions determined in the task. The rejected objects can participate in the *appeal procedure* (step 5 and 6). The whole procedure can be of iterative nature (after the 6 is the 2 step) and it may include the limits of: game time, iteration number or the number of individual adjustments.

A detailed explanation is required by the algorithm of *sub-system of the appeal simulation* (Fig. 3, step 5). It initiates the operation from taking the current values of the criteria vector $d_{kj}^{(t)}$ ($k = 1, 2, ..., r; j=1$) for the object W_t ($t = 1, 2, ..., n$). At the beginning it is possible to personalise the simulation settings. Modification applies to the threshold values on criteria $\varphi_k^{(t)}$, which are the borders of the correction resulting from the properties of the given object or the decision of the beneficiary. The change also subjects to the precision parameter $\eta_t > 0$, which value is a natural number, used to determine the partial value of the correction (so-called jump)

$$\delta_k^{(t)} = \frac{\left|\varphi_k^{(t)} - d_k^{(t)}\right|}{\eta_t}. \tag{6}$$

The decisive step of the simulation algorithm is the selection of strategies of seeking new criterion parameter values of the object. Three ways were used, each of them differs in the order of introducing amendments (jump values) to the elements of the criteria vector, namely: *strategy A* – according to the increasing preferences (annotating begins from the least important criterion), *strategy B* – according to the decreasing preferences (from the most important criterion), *strategy C* – collectively, single amendments are applied to all criteria.

After implementing every partial amendment to the criterion in strategies A, B or to the whole criterion vector within tactics C, the algorithm constructs the temporary multi-model matrix (*converter*) consisting of the partial models of current favourites and the analysed object and conducting the optimization. The simulation algorithm stops the operation, when the considered object (partial block of the multi-model) will be in the group of the best solutions (on the accepted list). Then, the located values of the criteria vector are presented, which can be confirmed, sent to the data form for the "manual" precision or ignore and repeat the simulation using another available strategy. In the worst case, optimization sessions of the simulation algorithm are repeated until the object achieves the determined border values on criteria.

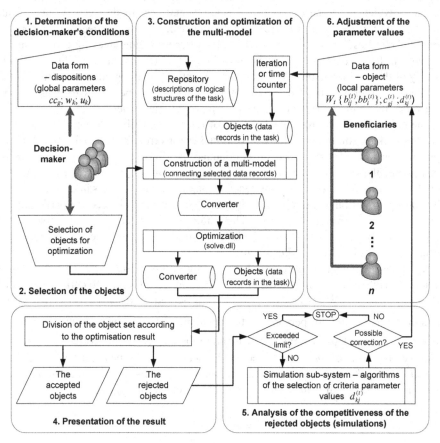

Fig. 3. Generating decision solutions with the iterative process of the multi-parameter object correction (source: [2])

The considered example of the multi-criteria assessment of the employees of the given company, which specializes in the design, manufacturing of steel structures [3]. Let us assume that the company's management organised a contest, in which five best locksmiths, from a total number of 17 people working in three branches, will get a cash bonus. The management cares not only for the improvement of performance. It propagates the policy of lifelong learning, improvement of: skills, knowledge and professional competence. The test procedure in the DSS 2.0 system included:

1) determination of the decision task template, criteria and preferences,
2) introduction of data on the employees (criterion parameter values),
3) optimization (Fig. 4), the emergence of five best assessed employees,
4) simulation of competitiveness of the rejected employees (Table 1).

Persons performing work are evaluated. There are three groups of criteria used during this operation, namely: a) *efficiency criteria* – relating to performance, e.g. the amount of pieces of product made in one month and the number of products with defects (d_1 – work performance, $w_1 = 0.45$), b) *eligibility criteria* – depending on the job; e.g.: work experience, familiarity with the devices, physical condition (d_2 – professional qualifications, $w_2 = 0.35$), c) *behavioural criteria* – e.g.: accountability, initiative, discipline (d_3 – professional attitudes, $w_3 = 0.15$). Additionally, *the extra skills* criterion was introduced in to the system for example: driver's license, certificates, courses, and foreign language skills (d_4 – additional skills, $w_4 = 0.05$).

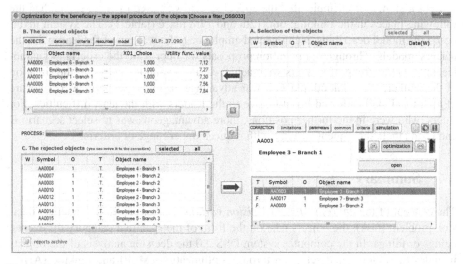

Fig. 4. The choice of five best assessed employees (source: the DSS 2.0 module of the appeal procedure of the beneficiaries – objects)

Table 1. Simulation results of fitters' competitiveness – correction of notes

Id	Object name	Main criteria	Values of the criteria [point]			
			Real	Simulated		
				A	B	C
AA0009	Employee 3 - branch 2	d1 - Work performance	8,0	8,0	8,2	8,1
		d2 - Professional qualifications	6,5	6,5	6,5	6,6
		d3 - Professional attitudes	6,8	6,8	6,8	6,9
		d4 - Additional skills	2,0	3,5	2,0	2,1
AA0003	Employee 3 - branch 1	d1 - Work performance	6,7	6,7	10,0	8,3
		d2 - Professional qualifications	3,8	5,9	4,0	5,4
		d3 - Professional attitudes	5,6	10,0	5,6	7,2
		d4 - Additional skills	7,0	10,0	7,0	8,6
AA0017	Employee 7 - branch 3	d1 - Work performance	3,6	3,6	10,0	7,3
		d2 - Professional qualifications	3,1	9,8	5,3	6,8
		d3 - Professional attitudes	3,1	10,0	3,1	6,8
		d4 - Additional skills	5,0	10,0	5,0	8,7

The criterion of productivity achieved the highest priority, as the data on the productivity of workers have the greatest impact on the system of rewards and punishments. All the weighting factors were determined by the Saaty's method [16] (the convergence coefficient CR for the criteria was 0,068).

In DSS 2.0 the fitter W_t (t = AA0001, AA0002, ..., AA0017) is represented by a simple mathematical model of partial linear programming, constructed on the basis of predefined template (according to Fig. 2). The template assumes that every fitter is represented by a binary variable of name "X01_Choice" and the vector of technical and economic parameters [c_1, c_3, d_1, d_2, d_3, d_4].

It was assumed that in the company at the end of the quarter, five best-rated employees would receive the bonus. At the beginning of the game the value cc_1 = 5 (Fig. 2) was set in the system and potential leaders were selected (Fig. 4, *B. The accepted objects*). Employees who have not been admitted to the preferred leaders were able to improve their ratings using *subsystem simulation of appeal* included DSS 2.0.

Let's assume, for example, that the appeal system was used by three fitters, had the following places in the ranking: 6th, 12th and 17th place (table 1). The system juxtaposed the model describing each rejected employee (who was out of the lead) with the leaders' models. Through optimization proposals for improving the criteria to guarantee entry to the group of winners were achieved. For example, analysing the situation of the worker, who took 6th place, it was advantageous to use scenario A. Disclosure of additional skills allowed him to move to the leaders with the least difficulties. For fitters, who were in the further positions, more advantageous was to select scenario B and C and their subsequent variants with reduced upper range of criteria.

5   Summary

The concept of the *analysis object (decision variant)* is abstract. In the practical decision problem (task), the objects take on the form of the specified category of: events, beings or things. In the computer system DSS 2.0 the decision analysis of the objects includes the bidding and selection through optimization (MLP) and ranking (AHP), grouping (ELECTRE TRI), quantitative and qualitative analysis. An example of the practical application of the decision support system is a group of tasks concerning the organisation of the tender for the purchase of services or products (the selection and bidding of offers: construction services, sales of cars, insurance, credit, etc.). Another category includes the tasks of the division of the financial resources for the particular economic goals (selection of proposals for funding: health service, municipal budgets, tasks covered by program and funds from the European Union, etc.). Individual tasks can involve the multi-criteria assessment of: offers, proposals, methods, technologies, companies, schools, development strategies, employees, students, financial reports, web portals, etc., where, additionally, through simulation we can find a strategy to improve the parameters for the low assessed variants.

References

1. Becker, J.: Architecture of information system for generating multi-criteria decision-making solution (part 1). In: Owsinski, J.W., Nahorski, Z., Szapiro, T. (eds.) Operational and systems research: decisions, the economy, human capital and quality (in Polish). Systems Research, vol. 64, pp. 103–112. Systems Research Institute, Polish Academy of Sciences, Warsaw (2008)

2. Becker, J.: Integration of knowledge sources in decision support system (methodological and design basics). The scientific monograph (in Polish). The Jacob of Paradyz University of Applied Sciences, Gorzow Wielkopolski (2015)

3. Becker, A., Becker, J.: Human resources analysis using a decision support system. Quantitative Methods in Economics **14**(1), 15–26 (2013)

4. Brans, J., Mareschal, B.: Promethee methods. In: Figueira, J., Greco, S., Ehrgott, M. (eds.) Multiple Criteria Decision Analysis: State of the Art Surveys, pp. 163–195. Springer, Berlin (2005)

5. Budzinski, R.: Methodology aspect in system processing of economical and financial data for enterprise (in Polish), vol. 446/372. Pub. House of Szczecin Univ., Szczecin (2001)

6. Budzinski, R., Becker, J.: Transformations of knowledge sources in decision support system. Journal of Automation, Mobile Robotics & Intelligent System **9**(2), 28–36 (2015)

7. De Almeida, A.T.: Multicriteria decision model for outsourcing contracts selection based on utility function and ELECTRE method. Computers & Operations Research **34**, 3569–3574 (2007)

8. Greco, S., Matarazzo, B., Słowiński, R.: Rough sets theory for multicriteria decision analysis. European J. of Operational Research **129**(1), 1–47 (2001)

9. Grzegorzewski, P.: The coefficient of concordance for vague data. Computational Statistics & Data Analysis **51**, 314–322 (2006)

10. Ma, J., Ruan, D., Xu, Y., Zhang, G.: A fuzzy-set approach to treat determinacy and consistency of linguistic terms in multi-criteria decision making. International J. of Approximate Reasoning **44**, 165–181 (2007)

11. Pawlak, Z.: Rough sets. Int. J. Computer and Information Sci. **11**, 341–356 (1982)

12. Piegat, A.: Are linguistic evaluations used by people of possibilistic or probabilistic nature? In: Rutkowski, L., Siekmann, J.H., Tadeusiewicz, R., Zadeh, L.A. (eds.) ICAISC 2004. LNCS (LNAI), vol. 3070, pp. 356–363. Springer, Heidelberg (2004)

13. Rao, R., Davim, J.: A decision-making framework model for material selection using a combined multiple attribute decision-making method. The International Journal of Advanced Manufacturing Technology **35**, 751–760 (2008)

14. Roy, B.: The outranking approach and the foundations of Electre methods. Theory and decision **31**, 49–73 (1991)

15. Roy, B., Bouyssou, D.: Aide multicritere a la decision: Methodes et cas (in French). Economica, Paris (1993)

16. Saaty, T.L.: Decision making with the analytic hierarchy process. International J. of Services Sciences **1**, 83–98 (2008)

17. Slowinski, R.: Ordinal regression approach to multiple-criteria ordering decision variants. In: Kulczycki, P., Hryniewicz, O., Kacprzyk, J. (eds.) Information techniques' in systems research (in Polish), pp. 315–337. Wyd. Naukowo-Techniczne, Warsaw (2007)

18. Wang, J.-J., Yang, D.-L.: Using a hybrid multi-criteria decision aid method for information systems outsourcing. Computers & Operations Research **34**, 3691–3700 (2007)

19. Zadeh, L.A.: From computing with numbers to computing with words – Form manipulation of measurements to manipulations of perceptions. IEEE Transactions on Circuits and Systems **45**, 105–119 (1999)
20. Zadeh, L.A.: Toward a perception-based theory of probabilistic reasoning with imprecise probabilities. Journal of Statistical Planning and Inference **105**, 233–264 (2002)
21. Zadeh, L.A.: Precisiated natural language. In: Mehler, A., Köhler, R. (eds.) Aspects of Automatic Text Analysis. Studies in Fuzziness and Soft Computing, vol. 209, pp. 33–60. Springer, Heidelberg (2006)

Supporting Cooperative Decisions with a Multicriteria Generalization of the Nash Solution

Lech Kruś[(✉)]

Systems Research Institute, Polish Academy of Sciences, Warsaw, Poland
krus@ibspan.waw.pl

Abstract. A decision situation is considered in which two decision makers negotiate cooperation conditions to realize a joint project. Each decision maker has his own set of criteria measuring results of the cooperation. The situation is modeled as the multicriteria bargaining problem. A special multiround mediation procedure is presented which can be implemented in a computer-based system. According to the procedure the system supports multicriteria analysis made by the decision makers and generates mediation proposals. The mediation proposals are derived on the basis of the original solution to the multicriteria problem, presented in the paper. The solution expresses preferences of the decision makers. It generalizes the classic Nash solution concept on the multicriteria case.

Keywords: Computer-based intelligent systems · Decision support · Multicriteria analysis · Cooperative games · Mediation

1 Introduction

The paper deals with computer intelligence problems related to construction of a computer-based system playing the role of a mediator in a bargaining process. The bargaining process is considered in the case of two decision makers discussing cooperation conditions to realize a joint project. The cooperation is possible if it is beneficial for both of them. It is assumed that each of the decision makers has his individual set of objectives which he would like to achieve. Achievements of the objectives are measured by given vectors of criteria, which are in general different for each decision maker. The criteria are conflicting in the case of the individual decision maker as well as between them. Each decision maker has also his individual preferences defined in the space of his criteria.

In the simplest buying - selling bargaining problem a buyer and a seller propose prices of a good trying to find a consensus. The consensus is possible if there exists an interval of prices beneficial for both sides, called as an agreement set. Typically, the positional negotiation are applied in this case. The positional negotiations frequently lead to an impasse, and the negotiations can not succeed even if the agreement set is not empty. To resolve the problem in this case,

© Springer International Publishing Switzerland 2015
M. Núñez et al. (Eds.): ICCCI 2015, Part II, LNCS 9330, pp. 193–202, 2015.
DOI: 10.1007/978-3-319-24306-1_19

but also in the case of more complicated negotiations, special procedures are applied with a mediation support. A mediator in the negotiations is an impartial outsider who tries to aid the negotiators in their quest to find a compromise agreement. The mediator can help with the negotiation process, but he does not have the authority to dictate a solution.

In the considered multicriteria bargaining process each of the two decision makers valuates variants of cooperations with use of his own vector of criteria. A compromise variant should be found which will be accepted by both sides despite the fact that the criteria are conflicting in the case of each decision maker as well between them. In practice of complicated international negotiations so called Single Negotiation Text Procedure is frequently applied. The procedure has been proposed by Roger Fisher during the Camp David negotiations to resolve an impasse which has occurred after several initial rounds of the positional negotiations (see Raiffa [14]). According to the procedure a negotiation process consists of a number of rounds. In each round a mediator prepares a package for the consideration of protagonists. Each package is meant as a single negotiation text to be criticized by protagonists then modified and remodified. Typically the negotiation process starts from the first single negotiation text which is far from the expectations of protagonists. The process is progressive for each of the protagonists.

A negotiation process in which a computer-based system plays the role of mediator is discussed in the paper in the case of the mentioned decision situation formulated as the multicriteria bargaining problem. The multicriteria bargaining problem presented in Section 2 is a generalization of the bargaining problem formulated and discussed in the classic game theory by many researchers, including (Nash [10,11], Raiffa [13], Kalai and Smorodinsky [3], Roth [15], Thomson [16], Peters [12], Moulin [9]) and others. In these papers many different solution concepts have been proposed and analyzed. In the classic theory the decision makers are treated as players playing the bargaining game and it is assumed that each of them has explicitly given utility function measuring his payoffs. The solution is looked for in the space of the utilities. In the multicriteria problem considered in this paper, the payoff of each decision maker (player) is measured by a vector of criteria and we do not assume that his utility function is given explicitly. The solution is looked for in the space being the cartesian product of the multicriteria spaces of the players. The solution concepts proposed in the classic theory do not transfer in a simple way to the multicriteria case. Looking for a solution in the multicriteria bargaining problem we have to consider jointly two decision problems: the first - the solution should be related to the preferences of each of the decision makers, and the second - the solution should fulfill fairness rules accepted by all of them. The bargaining process will succeed if the final cooperation conditions satisfy desirable benefits of each decision maker, measured by his criteria and valuated according to his individual preferences. In many situation, at the beginning of the bargaining process, decision makers can not be conscious of their preferences if they have not enough information about the attainable payoffs. Therefore an iterative process is required in which the decision makers can generate and collect such information.

The following Sections 3, 4, 5 present respectively proposals including: the mediation procedure, the unilateral analysis support and the formulation of the solution to the multicriteria bargaining problem, which is used to generate mediation proposals. The solution generalizes the Nash solution concept on the case of multicriteria payoffs of players.

This paper continues the line of research presented in the papers (Kruś and Bronisz [8], Kruś [7], [6], [5], [4]). The monograph [5] is devoted to computer-based methods supporting multicriteria cooperative decisions. It presents the methods applied in the case of the multicriteria bargaining problems, the multicriteria cooperative games with and without side payments, the cost allocation problems modelled with the use of the games. It summarizes results of the earlier papers. The paper [4] presents a proposal of the computer-based procedure supporting negotiations in the case of multicriteria bargaining problem with the use of the generalized Raiffa solution concept.

New results of this paper include a constructive proposal of the solution concept to the multicriteria bargaining problem which generalizes the classical Nash solution concept, derivation of the mediation proposals and the modified procedure with the application of this solution.

2 Multicriteria Bargaining Problem

Let us consider two decision makers negotiating conditions of possible cooperation.

Each decision maker i, $i = 1, 2$ has defined decision variables, denoted by a vector $x_i = (x_{i1}, x_{i2}, \ldots x_{ik^i})$, $x_i \in \mathbb{R}^{k^i}$, where k^i is a number of decision variables of decision maker $i = 1, 2$, and $\mathbb{R}^{k^i}$ is a space of his decisions.

Decision variables of all the decision makers are denoted by a vector $x = (x_1, x_2) \in \mathbb{R}^K$, $K = k^1 + k^2$, where $\mathbb{R}^K$ is the cartesian product of the decision spaces of the decision makers 1 and 2.

It is assumed that results of the cooperation are measured by a vector of criteria which is in general different for each decision maker. The criteria of the decision maker i, $i = 1, 2$, valuating his payoff are denoted by a vector $y_i = (y_{i1}, y_{i2}, \ldots y_{im^i}) \in \mathbb{R}^{m^i}$, where m^i is a number of criteria of the decision maker i, and $\mathbb{R}^{m^i}$ is a space of his criteria.

The criteria of all the decision makers are denoted by $y = (y_1, y_2) \in \mathbb{R}^M$, $M = m^1 + m^2$, where $\mathbb{R}^M$ is the cartesian product of the citeria spaces of all the decision makers.

We assume that a mathematical model is given describing the decision makers' payoffs as dependent on the decision variables. The model implemented in a computer-based system will be used to derive the payoffs of the decision makers for given variants of the decision variables.

Formally we assume that the model is defined by a set of admissible decisions $X_0 \subset \mathbb{R}^K$, and by a mapping $F : \mathbb{R}^K \to \mathbb{R}^M$ from the decision space to the space of the criteria. A set of attainable payoffs, denoted by $Y_0 = F(X_0)$ is defined in the space of criteria of all decision makers. However each decision maker has access to information in his criteria space only. In the space of criteria of i-th decision maker a set of his attainable payoffs Y_{0i}, can be defined, being an intersection of the set Y_0. The set of attainable payoffs of every decision maker depends on his set of admissible decisions and on the set of admissible decisions of other decision maker.

A partial ordering is introduced in the the criteria spaces in the standard way. Let $I\!R^m$ denote a space of criteria for an arbitrary number m of criteria. Each of m criterions can be maximized or minimized. However, to simplify the notation and without loss of generality we assume that decision makers maximize all their criteria.

A vector $y \in I\!R^m$ is **weakly Pareto optimal** (weakly nondominated) in set $Y_0 \subset I\!R^m$ if $y \in Y_0$ and does not exist $z \in Y_0$ such, that $z_i > y_i$ for $i = 1, 2 \ldots, m$. A vector $y \in I\!R^m$ is **Pareto optimal** (nondominated) in set $Y_0 \subset I\!R^m$ if $y \in Y_0$ and does not exist $z \in Y_0$, $z \neq y$ such, that $z_i \geq y_i$ for $i = 1, 2, \ldots, m$.

A bargaining problem with multicriteria payoffs of decision makers (multicriteria bargaining problem) can be formulated by a pair (S, d), where the element $d = (d_1, d_2) \in Y_0 \subset I\!R^M$ is a disagreement point, and the set S is an agreement set. The agreement set $S \subset Y_0 \subset I\!R^M$ is the subset of the set of the attainable payoffs dominating the disagreement point d. The agreement set defines payoffs attainable by all decision makers but under their unanimous agreement. If such agreement is not achieved, the payoffs of all decision makers are defined by the disagreement point d. We assume that the agreement set S is nonempty, compact and convex.

We assume, that each decision maker i, $i = 1, 2$, defines the vector $d_i \in I\!R^{m^i}$ as his reservation point in his space of criteria. Every decision maker, negotiating possible cooperation, will not agree for payoffs decreasing any component of the vector. A decision maker can assume the reservation point as the status quo point. He can however analyze some alternative options to the negotiated agreement and he define it on the basis of the BATNA concept presented in (Fisher, Ury [1]). The BATNA (abbreviation of Best Alternative to Negotiated Agreement) concept, is frequently applied in processes of international negotiations in a prenegotiation step. According to the concept, each side of negotiations should analyze possible alternatives to the negotiated agreement and select the best one according to its preferences. The best one is called as BATNA. It is the alternative for the side (decision maker), that can be achieved if the negotiations will not succeed.

A question arises, how each decision maker can be supported in the processes of decision analysis and in looking for the agreeable solution. The support should enable valuation of payoffs for different variants of his own decisions and the decisions of the second decision maker. It should also aid derivation of the agreeable, nondominated solution defining the payoffs of the decision makers in the agrement set. The solution should fulfil fair play rules such that it could be accepted by both the decision makers.

In this paper an interactive procedure is presented including multicriteria decision support of each decision maker. Within the procedure mediation proposals are derived using an idea of the Nash cooperative solution. The multicriteria decision support is made with the use of the reference point method developed by A.P. Wierzbicki [17], [18]. The Nash solution [10],[11] has been originally formulated under axioms describing the fair play distribution of cooperation benefits, that can be accepted by rational players. It has been proposed by Nash to the bargaining problem under assumptions of the scalar payoffs of players. It can not be applied directly in the multicriteria bargaining problem considered here. This paper presents a construction enabling application of this idea in the case of the multicriteria payoffs of decision makers.

3   Interactive Procedure Supporting Mediation Process

The procedure has been proposed under inspiration of the mentioned Single Negotiation Text Procedure frequently applied in the international negotiations. In the considered case the role of mediator is played by the computer based system.

The procedure is realized in some number of rounds $t = 1, 2, ..., T$. In each round t:

- each decision maker makes independently interactive analysis of nondominated payoffs in his multicriteria space of payoffs (the analysis is called further as unilateral) and indicates a direction improving his payoff in comparison to the disagreement point. The direction is selected by him according to his preferences as an effect of the multicriteria analysis.
- computer-based system collects improvement directions indicated by both decision makers and generates on this basis a mediation proposal d^t,
- decision makers analyze the mediation proposal and correct the preferred improvement directions, afterwards system derives next mediation proposal.

All mediation proposals d^t are generated on the basis of the improvement directions indicated by the decision makers and with application of an assumed solution concept of the multicriteria bargaining problem:
$d^t = d^{t-1} + \alpha^t [G^t - d^{t-1}]$, for $t = 1, 2, ...T$,
where $d^0 = d$, α^t is so called confidence coefficient assumed by the decision makers in the round t, G^t is a solution of the multicriteria bargaining problem.

The solution G^t is derived in the round t and defines payoffs of the decision makers. In this case a multicriteria solution concept is proposed which is a generalization of the Nash solution concept to the case of multicriteria payoffs of the decision makers.

Each decision makers can in each round reduce improvement of the payoffs (his own payoffs and at the same time payoffs of other decision maker) assuming respectively small value of the confidence coefficient α^t.

4   Unilateral Analysis

The unilateral analysis is made independently by each decision maker. It should lead the given decision maker i, $i = 1, 2$ to selection of the Pareto optimal element in the set S according to his in mind preferences. The element defines the required direction improving his current payoff.

The unilateral analysis is supported with the use of the reference point method and with application of the respective achievement function (see [17], [18]). The decision maker can obtain some number of the Pareto optimal points in the set S using this method and can select the preferred point.

Any Pareto optimal point $\overline{y}_i$ of the set S in the criteria space of the decision maker i $i = 1, 2$, can be derived as the solution of the following optimization problem:

$$\max_{x \in X_0} s(y_i, r_i), \tag{1}$$

where
$r_i \in I\!\!R^{m_i}$ is a reference point of the decision maker i in his space of criteria,
x is the vector of the decision variables,
$y_i = v_i(x)$ defines the vector of criteria of the decision maker i, as dependent on the decision variables x due to the mapping F, under additional constraints that the criteria of the second decision maker are on the level of his reservation point i.e. $y_{3-i} = d_{3-i}$,
$s(y_i, r_i)$ is the achievement function approximating order in the space $I\!\!R^{m_i}$.

The function $s(y_i, r_i) = \min_{1 \le k \le m_i} [a_k(y_{i,k} - r_{i,k})] + \rho \sum_{k=1}^{m_i} a_k(y_{i,k} - r_{i,k})$ states an example of the achievement function suitable in this case, where $a_k, 1 \le k \le m_i$ are scaling coefficients, and ρ is a relatively small number.

A representation of the Pareto frontier of the set S can be obtained by solving the optimization problem for different reference points r_i assumed by the decision maker i.

Fig. 1 illustrates how a given decision maker can generate and review his attainable nondominated payoffs. He assumes different reference points and then the system derives the respective Pareto optimal solutions. The reference points assumed by the decision maker and the Pareto optimal payoffs $\overline{y}_i$ derived by the system are stored in a data base, so that the decision maker can obtain a representation of the Pareto optimal frontier of the set S and can analyze it.

It is assumed that each decision maker i, $i = 1, 2$, finishing the multicriteria analysis, indicates his preferred nondominated payoff $\widehat{y}_i$ in his space of criteria. The payoff $\widehat{y}_1$ corresponds to the element $y^1 = (\widehat{y}_1, d_2) \in S$ in the case of the decision maker $i = 1$ and the payoff $\widehat{y}_2$ corresponds respectively to the element $y^2 = (d_1, \widehat{y}_2) \in S$ in the case of decision maker $i = 2$. The last elements are defined in the space of criteria of both the decision makers. The stage of unilateral analysis is finished when both the decision makers have indicated their preferred payoffs.

The unilateral analysis can be realized in different ways with respect to access to information available for decision makers. In the presented way it is assumed

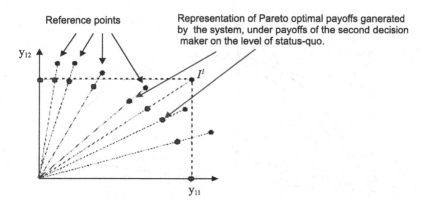

Reference points

Representation of Pareto optimal payoffs ganerated by the system, under payoffs of the second decision maker on the level of status-quo.

Fig. 1. Generation of the nondominated payoffs of the decision maker 1 for assumed reference points

that each decision maker makes unilateral analysis not knowing the criteria nor the reservation point of the second decision maker. Only the mediator has access to the full information. This information is used in calculations of the computer-based system. In general the decision makers have not permission to see the data introduced and generated by the others.

5 Derivation of the Mediation Proposal

A mediation proposal is derived by the system when both decision makers have indicated their preferred payoffs $\widehat{y}_1, \widehat{y}_2$ in their spaces of criteria and when respective points $y^1, y^2 \in S$ have been calculated by the system.

Let us construct a two dimensional hyperplane H^2 defined by points d, y^1, y^2 in the criteria space $\mathbb{R}^M$. Each point $y \in H^2$ may be defined as

$$y = d + a_1(y^1 - d) + a_2(y^2 - d).$$

Let A denote mapping from H^2 to $\mathbb{R}^2$ defined by $A(y) = A[d + a_1(y^1 - d) + a_2(y^2 - d)] = (a_1, a_2)$. A two person bargaining problem $(A(S^H), A(d))$ can be considered on the hyperplane H^2. Set $S^H = S \cap H^2$ in the problem.

A generalization of the Nash cooperative solution concept can be constructed using the hyperplane H^2. It is illustrated in Fig. 2. In this example the decision maker 1 has two criteria $y_{1,1}$ and $y_{1,2}$ respectively, the decision maker 2 has only one criterion $y_{2,1}$. Let the point y^1 be defined according to the preferences of the first decision maker. The preferred point y^2 of the second one is defined by the maximal attainable value of his payoff. The hyperplane H^2 is defined by the points d, y^1 and y^2. The arrows drown on the figure present the improvement directions leading to the nondominated payoffs selected by the decision makers. The Nash cooperative solution $y^N = f^N(S^H, d)$ to the bargaining problem (S^H, d) is defined as the point of the set S^H maximizing the product of the payoffs increases for the decision makers 1 and 2 on the hyperplane H^2.

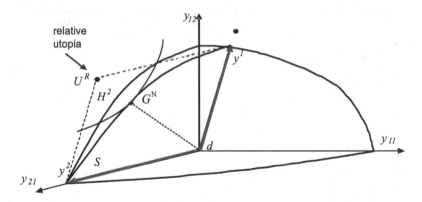

Fig. 2. Construction of the generalized Nash solution to the multicriteria bargaining problem

Let the preferences of decision makers are expressed by points y^1 i y^2. Then the final payoffs defined by this point fulfill the following axioms:

(A1) Pareto-optimality
$y^N = f^N(S^H, d)$ is Pareto-optimal in set S^H,

(A2) Individual rationality
For every bargaining game (S^H, d), $y^N = f^N(S^H, d) \geq d$.

(A3) Symmetry
We say, that bargaining problem (S^H, d) is symmetric, if $d_1 = d_2$ and $(x_1, x_2) \in S$, then $(x_2, x_1) \in S$. We say, that a solution fulfills symmetry property, if for symmetric (S^H, d) problem, $f_1^N(S^H, d) = f_2^N(S^H, d)$.

(A4) Independence of equivalent utility representation
Let L be a affine mapping, i.e. such that $Lx = (a_1 x_1 + b_1, a_2 x_2 + b_2)$ for any $x \in R^2$, where $a_i, b_i \in R, a_i > 0, i = 1, 2$. We say, that a solution is independent of equivalent utility representation, if $Lf^N(S^H, d) = f^N(LS^H, Ld)$.

(A5) Independence of irrelevant alternatives
Let (S^H, d) and (T^H, d) be bargaining problems such that $S^H \subset T^H$ and $f^N(T^H, d) \in S^H$. Then $f^N(S^H, d) = f^N(T^H, d)$.
The last axiom means that if the decision makers have agreed the solution $f^N(T^H, d)$ in the bargaining problem (T^H, d), then decreasing of the agreement set T^H to a set S^H which includes the solution, i.e. $f^N(T^H, d) \in S^H$, should not change the final payoffs of the decision makers.

It can be shown that for any bargaining problem (S^H, d) with the nonempty, compact and convex agreement set, there exists one and only one solution $f^N(S^H, d)$ of the form:

$$f_N(S^H, d) = \arg\max_{y \in S^H} \|y_1 - d_1\| \cdot \|y_2 - d_2\|,$$

satisfying the axioms A1 - A5.

$\|.\|$ is a distance measured on hyperplane H^2.

The axioms A1 - A5 can be treated as fair play rules satisfied by the mediation proposal constructed according to the Nash solution concept. The axiom A1 assures efficiency of the solution in the set S. The solution is individually rational according to the axiom A2. The axiom A3 means that the decision makers are treated in the same way. The axiom A4 prevent possible manipulation of the decision makers. Any decision maker will not benefit by changing scales measuring his payoffs.

The system derives the point $f_N(S^H, d)$ in the set S according to the above formula. The tentative mediation proposal is calculated improving the current status quo in the direction of the point, but the improvement is limited by the confidence coefficient. In the particular rounds of the mediation procedure the decision makers repeat unilateral multicriteria analysis. Each one can correct the preferred direction improving his payoffs, having more information about the form of the agreement set and knowing the tentative mediation proposal. Finally a sequence of mediation proposals is generated. It can be shown that the sequence converges to the Pareto optimal point in the agreement set and the payoffs defined by the point have the properties analogical to the properties of the classic Nash solution.

6   Conclusions

The iterative mediation procedure has been presented supporting decision analysis of two decision makers negotiating conditions of possible cooperation in realization of a join enterprize. Each of them valuates effects of the cooperation by his own different set of criteria. The multicriteria bargaining problem is formulated in the space being the cartesian product of the criteria spaces of the decision makers. The procedure consists of a sequence of rounds. In each round each decision maker makes multicriteria analysis looking independently for the preferred variants of cooperation. The preferred variants indicated by them are used to generate mediation proposals. The mediation proposals are derived on the basis of the proposed solution concept. This solution concept generalizes the classical Nash cooperative solution concept on the multicriteria case. The mediation proposals express the preferences of the decision makers but also satisfy the respective fair play rules. The rules are described by the axioms which are analogical as in the case of the Nash solution. It can be proved that the sequence of the mediation proposals leads to the Pareto optimal point in the agreement set. The procedure can be implemented in a computer-based system playing the role of a mediator in negotiations.

Different solution concepts to the multicriteria bargaining problem can be applied in the procedure to derive the mediation proposals. In particular the solutions based on the ideas of the Raiffa-Kalai-Smorodinsky and Lexicographic solution concepts have been proposed for the multicriteria bargaining problems and analyzed in the papers (Kruś and Bronisz [8], Kruś [7], [6], [5], [4]). The proposals relating to the Nash solution concept presented in this paper complete the previous results.

References

1. Fisher, R., Ury, W.: Getting to Yes. Hougton Mifflin, Boston (1981)
2. Imai, H.: Individual Monotonicity and Lexicographical Maxmin Solution. Econometrica **51**, 389–401 (1983)
3. Kalai, E., Smorodinsky, M.: Other Solutions to Nash's Bargaining Problem. Econometrica **43**, 513–518 (1975)
4. Kruś, L.: Computer-based support in multicriteria bargaining with the use of the generalized raiffa solution concept. In: Angelov, P., et al. (eds.) Intelligent Systems 2014. AISC, vol. 322, pp. 117–128. Springer, Heidelberg (2015)
5. Kruś, L.: Multicriteria cooperative decisions, methods of computer-based support (in Polish: wielokryterialne decyzje kooperacyjne, metody wspomagania komputerowego). Badania systemowe, vol. 70, p. 248 Warsaw, Systems Research Institute, Polish Academy of Sciences (2011)
6. Kruś, L.: Multicriteria decision support in bargaining, a problem of players's manipulations. In: Trzaskalik, T., Michnik, J. (eds.) Multiple Objective and Goal Programming Recent Developments, pp. 143–160. Physica verlag, Heidelberg (2001)
7. Kruś, L.: Multicriteria Decision Support in Negotiations. Control and Cybernetics **25**(6), 1245–1260 (1996)
8. Kruś, L., Bronisz, P.: Some new results in interactive approach to multicriteria bargaining. In: Wierzbicki, A.P., et al. (eds.) User Oriented Methodology and Techniques of Decision Analysis. Lecture Notes in Economics and Mathematical Systems, vol. 397, pp. 21–34. Springer, Berlin (1993)
9. Moulin, H.: Axioms of Cooperative Decision Making. Cambridge University Press, Cambridge (1988)
10. Nash, J.F.: The Bargaining Problem. Econometrica **18**, 155–162 (1950)
11. Nash, J.F.: Two-Person Cooperative Games. Econometrica **21**, 129–140 (1953)
12. Peters, H.: Bargaining Game Theory. Ph.D. Thesis, Catholic University of Nijmegen, The Nederlands (1986)
13. Raiffa, H.: Arbitration Schemes for Generalized Two-Person Games. Annals of Mathematics Studies (28), 361–387. Princeton (1953)
14. Raiffa, H.: The Art and Science of Negotiations. Harvard Univ. Press, Cambridge (1982)
15. Roth, A.E.: Axiomatic Model of Bargaining. Lecture Notes in Economics and Mathematical Systems, vol. 170. Springer, Berlin (1979)
16. Thomson, W.: Two Characterization of the Raiffa Solution. Economic Letters **6**, 225–231 (1980)
17. Wierzbicki, A.P.: On the Completeness and Constructiveness of Parametric Characterizations to Vector Optimization Problems. OR Spectrum **8**, 73–87 (1986)
18. Wierzbicki, A.P., Makowski, M., Wessels, J.: Model-based Decision Support Methodology with Environmental Applications. Kluwer Academic Press, Dordrecht (2000)

Performance Evaluation of Multi-UAV Cooperative Mission Planning Models

Cristian Ramirez-Atencia[1]([⊠]), Gema Bello-Orgaz[1], Maria D. R-Moreno[2], and David Camacho[1]

[1] Departamento de Ingeniería Informática, Universidad Autónoma de Madrid,
C/Francisco Tomás Y Valiente 11, 28049 Madrid, Spain
cristian.ramirez@inv.uam.es, {gema.bello,david.camacho}@uam.es
http://aida.ii.uam.es/
[2] Departamento de Automática, Universidad de Alcalá,
Carretera Madrid Barcelona, Km 33 600, 28871 Madrid, Spain
mdolores@aut.uah.es

Abstract. The Multi-UAV Cooperative Mission Planning Problem (MCMPP) is a complex problem which can be represented with a lower or higher level of complexity. In this paper we present a MCMPP which is modelled as a Constraint Satisfaction Problem (CSP) with 5 increasing levels of complexity. Each level adds additional variables and constraints to the problem. Using previous models, we solve the problem using a Branch and Bound search designed to minimize the fuel consumption and number of UAVs employed in the mission, and the results show how runtime increases as the level of complexity increases in most cases, as expected, but there are some cases where the opposite happens.

Keywords: Unmanned aircraft systems · Mission planning · Constraint satisfaction problems · Branch and bound

1 Introduction

Unmanned Aircraft Systems (UAS) have been used in many potential applications including monitoring coastal frontiers, road traffic, disaster management, agriculture, etc [6]. Nowadays, Unmanned Air Vehicles(UAVs) are controlled remotely from Ground Control Stations(GCSs) by humans operators who use legacy mission planning systems.

Mission planning for UAS deals with the actions that the vehicle must perform (loading/dropping a load, taking videos/pictures, etc.), typically over a time period. These planning problems can be solved using different methods such as Mixed-Integer Linear Programming (MILP) [14], Simulated Annealing (SA) [3], Auction Algorithms (AA) [8], etc. Usually, these methods are the best way to find the optimal solutions but, as the number of restrictions increases, the complexity grows exponentially because it is a NP-hard problem. Therefore, when the complexity of these models increases, it is quite usual to employ Heuristic Methods [16].

© Springer International Publishing Switzerland 2015
M. Núñez et al. (Eds.): ICCCI 2015, Part II, LNCS 9330, pp. 203–212, 2015.
DOI: 10.1007/978-3-319-24306-1_20

In the literature there are some attempts to implement UAS that achieve mission planning and decision making using Temporal Action Logic (TAL) for reasoning about actions and changes [4], Markov Decision Process (MDP) and dynamic programming algorithms [5], or hybrid Partial-Order Forward-Chaining (POFC) [7], among others. Other modern approaches formulate the mission planning problem as a Constraint Satisfaction Problem (CSP) [2], where the tactic mission is modelled and solved using constraint satisfaction techniques.

This work deals with multiple UAVs that must perform one or more tasks in different locations. The solution plans obtained should fulfill all the constraints given by the different components and capabilities of the UAVs involved. In previous works [11,12] a simple approach of the Multi-UAV Cooperative Mission Planning Problem (MCMPP) was modelled as a Constraint Satisfaction Optimization Problem (CSOP) along with and optimization function designed to minimize the fuel cost, the flight time and the number of UAVs needed, and solved with Branch & Bound (B&B). In this new work, we will redefine this model using 5 different levels of complexity to analyse the computational performance of these models when the complexity increases.

The rest of the paper is structured as follows: Section 2 shows basic concepts and definitions about CSPs. Section 3 describes how a UAV Mission is defined. Section 4 shows how the MCMPP is modelled as a CSP and the different levels of complexity to deal with. Section 5 explains the experiments carried out and the results obtained. Finally, the last section presents the final analysis and conclusions of this work.

2   Some Basis in Constraint Satisfaction Problems

Any CSP can be defined as a tuple $< V, D, C >$ where:

- A set of variables $V = v_1, ?, v_n$.
- For each variable, a finite set D_i (its domain) of possible values.
- And a set of constraints (C) restricting the values that variables can simultaneously take.

There are many studied methods to search for all the solutions in a CSP problem, such as Backtracking (BT), Backjumping (BJ) or Forward Checking (FC). These algorithms are usually combined with other techniques like consistency techniques (Node Consistency (NC), Arc Consistency (AC) or Path Consistency (PC)) to modify the CSP ensuring its local consistency conditions.

In most of cases, we have a goal test function that returns a numerical value (how good the solution is), and this special CSP in which we try to optimize the solutions is referred as CSOP. The most widely used method for finding optimal solutions is B&B.

A Temporal Constraint Satisfaction Problem (TCSP) is a particular class of CSP where variables represent times (time points, time intervals or durations) and constraints represent sets of allowed temporal relations between them [15]. Different classes of constraints are characterized by the underlying set of

Basic Temporal Relations (BTR). Most types of TCSPs can be represented with Point Algebra (PA), with BTR = $\{\varnothing, <, =, >, \leq, \geq, \neq, ?\}$.

In the related literature, Mouhoub [9] proved that on real-time or Maximal Temporal Constraint Satisfaction Problems(MTCSPs), the best methods for solving them are Min-Conflicts-Random-Walk (MCRW) in the case of under-constrained and middle-constrained problems, and Tabu Search (TS) and Steepest-Descent-Random-Walk (SDRW) in the over-constrained case. He also developed a temporal model, TemPro [10], which is based on interval algebra, to translate an application involving temporal information into a CSP.

3   Modelling a UAV Mission Planning

The MCMPP can be defined as a number n of tasks to accomplish for a team of m UAVs. There exists different kind of tasks, such as exploring a specific area or search for an object in a zone. These tasks can be carried out thanks to different sensors available by the UAVs performing the mission. In this approach, we consider several types of sensors: Electro-optical or Infra-red (EO/IR) cameras, Synthetic Aperture Radar (SAR), Signal Intelligence (SIGINT), etc. In addition, each task must be performed in a specific geographic *area* and in a specific *time interval*. This time interval could be specified with a start and end time for the task, or just with the duration of the task, so the start and end of the task must be computed in the planning.

On the other hand, the vehicles performing the mission has some features that must be taken into account in order to check if a mission plan is correct: their **initial positions**, their initial fuels, their **available sensors**, and one or more **flight profiles**. A vehicle's flight profile specifies at each moment its **speed**, its **fuel consumption ratio** and its **altitude** (unless it is a climb or descend profile).

When a task is assigned a vehicle, it is necessary to compute the **departure** time when the vehicle starts moving to the task. In addition, it is necessary to compute the **duration of the path** between the departure of the UAV and the start of the task. Finally, if a task is the last of the tasks assigned to a vehicle, then we must compute the **duration of the return** from this last task to the base (the initial position of the UAV).

In order to compute these durations, it is necessary to know which of the flight profiles (if there are more than one) of the UAV will be used. A UAV usually has, at least, three flight profiles: a **climb profile**, a **descend profile** and a **route profile**. Moreover, given that the flight profile provides the fuel consumption ratio, it will be possible to compute the fuel consumptions both in the path and the return. On the other hand, to compute the fuel consumption in the performance of the task, we must know the flight profile required by the sensor used in the task performance.

When start and end time of tasks are fixed, it is possible that the time when a vehicle finishes a task does not meet the time when the vehicle departs for the next task. In this case, we define this interval of time as the **loiter duration** for the second task. In this situation, the UAV must fly using the **minimum cost flight profile**, so the fuel consumption is minimum.

Finally, a mission could have some time and vehicle dependencies between different tasks. **Vehicle dependencies** consider if two tasks must be assigned the **same UAV** or **different UAVs**. Moreover, we consider **time dependencies** given by **Allen's Interval Algebra** [1].

4   Proposed CSP Model

Given the assumptions described in Section 3, we can consider several sets of variables in this kind of problem:

- Assignments (*assign*) of tasks to UAVs of the MCMPP.
- Orders (*order*), which define the order in which each UAV performs the tasks assigned to it. These variables are necessary when start and end times of tasks are not fixed.
- Path Flight Profiles (*fpPath*). These variables set the flight profile that the vehicle must take for the path performance. It is necessary to consider these variables just in the case that there are several route profiles.
- Return Flight Profiles (*fpReturn*), similar to the previous set of variables but for the return performance of each UAV.
- Task Flight Profile (*fpTask*). These variables set the flight profile that the vehicle must take to complete the task, which is required by the sensors used. It is necessary to consider these variables just in the case that the vehicle performing the task has several sensors that could perform that task.

In addition, it is necessary to define some extra variables that are computed in the propagation phase of the CSP solver. These variables are the time points (*departure*, *start* and *end*) and durations (*durPath*, *durTask*, *durLoiter*, *durReturn*) of the mission, as well as fuel consumptions (*fuelPath*, *fuelTask*, *fuelLoiter*, *fuelReturn*);

Now, we define the constraints of the CSP according to the level of complexity used. In this work, we have defined 5 levels of complexity, each one with a higher level of constraints.

4.1   CSP Model-1

This first level only uses sensor constraints, which are used to check if a UAV carries the required sensors to perform a task. Let $sensors(u)$ denote the sensors available for UAV u, and $sensors(t)$ the sensors that could perform the task t. The constraint to check it can be modelled as follows:

$$assign[i] = u \Rightarrow |sensors(t) \cap sensors(u)| > 0 \qquad (1)$$

In this level, the only variables that are considered are the assignments.

4.2   CSP Model-2

Here, we add order and temporal constraints that will check the time consistency of the model, transforming it into a TCSP. Consequently, this level will consider the order variables, and also all the secondary variables related to the time points and durations. We consider that each UAV has just one flight profile that will be used for the duration computations.

First of all, it is necessary to assure that the value of the order variable is less than the number of tasks assigned to the UAV performing that task. This constraint is modelled as Equation 2 shows:

$$assign[t] = u \Rightarrow order[t] < \sharp\{t \in T | assign[t] = u\} \tag{2}$$

and if two tasks are assigned the same UAV, they have different order values:

$$assign[i] = assign[j] \Rightarrow order[i] \neq order[j] \tag{3}$$

Then, it is necessary to assure that the start time of the task equals the sum of the departure time and the duration for the path:

$$departure[t] + durPath[t] = start[t] \tag{4}$$

and that end time is the sum of start time and the duration of the task:

$$start[t] + durTask[t] = end[t] \tag{5}$$

If tasks have fixed starts and end times, then it is necessary to compute the duration of the loiter as the difference between the end of a task and the departure for its consecutive task:

$$assign[i] = assign[j] \land order[i] = order[j] - 1 \Rightarrow durLoiter[j] = departure[j] - end[i] \tag{6}$$

Then, when two tasks are assigned to the same UAV, given their orders, the end time of the first task must be less or equal than the departure of the second:

$$assign[i] = assign[j] \land order[i] < order[j] \Rightarrow end[i] \leq departure[j] \tag{7}$$

Now, to compute the durations of the paths to the tasks, it is necessary to compute the distances from the UAV to the task $(d_{u \to t})$, between each pair of consecutive tasks $(d_{i \to j})$ and from the last task performed by a UAV to its initial position $(d_{t \to u})$. Given that these points are represented in geodesic coordinates, the distance between two points is computed using the Haversine formula with the latitude and longitude of the points.

With this and the speed v_u given by the flight profile of the vehicle, we can compute the duration of the path for the first task t performed by the vehicle u as $durPath[t] = \frac{d_{u \to t}}{v_u}$. On the other hand, for each pair of consecutive tasks i and j assigned to the same UAV, the duration of their path is computed as $durPath[j] = \frac{d_{i \to j}}{v_u}$. Finally, we compute the return duration for a UAV u, given its last planned task t, as $durReturn[u] = \frac{d_{t \to u}}{v_u}$.

4.3 CSP Model-3

In this level, we will add the fuel constraints, which check the fuel consumption for each UAV. The variables in this level remain the same as in Level 2.

First of all, in order to compute the fuel consumption in the performance of each task, it is necessary to know the flight profile required by the sensors to perform the task. For this, we will consider that all sensors require the same flight profile, which give us the speed $\overline{v_t}$ and the fuel consumption rate $\overline{fuelRate_t}$ for the performance of a task t.

With this, we can compute the fuel consumed by the UAV performing the task as $fuelTask[t] = durTask[t] \times \overline{v_t} \times \overline{fuelRate_t}$

On the other hand, the fuel consumed in the path for the first task t performed by a vehicle u is computed as $fuelPath[t] = d_{u \rightarrow t} \times fuelRate_u$. Moreover, the fuel consumed in the path between two consecutive tasks i and j is computed as $fuelPath[j] = d_{i \rightarrow j} \times fuelRate_u$.

Then, the fuel consumption for the loiter of a task, given that each vehicle has just one flight profile, is $fuelLoiter[t] = durLoiter[t] \times v_u \times fuelRate_u$.

Finally, we compute the return fuel consumption given the last task t performed by a UAV u as $fuelReturn[u] = d_{t \rightarrow u} \times fuelRate_u$.

With all this, we just have to sum all these fuel consumption values of a UAV and constraint it to be less than its initial fuel $fuel_u$:

$$\sum_{\substack{t \in T \\ assign[t]=u}} (fuelPath[t] + fuelTask[t]) + fuelReturn[u] < fuel_u \qquad (8)$$

4.4 CSP Model-4

In this level, we will consider that each UAV has four flight profiles: a climb profile, a descend profile, a minimum consumption profile and a maximum speed profile. In addition, each of the sensors will have a specific required flight profile, and not necessarily the same as assumed previously.

With this, all the duration and fuel consumption computations from previous levels must be recomputed considering the values given by the flight profiles selected in variables $fpPath$, $fpTask$ and $fpReturn$. On the computation of loiter fuel consumption, we must use the minimum consumption flight profile.

Following this, for the first task t performed by a UAV u, we check if its path profile must be climb or descend according to its increase or decrease of altitude. The altitude of the task is given by its task profile, so we have that:

$$\begin{cases} altitude(fpTask[t]) - altitude(u) > 0 \Leftrightarrow fpPath[t] = u.CLIMB \\ altitude(fpTask[t]) - altitude(u) < 0 \Leftrightarrow fpPath[t] = u.DESCEND \end{cases} \qquad (9)$$

On the other hand, for each pair of consecutive tasks i and j, a similar situation must be accomplished:

$$\begin{cases} altitude(fpTask[j]) - altitude(fpTask[i]) > 0 \Leftrightarrow fpPath[j] = u.CLIMB \\ altitude(fpTask[j]) - altitude(fpTask[i]) < 0 \Leftrightarrow fpPath[j] = u.DESCEND \end{cases}$$

$$(10)$$

On the other hand, for the return flight profile, given the last task t assigned to vehicle u, we must check if the return profile of u is climb or descend:

$$\begin{cases} altitude(u) - altitude(fpTask[t]) > 0 \Leftrightarrow fpReturn[u] = u.CLIMB \\ altitude(u) - altitude(fpTask[t]) < 0 \Leftrightarrow fpReturn[u] = u.DESCEND \end{cases}$$

$$(11)$$

4.5   CSP Model-5

Finally, in this level, we will add all the time and vehicle dependencies mentioned in Section 3. The time dependency constraints, for each pair of tasks i and j and according to Allen's Interval Algebra, are as follow:

$$i \quad precedes \quad j \Rightarrow \quad end[i] \leq start[j] \tag{12}$$

$$i \quad meets \quad j \Rightarrow end[i] = start[j] \tag{13}$$

$$i \quad overlaps \quad j \Rightarrow \begin{cases} start[i] \leq start[j] \\ end[i] \geq start[j] \\ end[i] \leq end[j] \end{cases} \tag{14}$$

$$i \quad starts \quad j \Rightarrow \begin{cases} start[i] = start[j] \\ end[i] \leq end[j] \end{cases} \tag{15}$$

$$i \quad during \quad j \Rightarrow \begin{cases} start[i] \geq start[j] \\ end[i] \leq end[j] \end{cases} \tag{16}$$

$$i \quad finishes \quad j \Rightarrow \begin{cases} start[i] \geq start[j] \\ end[i] = end[j] \end{cases} \tag{17}$$

$$i \quad equals \quad j \Rightarrow \begin{cases} start[i] = start[j] \\ end[i] = end[j] \end{cases} \tag{18}$$

On the other hand, the vehicle dependencies implies that some task must be performed by the same vehicle or by different vehicles:

$$sameUAV(i,j) \Rightarrow assign[i] = assign[j] \tag{19}$$

$$differentUAV(i,j) \Rightarrow assign[i] \neq assign[j] \tag{20}$$

4.6 Optimization Variables

Now, in order to solve the CSOP with B&B, we must define the optimization
function that this algorithm will use. Following a previous work [12], we will
consider three variables: the number of vehicles, the fuel consumption and the
flight time. To reckon them, first we compute the fuel consumed by each vehicle:

$$fuelConsumed[u] = \sum_{\substack{t \in T \\ assign[t]=u}} (fuelPath[t] + fuelTask[t] + fuelLoiter[t]) + fuelReturn[u]$$

(21)

and the flight time of each vehicle:

$$flightTime[u] = \sum_{\substack{t \in T \\ assign[t]=u}} (durPath[t] + durTask[t] + durLoiter[t]) + durReturn[u]$$

(22)

And then just compute the objective variables as:

- The number of UAVs used in the mission: $N_{uavs} = \sharp\{u \in U | \exists t \in T \quad assign[t] = u\}$.
- The total flight time of the mission: $flightTimeCost = \sum_{u \in U} flightTime[u]$.
- The total fuel consumption of the mission: $fuelCost = \sum_{u \in U} fuelConsumed[u]$.

Now, according to the previous work [12], we will use the function $0.9 \times N_{uavs} + 0.1 \times fuelCost$ with these problems. Nevertheless, in levels 1 and 2 it
is not possible to use this function because $fuelCost$ is not defined. Instead, we
will use N_{uavs} for Level 1, and $0.9 \times N_{uavs} + 0.1 \times flightTimeCost$ for Level 2.

5 Experimental Results

The Mission Scenario used in this experiment consists of 8 tasks and 5 UAVs
scattered throughout the map. Each UAV has different sensors and each task can
be performed by different sets of sensors, so there are several possible solutions.

On the other hand, we have fixed four time dependencies (of type $<$, m, f,
and $=$), and a same UAV vehicle dependency. Each UAV has been set with a
different amount of initial fuel and all departs from ground (altitude 0).

Now, we run B&B with this MCMPP modelled as a CSP using Gecode [13] on
a AMD FX-8350 8-Core 4.00GHz and 8GB RAM. For each level of complexity,
we run the problem twice: first, with start and end times of all tasks fixed, and
then with these unfixed, obtaining the run times shown in Table 1.

From the first three levels of complexity, we can see that adding complexity
implies adding some secondary variables (times and fuel) that are not iterated,
but computed in the propagation phase of the CSP. This is, the propagation
phase takes more time due to the variables that must be computed. On the
other hand, we can see that when times are unfixed, B&B takes much more time
to find the optimal solution, because of the order variables that must be iterated.

Table 1. Runtime of the MCMPP of 8 tasks and 5 UAVs, with each of the 5 levels of complexity and task times fixed and unfixed.

Level of complexity	Fixed Times	Unfixed Times
Level 1	20ms	-
Level 2	830ms	11.34s
Level 3	1.03s	1min 50s
Level 4	55.32s	6h 52min 45s
Level 5	56.05s	4h 29min 23s

Looking now at levels 4 and 5, we can see that they take much more time than the three first levels, because of the flight profile variables to iterate. The most interesting fact here is that, for unfixed times, level 4 spends more time than level 5. This is because adding the dependency constraints makes a little more propagation time but highly reduce the space of solutions, making it easier to the algorithm to found the optimal solution.

6   Conclusions

In this paper, we have presented a CSP model for Multi-UAV Mission Planning. The presented approach defines missions as a set of tasks to be performed by several UAVs with some capabilities. The problem is modelled according to 5 levels of complexity: Level 1 considers sensor constraints, Level 2 adds temporal constraints, Level 3 adds fuel constraints, Level 4 considers flight profiles and Level 5 add temporal and vehicle dependencies.

Using B&B, we have performed an experiment in order to measure the scalability of the problem in the different levels of complexity. We have considered two possible approaches: with fixed times and with unfixed times. Results showed that adding constraints that imply considering new variables always increase the runtime; but constraints that do not add new variables, as the dependency constraint added in Level 5, do not increase the runtime, but may decrease it.

In order to make these results more trustful, in further works we will consider different scenarios and topologies, so a more general conclusion would be obtained. Furthermore, we will use a Multiobjective model, such as the Multi-Objective Evolutionary Algorithm (MOEA), to find the Pareto Optimal Frontier (POF) of the optimization variables, instead of using a percentages function.

Acknowledgments. This work is supported by: Spanish Ministry of Science and Education under project TIN2014-56494-C4-4-P, Junta de Comunidades de Castilla-La Mancha under project PEII-2014-015-A, Comunidad Autónoma de Madrid under project CIBERDINE S2013/ICE-3095 and Savier Project (Airbus Defence & Space, FUAM-076915). The authors would like to acknowledge the support obtained from Airbus Defence & Space, specially from Savier Open Innovation project members: José Insenser, César Castro and Gemma Blasco.

References

1. Allen, J.F.: Maintaining knowledge about temporal intervals. Communications of the ACM, 832–843 (1983)
2. Barták, R.: Constraint programming: in pursuit of the holy grail. In: Week of Doctoral Students, pp. 555–564 (1999)
3. Chiang, W.C., Russell, R.A.: Simulated annealing metaheuristics for the vehicle routing problem with time windows. Annals of Operations Research **63**, 3–27 (1996)
4. Doherty, P., Kvarnström, J., Heintz, F.: A temporal logic-based planning and execution monitoring framework for Unmanned Aircraft Systems. Autonomous Agents and Multi-Agent Systems **19**(3), 332–377 (2009)
5. Fabiani, P., Fuertes, V., Piquereau, A., Mampey, R., Teichteil-Konigsbuch, F.: Autonomous flight and navigation of VTOL UAVs: from autonomy demonstrations to out-of-sight flights. Aerospace Science and Technology **11**(2–3), 183–193 (2007)
6. Kendoul, F.: Survey of advances in guidance, navigation, and control of unmanned rotorcraft systems. J. Field Robot. **29**(2), 315–378 (2012)
7. Kvarnström, J., Doherty, P.: Automated planning for collaborative UAV systems. In: Control Automation Robotics & Vision, pp. 1078–1085, December 2010
8. Leary, S., Deittert, M., Bookless, J.: Constrained UAV mission planning: a comparison of approaches. In: 2011 IEEE International Conference on Computer Vision Workshops (ICCV Workshops), pp. 2002–2009, November 2011
9. Mouhoub, M.: Solving temporal constraints in real time and in a dynamic environment. Tech. Rep. WS-02-17. AAAI (2002)
10. Mouhoub, M.: Reasoning with numeric and symbolic time information. Artif. Intell. Rev. **21**(1), 25–56 (2004)
11. Ramirez-Atencia, C., Bello-Orgaz, G., R-Moreno, M.D., Camacho, D.: A simple CSP-based model for unmanned air vehicle mission planning. In: IEEE International Symposium on INnovations in Intelligent SysTems and Application, pp. 146–153 (2014)
12. Ramírez-Atencia, C., Bello-Orgaz, G., R-Moreno, M.D., Camacho, D.: Branching to find feasible solutions in unmanned air vehicle mission planning. In: Corchado, E., Lozano, J.A., Quintián, H., Yin, H. (eds.) IDEAL 2014. LNCS, vol. 8669, pp. 286–294. Springer, Heidelberg (2014)
13. Schulte, C., Tack, G., Lagerkvist, M.Z.: Modeling and Programming with Gecode (2010). http://www.gecode.org/
14. Schumacher, C., Chandler, P., Pachter, M., Pachter, L.: UAV Task Assignment with Timing Constraints via Mixed-Integer Linear Programming. Tech. rep, DTIC Document (2004)
15. Schwalb, E., Vila, L.: Temporal constraints: A survey. Constraints **3**(2–3), 129–149 (1998)
16. Tang, L., Zhu, C., Zhang, W., Liu, Z.: Robust mission planning based on nested genetic algorithm. In: 2011 Fourth International Workshop on Advanced Computational Intelligence (IWACI), pp. 45–49, October 2011

A Hybrid Distance-Based and Naive Bayes Online Classifier

Joanna Jędrzejowicz[1]($\boxtimes$) and Piotr Jędrzejowicz[2]

[1] Institute of Informatics, Gdańsk University,
Wita Stwosza 57, 80-952 Gdańsk, Poland
jj@inf.ug.edu.pl
[2] Department of Information Systems, Gdynia Maritime University,
Morska 83, 81-225 Gdynia, Poland
pj@am.gdynia.pl

Abstract. The paper combines distance-based weak classifiers constructed using kernel fuzzy clustering technique with the naive Bayes algorithm. Resulting hybrid online ensemble is validated through computational experiment involving a number of datasets often used for testing data streams mining algorithms.

Keywords: Online learning · Kernel based fuzzy clustering

1 Introduction

According to [10] learning from data streams is a pervasive area of increasing importance. Typically, stream learning algorithms use decision models that are evolving and adapting to changes in the environment generating the data. Algorithms capable of dealing with data streams and a large amount of data, possibly with presence of the concept drift, are usually referred to as online or incremental. Review of such algorithms can be found, for example, in [5] and [14]. In the online approach a classifier is induced from the available training set. However, in contrast to a traditional approach, there is also some adaptation mechanism providing for a classifier evolution after the classification task has been initiated and started. In each subsequent round a class label of the incoming instance is predicted and afterwards information as to whether the prediction was correct or not, becomes available. Based on this information adaptation mechanism may decide to leave a classifier unchanged, or modify it, or induce a new one. Main problem with designing online classification algorithms is a need to find a compromise between memory and time complexity (both should be low) and classification accuracy (which should be high). In this paper a hybrid distance-based and naive Bayes on-line classifier based on kernel clustering is considered. The approach is another variant of the earlier proposed ideas where a family of the online distance-based classifiers based on fuzzy C-means clustering were suggested [7]. Later on the approach was extended by adding an implementation of the rotation forest algorithm [8]. The paper is organized as follows. The next

© Springer International Publishing Switzerland 2015
M. Núñez et al. (Eds.): ICCCI 2015, Part II, LNCS 9330, pp. 213–222, 2015.
DOI: 10.1007/978-3-319-24306-1_21

section describes kernel based fuzzy clustering. In Section 3 the proposed online classification algorithm is discussed. Section 4 presents results of the validating experiment. Final section contains conclusions and suggestions for future research.

2   Kernel Based Fuzzy Clustering

Basic clustering methods are assigned into two groups: crisp and fuzzy. One of the most used fuzzy methods is fuzzy C-means clustering (FCM) [3]. FCM considers each cluster as a fuzzy set with membership function measuring the possibility for each data to belong to a given cluster. The method proved useful overcoming some drawbacks of crisp C-means clustering. Recently, to deal with over-lapping and noisy data kernel methods have been applied to fuzzy clustering ([4], [9], [20]) and the kernelized version of FCM is referred to as kernel-based fuzzy C-means clustering. Graves and Pedrycz [6] give a broad report on kernel methods applied to clustering. The basic idea of kernel-based fuzzy C-means clustering is to transform the input data (feature space) into a higher dimensional kernel space via a non-linear mapping Φ which increases the possibility of linear separability of the patterns in kernel space and then allows for fuzzy C-means clustering in the feature space. As noted in [6], the kernel method uses the fact that product in the kernel space can be expressed by a Mercer kernel K given by

$$K(\mathbf{x}, \mathbf{y}) \equiv \Phi(\mathbf{x})^T \Phi(\mathbf{y}).$$

This allows to replace the computation of distances in the kernel space by a Mercer kernel function, which is known as a 'kernel trick'. Two variants of most common kernel functions are:

- gaussian kernel function $K(\mathbf{x}, \mathbf{y}) = \exp \frac{-dist^2(\mathbf{x}, \mathbf{y})}{\sigma^2}$ for $\sigma^2 > 0$,
- polynomial kernel function: $K(\mathbf{x}, \mathbf{y}) = (\mathbf{x}^T \mathbf{y} + \Theta)^p$ where $\Theta \geq 0$, and p - natural, are parameters.

Following the terminology of [6], two classes of kernel-based fuzzy clustering algorithms are considered: KFCM-F - kernel-based fuzzy C-means clustering with centroids in feature space, and KFCM-K - kernel-based fuzzy C-means clustering with centroids in kernel space. The online algorithm considered in the paper makes use of KFCM-K.

In what follows, it is assumed that $D = \{x_1, \ldots, x_M\} \subset R^N$ is a finite collection of data (so called feature, or attribute vectors) to be clustered into nc clusters. Clusters are characterized by centroids c_i and fuzzy partition matrix $U = (u_{ij})$ of size $nc \times M$, where u_{ik} is the degree of membership of x_k in cluster i. Parameter $m \in N$ is a fixed natural, most often equal 2. The standard constraints for fuzzy clustering are assumed:

$$0 \leq u_{ik} \leq 1, \quad \sum_{i=1}^{nc} u_{ik} = 1 \text{ for each } k, \quad 0 < \sum_{k=1}^{M} u_{ik} < M \text{ for each } i. \quad (1)$$

The partition of data D into clusters is performed in accordance with the maximal value of membership degree:

$$clust(x_k) = \arg max_{1 \leq j \leq nc} u_{jk} \qquad (2)$$

2.1 KFCM-K Algorithm

For KFCM-K clustering it is assumed that prototypes v_i are located in the kernel space and centroids further need to be approximated by an inverse mapping $c_i = \Phi^{-1}(v_i)$ to the feature space. The objective function to be minimized is:

$$J_m = \sum_{i=1}^{nc} \sum_{k=1}^{M} u_{ik}^m \, dist^2(\Phi(x_k), v_i)$$

Optimizing J_m with respect to v_i gives:

$$v_i = \frac{\sum_{k=1}^{M} u_{ik}^m \Phi(x_k)}{\sum_{k=1}^{M} u_{ik}^m}$$

which leads to the formula for partition matrix

$$u_{ik} = \frac{1}{\sum_{j=1}^{nc} \left(\frac{dist(\Phi(x_k), v_i)}{dist(\Phi(x_k), v_j)} \right)^{\frac{2}{m-1}}} \qquad (3)$$

where

$$dist^2(\Phi(x_k), v_i) = K(x_k, x_k) - 2 \frac{\sum_{j=1}^{M} u_{ij}^m K(x_k, x_j)}{\sum_{j=1}^{M} u_{ij}^m} \qquad (4)$$
$$+ \frac{\sum_{j=1}^{M} \sum_{l=1}^{M} u_{ij}^m u_{il}^m K(x_j, x_l)}{\left(\sum_{j=1}^{M} u_{ij}^m \right)^2}$$

Centroids c_i are approximated in the feature space using the minimization of

$$V = \sum_{i=1}^{nc} dist(\Phi(c_i), v_i)$$

which in case of gaussian kernel gives:

$$c_i = \frac{\sum_{k=1}^{M} u_{ik}^m \cdot K(x_k, c_i) \cdot x_k}{\sum_{k=1}^{M} u_{ik}^m \cdot K(x_k, c_i)} \qquad (5)$$

and for the polynomial kernel:

$$c_i = \frac{\sum_{k=1}^{M} u_{ik}^m (x_k^T c_i + \Theta)^{p-1} x_k}{(c_i^T c_i + \Theta)^{p-1} \sum_{j=1}^{M} u_{ij}^m} \qquad (6)$$

The algorithm of kernel based fuzzy C-means clustering is shown as **Algorithm 1**.

Algorithm 1. Kernel-based fuzzy C-means clustering

Input: data D, kernel function K, number of clusters nc,
Output: fuzzy partition U, centroids $\{c_1, \ldots, c_{nc}\}$
 1: initialize U to random fuzzy partition satisfying (1)
 2: **repeat**
 3:   update U according to (3) and (4)
 4: **until** stopping criteria satisfied or maximum number iterations reached
 5: **repeat**
 6:   update centroids according to (5) for gaussian kernel and (6) for polynomial
      kernel
 7: **until** maximum number iterations reached

2.2   Fixing the Number of Clusters

As noted by [21] an additional advantage of using kernel functions is the possibility to determine the number of clusters based on significant eigenvalues of a matrix determined by the kernel function applied to feature vectors. In the sequel

---

Algorithm 2. Number of clusters estimation

Input: data D with M feature vectors, kernel function K, threshold δ
Output: number of clusters nc
 1: let $K_{ij} = K(x_i, x_j)$ be the quadratic matrix of size $M \times M$,
 2: calculate the eigenvalues of matrix (K_{ij})
 3: nc$\leftarrow$ number of eigenvalues exceeding δ
 4: **return** nc

---

it is assumed that kernel-based fuzzy clustering is preceded by **Algorithm 2** to fix the number of clusters.

2.3   Clustering for Classification Using Different Metrics

For the online algorithms considered in the paper two measures are applied to classify unseen feature vectors. The first one is connected with clustering partition. Suppose that the fuzzy algorithm with nc -number of clusters, was applied to a set of feature vectors $X = \bigcup_{c \in C} X^c$. As a result one obtains, for each $c \in C$, centroids set Cnt^c of size nc and using the partition matrix with application of (2), the partition into clusters $X^c = \bigcup_{j=1}^{nc} X_j^c$.

To classify new vector $r \notin X$ the distance from r to any element from cluster X_j^c is calculated. Then the distances are sorted in a non-decreasing order so that l_1^{cj} points to the element nearest to r, l_2^{cj} - to the second nearest etc. The coefficient S_{cj}^x measures the mean distance from x neighbors:

$$S_{cj}^x = \frac{\sum_{i=1}^{x} dist(r, l_i^{cj})}{x} \tag{7}$$

For each class $c \in C$ and each cluster $X_j^c \subset TD^c$ the coefficient S_{cj}^x is calculated. Using the partition, the row r is classified as class c, for which the value (7) is minimal:

$$class(r) = arg \min_{c \in C, j \leq nc} S_{cj}^x \qquad (8)$$

Note that considered algorithms are parametrized by different metrics $dist(x, y)$.

Algorithm 3. Classification based on fuzzy clustering partition

---

Input: partition of learning data $X = \bigcup_{c \in C} \bigcup_{j \leq nc} X_j^c$, data row $r \notin X$, number of neighbours x
Output: class cl for row r
 1: calculate S_{cj}^x for all $c \in C$, $j \leq nc$ according to (7)
 2: $cl = arg \min_{c \in C, j \leq nc} S_{cj}^x$
 3: **return** class cl

Since some of most popular metrics (Euclidean, Manhattan) do not handle well non-continuous input data, a variety of metrics is considered. What is more, benchmark sets often contain both nominal (or, symbolic) and continuous features which makes heterogeneous metrics well suited. The following metrics were used (and compared) in the experiments, see [19]. For s, t being feature vectors of dimension n we have:

- Euclidean metrics: $d_e(s,t) = \sqrt{\sum_{k=1}^{n}(s_k - t_k)^2}$
- Manhattan metrics: $d_m(s,t) = \sum_{k=1}^{n} \mid s_k - t_k \mid$
- Chebychev: $d_{ch}(s,t) = \max_{1 \leq k \leq n} \mid s_k - t_k \mid$
- Camberra: $d_c(s,t) = \sum_{k=1}^{n} \frac{|s_k - t_k|}{|s_k + t_k|}$
- Heterogeneous Euclidean - Overlap: $HEOM(s,t) = \sqrt{\sum_{k=1}^{n} g_k^2(s_k, t_k)}$, where

$$g_k(s_k, t_k) = \begin{cases} overlap(s_k, t_k) & \text{if k is nominal} \\ rn_k(s_k, t_k) & \text{otherwise} \end{cases}$$

$$overlap(s_k, t_k) = \begin{cases} 0 & s_k = t_k \\ 1 & \text{otherwise} \end{cases}$$

$$rn_k(s_k, t_k) = \frac{\mid s_k - t_k \mid}{\max_k - \min_k}$$

  where $\max_k$ ($\min_k$) is the maximal (minimal) value of the attribute k in the training set.
- Heterogeneous Value Difference metric: $HVDM(s,t) = \sqrt{\sum_{k=1}^{n} g_k^2(s_k, t_k)}$, where

$$g_k(s_k, t_k) = \begin{cases} vdm_k(s_k, t_k) & \text{if k is nominal} \\ \frac{|s_k - t_k|}{4 \cdot \sigma_k} & \text{otherwise} \end{cases}$$

vdm_k - the value difference metric is appropriate for values of a nominal attribute k

$$vdm_k(x,y) = \sum_{c=1}^{C} \mid \frac{N_{k,x,c}}{N_{k,x}} - \frac{N_{k,y,c}}{N_{k,y}} \mid$$

where $N_{k,x,c}$ is the number of instances in the training set that have values x for attribute k and class c, $N_{k,x} = \sum_{c=1}^{C} N_{k,x,c}$ and C is the number of classes; for the continuous k attribute σ_k stands for the standard deviation
- Bray Curtis:

$$d_{BC}(s,t) = \sum_{k=1}^{n} \mid s_k - t_k \mid$$

- cosine metrics

$$d_{cos}(s,t) = 1 - \frac{s \cdot t}{\mid s \mid \cdot \mid t \mid} = 1 - \frac{\sum_{k=1}^{n} s_k \cdot t_k}{\sqrt{\sum_{k=1}^{n} s_k^2} \cdot \sqrt{\sum_{k=1}^{n} t_k^2}}$$

- discrete metric $d_d(s,t) = \sum_{k=1}^{n} g(s_k, t_k)$, where

$$g(s_k, t_k) = \begin{cases} 0 \text{ if } (s_k < 0.5) \wedge (t_k < 0.5) \\ \quad \vee (s_k \geq 0.5) \wedge (t_k \geq 0.5) \\ 1 \text{ otherwise} \end{cases}$$

3   Online Algorithm

The online algorithm works in rounds. In each round a chunk of data is classified by an ensemble of classifiers which consists of L (L - parameter) classifiers based on kernel clustering and one naive Bayes classifier. For each of L classifiers a distance metric is fixed and kernel based fuzzy clustering is performed on a fixed chunk of data (learning step). In the next step (classification) for each feature vector r from the next chunk of data, each of l classifiers for $l = 1, \ldots, L$ applies **Algorithm 3** to assign class c_l^r. Naive Bayes classification is applied to obtain class cB^r. Finally, voting with weights applied to $c_1^r, \ldots, c_L^r, cB^r$ is performed. Weights are assigned to classifiers, all are initially equal 1.0 and are modified after all the data from the chunk were classified. The idea is to boost those classifiers that made correct decisions. Let $w(i)$ stand for the weight of the ith classifier, for $i = 1, \ldots, L + 1$. After each chunk all weights are modified as follows:

$$w(i) := w(i) + \frac{\text{number of vectors correctly classified by ith classifier}}{\text{chunk size}} - 0.5 \quad (9)$$

In case when the weight is negative after the modification, it gets value 0. The number of rounds is proportional to the size of data divided by the size of the chunk. The details are in **Algorithm 4**.

Remark. In **Algoritm 1, 2, 3** data set stands for a subset of R^N, where N is the number of features and in **Algorithm 4** data also contains class value, in order to check the classification quality.

Algorithm 4. Online algorithm DB-NB-OL

Input: data D, chunk size ch, list of L metrics
Output: qc - quality of online classification
 1: initialize $correctClsf \leftarrow 0$
 2: initialize weights $w(i) \leftarrow 0$ for $i = 1, \ldots, L + 1$
 3: $dataClust \leftarrow$ first ch rows from D
 4: $dataClassif \leftarrow$ next ch rows from D
 5: **while** not all feature vectors in D considered yet **do**
 6:     **for** $l = 1$ to L **do**
 7:         apply **Algorithm 1** with lth metric to cluster $dataClust$ into the partition X^l
 8:     **end for**
 9:     **for all** feature vector $r \in dataClassif$ **do**
10:         **for** $l = 1$ to L **do**
11:             assign class label c_l^r to r according to **Algorithm 3**, using partition X^l
12:         **end for**
13:         use Bayes naive classifier with training data $dataClust$ to obtain class cB^r
14:         perform voting with weights w on $c_1^r, \ldots, c_L^r, cB^r$ to assign class c to r
15:         **if** c is a correct class label **then**
16:             $correctClsf \leftarrow correctClfs + 1$
17:         **end if**
18:     **end for**
19:     modify weights w according to (9)
20:     $dataClust \leftarrow dataClassif$
21:     $dataClassif \leftarrow$ next ch rows from D
22: **end while**
23: $qc \leftarrow \frac{correctClfs}{|D| - ch}$

---

4 Validating Experiment Results

To validate the proposed approach computational experiment has been carried out. It has involved 13 datasets shown in Table 1.

In Table 2 classification accuracies averaged over 25 runs of the proposed hybrid distance-based and naive Bayes online classifier (denoted DB-NB-OL) are shown. These accuracies are compared with recent literature reported results. In all cases it has been assumed that training sets consisted of 20 instances with known class labels and that further instances have been arriving in chunks of 10 instances for smaller datasets with up to 5000 instances, and in chunks of 20 instances for large datasets, respectively. In all cases the number of considered neighbors x (see equation (7)) has been set to 5.

Results generated by DB-NB-OL are fairly stable with standard deviation for all cases staying in the range between 0 and 1.4%. Computation times depend mostly on number of instances to be classified, size of the chunk and number of attributes. Assuming chunk size of 10 and number of attributes not greater than 30, computation time does not exceed 1 s. per 1000 instances on a standard PC.

Table 1. Benchmark datasets used in the experiment

Dataset	Instances	Attributes	Classes	Dataset	Instances	Attributes	Classes
Banana [11]	5300	16	2	Bank M. [1]	4522	17	2
Breast [1]	263	10	2	Chess [22]	503	9	2
Flare Solar[1]	1066	12	6	Heart [1]	303	14	2
Image [1]	2086	19	2	Poker - Hand [16]	14179	11	10
Luxembourg [22]	1901	32	2	Spam [1]	4601	58	2
Thyroid [1]	7000	21	3	Twonorm [11]	7400	21	2
Waveform [1]	5000	41	2				

Table 2. Classification accuracies

Dataset	Accuracy	
	DB-NB-OL	Recent
Banana	87,8	89,3 (Inc SVM [17] 2013)
Bank M.	88,1	86,9 (LibSVM [18] 2013)
Breast	74,7	72,2 (IncSVM[17] 2013)
Chess	76,8	71,8 (EDDM [22] 2011)
Heart	79,2	83,8 (IncSVM [17] 2013)
Image	95,8	94,21 (SVM, [13] 2012)
Poker-Hand	54,9	52,0 (LWF [12] 2014)
Flare Solar	66,1	64,05 (FPA [17] 2013)
Spam	79,8	85,5 (K-a graph [2] 2013)
Thyroid	92,5	95,8 (LibSVM [18] 2013)
Twonorm	95,4	97,6 (FPA [17] 2013)
Waveform	78,2	86,7 (Inc SVM [17] 2014)
Luxembourg	86,7	85,62 (LCD [15] 2013)

5   Conclusions

The paper contributes through combining known algorithms including distance-based classifiers and naive Bayes approach into an online hybrid classification ensemble, which is stable, effective in terms of time and memory complexity, and offers satisfactory accuracy while carrying-out the classification tasks. Possible improvement in terms of the time-wise efficiency could be expected through transferring computation to a parallel environment which will be the focus of future research.

References

1. Asuncion, A., Newman, D.J.: UCI Machine Learning Repository. University of California, School of Information and Computer Science (2007). http://www.ics. uci.edu/mlearn/MLRepository.html
2. Bertini, J.R., Zhao, L., Lopes, A.: An Incremental Learning Algorithm Based on the K-associated Graph for Non-stationary Data Classification. Information Sciences **246**, 52–68 (2013)
3. Bezdek, J.C.: Pattern Recognition with Fuzzy Objective Function Algorithms. Kluwer Academic Publishers (1981)
4. Chiang, J.H., Hao, P.Y.: A new kernel-based fuzzy clustering approach: support vector clustering with cell growing. IEEE T. Fuzzy Systems **11**(4), 518–527 (2003)
5. Gaber, M.M., Zaslavsky, A., Krishnaswamy, S.: Data stream mining. In: Maimon, O., Rokach, L. (eds.) Data Mining and Knowledge Discovery Handbook, Part 6, pp. 759–787 (2010)
6. Graves, D., Pedrycz, W.: Kernel-based Fuzzy Clustering and Fuzzy clustering: A Comparative Experimental Study. Fuzzy Sets and Systems **161**(4), 522–543 (2010)
7. Jędrzejowicz, J., Jędrzejowicz, P.: Online classifiers based on fuzzy C-means clustering. In: Bădică, C., Nguyen, N.T., Brezovan, M. (eds.) ICCCI 2013. LNCS, vol. 8083, pp. 427–436. Springer, Heidelberg (2013)
8. Jędrzejowicz, J., Jędrzejowicz, P.: A family of the online distance-based classifiers. In: Nguyen, N.T., Attachoo, B., Trawiński, B., Somboonviwat, K. (eds.) ACIIDS 2014, Part II. LNCS, vol. 8398, pp. 177–186. Springer, Heidelberg (2014)
9. Li, Z., Tang, S., Xue, J., Jiang, J.: Modified FCM Clustering Based on Kernel Mapping. Proc. SPIE **4554**, 241–245 (2001)
10. Lopes, N., Ribeiro, B.: Machine Learning for Adaptive Many-core Machines: A Practical Approach, Studies in Big Data 7. Springer International Publishing (2015)
11. Machine Learning Data Set Repository (2013). http://mldata.org/repository/tags/data/IDA_Benchmark_Repository/
12. Mena-Torres, D., Aguilar-Ruiz, J.S.: A Similarity-based Approach for Data Stream Classification. Expert Systems with Applications **41**, 4224–4234 (2014)
13. Moreno-Torres, J.G., Sáez, J.A., Herrera, F.: Study on the Impact of Partition-Induced Dataset Shift on k-Fold Cross-Validation. IEEE Trans. Neural Netw. Learning Syst. **23**(8), 1304–1312 (2012)
14. Pramod, S., Vyas, O.P.: Data Stream Mining: A Review on Windowing Approach. Global Journal of Computer Science and Technology Software & Data Engineering **12**(11), 26–30 (2012)
15. Turkov, P., Krasotkina, O., Mottl, V.: Dynamic programming for bayesian logistic regression learning under concept drift. In: Maji, P., Ghosh, A., Murty, M.N., Ghosh, K., Pal, S.K. (eds.) PReMI 2013. LNCS, vol. 8251, pp. 190–195. Springer, Heidelberg (2013)
16. Waikato (2013). http://moa.cms.waikato.ac.nz/datasets/
17. Wang, L., Ji, H.-B., Jin, Y.: Fuzzy Passive-Aggressive Classification: A Robust and Efficient Algorithm for Online Classification Problems. Information Sciences **220**, 46–63 (2013)

18. Wisaeng, K.: A Comparison of Different Classification Techniques for Bank Direct Marketing. International Journal of Soft Computing and Engineering **3**(4), 116–119 (2013)
19. Wilson, D.R., Martinez, T.R.: Improved Heterogeneous Distance Functions. Journal of Artificial Intell. Research **6**, 1–34 (1997)
20. Zhang, D., Chen, S.: Clustering Incomplete Data Using Kernel-Based Fuzzy C-means Algorithm. Neural Processing Letters **18**(3), 155–162 (2003)
21. Zhang, D., Chen, S.: Fuzzy clustering using kernel method. In: Proc. International Conference on Control and Automation ICCA, Xiamen, China, pp. 162–163 (2002)
22. Žliobaite, I.: Combining Similarity in Time and Space for Training Set Formation under Concept Drift. Intelligent Data Analysis **15**(4), 589–611 (2011)

Cluster-Dependent Feature Selection
for the RBF Networks

Ireneusz Czarnowski[(✉)] and Piotr Jędrzejowicz

Department of Information Systems, Gdynia Maritime University,
Morska 83 81-225, Gdynia, Poland
{irek,pj}@am.gdynia.pl

Abstract. The paper addresses the problem of a radial basis function network initialization with feature section. Main idea of the proposed approach is to achieve the reduction of data dimensionality through feature selection carried-out independently in each hidden unit of the RBFN. To select features we use the so called cluster-dependent feature selection technique. In this paper three different algorithms for determining unique subset of features for each hidden unit are considered. These are RELIEF, Random Forest and Random-based Ensembles. The processes of feature selection and learning are carried-out by program agents working within a specially designed framework which is also described in the paper. The approach is validated experimentally. Classification results of the RBFN with cluster-dependent feature selection are compared with results obtained using RBFNs implementations with some other types of feature selection methods, over several UCI datasets.

Keywords: RBF networks · Feature selection · Multi-agent systems

1 Introduction

Radial basis function networks (RBFNs), introduced by Bromhead and Lowe [6], are a popular type of the feedforward networks. Performance of the RBF network depends on numerous factors. The basic problem when designing an RBFN is to select an appropriate number of radial basis functions, i.e. a number of the hidden units. Deciding on this number results in fixing the number of clusters composed from training data, as well as their centroids. When the size of the network will be too large, the generalization ability of such a large RBF may be poor. Smaller RBF networks may have better generalization ability, but too small will perform poorly. So, determination of the sufficient number of hidden units requires some a priori knowledge. Incorrect input neurons or poorly located RBF centers can induce bias to the fitted network model. Another factor influencing performance of the RBFN involves values of parameters, such as the shape parameter of the radial basis function. Another problem concerns connection weights as well as the method used for learning the input-output mapping. Both have a direct influence on the RBFN performance, and both, traditionally, need to be somehow set.

© Springer International Publishing Switzerland 2015
M. Núñez et al. (Eds.): ICCCI 2015, Part II, LNCS 9330, pp. 223–232, 2015.
DOI: 10.1007/978-3-319-24306-1_22

In general, designing the RBFN involves two stages: initialization and training. At the first stage the RBF parameters, including the number of centroids with their locations and the number of the radial basis functions with their parameters, need to be calculated or somehow induced. At the second stage weights used to form a linear combination of the outputs of the hidden neurons are estimated.

Hence, designing a RBFN is neither simple nor straightforward task and the problem is how to determine key parameters of the RBFN-based classifier. Different approaches for setting the RBF initialization parameters have been proposed in the literature. However, none of the approaches proposed, so far, for RBF network design can be considered as superior and guaranteeing optimal results in terms of the learning error reduction or increased efficiency of the learning process.

Improving the RBF network performance including its generalization ability may be also achieved through data dimensionality reduction (DDR). DDR can be can be based on instance or feature reduction or both. Data dimensionality reduction can reduce the computation burden in case, where, for example, RBF neural network to classify data is constructed. In [7] it was shown that using different feature masks for different radial basis functions can improve RBFs performance. Such masks can be produced applying some clustering techniques for feature selection. In [7] process of feature selection supported by clustering was named cluster-dependent feature selection.

In [7] an agent-based population learning algorithm has been used to select cluster centroids and to determine the independent set of features for each cluster. This paper extends the idea presented in [7]. The main goal of the paper is to validate different approaches to selecting independent set of features for each cluster, when the centroids are selected using the agent-based population learning algorithm. The basic question is whether feature selection carried-out for each cluster independently using selected, well-known approaches, i.e. RELIEF, Random Forest and Random-based Ensembles, can improve the quality of the RBFN-based classifier in comparison to the approach presented in [7]. In this paper the agent-based population learning algorithm has been also used to set output weights of the RBFN.

The paper is organized as follows. A general motivation for the feature selection with a description of selected techniques dedicated for RBFNs is included in Section 2. Section 3 provides some details of data processing procedure within RBFNs. The proposed methodology, based on using the agent-based population learning algorithm, is presented in Section 4. Section 5 provides details on the computational experiment setup and discusses its results. Finally, the last section contains conclusions and suggestions for future research.

2   Data Dimensionality Reduction and Cluster-dependent Feature Selection

Data dimensionality reduction process aims at identifying and eliminating irrelevant and redundant information. DDR carried-out without losing extractable information is considered as an important approach to increasing the effectiveness of the learning process.

In practice the data dimensionality reduction is concerned with selecting informative instances and features in the training dataset which can be achieved, through instances or feature selection, or both simultaneously, or through prototype extraction. The last one can include the process of feature construction, where decrease in the number of features is obtained by creating new features on basis of some transformation of the original feature set [3]. Example technique for feature construction is the Principal Component Analysis (PCA) [19].

In general, the aim of feature selection is to select the best subset of features from the original set in such manner that the performance of classifier built on the set of features is better or at least not worse than a classifier built on the original set. The feature selection approaches can be grouped into tree classes. Two main ones include filter and wrapper methods. The third approach comes under the umbrella of the so-called ensemble methods. The filter methods choose the predefined feature set optimizing some assumed evaluation function like, for example, distance, information level, dependency or consistency. The wrapper methods use the learning algorithm performance as the evaluation function and the classifier error rate is considered as the feature selection criterion. Finally, another type of feature selection method is called embedded feature selection. In this case selection process is carried-out within the classification algorithm itself.

In this paper we focus on class-independent feature selection, class-dependent feature selection and cluster-dependent feature selection. In the class dependent approach a unique subset of features is used for each class, while in case of class independent case we have a uniform set of features for all classes [14], [21].

An example of the class-independent feature selection for RBFN has been investigated in [15]. In [21] a class-dependent feature selection for discriminating all considered classes by selecting heterogeneous subset of features for each class, has been proposed. In [21], it has been shown that the class-dependent feature selection can help to select most distinctive feature sets for classifying different classes and that the class-dependent approaches can improve classification accuracy as compared to class independent feature selection methods.

In [14] a RBF classifier with class-dependent features selection was proposed. For different groups of hidden units corresponding to different classes in the RBF neural network, different feature subsets have been selected as inputs. It should be also noted that RBFNs based on class-dependent feature selection leads to a heterogeneous networks.

Cluster-dependent feature selection is a special case of the class-dependent selection. This type of feature selection was introduced in the previous paper of authors [12]. The proposed approach can help to identify the feature subsets that accurately describe clusters. According to [12], *this improves the interpretability of the induced model, as only relevant features are involved in it, without degrading its descriptive accuracy.*

The main feature of the approach discussed in [7] is that different feature masks are used for different kernel functions. Thus each Gaussian kernel function of the RBF neural network is active for a subset of patterns. This results in obtaining a heterogeneous activation functions. It also was shown, that heterogeneous network uses different type of decisions at each stage, enabling discovery of the most appropriate bias for the data. In [7] the agent-based population learning algorithm was used, at the first step, to selection of cluster centroids, and, at the next step, for determining an independent set of features for each cluster.

3   RBF Network Background

Output of the RBF network is a linear combination of outputs of the hidden units, i.e. a linear combination of the nonlinear radial basis function generating approximation of the unknown function. In case of the classification problems the output value is produced using the sigmoid function with a linear combination of the outputs of the hidden units as an argument.

In a conventional RBF neural network output function has the following form:

$$f(x) = \sum_{i=1}^{M} w_i G_i(x, c_i, p_i) = \sum_{i=1}^{M} w_i G_i(\|x - c_i\|, p_i) ,$$

where M defines the number of hidden neurons, G_i denotes a radial basis function associated with i-th hidden neuron, p_i is a vector of parameters, which may include locations of centroids, dispersion or other parameters describing the radial function, x is an n-dimensional input instance ($x \in \Re^n$), $\{c_i \in \Re^n : i = 1,..., M\}$ are cluster's centers, $\{w_i \in \Re^n : i = 1,..., M\}$ represents the output layer weights and $\|\cdot\|$ is the Euclidean norm.

For a RBFN-based classifier with a class-dependent features the RBF output function has the following form:

$$f(x) = \sum_{i=1}^{K} \sum_{j=1}^{k_i} w_{ij} G_{ij}(\|x \cdot g_i - c_i \cdot g_i\|, p_{ij}) .$$

For a RBF network with cluster-dependent features the RBF output function has the following form:

$$f(x) = \sum_{i=1}^{M} w_i G_i(\|x \cdot g_i - c_i \cdot g_i\|, p_i),$$

where g_i is the feature mask of i-th hidden neuron or class i, K is the number of classes, k_i is the number of hidden units serving i-th class

4   Proposed Methodology

The paper deals with the problem of RBF networks with cluster-dependent features. Hence we aim at finding the optimal structure of the RBF network with respect to producing clusters, with their centroids, and determining the best feature subset for representing each cluster (i.e. for each hidden units). In this paper process of searching for the output weights values is integrated with the process of finding the optimal structure of the RBF network.

Since the RBF neural network designing belongs to the class of computationally difficult combinatorial optimization problems [11] the practical way to solve this task is to apply one of the known metaheuristics. In the earlier paper authors implemented the agent-based framework to solve the RBFN design problem [8]. This paper proposes to extend the above mentioned approach to obtain a more flexible tool for designing RBF networks.

4.1   RBF Network Initialization and Clusters Generation

Under the proposed approach clusters are produced at the first stage of the initialization process. Clusters are generated using the procedure based on the similarity coefficients calculated in accordance with the scheme proposed in [9]. Clusters contain instances with identical similarity coefficient and the number of clusters is determined by value distribution of this coefficient. Thus the clusters are initialized automatically, which also means that the number of radial basis function is initialized automatically. Next from thus obtained clusters of instances centroids are selected. An agent-based algorithm with a dedicated set of agents is used to locate centroids within clusters.

4.2   Implementation of the Agent-based Population Learning Algorithm

The main feature of the proposed agent-based population learning algorithm (PLA) is its ability to select centroids, other important parameters of transfer function and estimate values of output weights of the RBFN in cooperation between agents. Searches performed by each agent and information exchange between agents which is a form of cooperation, are carried-out in parallel.

Main feature of the approach is integration of the initialization and training processes which are carried-out in parallel. The discussed approach has been already studied in [8]. It is based on the agent-based population learning algorithm, originally introduced in [2]. Within this algorithm a set of agents cooperate and exchange information forming an asynchronous team of agents (A-Team – see, for example, [18]).

In the discussed case an A-Team consists of agents which execute various improvement procedures aiming at achieving better value of the fitness function. They cooperate through exchanging feasible solutions of the RBF network designing problem. Most important assumptions behind the approach can be summarized as follows:
- Shared memory of the A-Team is used to store a population of solutions to the RBFN design problem.
- A solution is represented by a string consisting of three parts for representing, respectively, numbers of instances selected as centroids, numbers of features selected for different kernel functions (i.e. chosen to represent the clusters) different for each class or same for all classes and weights of connections between neurons of the network.
- The initial population is generated randomly.
- Initially, numbers representing instances, feature numbers and weights are generated randomly.
- Each solution from the population is evaluated in terms of the RBFN performance and the value of its fitness is calculated. The evaluation is carried-out by estimating classification accuracy or error approximation of the RBFN, which is initialized using prototypes, set of the selected features and set of weights as indicated by the proposed approach.

The RBFN design problem is solved using three groups of optimizing agents:
- agents executing procedures for centroid selection. These involves two local search procedures, one simple random search and another, being an implementation of the tabu search for the prototype selection (these procedures are described in a more detailed manner in the earlier authors' paper [8]),

- agents that are implementation of procedures for feature selection (the implement-ed procedure for feature selection will be discussed in the next section),
- agents with procedures dedicated for estimation of the output weights (the proce-dures has been discussed in [8]).

The pseudo-code of the proposed agent-based population learning approach is shown as Algorithm 1.

Algorithm 1. Agent-based PLA for the RBF networks design

```
Generate clusters using the procedure based on the similarity
coefficient
Generate an initial population of solutions (individuals) and
store them in the common memory
Implement agent-based improvement procedures
Activate optimization agents, which execute improvement
procedures
    while (stopping criterion is not met) do {in parallel}
        Read randomly selected individual from the common memory
        Execute improvement algorithm
        Store individual back in the common memory
    end while
Take the best solution from the population as the result
```

5   Computational Experiment

This section contains results of several computational experiments carried-out with a view to evaluate the performance of the proposed approach. In particular, the reported experiments aimed at evaluating quality of the RBF-based classifiers with cluster-dependent feature selection. The performance of the RBF-based classifiers initialized using class-independent and class-dependent features is also analysed.

Experiments aimed also at answering the question whether the choice of the feature selection method influences performance of the RBF-based classifier. In the reported experiments the following feature selection methods have been used and implemented as agent-executed procedures:

- RELIEF [17], which is a popular algorithm belonging to the filter class methods with feature selection based on ranking of features.
- Random forest (RF) [5], which is an ensemble approach where the CART algo-rithm has been used as a base classifier. Its implementation has been adopted form [4].
- Rotation-based ensemble (RE) [16], which is a feature construction technique, where the idea is to group the features, and for each group to apply an axis rota-tion. The transformation process is based on Principle Component Analysis.

In this paper the above techniques have been implemented for designing RBFN-based classifiers and combined with:

- cluster-dependent feature selection – the combination has been denoted further on as $ABRBF_{CD\ RELIEF}$, $ABRBF_{CD\ RF}$, $ABRBF_{CD\ RE}$, respectively,

- class-dependent feature selection – the combination has been denoted further on as ABRBF$_{CLD\ RELIEF}$, ABRBF$_{CLD\ RF}$, ABRBF$_{CLD\ RE}$, respectively,
- class-independent feature selection - the combination has been denoted further on as ABRBF$_{CLI\ RELIEF}$, ABRBF$_{CLI\ RF}$, ABRBF$_{CLI\ RE}$, respectively.

The discussed approaches have been also compared with two earlier version of the approach, where the agent-based PLA has been applied to RBF initialization and to the class-dependent feature selection - denoted originally as ABRBFN_FS 1 (see: [7]), for the first version, and to the cluster-dependent feature selection - denoted originally as ABRBFN_FS 2 (see: [7]), for the second version. Please also note that in [7] these two version of the algorithm have been trained using the backpropagation algorithm.

Evaluation of the proposed approaches and performance comparisons are based on the classification and the regression problems. In both cases the proposed algorithms have been applied to solve respective problems using several benchmark datasets obtained from the UCI Machine Learning Repository [1]. Basic characteristics of these datasets are shown in Table 1.

Each benchmark problem has been solved 50 times, and the experiment plan involved 10 repetitions of the 10-cross-validation scheme. The reported values of the goal function have been averaged over all runs. The goal function, in case of the classification problems, was the correct classification ratio – accuracy (Acc). The overall goal function for regression problems was the mean squared error (MSE) calculated as the approximation error over the test set.

Parameter settings for computations involving the agent-based population learning algorithm are shown in Table 2. Values of some parameters have been set arbitrarily in the trials and errors procedure. We use the same parameter settings as it has been done in [8].

The Gaussian function has been used for computation within the RBF hidden units. The dispersion of the Radial function has been calculated as a double value of the minimum distance between basis functions [6].

Table 1. Datasets used in the experiment

Dataset	Type of problem	Number of instances	Number of attributes	Number of classes	Reported results
Forest Fires	Regression	517	12	-	-
Housing	Regression	506	14	-	-
WBC	Classification	699	9	2	97.5% [10] (Acc.)
ACredit	Classification	690	15	2	86.9% [10] (Acc.)
GCredit	Classification	999	21		77.47% [13] (Acc.)
Sonar	Classification	208	60	2	97.1% [10] (Acc.)
Diabetes	Classification	768	9	2	77.34% [10] (Acc.)

Table 2. Parameter settings for agent-based algorithm in the reported experiment

Parameter	
Max number of iteration during the search	500
Max number of epoch reached in RBF network training	1000
Population size	60
Probability of mutation for the standard and non-uniform mutation (p_m, p_{mu})	20%
Range values for left and right slope of the transfer function	[-1,1]

The proposed A-Team has been implemented using the middleware environment called JABAT [2]. The RELIEF, RF, RE algorithms have been implemented using the WEKA classes [19].

Table 3 shows mean values of the classification accuracy of the classifiers (Acc) and the mean squared error of the function approximation models (MSE) obtained using the RBFN architecture initialized by the proposed approach. Table 3 also contains results obtained by the other well-known approaches including MLP, Multiple linear regression, SVR/SVM and C4.5

Table 3. RBFN results obtained for different feature selection techniques compared with performance of some different classification models

Problem	Forest fires	Housing	WBC	ACredit	GCredit	Sonar	Diabetes
Algorithm	MSE				Acc. (%)		
ABRBF$_{CD\ RELIEF}$	2.02	36.84	92.34	85.42	68.23	81.5	75.47
ABRBF$_{CD\ RF}$	2.11	37,12	96.2	82.21	**72.45**	81.71	72.52
ABRBF$_{CD\ RE}$	2.01	**34.07**	94.04	84,06	71.9	74.56	74.61
ABRBF$_{CLD\ RELIEF}$	2.25	41.52	92.53	83.42	68.4	82.08	71.84
ABRBF$_{CLD\ RF}$	2.09	39.49	91.4	80.94	69.82	80.46	73.6
ABRBF$_{CLD\ RE}$	1.99	35.11	93.71	84.45	70.47	78.39	73.59
ABRBF$_{CLI\ RELIEF}$	2.02	40,78	90.26	83.8	69.68	76.84	67.42
ABRBF$_{CLI\ RF}$	2,13	40.4	92.84	83.54	68.12	75.42	68.4
ABRBF$_{CLI\ RE}$	2.01	37.45	92,62	84.61	69.42	80.21	75.87
ABRBF_FS 1	2.08	34.17	95.02	83.4	70.08	81.71	**76.8**
ABRBF_FS 2	**1.98**	34.71	96.11	**86.13**	71.31	83.14	75.1
Neural network – MLP	2.11[20]	40.62[20]	96.7[10]	84.6[10]	77.2[10]	**84.5**[10]	70.5[10]
Multiple linear regression	2.38[20]	36.26[20]	-	-	-	-	-
SVR/SVM	1.97[20]	44.91[20]	**96.9**[10]	84.8[10]	-	76.9[10]	75.3[10]
C 4.5	-	-	94.7[10]	85.5[10]	70.5	76.9[10]	73.82[13]

From Table 3, one can observe that cluster-dependent feature selection outperforms other approaches. Second best proved to be RBFNs with class-dependent feature selection. The worst result has been obtained using class-independent model for feature selection which is true independently from algorithm used for feature selection. It can be also observed that the best results have been obtained for Rotation-based ensemble. Rotation-based ensemble approach with the cluster-dependent feature selection model has been superior to other compared methods, i.e. Random Forest and RELIEF. However, the RE approach have not been able to provide better generalization ability as compared to the cluster-dependent feature selection using the agent-based PLA. It can be also observed that the RBFN classifier with cluster-dependent feature selection can be considered as superior to the other methods including MLP, Multiple linear regression, SVR/SVM and C4.5.

6   Conclusions

The paper investigates an influence of feature selection technique on the quality and performance of RBF networks with cluster-dependent feature selection. The agent-based population learning algorithm is used for designing the RBFNs with feature selection. The agent-based RBFN approach has several properties which make the RBFN design more flexible. Among them are: automatically initialized structure of the RBFNs, modification of location of centroids during the training process and possibility of producing a heterogeneous structure of the network. The last one property has been widely studied in this paper and it has been shown that the approach is also flexible with respect to the applying different techniques for feature selection. The agent-based RBFN approach is open to the implementation of various techniques for feature selection including: filter, wrapper or ensemble methods, and is also open for the implementation of feature extraction techniques.

The computational experiment results show that not all feature selection techniques are equally suitable and do not guarantee a similar performance. This fact should be taken into consideration while designing a heterogeneous structure of the network. Nevertheless, the experimental results showed that cluster-dependent feature selection is a promising approach to designing of heterogeneous structure of RBFN. Future research will focus on studying properties of this feature selection model.

References

1. Asuncion, A., Newman, D.J.: UCI Machine Learning Repository. Irvine, CA: University of California, School of Information and Computer Science (2007). http://www.ics.uci.edu/~mlearn/MLRepository.html (accessed June 24, 2009)
2. Barbucha, D., Czarnowski, I., Jędrzejowicz, P., Ratajczak-Ropel, E., Wierzbowska, I.: e-JABAT - an Implementation of the web-based a-team. In: Nguyen, N.T., Jain, I.C. (eds.) Intelligent Agents in the Evolution of Web and Applications. Studies in Computational Intelligence, vol. 167, pp. 57–86. Springer, Heidelberg (2009)

3. Bezdek, J.C., Kuncheva, L.I.: Nearest Prototype Classifier Design: An Experimental Study. International Journal of Intelligence Systems **16**(2), 1445–1473 (2000)
4. Botsch, M., Nossek, J.A.: Construction of interpretable radial basis function classifier based on the random forest kernel. In: Proceedings of IEEE World Congress on Computational Intelligence, IEEE International Joint Conference on Neural Network, Hong Kong, pp. 220–227 (2008)
5. Breiman, L.: Random forests. Machine Learning **45**, 5–32 (2001)
6. Broomhead, D.S., Lowe, D.: Multivariable Functional Interpolation and Adaptive Networks. Complex Systems **2**, 321–355 (1988)
7. Czarnowski, I., Jędrzejowicz, P.: An approach to RBF initialization with feature selection. In: Plamen, P., et al. (eds.) Intelligent Systems 2014. AISC, vol. 322, pp. 671–682. (2015)
8. Czarnowski, I., Jędrzejowicz, P.: Designing RBF Networks Using the Agent-Based Population Learning Algorithm. New Generation Computing **32**(3–4), 331–351 (2014)
9. Czarnowski, I.: Cluster-based Instance Selection for Machine Classification. Knowledge and Information Systems **30**(1), 113–133 (2012)
10. Datasets used for classification: comparison of results. In: directory of data sets. http://www.is.umk.pl/projects/datasets.html. (accessed September 1, 2009)
11. Gao, H., Feng, B.-q., Hou, Y., Zhu, L.: Training RBF neural network with hybrid particle swarm optimization. In: Wang, J., Yi, Z., Żurada, J.M., Lu, B.-L., Yin, H. (eds.) ISNN 2006. LNCS, vol. 3971, pp. 577–583. Springer, Heidelberg (2006)
12. Grozavu, N., Bennani, Y., Lebbah, M.: Cluster-dependent feature selection through a weighted learning paradigm. In: Guillet, F., Ritschard, G., Zighed, D.A., Briand, H. (eds.) Advances in Knowledge Discovery and Management. SCI, vol. 292, pp. 133–147. Springer, Heidelberg (2010)
13. Jędrzejowicz, J., Jędrzejowicz, P.: Cellular GEP-induced classifiers. In: Pan, J.-S., Chen, S.-M., Nguyen, N.T. (eds.) ICCCI 2010, Part I. LNCS, vol. 6421, pp. 343–352. Springer, Heidelberg (2010)
14. Wang, L., Fu, X.: Evolutionary Computation in Data Mining. Studies in Fuzziness and Soft Computing, vol. 163, pp. 79–99 (2005)
15. Novakovic, J.: Wrapper approach for feature selections in RBF network classifier. Theory and Applications of Mathematics & Computer Science **1**(2), 31–41 (2011)
16. Rodriguez, J.J., Maudes, J.M., Alonso, J.C.: Rotation-based ensembles of RBF networks. In: Proceedings of the European Symposium on Artificial Neural Networks, Bruges, Belgium, pp. 605–610 (2006)
17. Sun, Y., Wu, D.: A RELIEF based feature extraction algorithm. In: Proceedings of the SIAM International Conference on Data Mining, pp. 188–195 (2008)
18. Talukdar, S., Baerentzen, L., Gove, A., de Souza, P.: Asynchronous Teams: Co-operation Schemes for Autonomous, Computer-Based Agents. Technical Report EDRC 18-59-96, Carnegie Mellon University, Pittsburgh (1996)
19. Witten, I.H., Merz, C.J.: Data Mining: Practical Machine Learning Tools and Techniques, 2nd edn. Morgan Kaufmann (2005)
20. Zhang, D., Tian, Y., Zhang, P.: Kernel-based nonparametric regression method. In: Proceedings of the IEEE/WIC/ACM International Conference on Web Intelligence and Intelligent Agent Technology, pp. 410–413 (2008)
21. Zhu, W., Dickerson, J.A.: A Novel Class Dependent Feature Selection Methods for Cancer Biomarker Discovery. Computers in Biology and Medicine **47**, 66–75 (2014)

ANST: Special Session on Advanced Networking and Security Technologies

SAME: An Intelligent Anti-malware Extension for Android ART Virtual Machine

Konstantinos Demertzis[(✉)] and Lazaros Iliadis[(✉)]

Democritus University of Thrace, 193 Pandazidou st. 68200, Orestiada, Greece
{kdemertz,liliadis}@fmenr.duth.gr

Abstract. It is well known that cyber criminal gangs are already using advanced and especially intelligent types of Android malware, in order to overcome the out-of-band security measures. This is done in order to broaden and enhance their attacks which mainly target financial and credit foundations and their transactions. It is a fact that most applications used under the Android system are written in Java. The research described herein, proposes the development of an innovative active security system that goes beyond the limits of the existing ones. The developed system acts as an extension on the ART (Android Run Time) Virtual Machine architecture, used by the Android Lolipop 5.0 version. Its main task is the analysis and classification of the Java classes of each application. It is a flexible intelligent system with low requirements in computational resources, named Smart Anti Malware Extension (SAME). It uses the biologically inspired Biogeography-Based Optimizer (BBO) heuristic algorithm for the training of a Multi-Layer Perceptron (MLP) in order to classify the Java classes of an application as benign or malicious. SAME was run in parallel with the Particle Swarm Optimization (PSO), Ant Colony Optimization (ACO) and Genetic Algorithm (GA) and it has shown its validity.

Keywords: Android malware · Java Class File Analysis (JCFA) · ART virtual machine · Multi-Layer Perceptron (MLP) · Biogeography-Based Optimizer (BBO) · Bio-inspired optimization algorithms

1 Introduction

1.1 Android Malware

Advanced generations of Android malware, often appear as legitimate social networking or banking applications. They are even involved in security systems in order to enter mobile devices and steel crucial data like the Two-Factor Authentication (2FA) sent by the banks. In this way the offer the cyber criminals the chance to have access in accounts despite the existence of additional protection level. Also this type of malware is characterized by a series of upgraded characteristics allowing the attackers to change the device control between HTTP and SMS, regardless the availability of Internet connection. They can also be used for the development of portable botnets and for spying on their victims. [1].

M. Núñez et al. (Eds.): ICCCI 2015, Part II, LNCS 9330, pp. 235–245, 2015.
DOI: 10.1007/978-3-319-24306-1_23

1.2   Android Package (APK)

An Android application is usually written in Java and it is compiled by an SDK tool, together with all of the source files in an Android package (.APK file). Such a typical application includes the Android Manifest.xml file and the classes' .dex (a file that contains the full byte code that is interpreted by the Java Virtual Machine (JVM) [2].

1.3   ART JVM

According to the changes in the Android code after the release of the Lolipop 5.0 Android OS (release date 12/11/2014) the ART JVM replaced Dalvik JVM. Dalvik JVM interpreted the Android applications in "machine language" (ML) so that they can be executed by the device processing unit only during the application execution (Just-In-Time (JIT) compiler). On the other hand, ART creates and stores the interpreted ML during the installation of the application. This is done so that it can be available all the time in the operating system (Ahead-Of-Time (AOT) compiler). ART uses the same encoding with Dalvik, by keeping the .dex files as parts of the .APK files, but it replaces the .odex files (optimized .dex files) with the corresponding .ELF ones (Executable and Linkable Format) [2]. A comparison of Dalvik and ART JVM architectures can be seen in the figure 1.

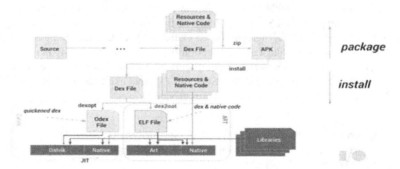

Fig. 1. A comparison of Dalvik and ART JVM architectures (photo by anandtech.com)

1.4   Java Class File Analysis (JCFA)

Generally the architecture of a Java application is described as follows: The source code Java files (.java) are compiled to byte code files (.class) which are platform independent and they can be executed by a JVM just like ART. The classes are organized in the .java files with each file containing at least one public class. The name of the file is identical to the name of the contained public class. The ART loads the classes required to execute the Java program (class loader) and then it verifies the validity of the byte code files before execution (byte code verifier) [3]. The JCFA process includes also the analysis of the classes, methods and specific characteristics included in an application.

1.5   Proposed System

This research paper, introduces advanced Artificial Intelligence (AI) methods, applied on specific parameters and data (obtained after the JCFA process) in order to perform binary classification of the classes comprising an application, in benign or malicious. More specifically, it describes the development of the SAME system, which acts as an extension of the ARTJVM. The SAME employs the Biogeography-Based Optimizer in order to train a MLP which classifies the Java classes of an application successfully in benign or malicious.

It is really important that this is achieved by consuming minimum computational resources. The proposed system enhances the Android operating system with the JCFA process.

In a second stage a comparative analysis with other timely methods like the Particle Swarm Optimization (PSO), Ant Colony Optimization (ACO) and Genetic Algorithm (GA) was performed with encouraging results.

1.6   Literature Review

It is a fact that a lot and significant work has been published in the literature, in applying machine learning (ML) techniques, using features derived from both static [4][5][6] and dynamic [7] analysis to identify malicious Android applications [8][9]. Yerima et al. [10] proposed a parallel machine learning approach for early detection of Android malware by utilizing several classifiers with diverse characteristics. Also, in [11], PUMA (Permission usage to detect malware in Android) detects malicious Android applications through machine-learning techniques by analyzing the extracted permissions from the application itself. Dini et al. [12] proposed a Multi-level Anomaly Detector for Android Malware (MADAM) system in order to monitors Android at the kernel-level and user-level to detect real malware infections using ML techniques to distinguish between standard behaviors and malicious ones.

On the other hand Dan Simon [13] employed the BBO algorithm on a real-world sensor selection problem for aircraft engine health estimation. Panchal et al. [14] proposed a Biogeography based Satellite Image Classification system and Lohokare et al. [15] demonstrated the performance of BBO for block motion estimation in video coding. Ovreiu et al. [16] trained a neuro-fuzzy network for classifying P wave features for the diagnosis of cardiomyopathy. In this work we employ 11 standard datasets to provide a comprehensive test bed for investigating the abilities of BBO in training MLPs. Finally Mirjalili et al. [17] proposed the use of the Bio-geography-Based Optimization (BBO) algorithm for training MLPs to reduce the problems of entrapment in local minima, convergence speed and sensitivity to initialization.

1.7   Innovation of the SAME Project

A basic innovation of the system described herein is the inclusion of a machine learning approach as an extension of the ARTJVM used by the Android OS. This join with the JCFA and the fact that the ARTJVM resolves Ahead-Of-Time all of the dependencies during the loading of classes, introduces Intelligence in compiler level. This fact enhances the defensive capabilities of the system significantly. It is important that the dependencies and the structural elements of an application are checked before its installation enabling the malware cases.

An important innovative part of this research is related to the choice of the independent parameters, which was done after several exhaustive tests, in order to ensure the maximum performance and generalization of the algorithm and the consumption of the minimum resources. For example it is the first time that such a system does not consider as independent parameters, the permissions required by an application for her installation and execution, unlike all existing static or dynamic malware location analysis systems so far.

Finally, it is worth mentioning that the BBO optimization algorithm (popular for engineering cases) is used for the first time to train an Artificial Neural Network (ANN) for a real information security problem.

2   Architecture of the SAME

The architectural design of the SAME introduces an additional functional level inside the ARTJVM, which analyzes the Java classes before their loading and before the execution of the Java program (class loader).

The introduction of the files in the ARTJVM, always passes from the above level, where the check for malicious classes is done. If malicious classes are detected, decisions are done depending on the accuracy of the classification. If the accuracy is high, then the decisions are done automatically, otherwise the actions are imposed by the user regarding the acceptance or rejection of the application installation. In the case that the classes are benign the installation is performed normally and the user is notified that this is a secure application. The proposed architecture is presented in the following figure 2.

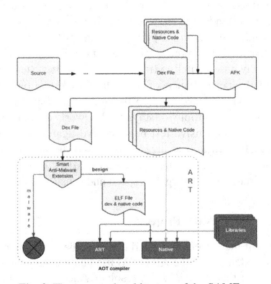

Fig. 2. The proposed architecture of the SAME

Trying a comprehensive analysis of the way the above architecture works, we clearly realize that in the proposed method, the detection and disposal of the malware is done in dead time, before its installation in the cell phone. The relative security systems presented in 1.6, analyze the operation mode of an application, based on characteristics obtained by static and dynamic analysis, by the control method of the Android Permissions and by the anomalies that can emerge in the operation of the cell phone. Of course the above process aims to spot the malware. This innovation enhances the energetic security of the Android mobile phones significantly. It also creates new perspectives in the design architecture of the operating systems, which adopt smart defense mechanisms against sophisticated attacks.

3 Dataset

3.1 Features Extraction

The dataset was constructed after using 2010 Android applications, from which 1160 were benign and were selected from the official Google Play store and 850 were malicious originating from the Drebin dataset [18][19]. The process of feature extraction was done based on the Python language, combined with the JSON technique for representing simple data structures as described in [20].

During the extract out the features process, the extracted features were related to access_flags, statistical class data and finally to methods that determine the structure of the application. The full list of the 24 features, including the class (Benign or Malicious) is presented in the following table 1.

Table 1. Extracted features

ID	Feature Name	ID	Feature Name
1	acc_abstract	13	J_class_name_slash_count
2	acc_annotation	14	J_class_name_uppercase_count
3	acc_enum	15	constant_pool_count
4	acc_final	16	entropy
5	acc_interface	17	majorversion
6	acc_public	18	method_name_digit_count
7	acc_super	19	method_name_lowercase_count
8	acc_synthetic	20	method_name_uppercase_count
9	ap_count	21	methods_count
10	J_class_name_digit_count	22	minor version
11	J_class_name_length_count	23	size
12	J_class_name_lowercase_count	24	class (Benign or Malicious)

3.2 Features Selection

As it has already been mentioned, the basic idea behind the SAME development was the use of the minimum computational resources and computational power, without making any compromise related to the performance. Thus, an effort has been made to determine the features' vector with the well-known Principal Components Analysis (PCA) method. This was done in order for the new linear combinations to contain the biggest part of the variance of the initial information. This method was not selected because the results were not as good as expected. More specifically the mean accuracy of the algorithm was reduced by 5%.

The next step was the use of correlation analysis. In fact, we tested features' vectors that had higher correlation with the class, regardless their mutual correlation and also we tested features that had higher correlation with the class and high correlation between them. Also we tested vectors based on the cost-sensitive classification by using the cost-matrix. Additionally vectors were tested by estimating the value of each feature based on the information gain for each class (Information Gain Attribute Evaluation). Finally the optimal subset was selected based on a genetic algorithm which optimized the classification error for the training and testing data in correlation with the value of each feature [21].

3.3   Proposed Dataset

The number of parameters in the final determined dataset (after the features selection process) was reduced by 52% whereas the reduction in the total average performance of the algorithm (comparing with the performance when 24 parameters were used) was as high as 0.3% and it can be considered as insignificant. The 12 features used in the final dataset are described in the following table 2.

Table 2. Selected Features

ID	Feature Name	Interpretation	
1	acc_abstract	Declared abstract; must not be instantiated.	The value of the
2	acc_public	Declared public; may be accessed from outside its package	access_flags item is a mask of flags used to denote
3	acc_synthetic	Declared synthetic; not present in the source code	access permissions to and properties of this class or inter- face
4	J_class_name_length	The length of the class	
5	J_class_name_digit_ count	The number of the digits of the class	
6	J_class_name_lowercase_count	The number of the lower case of the class	
7	J_class_name_uppercase_count	The number of the uppercase of the class	
8	J_class_name_slash_count	Thenumber of the slash (/) characters in the name of the class	
9	constant_pool_count	The constant_pool is a table of structures representing various string constants, class and interface names, field names, and other constants that are referred to within the Class File structure and its substructures. The value of the constant_pool_count item is equal to the number of entries in the constant_pool table plus one.	
10	methods_count	Each value in the methods table must be a method_info structure giving a complete description of a method in this class or interface. The value of the methods_count item gives the number of method_info structures in the methods table.	
11	entropy	The Entropy value was estimated for each class as degree of uncertainty, with the highest values recorded for the malicious classes.	
12	class	Benign or Malicious	

4   Methods and Materials

4.1   MLP and Heuristic Optimization Methods

A common form of ANN is the MLP with three layers (Input, Hidden and Output). Various heuristic optimization methods have been used to train MLPs such as GA, PSO and ACO algorithms. Generally, there are three heuristic approaches used to train or optimize MLPs. First, heuristic algorithms are used to find a combination of weights and biases which provide the minimum error for a MLP. Second, heuristic algorithms are employed to find the optimal architecture for the network related to a specific case. The last approach is to use an evolutionary algorithm to tune the parameters of a gradient-based learning algorithm, such as the learning rate and momentum [22]. In the first method, the architecture does not change during the learning process. The training algorithm is required to find proper values for all connection weights and biases in order to minimize the overall error of the MLP. The most important parts of MLPs are the connection weights and biases because these parameters define the final values of output. Training an MLP involves finding optimum values for weights and biases in order to achieve desirable outputs from certain given inputs. In this study, the BBO algorithm is applied to an MLP using the first method.

4.2   Biogeography-Based Optimizer

It is a biologically inspired optimization algorithm based on biogeography, which studies the development and evolution of various species in neighboring areas called "islands". Each island is a solution to the problem whereas the species are the parameters of each solution. Each solution is characterized by variables called suitability indices (Suitability Index Variables-SIVs) [13] which define the Habitat Suitability Index-HSI [13]. The areas with the high suitability values offer the best solutions. Each solution becomes optimal by taking characteristics or species from other areas through the immigration mechanism. The algorithm is terminated either after a predefined number of iterations or after achieving a target. An essential difference with the corresponding biologically inspired optimization algorithms is that in the BBO there is no reproduction or offspring concept. There are no new solutions produced from iteration to iteration, but the existing ones are evolving. When there are changes or new solutions the parameters are notified by exchanging characteristics through the immigration [13].

4.3   BBO for Training the Proposed MLP

By introducing a .dex file to the ART the 11 features described in chapter 3.3. are exported, which comprise the independent parameters of the problem, used as independent input to the MLP. In the hidden layer 9 neurons are used, according to the following empirical function 1 [23]:

$$\left(\frac{2}{3}*\text{Inputs}\right) + \text{Outputs} = \left(\frac{2}{3}*11\right) + 2 = 9 \tag{1}$$

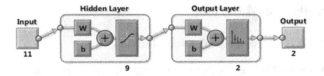

Fig. 3. Architecture of proposed MLP

The output vector of the MLP comprises of the two potential classification values benign or malicious. The architecture of the MLP is shown in the figure 3.

The proposed method starts by generating a random set of MLPs based on the defined number of habitats (BBO employs a number of search agents called habitats which are analogous to chromosomes in GAs.). Each MLP corresponds to a habitat, and each weight/bias corresponds to habitants in the habitat. After the initialization step, the MSE for each MLP is calculated by Eq. (2) [17]:

$$\text{\hspace{3cm}}\tag{2}$$

Where q is the number of training samples, m is the number of outputs,   is the desired output of the ith input unit when the kth training sample is used and   is the actual output of the ith input unit when the kth training sample appears in the input.

The next step is to update emigration, immigration, and mutation rates by Eqs. (3) and (4), respectively [17].

$$\text{\hspace{2cm}}\tag{3},$$

$$\text{\hspace{2cm}}\tag{4}$$

Where n is the current number of habitants, N is the allowed maximum number of habitants, which is increased by HSI (the more suitable the habitat, the higher the number of habitants), E is the maximum emigration rate, and I indicates the maximum immigration rate [17].

Afterwards, the MLPs are combined based on the emigration and immigration rates. Each MLP is then mutated based on its habitat's mutation rate.

The last step of the proposed method is elite selection, in which some of the best MLPs are saved in order to prevent them from being corrupted by the evolutionary and mutation operators in the next generation. These steps (from calculating MSE for every MLP to elitism) are iterated until satisfaction of a termination criterion. The flow chart of the proposed method can be seen in the figure 4.

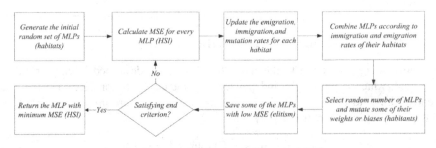

Fig. 4. Flow chart of the proposed method [17]

5   Comparative Analysis and Results

Regarding the overall efficiency of the methods, the results show that the BBO is highly suitable for training MLPs and has much better generalization performance and more accurate classification output from the other compared algorithms (the initial parameters of compared algorithms described in the [17]). The detailed accuracy by class comparison of the other bio-inspired algorithms is shown in table 3. In this case the hold out approach was used (70% Training, 15% Validation and 15% Testing).

Table 3. Comparison between BBO,PSO, ACO and GA algorithms

Classifier	ACC	MSE	TP	FP	Precision	F-measure	ROC	Class
MLP-BBO	96.7%	0.4227	0.955	0.025	0.966	0.960	0.973	Malicious
			0.975	0.045	0.967	0.971	0.973	Benign
MLP-PSO	89.4%	0.5550	0.863	0.083	0.889	0.867	0.886	Malicious
			0.917	0.137	0.898	0.908	0.886	Benign
MLP-ACO	72.1%	0.6738	0.653	0.222	0.716	0.683	0.712	Malicious
			0.778	0.347	0.724	0.750	0.712	Benign
MLP-GA	93.8%	0.4931	0.919	0.047	0.936	0.927	0.937	Malicious
			0.953	0.081	0.940	0.947	0.937	Benign

According to this comparative analysis, it appears that BBO is highly suitable for training MLPs. This algorithm successfully reduces the problem of entrapment in local minima in training MLPs, with very fast convergence rates. These improvements are accompanied by high classification rates and low test errors as well. The reasons for the better performance of BBO are as follows [17]:

- Varying values of emigration and immigration rates provide diverse information exchange behavior between habitats and consequently improve exploration.
- Over the course of generations, the HSI of all habitats are improved since habitants living in high-HSI habitats tend to migrate to the low-HSI habitats. This guarantees the convergence of BBO.
- Migration operators emphasize exploration and consequently prevent BBO from easily getting trapped in local minima.
- Different mutation rates keep habitats as diverse as possible.
- Elitism assists BBO to save and retrieve some of the best solutions, so they are never lost.

The final conclusion is that the proposed method has proven to be reliable and efficient and has outperformed at least for this dataset the other approaches.

6   Discussion – Conclusions

The Android system is very popular and thus is one of the main attack targets of the cyber criminals. This research paper presents an innovative, timely and effective AI system applied successfully in the field of Information systems' security. It is a Smart anti-Malware Extension for Android ARTJVM, which introduces intelligence to the compiler and classifies malicious Java classes in time in order to spot the Android malwares. This is done by applying the JCFA approach and based on the effective BBO optimization algorithm, which is used to train a MLP. The most significant innovation of this methodology is that it uses embedded AI algorithms in compiler of the Android system that ensure mobile security.

Future research could involve its extension under a hybrid scheme, which will combine semi supervised methods and online learning for the trace and exploitation of hidden knowledge between the inhomogeneous data that might emerge. Also, the SAME system could be improved by optimizing further the BBO parameters. Another direction would be the use of the BBO in the training of other ANN models like the RBF NN, Cascade NN. Finally the ultimate challenge would be to test the same model in an ensemble approach.

References

1. https://blog.malwarebytes.org/
2. http://developer.android.com/
3. http://docs.oracle.com/en/java/
4. Scandariato, R., Walden, J.: Predicting vulnerable classes in an android application (2012)
5. Shabtai, A., Fledel, Y., Elovici, Y.: Automated static code analysis for classifying android applications using machine learning. In: Conference on CIS, pp. 329–333. IEEE (2010)
6. Chin, E., Felt, A., Greenwood, K., Wagner, D.: Analyzing inter-application communication in android. In: 9th Conf. on Mobile Systems, Applications, and Services, pp. 239–252. ACM (2011)
7. Burguera, I., Zurutuza, U., Nadjm-Tehrani, S.: Crowdroid: behavior-based malware detection system for android. In: 1st ACM Workshop on SPSM, pp. 15–26. ACM (2011)
8. Glodek, W., Harang, R.R.: Permissions-based detection and analysis of mobile malware using random decision forests. In: IEEE Military Communications Conference (2013)
9. Demertzis, K., Iliadis, L.: Bio-inspired hybrid intelligent method for detecting android malware. In: Proceedings of 9th KICSS 2014, Limassol, Cyprus (2014). ISBN: 978-9963-700-84-4
10. Yerima, Y.S., Sezer, S., Muttik, I.: Android malware detection using parallel machine learning classifiers. In: Proceedings of 8th NGMAST, September 10-14, 2014
11. Sanz, B., Santos, I., Laorden, C., Ugarte-Pedrero, X., Bringas, P.G., Álvarez, G.: PUMA: permission usage to detect malware in android. In: Herrero, Á., Snášel, V., Abraham, A., Zelinka, I., Baruque, B., Quintián, H., Calvo, J.L., Sedano, J., Corchado, E. (eds.) Int. Joint Conf. CISIS'12-ICEUTE'12-SOCO'12. AISC, vol. 189, pp. 289–298. Springer, Heidelberg (2013)
12. Dini, G., Martinelli, F., Saracino, A., Sgandurra, D.: MADAM: a multi-level anomaly detector for android malware. In: Kotenko, I., Skormin, V. (eds.) MMM-ACNS 2012. LNCS, vol. 7531, pp. 240–253. Springer, Heidelberg (2012)
13. Simon, D.: Biogeography-based optimization. IEEE TEC **12**, 702–713 (2008)
14. Panchal, V.K., Singh, P., Kaur, N., Kundra, H.: Biogeography based Satellite Image Classification. Journal of Computer Science and Information Security **6**(2) (2009)

15. Lohokare, M.R., Pattnaik, S.S., Devi, S., Bakwad, K.M., Jadhav, D.G.: Biogeography-Based Optimization techniquefor Block-Based Motion Estimation in Video Coding CSIO 2010 (2010)
16. Ovreiu, M., Simon, D.: Biogeography-based optimization of neuro-fuzzy system parameters for diagnosis of cardiac disease. In: Genetic and Evolutionary Computation Conference (2010)
17. Mirjalili, S., Mirjalili, S.M., Lewis, A.: Let a biogeography-based optimizer train your Multi-Layer Perceptron. Elsevier (2014). http://dx.doi.org/10.1016/j.ins.2014.01.0380020-0255/
18. Arp, D., Spreitzenbarth, M., Huebner, M., Gascon, H., Rieck, K.: Drebin: efficient and explainable detection of android malware in your pocket. In: 21st NDSS, February 2014
19. Spreitzenbarth, M., Echtler, F., Schreck, T., Freling, F.C., Hoffmann, J.: Mobile sandbox: looking deeper into android applications. In: 28th SAC 2013
20. https://github.com/ClickSecurity/data_hacking/java_classification/
21. Hall, M., Eibe, F., Holmes, G., Pfahringer, B., Reutemann, P., Witten, H.I.: The WEKA Data Mining Software: An Update. SIGKDD Explorations **11** (2009)
22. Iliadis, L.: Intelligent Information Systems and applications in risk estimation. Stamoulis Publication, Thessaloniki (2008). ISBN 978-960-6741-33-3 A
23. Heaton, J.: Introduction to Neural Networks with Java, 1st edn. (2008). ISBN: 097732060X

Endpoint Firewall for Local Security Hardening in Academic Research Environment

Ladislav Balik[1], Josef Horalek[1(✉)], Ondrej Hornig[1], Vladimir Sobeslav[1], Rafael Dolezal[1,2], and Kamil Kuca[1,2]

[1] Department of Information Technologies, Faculty of Informatics and Management, University of Hradec Kralove, Hradec Kralove, Czech Republic
{ladislav.balik,josef.horalek,ondrej.hornig, vladimir.sobeslav,rafael.dolezal,kamil.kuca}@uhk.cz
[2] Biomedical Research Center, University Hospital Hradec Kralove, Hradec Kralove, Czech Republic

Abstract. This article presents a security system proposal, providing a low-level endpoint security and network activity monitoring. Its focus is to provide a necessary information for local administrators, who does not necessarily have the knowledge of networking infrastructure or access to it, according to the security policies of a parent organization. This paper presents a system designed for academic research environments, where it serves as a tool for an extended security in protection of sensitive data used in research and development against the local and remote threats.

Keywords: Firewall · Endpoint · Local security · Packet inspection · Iptables · Advfirewall · Java firewall · Research data security

1 Introduction

Security, privacy and data protection are becoming more and more important topics in networking. In the academic environment, where new inventions and cost-extensive experiment outputs are stored, all these security topics are on the top of interest. Every local network perimeter must be secured by at least one border network element, such as router, firewall or IPS. Local policies may be hardened by a distributed software firewall, which protects local end devices from the remote and local attackers in the same time. In general, the use of multiple, highly specialized and single purpose devices can produce far better results than the versatile one.

The environment of academic and business computer network usually differs dramatically. Within one academic organization, there can be multiple teams with the same field of interest. This highly concurrent environment emphasize a need to enforce security and protect privacy, especially when focused on local employees in the network. Furthermore these teams usually operate under one of the security policies managed by parent organization [1].

© Springer International Publishing Switzerland 2015
M. Núñez et al. (Eds.): ICCCI 2015, Part II, LNCS 9330, pp. 246–255, 2015.
DOI: 10.1007/978-3-319-24306-1_24

The concept of network firewall, that is usually deployed and managed by the parent organization, enforces security on the network borders and is not able to monitor, reveal nor stop potential attackers within the local networks. This leaves inner networks vulnerable especially to the man-in-the-middle attacks. Besides these vulnerabilities, security policies of the parent organization often does not address special needs of before mentioned research teams.

Specialized security requirements of the research teams can be addressed by adding another post, where security is enforced. This post should be administered directly by individual research teams. The network environment does not allow to utilize another central point of security enhancement, therefore the only possible placement left are endpoint devices. This approach requires to individually secure every endpoint, such as computers, mainframes, servers, mobile devices, etc. Fig. 1 describes a topology of typical academic environment with local and border firewall implemented.

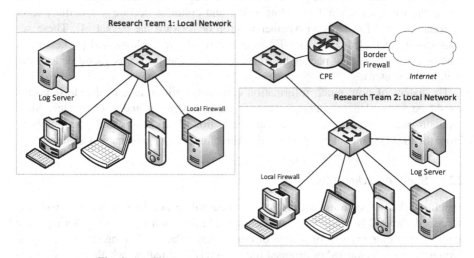

Fig. 1. Academy environment with distinct research groups.

Locally administered firewall with a strictly configured rules for network communication can possibly protect old or obsolete unpatched operating systems and services [2], because it can provide access only for a defined group of users (or devices). Such devices should not occur in any network, but in academic environment there is an occasional need for application of specific version, or an application, which is no more maintained neither updated.

Another benefit of distributed firewalling is a gradual elimination of threats on different levels, which is connected with a distributed computational load on different machines. The load distribution brings a new aspect of backward information distribution, mainly about actions taken to increase security. If a distributed environment is needed, a functional logging and central heartbeat algorithms may be applied [3].

2   State of the Art

Network security problems, arising gradually in time, has to be overtaken by appropriately dimensioned firewall appliance. One solution may not successfully solve all the possible problems. Combination of more firewalling approaches brings together more methods for threats detection, while the complexity of such system increases significantly. Most common types of today's firewalls are described below.

2.1   Packet Filters

Packet filters represents the most common and basic form of a firewall. The Basic firewall principle is to compare each packet header information from every monitored interface to packet forwarding rules [1]. These records are stored in table structure, where the rows represent forwarding rules, and columns declare a specific packet information used for decision whether to drop or forward the packet [4]. These actions can be extended with a modification action. More sophisticated packet filters usually allow to modify packet information before the packet is forwarded (either switched or routed).

The extent of how much information can packet filters use, to decide if packet should be forwarded or dropped depends on its operation and performance. Most common packet filters use source and destination IP addresses and port numbers. From the ISO/OSI model perspective, third and fourth layer information are used.

2.2   Stateful Packet Filters

Stateful packet inspection is an evolution of original packet filtering method, which is described above. The difference between stateful inspection and packet filtering is the ability to monitor firewall or a state of connections. Packets belonging to these connections can be forwarded or dropped based on the information, whether it belongs to a previously established traffic flow. Stateful packet inspection (SPI) deny all new incoming connections from an untrusted zone (or interface) by default. Connections from a trusted zone (e.g. local network) are matched to a firewall rules and only connections belonging to these flows are permitted from an outside network. This mechanism simplifies the rule creation, because it is necessary to define rules only in one direction. Rules for packets from an outside networks are created dynamically [5].

2.3   Advanced Firewalling and Inspection Techniques

Application firewall, sometimes called as proxy firewall, provides similar functionality as a packet filter, but adds additional capabilities in traffic inspection. Application firewall is able to analyze header information from first up to the seventh layer of ISO/OSI reference model. This method inspects largely deployed Internet protocols such as HTTP, FTP, POP3, SMTP and protocols for instant messaging, where application firewall is able to filter traffic based on URI information, email sender address, etc. [1]. This concept of firewall is often implemented as an appliance designed to process one or few specific protocols, therefore it is convenient to pass only the permitted traffic through this firewall. This type of application and its deployment is then often called proxy firewall.

Another logical step in firewall development is to add a possibility of deep packet inspection or packet flow inspection. This approach enables firewall to process not only the header information, but also the data portion of PDU. Inspection of data traffic allows a detection of masked or tunneled communication. Another yet important improvement of such principle is to integrate a methods for attack detection based on signatures of previously discovered intrusions. Signature database matching is usually done on the basis of hash matching, due to easy wire-speed hashing capabilities of specialized hardware [6].

More complex, but also more resource-demanding approach is to inspect those packets that belong together in to the same data stream. Larger caching and better protocol knowledge are necessary. It is common, that network attackers use the sequences of packets that can itself act as legitimate traffic, but together present a security threat to vulnerable systems. Monitoring and analyzing of such packet streams enables to identify even segmented types of attacks [7]. This method is used by IDS, IPS and some advanced hybrid antivirus applications. This approach, as written above, is very source-demanding and its utilization is usually planned on separate hardware-accelerated devices on the border of local network segment.

In modern cloud-oriented solutions, it is possible to use a cloud-based firewall appliance for site protection against remote attackers in addition to endpoint or border firewall solution [8].

Such sophisticated system also brings a new type of risk. As it is made up of many components and modules, which may be manufactured by many different vendors, security risk of single component failure is therefore ameliorated. Critical errors of a single module may significantly affect the whole system and generate new types of vulnerabilities. In this example, one software component failure (multiplied by number of components in every system) may open new security hole for potential attacker. The same problem may be produced by misconfiguration of sophisticated appliance [9]. While such large well known system and its version is detected, exploit sites or tools for exactly that version is used by attacker. Compared to simple stateful packet filter, this improvements increase the security significantly, but mostly at the expense of system overall slow-down.

3   System-Based Packet Analysis

Behavior of the smart firewall appliance is driven by system events. It has to dynamically adapt forwarding rules to new situations, potential attacks and abnormal traffic flows. Multiple incorrect login attempts, port scanning, incomplete or split TCP handshakes are examples of these situations [10]. All these events may indicate some type of a network attack towards a running service. Almost every running service creates some kind of log, whether in a syslog compatible form, or simply as a text file, where every row stores an information about one accident or event.

Self-regulating or learning from data flows through firewall may be considered as smart learning appliance, formerly described in [11]. As shown in Fig. 2 below, self-learning in firewall software may occur at regular intervals based on refresh interval from amount of data captured during firewall function [4].

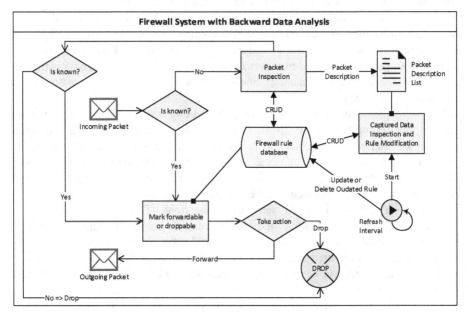

Fig. 2. Self-learning firewall application with backward data analysis.

Two possible approaches of proactive firewall behavior are traffic monitoring that requires to process traffic in near real-time basis or monitor logs of known services running on protected systems in regular intervals. First of these methods is more re-source-demanding and slows down the communication passed through the firewall. Self-repeating high resource-demanding operations can be offloaded to application specific integrated circuits (ASIC), but one of their limitations is dependency on standard structure of analyzed data [2]. Furthermore, its ability to monitor and analyze encrypted traffic (such as encrypted variants of protocols mentioned above – e.g. HTTPS, IPsec, etc.) is rather limited to certain signatures.

Second approach, log monitoring, brings even more accurate information directly from the source of possible problem, service itself respectively. Regular checking intervals in this approach are crucial. Longer intervals reduce hardware requirements, but extends a possible attack window. If too short checking interval is configured, hardware can be utilized more than in online solution as described above. Balanced time schedule must be created and has to be proven on particular secured system, because of expected differences between testing and production environment. Downfall of this method is that has to be considered in general orientation of log messages. By design, they are not intended as a source for network communication analysis [12]. One possible overcome of this downfall is to implement special agent, that directly communicates with the running services and provides firewall with additional network oriented information.

The detection methods are relatively obvious, connected to firewall-update mechanism, they are able to create a new security layer. Attackers, that exceed defined threshold of login attempts or opened connection, may be blocked as in Fail2Ban principle with new permanent or time-limited firewall rule. Definable time of ban or permanent ban may be chosen to discourage a potential attacker. This approach is also useful in today's very common DoS and advanced DDoS attacks [13] and is also efficient against the brute force password cracking [14].

Another detection method is based on heuristic analysis of the amount and type of traffic that is generated by different services. This method is capable to detect different hijacking attacks that are difficult to eliminate by static rule based restrictions. Unfortunately possibilities to pull out such information from logs are rather limited. In this situation, detection based on traffic capture and analysis method can produce dramatically better results [14].

4   Pilot Application

To address the specifics of an environment, in which application should be used, multiplatform open Java standard was chosen for an implementation. The advantage of Java operating system interoperability enables much wider usage. The following application example (Fig. 3) is compiled with the Linux based module, but it is possible to switch this module with another operating system. There is a possibility to use module designed for Microsoft Windows, which utilizes wide spread AdvFirewall functionality (presented with Windows Vista operating system). Standard Linux implementation utilizes iptables functionality. Beside this, libpcap and jNetPCap are used for traffic capturing like in [5].

The developed firewall application uses multithread design with a main thread and in addition, 5 supporting threads were implemented. Every thread has its own isolated responsibility for specific task. This principle is described in various design patterns that support code manageability and interoperability [15].

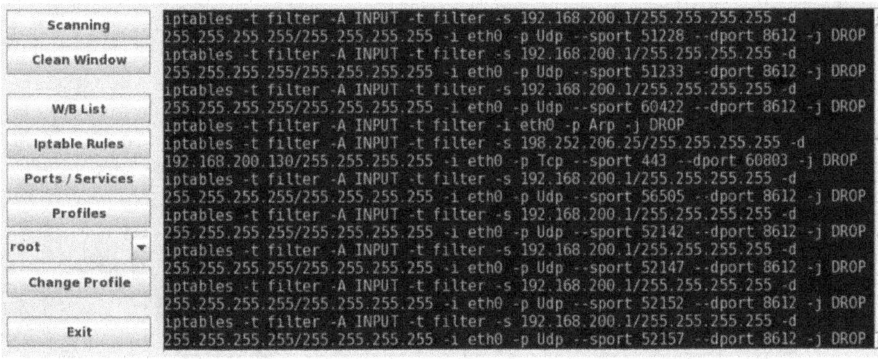

Fig. 3. Log output of packet filtering in application GUI.

4.1   Firewall Application Design

Main program thread provides a simple GUI for user (it is shown on Fig. 4 below). It also ensures communication with operating systems via standard shell and management of files that application use.

Primary thread so-called "Snooper" manage the application focus on oriented behavior thus, is the thread that grants access to data flows that are represented by data packets on communicating interfaces. It is designed to capture and describe both incoming and outgoing packets. Monitored interfaces are one section of the user definable configuration of application, which depends on selected user profile. Captured data are written to specific lists for every chosen interface.

Second, "Enforcement", thread represents a part of application that manages analysis. It is responsible for scanning lists of captured messages after a configured amount of time is elapsed. If needed, this thread performs a deep packet analysis. After the packet is analyzed, this thread may create a new firewall rule based on measured data and predefined action. Furthermore it is automatically installed into chosen firewall which can cause an immediate packet processing change.

Because of detached reading and writing actions within the same list of objects in the same time, the third thread is introduced. Its aim is to switch the buffers used by first and second thread for its functionality. It also manages the user context switching between two different profiles, that every two users, defined in this system, may have. In this application this thread is called "Rotor".

Last two supporting threads are used for communication between instances of the application running on different nodes within local network. First of them ("Sender" thread) is fulfilling the role of application interface for creating and sending update information messages to other computers in the same network segment. The second one ("Updater" thread) is receiving updates from other computers, which are then processed and applied to the configuration of local firewall. Features of this two application threads may not be in use mandatory.

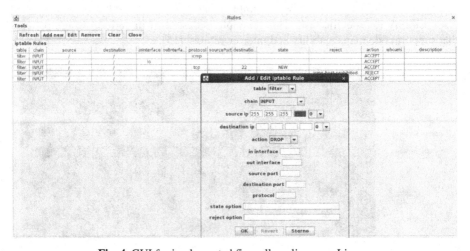

Fig. 4. GUI for implemented firewall appliance on Linux.

4.2   Resulting Application Behavior

From the first look, the thread synchronization was identified as a main goal to achieve a stable firewall application. During implementation and mainly in testing phase of the project it was revealed that the threads mentioned above can work simultaneously, but unsynchronized. This is caused by using Rotor thread, which ensures, that in the same time, there won't be two different threads acceding to the same list. This finding was the main key to simplify application design in one hand and speed up the whole application on the other hand. Due to this fact, this application is a typical example of computer software, which does not enhance user experience, but only slow down the whole system from user perspective.

As mentioned earlier, the goal of application design was to simplify and automate list-based firewall configuration (iptables in Linux) [16]. In the picture above, the list of rules is shown. There is a possibility to manually reconfigure or delete a rule that is configured earlier manually or that is earlier learned via described automated processes.

4.3   Application Performance

The developed firewall software was tested in education classroom environment with standard traffic within the classes. Some piece of information grabbed from network are presented by Fig. 5. CPU usage is negligible, unlike the use of memory, which can be considered as medium-duty and as a possible problem in long-term usage.

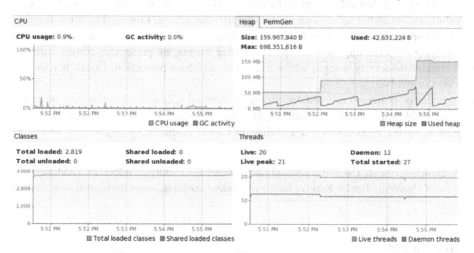

Fig. 5. Resource utilization on Linux machine during testing.

4.4   Future Works

In future works, it is planned to implement central server, thus the point to exchange information about firewall rules. In addition, there is a possibility to delegate the decision power for grey-zone traffic flows to central point which can have more information about flows from other stations. If this, possibly harmful, traffic is recognized on more computer systems, new firewall rule can be added to all nodes by the server in a shorter time than if every single node has to decide based on its own self-cached data.

One of possible future goals is to optimize the memory assignment, its utilization and thus, overall performance of developed application.

Different approach to utilization and processing of firewall rules is presented by so-called Tree-Rule firewall [16]. In connection with the rise of number of rules, brought by long-term use or as a result of synchronization, tree-rule approach is a way to optimize performance. With an implementation of tree-rule based firewall instead of a list-rule, there is also a possibility to extend usability of this system to high speed server nodes, server farms and private cloud environments with a high network utilization [17].

5   Conclusion

This paper presented the endpoint security problematics with a focus on non traditional environment, the academic area. Firewall design addresses multiple specifics of academic researchers end devices and presents a possible mechanisms of protection against the most common and critical threats in such surroundings. It also presents possibly the new approach to elimination of some types of vulnerabilities on outdated systems.

This type of firewall design does not intend to substitute commonly used firewall solutions, but rather focuses on overall security hardening. Proposed solution is suitable for protecting different types of research data, primarily against local attackers and for security hardening of weak points. Better results, as presented, comes with the hierarchical security design, where every part of chain protection. Local, border and cloud is involved.

Security policies enforced by proposed firewall solution are presented in easily transferable form, enabling central rule collection, cooperation of different instances running on different nodes and usability for further research.

Acknowledgement. The authors would like to thank the Faculty of Informatics and Management of University Hradec Kralove for the continuous cooperation and the support over the past years.

References

1. Greenwald, M., Singhal, S.K., Stone, J.R., Cheriton, D.R.: Designing an academic firewall: policy, practice, and experience with SURF. In: Network and Distributed System Security, pp. 79–92, February 22–23, 1996
2. Moon, C.-S., Kim, S.-H.: A study on the integrated security system based real-time network packet deep inspection. International Journal of Security and its Applications **8**, 113–122 (2014)

3. Hamelin, M.: Preventing firewall meltdowns. Network Security **2010**(6), 15–16 (2010). doi:10.1016/S1353-4858(10)70083-0
4. Hamed, H., Al-Shaer, E.: Dynamic rule-ordering optimization for high-speed firewall filtering. In: ASIACCS 2006 Proceedings of the 2006 ACM Symposium on Information, computer and communications security, pp. 332–342 (2006). doi:10.1145/1128817. 1128867
5. Mishra, A., Agrawal, A., Ranjan R.: Artificial Intelligent Firewall. In: ACAI 2011 Proceedings of the International Conference on Advances in Computing and Artificial Intelligence, pp. 204–207 (2011). doi:10.1145/2007052.2007094
6. Dubrawsky, I.: Firewall Evolution - Deep Packet Inspection. Infocus (July 2003)
7. Zhang, R., Qian, D., Ba, C., Wu, W., Guo, X.: Multi-agent based intrusion detection architecture. In: Computer Networks and Mobile Computing, pp. 494–501 (2001)
8. Dilley, J.A., Laghate, P., Summers, J., Devanneaux, T.: Cloud Based Firewall System And Service (2015). US Patent No.: 20150089582
9. Wool, A.: Trends in firewall configuration errors: Measuring the holes in swiss cheese. IEEE Internet Computing **14**(4), 58–65 (2010)
10. Beardsley, T., Qian, J.: The TCP Split Handshake: Practical Effects on Modern Network Equipment. Network Protocols and Algorithms **2** (2010)
11. Butler, D.L., Winne, P.H.: Feedback and Self-Regulated Learning: A Theoretical Synthesis. Review of Educational Research **65**(3), 245–281 (1995). doi:10.3102/003465430 65003245
12. Vaarandi, R.: Simple Event Correlator for real-time security log monitoring. Hakin9 Magazine **1**(6), 28–39 (2006)
13. Manolache, F.B., Hou, Q., Rusu, O.: Analysis and prevention of network password guessing attacks in an enterprise environment. In: RoEduNet Conference 13th Edition: Networking in Education and Research Joint Event RENAM 8th Conference, pp. 1–7, September 11–13, 2014
14. Waldvogel, M., Kollek, J.: SIEGE: Service-Independent Enterprise-GradE protection against password scans (2014)
15. Czajkowski, G., Daynès, L.: Multitasking without compromise: a virtual machine evolution. In: Proceedings of the 16th ACM SIGPLAN Conference on Object-oriented Programming, Systems, Languages, and Applications, vol. 36(11), pp. 125–138 (2001). doi:10.1145/504282.504292
16. He, A., Chomsiri, T., Nanda, P., Tan, Z.: Improving cloud network security using the Tree-Rule firewall. Future Generation Computer Systems **30**, 116–126 (2014)
17. Cimler, R., Matyska, J., Sobeslav, V.: Cloud based solution for mobile healthcare application. In: Proceedings of the 18th International Database Engineering & Applications Symposium on - IDEAS 2014. doi:10.1145/2628194.2628217

Fuzzy Granular Classifier Approach for Spam Detection

Saber Salehi[1], Ali Selamat[1,2($\boxtimes$)], Ondrej Krejcar[3], and Kamil Kuca[3]

[1] UTM-IRDA Digital Media Centre, Universiti Teknologi Malaysia,
81310, Skudai, Johor, Malaysia
aselamat@utm.my
[2] Faculty of Computing, Universiti Teknologi Malaysia, 81310, Skudai, Johor, Malaysia
[3] Faculty of Informatics and Management, Center for Basic and Applied Research,
University of Hradec Kralove, Rokitanskeho 62 500 03, Hradec Kralove, Czech Republic
ondrej@krejcar.org, kamil.kuca@uhk.cz

Abstract. Spam email problem is a major shortcoming of email technology for computer security. In this research, a granular classifier model is proposed to discover hyperboxes in the geometry of information granules for spam detection in three steps. In the first step, the k-means clustering algorithm is applied to find the seed_points to build the granular structure of the spam and non-spam patterns. Moreover, applying the interval analysis through the high homogeneity of the patterns captures the key part of the spam and non-spam classifiers' structure. In the second step, PSO algorithm is hybridized with the k-means to optimize the formalized information granules' performance. The proposed model is evaluated based on the accuracy, misclassification and coverage criteria. Experimental results reveal that the performance of our proposed model is increased through applying Particle Swarm Optimization and fuzzy set.

Keywords: Spam detection · Hyperbox geometry of classifiers · Granular classifier · Membership functions · Particle Swarm Optimization · K-means clustering algorithm

1 Introduction

Email is a powerful communication channel in a human life to improve group communications. This technology has great impact on human life, national development and business in a positive growth [1]. Due to popularity, low cost and fast delivery characteristics of email, this media is prone to misuse. Spammers abuse email's characteristics to send unsolicited messages (spam) to fill users' accounts and waste network bandwidth, memory and server capacity in ISPs [2]. Spam email problem is a major shortcoming of this technology for computer security. Hence, theoretical and applied research on spam filtering becomes fundamental to eliminate spam. In this regard, classification of email is an important and growing task that concluded from the spam email's menace [1, 3]. Several spam filtering techniques have been introduced by using machine learning approaches including SVM [4], GA [5], AIS [1, 6, 7], case based technique [8] and ANN [9] to combat spam messages.

© Springer International Publishing Switzerland 2015
M. Núñez et al. (Eds.): ICCCI 2015, Part II, LNCS 9330, pp. 256–264, 2015.
DOI: 10.1007/978-3-319-24306-1_25

To discover hyperboxes for spam detection, a granular classifier model is proposed in this research as follows: (i) To build the centers of clusters that called seed_points [11] of the hyperboxes in spam and non-spam classes, the k-means clustering algorithm [10] is applied on the emails. In this regards, the interval analysis through the high homogeneity of the patterns is applied to capture the key part of the spam and non-spam classifiers' structure. This paper is structured as follows. The related works for spam detection is presented in section 2. The proposed model to discover hyperboxes in spam and non_spam dataset, and noise points in the geometry of the information granules is described in section 3. In section 4, we evaluate the proposed model for spam detection based on the coverage, misclassification and accuracy criteria. Finally, the conclusion of this paper is presented in section 5.

2   Related Works

In this section, a review of recent studies for spam detection is presented. Idris and Selamat [1] proposed model was hybridized the NSA [13] with PSO. The proposed model was inspired by AIS [14] model with the adaptive nature of NSA to generate random detectors. Then, PSO was applied to enhance the random generation phase of NSA. Santos et al [15] proposed a semantic-aware model by focusing on synonyms [16] among several linguistic phenomena in natural languages [17] for spam detection. They applied their model on the information of body and subject of the email messages to detect the internal semantics of spam messages. The proposed model removes words that add noise to the model and do not prepare any semantic information [18] by the enhanced Topic-based Vector Space Model (eTVSM). Carlos et al. [19] considered the spam detection problem through the anomaly detection systems inspiration. This approach determines the email is spam or non-spam by comparing word frequency features with a dataset. The dataset usually assumed as legitimate emails. If the significant deviation is found from what is considered typical, that email is recognized as a spam. Zhang et al. [20] proposed a hybrid method of the wrapper-based feature selection method and C4.5 decision tree algorithm to reduce the false positive error in email classification. Moreover, different weights were given to the false positive and false negative errors by cost matrix. In addition, out-of sample error was reduced with K-fold cross validation. Finally, they used binary PSO with mutation operator (MBPSO) for the search strategy of the wrapper. DeBarr and Wechsler [21] proposed two alternative forms of random projections as Random Project and Random Boost for spam detection that can be used with a small number of patterns for training. Random Project employed a random projection matrix to produce linear combinations of input features. Moreover, Random Boost used random feature selection through the combination of Logit Boost and Random Forest to enhance the performance of the Logit Boost algorithm.

3   Proposed Model

In this section, the proposed model is introduced for spam detection by creating the hyperboxes in the geometry of information granules. Each step of the proposed model is outlined on the following sentences: (i) find the seed_points to build the granular structure of the spam and non-spam patterns by applying the k-means clustering algorithm in the training phases; (ii) enhance the performance of the found seed_points by hybridizing the k-means and PSO; (iii) apply the fuzzy sets to solve the noise points problem by calculating the membership grades of them. There are some relevant definitions in [11] to construct the granular structure.

3.1   First Step: The Design of the Spam and Non_spam Classifiers

In this section, the procedure of designing the spam and non_spam classifiers by k-means clustering algorithm is presented. Based on the density-based cluster and noise points definitions [11], the seed_points of the unsupervised hyperboxes are built to discover the information granules for creating the clusters and noise points in spam and non_spam dataset. In this step, interval analysis is applied to build the sets as the key part of the classifier's structure on the high homogeneity of the patterns. The significant point in the granular structure is that the density of patterns in the discovered clusters is more than the outside of that clusters based on the predefined parameters as '*MinPoints*' and '*EPS*' [11]. The process of designing the classifiers to establish the decision boundaries and the geometry of the feature space is illustrated on the following paragraph. In the beginning, the spam and non_spam datasets are split in the training phase to generate granular classifiers. One of the capabilities of the granular computing is that each granule plays a role of classifier [11]. Therefore, the computing time is more than the traditional classifiers, specifically for loading big data. In addition, the program consumes a lot of memory and Java Heap Space Error is occurred during the running of the program. In order to solve this problem, the training datasets were split to 8 sections to discover the hyperboxes.

Algorithm: hyperbox generation
Input:
      Tr// Training datasets (70% of total datasets)
Output: hyperboxes

Algorithm:
    Tr $\{tr_1, tr_2, ..., tr_n\}$
    Initial: visited points *vPts*
    For each $tr_i \in Tr$ do
      If1 $tr_i \not\in vPts$ then
        vPts tr_i
        $N_{EPS}(tr_i)$ calculate the EPS-neighborhood of tr_i
            If2 $N_{EPS}(tr_i) \geq$ MinPts then
                hyp   create new hyperbox
                add *hyp* to the cluster
            End If2
      End If1
    Next *tr*

Fig. 1. Hyperbox generation algorithm

Fig. 1 represents the algorithm of hyperbox generation. The hyperboxes is found according to the EPS-neighborhood [11] of each pattern in the training patterns. Therefore, all density-reachable [11] patterns from that pattern are returned to create a new hyperbox 'H_i' with respect to 'EPS' and '$MinPoints$'. In addition, the new cluster 'C' is created for the spam or non-spam class label for the new hyperbox 'H_i' in each class label. Otherwise, the new hyperbox is added to the existing cluster. Each hyperbox belongs to a single class. In the next step, the aforementioned process is also applied on all unvisited points in 'H_i' to create new hyperbox 'H_{i+1}' and add it to cluster 'C'. This process is applied iteratively to collect all directly density-reachable [11] points until there is not any point remains for adding to the current cluster 'C'. Hyperboxes finding procedure is performed on all of the spam and non_spam patterns in training phase separately to discover clusters and noise points. Formula 1 is used to as a distance formula in k-means clustering algorithm [10].

$$dis(x,y) = \sum_{i=1}^{n} |x_i - y_i|^2 \tag{1}$$

where 'x' and 'y' denote the vectors of two datasets and 'x_i' and 'y_i' denote the attribute of them. 'n' presents the number of attributes.

3.2   Second Step: PSO as a Tool of Optimization

In the first step, the hyperboxes of spam and non_spam are created. In this step, the PSO algorithm is applied to enhance the performance of seed_points for classification. In fact, the PSO expands the dimension of hyperbox to cover more points. In this situation, the size of the hyperbox does not change through the PSO algorithm with invariant value of 'EPS' value. The details of expanding the hyperboxes are described in the following subsections.

Seed_Points Initialization: The seed_points of spam and non_spam classes and the training dataset belonging to the aforementioned classes are separated. Then, the PSO algorithm is applied on spam seed_points by spam training dataset in order to expand the size of the hyperboxes by changing the seed_points. This process is also applied on non_spam seed_points among the non_spam training dataset.

Initial State of PSO: There are some parameters as position, velocity, 'p_i' and 'p_g' that must be initialized and updated in the initial state and updating process, respectively, in order to earn all possible solutions in the search space. The position and velocity list are initialized with seed_points and zero value, respectively. Based on the distance formula 1, the 'p_i' (i = 1, 2, ... , n) list is defined as a container of a minimum distance of creating seed_points; in which 'i' and 'n' are denoted the ith particle and the total number of seed_points (population size), respectively. In addition, a minimum value of 'p_i' list is assigned for the 'p_g' value as a global best particle in each iteration.

PSO Procedure: Based on the defined token rate, restricted numbers of attributes are chosen to update the position and velocity in each iteration. 10% of total attributes are defined to select the random attributes for token rate in this process. Furthermore, 500 times and 40% are assigned for iteration number and total rate. Formula 2 represents the update formula of seed_point's velocity.

$$V_i^{t+1} = wV_i^t + c_1 r_1 (p_i + X_i^t) + c_2 r_2 (p_g + X_i^t) \tag{2}$$

where 'i' variable denotes the 'ith' seed_points in the datasets space (swarm). 'r_1' and 'r_2' variables are the random values in velocity updating processes. 'p_i' and 'p_g' are denoted the location of the best problem solution vector found by 'i' and the location of the best particle (seed_point) among all the particles (seed_points) in the population, respectively. 'w', 'c_1' and 'c_2' (user-supplied coefficients parameters) default values was defined as '1.2' for 'w' and '0.5' for 'c_1' and 'c_2' variables. 'w' ensures convergence of the PSO algorithm. Then, the updated velocity (V_i^{t+1}) is added to the current position (X_i^t) (which assumed as an attribute of seed_point) to achieve the updated position (X_i^{t+1}) as follows:

$$X_i^{t+1} = X_i^t + V_i^{t+1} \tag{3}$$

Fitness Evaluation: The fitness value of modified seed_points in each iteration must be calculated after applying the PSO. Then, the achieved value is compared with the fitness value of last seed_point to estimate the improvement of modified seed_point's performance. The process of calculating the fitness value is described on the following: (i) calculate the distance summation of the last seed_point (*laSeed*) and modified seed_point (*moSeed*) with the existing points in the hyperbox as presented in formula 4 and 5, respectively. The maximization of fitness value is achieved through the minimization of G'; So, (ii) formula 6 and 7 present the fitness value of '*laSeed*' and '*moSeed*', respectively. Improvement of '*moSeed*' performance is resulted from the increasing fitness value. In this situation, the '*moSeed*' is substituted with the '*laSeed*'.

$$G = \sum_{i=1}^{n} d(laSeed, x_i) \tag{4}$$

$$G' = \sum_{i=1}^{n} d(moSeed', x_i) \tag{5}$$

$$Fit = \left(\frac{1}{G}\right) \tag{6}$$

$$Fit' = \left(\frac{1}{G'}\right) \tag{7}$$

3.3   Third Step: Computing of Class Membership

A considerable degree of patterns is covered with the structure of granular classifier in the first and second phases. On the other hand, there are some patterns that do not placed in any of the created clusters and known as noise points [11]. The capability of membership function of a fuzzy set was used to improve the geometry of the classifier by identifying the belongingness of noise points. The discovered noise points could be located close or far from the created clusters. The membership degree of noise point, as shown '*noise*', is presented here by 'u_i'. The way of computing the class membership is explained on the following: (i) calculate the distance of '*noise*' and seed_points of each cluster by formula 1; (ii) calculate the average distance of achieved distances in step (i) by formula 8, which 'c' and 'n' indicate the total number of created clusters and seed_points, respectively; (iii) calculate the 'u_i' of '*noise*' in each cluster 'C_i' among the created clusters 'C_j' through the Lagrange multipliers technique by formula 9.

$$avg(noise, C_i) = \frac{dis(noise, S_1) + dis(noise, S_2) + \cdots + dis(noise, S_n)}{n} \qquad (8)$$

$$u_i = \frac{1}{\sum_{j=1}^{c}(avg(noise, C_i) \div avg(noise, C_j))^2} \qquad (9)$$

4   Experimental Studies

In this section, the results of our proposed method are presented and analyzed on testing datasets. The Spam-base dataset [22] is used to evaluate and compare the performance of the proposed approach for spam detection. It contains 4601 instances of spam and non_spam email messages which 39.4% being spam. Each instance is characterized by 57 attributes and labeled with the '0' and '1': in which '0' and '1' consider the email is non-spam or spam, respectively. In the following, the performance evaluation criteria that are used in this study are presented. The proposed model is applied based on 10 runs for new training and testing datasets in each run. The proposed model is implemented using Eclipse Java JDK 8 on a DELL workstation with 3.30 GHz CPU, 3.41 GB memory, where the operating system is WINDOWS 7.

4.1   Evaluation Criteria

Performance of the proposed approach is considered based coverage, misclassification error and accuracy. Table 1 represents the evaluation criteria and their interpretations, which are used in this study: where '$NoCp$' denotes the number of emails covered by proposed model, '$NoMe$' denotes the number of misclassified emails, 'NoE' denotes total number of emails, 'TP' denotes the number of non-spam emails that recognized correctly, 'TN' denotes the number of spam that recognized correctly, 'FN' denotes the number of spam that recognized wrongly, and 'FP' denotes the number of non-spam that recognized wrongly.

Table 1. Performance measures

Performance measure	Formula	Interpretation
Coverage	$Cov = \dfrac{NoCe}{NoE}$	Measures the number of emails cover by the proposed granular classifier model for classification
Misclassification error	$MissErr = \dfrac{NoMe}{NoE}$	Measures the fraction of emails did not cover by the proposed granular classifier model for classification
Accuracy	$Acc = \dfrac{TP + TN}{NoE}$	Measures the effectiveness of the granular classifier model to classify the emails correctly

4.2   Experimental Results

In this part, the result of the proposed model is presented and analyzed to determine its robustness in spam filtering. This model is applied based on 10 runs each is based on new training-testing datasets. The compared models used the same number of runs, where their average results are presented in various '*MinPts*' and '*EPS*' values. As it was mentioned, the proposed granular computing classifier consumes a lot of memory and Java Heap space Error occurs for the Spam-base datasets. Therefore, the training datasets in each run was split to 8 sections to discover the hyperboxes for training in order to solve this problem.

Table 2 and 3 represent the achieved results in various '*MinPts*' and '*EPS*' values based on aforementioned evaluation criteria. In Table 3, the achieved best results are presented in bold. Table 3 shows the best coverage, misclassification error, accuracy in second phase and accuracy in third phase are '94.42', '0.004', '92.55' and '94.62', respectively; in which '*MinPts*' and '*EPS*' values are '3' and '0.7'. Analysis of accuracy results in the first, second and third steps with various '*MinPts*' and '*EPS*' values show that the accuracy results are increased by applying the PSO algorithm in most of the cases. In addition, the accuracy results are increased after applying the fuzzy sets to solve the noise point's problem.

Table 2. Result of the proposed model in the first step for the Spam-base datasets (mean ± standard deviation)

| MinPoints** | EPS* | First phase | | | |
		Hyperbox No	Coverage (%)	Misclassification error(%)	Accuracy
2	0.5	2473.9±14.8446	85.35±1.0415	0.0106±0.0007	82.44±0.0108
2	0.6	2784.2±13.9445	93.39±0.3445	0.0047±0.0002	90.06±0.0064
3	0.5	2378.9±24.9572	83.22±1.466	0.0121±0.001	80.97±0.0154
3	0.6	2714±11.1243	91.43±0.9369	0.0062±0.0006	88.61±0.0082
3	0.7	2888.2±8.9833	94.4±0.5285	0.004±0.0003	92.55±0.0073
4	0.5	2307.9±17.2369	82.81±1.2333	0.0124±0.0008	80.55±0.0094
4	0.6	2673.7±9.7346	89.55±0.6599	0.0075±0.0004	87.89±0.0073
4	0.7	2838.7±15.45194	94.09±0.4877	0.0042±0.0003	92.34±0.0062

Table 3. Result of the proposed model in the second and third step for the Spam-base datasets (mean ± standard deviation)

| MinPoints** | EPS* | Second phase | | | Third phase |
		Coverage (%)	Misclassification error(%)	Accuracy	Accuracy
2	0.5	85.41±1.1096	0.0105±0.0008	82.49±0.0104	87.81±0.0084
2	0.6	93.44±0.3734	0.0047±0.0002	90.07±0.0064	92.41±0.0084
3	0.5	83.23±1.4547	0.0121±0.001	81.02±0.0153	86.85±0.0118
3	0.6	91.4637±0.9612	0.0061±0.0006	88.65±0.0083	91.47±0.0086
3	0.7	**94.42±0.5342**	**0.004±0.0003**	**92.55±0.0071**	**94.62±0.0086**
4	0.5	82.8405±1.2312	0.0124±0.0008	80.62±0.0093	87.15±0.0084
4	0.6	89.56±0.672	0.0075±0.0004	87.9±0.0072	90.99±0.0058
4	0.7	94.1±0.4917	0.0042±0.0003	92.35±0.0061	94.25±0.0076

5   Conclusion

Email technology has profound impacts on human life, growth of business and national development in a positive path. Spam email problem is a major shortcoming of this technology for computer security. In this study, we consider granular classifier model to create the hyperboxes in the geometry of information granules for spam detection in three steps. In the first step, the k-means clustering algorithm is applied to build the centers of clusters that called seed_points of the hyperboxes in spam and non-spam classes. In this step, interval analysis is applied to build the sets as the key part of the classifier's structure on the high homogeneity of the patterns. In the second step, PSO algorithm is hybridized with the k-means to enhance the performance of hyperboxes. In fact, the PSO expands the dimension of hyperbox to cover more points in related classes within the seed_points optimization. In the third step, the noise point belongingness is solved by applying the fuzzy membership grades to cover the universe in classification. We evaluate the proposed model for spam detection based on the coverage, misclassification and accuracy criteria. The achieved results show the performance of the proposed model in the first step is increased by applying PSO and fuzzy set in the second and third steps, respectively.

Acknowledgement. The Universiti Teknologi Malaysia (UTM) and Ministry of Education Malaysia under Research University grants 02G31, and 4F550 are hereby acknowledged for some of the facilities that were utilized during the course of this research work. This work and the contribution were supported by project "SP-103-2015 - Smart Solutions for Ubiquitous Computing Environments" Faculty of Informatics and Management, University of Hradec Kralove, Czech Republic.

References

1. Idris, I., Selamat, A.: Improved email spam detection model with negative selection algorithm and particle swarm optimization. Applied Soft Computing **22**, 11–27 (2014)
2. Salcedo-Campos, F., Díaz-Verdejo, J., García-Teodoro, P.: Segmental parameterisation and statistical modelling of e-mail headers for spam detection. Information Sciences **195**, 45–61 (2012)
3. Méndez, J.R., Reboiro-Jato, M., Díaz, F., Díaz, E., Fdez-Riverola, F.: Grindstone4Spam: An optimization toolkit for boosting e-mail classification. Journal of Systems and Software **85**, 2909–2920 (2012)
4. Bouguila, N., Amayri, O.: A discrete mixture-based kernel for SVMs: application to spam and image categorization. Information Processing & Management **45**, 631–642 (2009)
5. Salehi, S., Selamat, A., Bostanian, M.: Enhanced genetic algorithm for spam detection in email. In: 2011 IEEE 2nd International Conference on Software Engineering and Service Science (ICSESS), pp. 594–597. IEEE (2011)
6. Guzella, T.S., Mota-Santos, T.A., Uchôa, J.Q., Caminhas, W.M.: Identification of SPAM messages using an approach inspired on the immune system. Biosystems **92**, 215–225 (2008)

7. Salehi, S., Selamat, A.: Hybrid simple artificial immune system (SAIS) and particle swarm optimization (PSO) for spam detection. In: 2011 5th Malaysian Conference in Software Engineering (MySEC), pp. 124–129. IEEE (2011)
8. Fdez-Riverola, F., Iglesias, E.L., Díaz, F., Méndez, J.R., Corchado, J.M.: SpamHunting: An instance-based reasoning system for spam labelling and filtering. Decision Support Systems **43**, 722–736 (2007)
9. Özgür, L., Güngör, T., Gürgen, F.: Adaptive anti-spam filtering for agglutinative languages: a special case for Turkish. Pattern Recognition Letters **25**, 1819–1831 (2004)
10. Velmurugan, T.: Performance based analysis between k-Means and Fuzzy C-Means clustering algorithms for connection oriented telecommunication data. Applied Soft Computing **19**, 134–146 (2014)
11. Salehi, S., Selamat, A., Fujita, H.: Systematic mapping study on Granular computing. Knowledge-Based Systems (2015)
12. Pedrycz, W., Park, B.-J., Oh, S.-K.: The design of granular classifiers: A study in the synergy of interval calculus and fuzzy sets in pattern recognition. Pattern Recognition **41**, 3720–3735 (2008)
13. Forrest, S., Perelson, A.S., Allen, L., Cherukuri, R.: Self-nonself discrimination in a computer. In: 2012 IEEE Symposium on Security and Privacy, pp. 202–202. IEEE Computer Society (1994)
14. Oda, T., White, T.: Developing an immunity to spam. In: Cantú-Paz, E., et al. (eds.) GECCO 2003. LNCS, vol. 2723. Springer, Heidelberg (2003)
15. Santos, I., Laorden, C., Sanz, B., Bringas, P.G.: Enhanced topic-based vector space model for semantics-aware spam filtering. Expert Systems with applications **39**, 437–444 (2012)
16. Carnap, R.: Meaning and synonymy in natural languages. Philosophical Studies **6**, 33–47 (1955)
17. Polyvyanyy, A.: Evaluation of a novel information retrieval model: eTVSM. Master's thesis, Hasso Plattner Institut (2007)
18. Salton, G., McGill, M.J.: Introduction to modern information retrieval (1983)
19. Laorden, C., Ugarte-Pedrero, X., Santos, I., Sanz, B., Nieves, J., Bringas, P.G.: Study on the effectiveness of anomaly detection for spam filtering. Information Sciences **277**, 421–444 (2014)
20. Zhang, Y., Wang, S., Phillips, P., Ji, G.: Binary PSO with mutation operator for feature selection using decision tree applied to spam detection. Knowledge-Based Systems **64**, 22–31 (2014)
21. DeBarr, D., Wechsler, H.: Spam detection using random boost. Pattern Recognition Letters **33**, 1237–1244 (2012)
22. Asuncion, A., Newman, D.: UCI machine learning repository (2007)

A Recent Study on the Rough Set Theory in Multi-Criteria Decision Analysis Problems

Masurah Mohamad[1,2], Ali Selamat[1,2(✉)], Ondrej Krejcar[3], and Kamil Kuca[3]

[1] UTM-IRDA Digital Media Center of Excellence,
Universiti Teknologi Malaysia, 81310, Johor, Malaysia
`masur480@perak.uitm.edu.my`, `aselamat@utm.my`
[2] Faculty of Computing, Universiti Teknologi Malaysia, 81310, Johor, Malaysia
[3] Faculty of Informatics and Management, Center for Basic and Applied Research,
University of Hradec Kralove, Rokitanskeho 62 500 03, Hradec Kralove, Czech Republic
`ondrej@krejcar.org`, `kamil.kuca@uhk.cz`

Abstract. Rough set theory (RST) is one of the data mining tools, which have many capabilities such as to minimize the size of an input data and to produce sets of decision rules from a set of data. RST is also one of the great techniques used in dealing with ambiguity and uncertainty of datasets. It was introduced by Z. Pawlak in 1997 and until now, there are many researchers who really make use of its advantages either to make an enhancement of the RST or to apply in various research areas such as in decision analysis, pattern recognition, machine learning, intelligent systems, inductive reasoning, data preprocessing, knowledge discovery, and expert systems. This paper presents a recent study on the elementary concepts of RST and its implementation in the multi-criteria decision analysis (MCDA) problems.

Keywords: Rough Set Theory · Multi-Criteria Decision Analysis · Multi-Criteria Decision Making

1 Introduction

Rough set theory (RST) is a mathematical method which was introduced by Z. Pawlak in 1982 to deal with the ambiguity and uncertainty of data [1], [2]. Later, in 1997, RST had been extended to deal with many application areas especially in decision analysis. RST was commonly employed as a feature selection tool, attribute or knowledge reduction, rule extraction, uncertainty reasoning, granular computing, derivation of decision table in data mining, pattern identification and also in machine learning area [3]–[5]. Furthermore, due to many advantages of RST, researchers recently constructed many experimental works by integrating other theories such as genetic algorithm, neural network, fuzzy set and others with RST [2], [3]. These novelty works had contributed to many areas especially in reasoning, optimization, and classification tasks when imprecision and uncertainty of a data is involved in the decision process [1], [5]–[7]. RST is a beneficial tool compared to other techniques whereby it can simplify a work with quantitative and qualitative attributes in parallel

© Springer International Publishing Switzerland 2015
M. Núñez et al. (Eds.): ICCCI 2015, Part II, LNCS 9330, pp. 265–274, 2015.
DOI: 10.1007/978-3-319-24306-1_26

without any preliminary information required [8]. For instance, when RST is used as a rule extractor in the classification of task, it is not necessary to provide the probability distribution before the data is classified by RST.

As mentioned in [1], RST has contributes to many decision analysis applications including multi-criteria decision analysis (MCDA) problems although it has the ability to produce sets of decision rules based on the predefined criteria during the process of decision-making. Thus, it is one of the most efficient MCDA tools in helping the decision maker in making a good decision. MCDA is a decision-making method which assists any decision makers in making a decision that involves multiple inconsistencies of decision criteria [9]. An MCDA method is also known as an MCDM method which has been widely applied and developed since 1970s [10]. There are six types of problems regarding MCDA framework, namely choice problem, sorting problem, ranking problem, description problem, elimination problem and design problem [8].

Consequently, to appreciate the advantages of RST, this paper presents a recent study on the implementation of RST in MCDA problems. The core structure of this paper is to concisely discuss the desired RST and its application in different MCDA problems. Secondly, the discussion also includes the basic concept of MCDA. Thirdly, this paper will describe the techniques used in solving MCDA problems within the RST scope. The discussion will be followed by the explanation regarding the approaches applied by the decision maker in creating the decision result using RST. Finally, this paper will demonstrate the statistic of RST usage over the domain application area. In addition, a future rough set research direction will be highlighted as the contribution to this study at the end the section of discussion.

2   Rough Set Theory (RST)

This section will discuss the elementary part of RST and the way it is implemented to any application area.

2.1   Basic Terminology of RST

Data is a group of raw facts that built up information. Data is different among each other based on the dimension and complexity [2]. Hence, it is difficult to manage and analyze the data without having good data administration officers, techniques, theories, or tools. One of the beneficial theories that really can help in managing the size and complexity of data is called RST. The philosophies of RST are to minimize the data size and also to deal with uncertainties of the data [3]–[5]. RST will construct a set of rules that provides useful information from uncertain and inconsistent data [11]. The theory behind this powerful mathematical tool is grounded by a classification task. Objects are categorized according to the similar (indiscernible) information which is also called as elementary set. These indiscernible objects are unnoticeable when the information of the objects are viewed. The elementary sets are denoted as crisp (precise) sets when a number of elementary sets are united together, otherwise the set is denoted as rough (imprecise) set [12].

Based on the fundamental concept of RST mentioned above, it shows that rough set will use approximation spaces and searching strategies as the fundamental tools to estimate an uncertain information [5] and also to deal with redundant data. This rough set indiscernibility relation theory has been generated by the knowledge from the objects of interest whereas some objects of the interest can not be discriminated. The theory of approximation is considered accomplished when Boolean reasoning and approximate Boolean reasoning are constructed during the process of searching approximation spaces or in specific assignment, searching for appropriate features, discretization [13], and symbolic value grouping.

2.2   Rough Set in Decision Analysis

RST is an independent method where it partly covers many other theories [5]. The special feature of a rough set is, it does not need any fundamental or additional information in analyzing data. Therefore, it is always preferred and selected as one of the methods which can be implemented in many areas mainly in cognitive sciences and artificial intelligence such as in decision analysis [5],[14], machine learning [13], intelligent systems, inductive reasoning, pattern recognition [15], mereology, image processing, signal analysis, knowledge discovery, and expert systems [2], [7], [16], [17]. In real world implementation, a decision problem is presented using a decision table. To come out with the best decision, all conditions or rules of attributes need to be analyzed. There are two basic consideration highlighted during the process of decision analysis. Firstly, the decision should be described in the context of the environment in which it is applied. Secondly is to provide the instructions in making a decision based on the specific problem area. The instruction is basically referred to the decision rules, which is the result of the derivation table. In order to deal with these two issues, RST will provide its mathematical tools to construct minimum sets of decision rules from the problem domain [1].

2.3   Rough Set in Decision Analysis

RST is an independent method where it partly covers many other theories [5]. The special feature of a rough set is, it does not need any fundamental or additional information in analyzing data. Therefore, it is always preferred and selected as one of the methods which can be implemented in many areas mainly in cognitive sciences and artificial intelligence such as in decision analysis [5],[14], machine learning [13], intelligent systems, inductive reasoning, pattern recognition [15], mereology, image processing, signal analysis, knowledge discovery, and expert systems [2], [7], [16], [17].

3   Decision Analysis

Technically, a decision analysis is defined as a systematic process of analyzing the complexities of decision problems based on the given criteria. The essential aspects of decision analysis that might be applied to all decision problems are: i) an observation is needed to solve some objectives, ii) one of the listed alternatives must be selected,

iii) the combination of significance and alternatives are different, iv) vagueness is always related to the significance of each alternative, and v) the possible significance are not all equally valued [18]. According to Chai and Liu, a decision analysis is a process of providing a recommendation such as actions, alternatives, or candidates to a certain problem domain by analyzing some criteria (attributes, alternatives, actions or variables) [19]. Decision analysis is divided into two main categories, which referred to the type of feature either in preference - ordered, or not. If the type of a feature contains preference-ordered of attributes or called as criteria, it is categorized as MCDA while another category is called as multiple attributes decision analysis (MADA) because it contains attributes that are not preference-ordered. Researchers are always attracted to investigate the certainty and risk problem instead of the ambiguity. The ambiguity problem is difficult to work with rather than certainty and risk problem. More efforts might be used in investigating the problem because there are a lot of uncertainties, imprecise and vagueness data to analyze. Moreover, the process of analyzing uncertain data becomes more complicated when dealing with multiple attributes or multiple criteria or multiple event problems. The existing decision analysis tools are unable to support this problem and gain the expected outcomes. Therefore an ambiguity decision model were proposed and being investigated to solve this uncertain multiple criteria decision problems [20]. On top of that, this paper also takes an initiative to explore the environment of multiple criteria analysis in real world implementation. The following sections will discuss on the elementary of MCDA, the most common techniques used in solving the MCDA problems and the implementation of RST in solving MCDA problems.

3.1   Multi-criteria Decision Analysis (MCDA)

MCDA is a method that assists decision makers to deal with multiple inconsistent criteria when making decisions. It helps the decision maker to construct an aggregation model (preference model) by referring to the preference information provided by the decision maker [21]. Preference information can be divided into two types, direct and indirect. Direct preference information referred to the parameters which values had been specified such as performance and ranking of a student for each subject [14], [21], [22], or performance and financial background of a supplier [9][23]. While an indirect preference referred to the parameters that are specified from decision maker's general judgment and knowledge [24][25].

Decision-maker (DM) preferred indirect preference rather than direct preference. There are several reasons why the direct preference is not preferable. Firstly, this process may let the DM feels uncomfortable during the interview session. Secondly the information acquired from the DM is quite difficult to be defined, and lastly this type of preference requires a lot of effort in reasoning activity to construct the required information [21]. Moreover, direct preference may cause difficulty to create the DM judgment if the size of criteria increases [26]. MCDA helps the DM in making good decisions by dividing or aggregating all the possible evaluations criteria into relevant evaluations criteria. As stated in [27], criteria may always relate to a measurable attribute either in quantitative or qualitative scale. This measurable attribute is used to evaluate the performance of each criterion or condition for a desired problem.

There are several common problems regarding MCDA, which need to be identified before the decision-making process is conducted. The problems are ranking, choice, sorting, description, elimination, design, and criteria analysis [9], [19]. The most preferred methods by all researchers and practitioners are ranking and followed by classification and other category. Normally many researchers and practitioners applied a single decision analysis method in problem-solving to avoid complexity. However, MCDA method has been widely used in solving problems either with uncertain or precise criteria and alternatives [28].

3.2   Multiple Criteria Decision Making (MCDM) Methods

This section will highlight several MCDM methods, which are practically applied in various fields. MCDM methods are divided into 2 divisions. First, the methods that are used for attributes evaluation performance. Second, the methods which are used to combine all available attributes evaluations in order to get a simplified attributes evaluation [27]. Science management and operations research fields [29] have listed eleven most popular MCDM methods. The methods are multi-attribute utility theory (MAUT), analytic hierarchy process (AHP), fuzzy set theory (FST), case-based reasoning (CBR), data envelopment analysis (DEA), simple multi-attribute rating technique (SMART), goal programming (GP), elimination and choice expressing the reality (ELECTRE), preference ranking organization method for enrichment of evaluation (PROMETHEE), simple additive weighting (SAW) and technique for order of preference by similarity to ideal solution (TOPSIS). Among these eleven MCDM methods, only five methods considered the uncertainty and vagueness of data or criteria. The methods are MAUT, FST, ELECTRE, PROMETHEE and TOPSIS. These five methods are compared to DRSA where DRSA also has its own weaknesses as mentioned in [30]. DRSA is specifically and the rough set is generally not capable of handling the ranking problem successfully [31]. Cinelli et al [30] also proved that DRSA is the most flexible and intelligent among all highlighted methods.

Recently, many researchers have proposed an enhancement or hybrid MCDM methods due to the advantages and disadvantages of each method and also to suit with a certain problem [19], [29]. For instance, interval value fuzzy ANP (IVF-ANP) [25] is developed to solve MCDM problems by the determined weights of criteria, while rough AHP is proposed to measure the system performance [32], [33], and to solve the problem of selection. Fuzzy AHP is integrated [34], [35], and rough-neuro fuzzy [36] is developed for pattern classification. The purpose of these hybrid methods is to overcome the limitation of the original proposed method in making decisions. For instance, rough AHP is applied to reduce the bias criteria in decision-making process [32].

3.3   Rough Set Theory in MCDM

As mentioned in previous section, the theory of approximation, which is owned by rough set, helps to solve many problems. Therefore, with the ability to deal with uncertain and vagueness data, RST is known as one of the available methods preferred by researchers in solving MCDM problems. However, the method which was proposed by Pawlak in 1982 was not really successful in handling data and does not consider the attributes with preference-ranked domain such as criteria, choice and selections [31], [37], [38]. It may generate impure results (noise or error) during the decision analysis task and restricted only to the classification of work [39]. This noise problem is improved by Ziarko [40] who proposed an improved rough set model named variable precision rough set model (VPRS) [40], [41]. The VPRS model enhanced the noise problem of the original RST by allowing a number of misclassification tasks whereas an error factor issue is added to the basic RST. Some of the VPRS implementations are VPRS as a feature space partition maker for the training data [42] and the combination of VPRS with dominance relation for MCDA method [41].

Later on, a new extension of RST named dominance-based rough set approach (DRSA) is introduced by Greco in MCDM process [14], [21], [43]–[45] in dealing with decision makers' preference information. Many decision problems applied DRSA in making decision such as formulating the airline service strategies [39], highway asset management [44], location for a waste incinerator selection [46], and executive competences analysis [17]. Besides, Greco also proposed an enhancement for DRSA which is called as variable consistency dominance-based rough set approach (VC-DRSA) to restructure the inconsistencies of preferences relations (preference information) given by the decision maker [17], [22]. It means that VC-DRSA has the ability to fully access the dominance relations from the information table and deals with the inconsistency of large data [47]. For instance, VC-DRSA helps the decision maker in airline service management to formulate an easy understanding of decision rules by using natural language. Other than that, VC-DRSA has been widely used in data mining analysis such as in water quality analysis [11], and supplier selection problems [48] as well as acts as a rule extractor or as a procedures' maker for rule initiation from dominance- based rough approximation [11], [14], [43]. Another method which is derived from RST is believable rough set approach (BRSA) proposed by J. Chai [48] which has an additional feature as a rule extractor in uncertain information and it is proven that the method performed very well in supplier selection case problem. The following figures summarize all the implementation of RST methods in MCDA research problems for the past five years as mentioned in the previous section. Thirteen methods inspired by RST were identified and applied in the empirical studies proposed by researchers in various different application fields. The case study is divided into five main MCDM problem categories; namely choice, ranking, classification, sorting and others in which others category represented the description, elimination and design categories. Fig. 1 below presents the number of article that have applied RST and its families. According to Fig. 2, which is the conclusion of Fig. 1, the most preferred method of MCDM RST is DRSA where it is tested and applied in all application areas and followed by VC-DRSA.

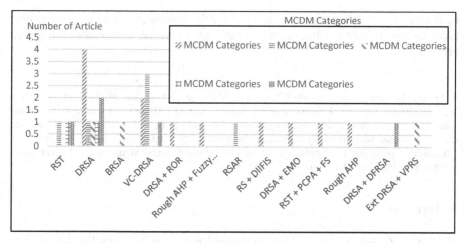

Fig. 1. The number of articles, which applied RST in the research works.

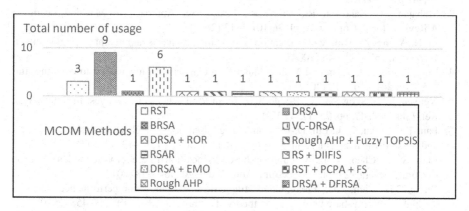

Fig. 2. The number of MCDM Rough Set methods' used in 2010 until 2015.

4   Conclusion

It is well known that RST is applied when there is uncertain or ambiguous data need to be handled and solved. This paper discusses a recent study of the implementation of RST in MCDA problems. The major concern of this study is the basic concept of the rough set itself, the extended version of RST and also the area where these theories are being implemented. There are four major MCDM categories, namely ranking, classification, sorting and choice, which are always being selected by the researchers in solving MCDA problems. Based on the analysis, the most popular of MCDM category is ranking problem and followed by classification problem. As illustrated in Fig. 2, the most selected rough set method in solving MCDA problem is DRSA where the total number of usage is 9. It is followed by VC-DRSA; where VC-DRSA is the improvement of DRSA. with uncertainty information in MCDA area. Moreover, further enhancement of RST methods or even an integration between other theories should be well thought-out by the researchers in order to solve not only for the ranking problem but also for other problems which have not been explored yet such as choice, elimination, and design problems.

Acknowledgement. The Universiti Teknologi Malaysia (UTM) and Ministry of Education Malaysia under Research University grants 02G31, and 4F550 are hereby acknowledged for some of the facilities that were utilized during the course of this research work. This work and the contribution were partially supported by project "SP-103-2015 - Smart Solutions for Ubiquitous Computing Environments" Faculty of Informatics and Management, University of Hradec Kralove, Czech Republic.

References

1. Pawlak, Z.: Rough set approach to knowledge-based decision support. Eur. J. Oper. Res. **99**, 48–57 (1997)
2. Mahajan, P., Kandwal, R., Vijay, R.: Rough Set Approach in Machine Learning: A Review. Int. J. Comput. Appl. **56**(10), 1–13 (2012)
3. Li, R., Wang, Z.: Mining classification rules using rough sets and neural networks. Eur. J. Oper. Res. **157**, 439–448 (2004)
4. Lin, G., Liang, J., Qian, Y.: Multigranulation rough sets: From partition to covering. Inf. Sci. (Ny) **241**, 101–118 (2013)
5. Nguyen, H.S., Skowron, A.: Rough Sets: From Rudiments to Challenges. In: Intell. Syst. Ref. Libr., vol. 42, pp. 75–173 (2013)
6. Fan, T.-F., Liau, C.-J., Liu, D.-R.: Dominance-based fuzzy rough set analysis of uncertain and possibilistic data tables. Int. J. Approx. Reason. **52**(9), 1283–1297 (2011)
7. Liao, S.-H., Chen, Y.-J.: A rough set-based association rule approach implemented on exploring beverages product spectrum. Appl. Intell. **40**, 464–478 (2013)
8. Wang, C.H., Chin, Y.C., Tzeng, G.H.: Mining the R&D innovation performance processes for high-tech firms based on rough set theory. Technovation **30**(7–8), 447–458 (2010)
9. Ishizaka, A., Pearman, C., Nemery, P.: AHPSort: an AHP-based method for sorting problems. Int. J. Prod. Res. **50**, 4767–4784 (2012)
10. Wu, W., Kou, G., Peng, Y., Ergu, D.: Improved AHP-group decision making for investment strategy selection. Technol. Econ. Dev. Econ. **18**(2), 299–316 (2012)
11. Karami, J., Alimohammadi, A., Seifouri, T.: Water quality analysis using a variable consistency dominance-based rough set approach. Comput. Environ. Urban Syst. **43**, 25–33 (2014)
12. Pawlak, Z.: Rough set theory and its applications. J. Telecommun. Inf. Technol. **29**, 7–10 (1998)
13. Ali, R., Siddiqi, M.H., Lee, S.: Rough set-based approaches for discretization: a compact review (2015)
14. Błaszczy, J., Greco, S., Matarazzo, B., Słowi, R.: jMAF - Dominance-Based Rough Set Data, pp. 185–209
15. Hu, Y.C.: Rough sets for pattern classification using pairwise-comparison-based tables. Appl. Math. Model. **37**(12–13), 7330–7337 (2013)
16. Liang, J., Wang, F., Dang, C., Qian, Y.: An efficient rough feature selection algorithm with a multi-granulation view. Int. J. Approx. Reason. **53**(6), 912–926 (2012)

17. Vol, F., Computing, O.F., No, D.S., Ciznicki, M., Kurowski, K., We, J.: Evaluation of Selected Resource Allocation Many-Core Processors and Graphics. **3** (2014)
18. Keeney, R.L.: Decision Analysis: An Overview. Operations Research **30**, 803–838 (1982)
19. Chai, J., Liu, J.N.K.: Dominance-based decision rule induction for multicriteria ranking. Int. J. Mach. Learn. Cybern. **4**, 427–444 (2013)
20. Borgonovo, E., Marinacci, M.: Decision analysis under ambiguity. Eur. J. Oper. Res. **000**, 1–14 (2015)
21. Greco, S., Słowiński, R., Zielniewicz, P.: Putting Dominance-based Rough Set Approach and robust ordinal regression together. Decis. Support Syst. **54**, 891–903 (2013)
22. Szeląg, M., Greco, S., Słowiński, R.: Variable consistency dominance-based rough set approach to preference learning in multicriteria ranking. Inf. Sci. (Ny) **277**, 525–552 (2014)
23. Chai, J., Liu, J.N.K., Ngai, E.W.T.: Application of decision-making techniques in supplier selection: A systematic review of literature. Expert Syst. Appl. **40**(10), 3872–3885 (2013)
24. Kavita, Yadav, S.P., Kumar, S.: A Multi-criteria Interval-valued intuitionistic fuzzy group decision making for supplier selection with TOPSIS method. In: Sakai, H., Chakraborty, M.K., Hassanien, A.E., Ślęzak, D., Zhu, W. (eds.) RSFDGrC 2009. LNCS, vol. 5908, pp. 303–312. Springer, Heidelberg (2009)
25. Vahdani, B., Hadipour, H., Tavakkoli-Moghaddam, R.: Soft computing based on interval valued fuzzy ANP-A novel methodology. J. Intell. Manuf. **23**, 1529–1544 (2012)
26. Fernandez, E., Lopez, E., Bernal, S., Coello Coello, C., Navarro, J.: Evolutionary multiobjective optimization using an outranking-based dominance generalization. Comput. Oper. Res. **37**(2), 390–395 (2010)
27. Durbach, I.N., Stewart, T.J.: Modeling uncertainty in multi-criteria decision analysis. Eur. J. Oper. Res. **223**(1), 1–14 (2012)
28. Chakhar, S., Saad, I.: Dominance-based rough set approach for groups in multicriteria classification problems. Decis. Support Syst. **54**(1), 372–380 (2012)
29. Velasquez, M., Hester, P.T.: An Analysis of Multi-Criteria Decision Making Methods. Int. J. Oper. Res. **10**(2), 56–66 (2013)
30. Cinelli, M., Coles, S.R., Kirwan, K.: Analysis of the potentials of multi criteria decision analysis methods to conduct sustainability assessment. Ecol. Indic. **46**, 138–148 (2014)
31. Huang, B., Wei, D., Li, H., Zhuang, Y.: Using a rough set model to extract rules in dominance-based interval-valued intuitionistic fuzzy information systems. Inf. Sci. (Ny) **221**, 215–229 (2012)
32. Aydogan, E.K.: Performance measurement model for Turkish aviation firms using the rough-AHP and TOPSIS methods under fuzzy environment. Expert Syst. Appl. **38**, 3992–3998 (2011)
33. Lee, C., Lee, H., Seol, H., Park, Y.: Evaluation of new service concepts using rough set theory and group analytic hierarchy process. Expert Syst. Appl. **39**(3), 3404–3412 (2012)
34. Dutta, M., Husain, Z.: An application of Multicriteria Decision Making to built heritage. The case of Calcutta. J. Cult. Herit. **10**, 237–243 (2009)
35. Devi, K., Yadav, S.P.: A multicriteria intuitionistic fuzzy group decision making for plant location selection with ELECTRE method. Int. J. Adv. Manuf. Technol. **66**, 1219–1229 (2013)
36. Lee, C.S.: A rough-fuzzy hybrid approach on a Neuro-Fuzzy classifier for high dimensional data. In: Proc. Int. Jt. Conf. Neural Networks, pp. 2764–2769 (2011)
37. Chai, J., Liu, J.N.K.: Class-based rough approximation with dominance principle. In: Proc. - 2011 IEEE Int. Conf. Granul. Comput. GrC 2011, pp. 77–82 (2011)
38. Tzeng, K.S.G.: A decision rule-based soft computing model for supporting financial performance improvement of the banking industry (2014)

39. Liou, J.J.H., Yen, L., Tzeng, G.H.: Using decision rules to achieve mass customization of airline services. Eur. J. Oper. Res. **205**(3), 680–686 (2010)
40. Hu, M., Shen, F., Chen, Y., Wang, J.: Method of multi-attribute decision analysis based on rough sets dealing with grey information. In: 2011 IEEE Int. Conf. Syst. Man, Cybern., no. 90924022, pp. 1457–1462 (2011)
41. Hu, M., Shen, F., Chen, Y.: A multi-attribute decision analysis method based on rough sets dealing with uncertain information. In: Proc. 2011 IEEE Int. Conf. Grey Syst. Intell. Serv., pp. 576–581 (2011)
42. Miao, D., Duan, Q., Zhang, H., Jiao, N.: Rough set based hybrid algorithm for text classification. Expert Syst. Appl. **36**(5), 9168–9174 (2009)
43. Błaszczyński, J., Greco, S., Słowiński, R.: Inductive discovery of laws using monotonic rules. Eng. Appl. Artif. Intell. **25**, 284–294 (2012)
44. Augeri, M.G., Colombrita, R., Greco, S., Lo Certo, A., Matarazzo, B., Slowinski, R.: Dominance-Based Rough Set Approach to Budget Allocation in Highway Maintenance Activities. J. Infrastruct. Syst. **17**(June), 75–85 (2011)
45. Greco, S., Matarazzo, B., Słowinski, R.: Interactive Evolutionary Multiobjective Optimization using Dominance-based Rough Set Approach (2010)
46. Phillips, L.D.: How Raiffa's RAND memo led to a multi-criteria computer program **23**(July 2013), 3–23 (2006)
47. Liou, J.J.H.: Variable Consistency Dominance-based Rough Set Approach to formulate airline service strategies. Appl. Soft Comput. J. **11**(5), 4011–4020 (2011)
48. Chai, J., Liu, J.N.K.: A novel believable rough set approach for supplier selection. Expert Syst. Appl. **41**(1), 92–104 (2014)
49. Capotorti, A., Barbanera, E.: Credit scoring analysis using a fuzzy probabilistic rough set model. Comput. Stat. Data Anal. **56**(4), 981–994 (2012)

Network Visualization Survey

Ales Komarek, Jakub Pavlik, and Vladimir Sobeslav[✉]

Faculty of Informatics and Management, University of Hradec Kralove,
Rokitanskeho 62, Hradec Kralove, Czech Republic
{ales.komarek,jakub.pavlik.7,vladimir.sobeslav}@uhk.cz
http://uhk.cz/

Abstract. This paper compares modern approaches to draw complex graph data that create create compelling visualizations. Graphs are used to represent more and more complex systems that are used across various scientific domains. Some graph visualizations are used to model network topologies or service architectures in information service fields, model genomes in biomedicine or complex molecules in chemistry. Visualizations of graph data play important role in interpreting the meaning of graphed data. The adage *"A picture is worth a thousand words"* generally refers to the notion that a complex idea can be conveyed with just a single still image and that is actually the main reason for creating visualisations of any kind. Graph drawings from all domains follow the same rules covered by the graph theory. This work outlines different layout strategies used for drawing graph data. Tested graph drawing layouts are compared by standard quality measures to determine their suitability for various areas of usage.

Keywords: Graph · Graph theory · Data flow · Visualization

1 Introduction

Growing number of systems are using various visualization techniques to display data it holds. Business intelligence systems use graphs to simplify data and graph drawings are interface for the users to better comprehend the data. [2,5,14] This paper's goal is to present and compare traditional and modern visualisation techniques and determine what kind of graph data it can represent. Main areas of focus for graph drawings are following knowledge domains.

1. Network topologies with flows
2. Service oriented architectures

The section 2 of the paper summarizes graph theory therms and principles relevant to the graph drawing. [1] This chapter also lists quality measures that are used for graph layout quality evaluations.

Chapter 3 introduces various graph drawing layouts that can be used for actual graph visualisations. Covered visualisation techniques are traditional force-directed algorithms, adjacency matrices or arc diagrams, [7,12] as well

© Springer International Publishing Switzerland 2015
M. Núñez et al. (Eds.): ICCCI 2015, Part II, LNCS 9330, pp. 275–284, 2015.
DOI: 10.1007/978-3-319-24306-1_27

as new approaches to graph layouts like hive plots, chord diagrams, Sankey diagrams and Pivot graphs. [8,9,13] Adjacency matrix is special as it displays graph edges as the graph's matrix coordinates instead of lines as the other graph drawings.

Following chapter shows differences of these visualisation techniques by given parameters. All of the visualisations were tested with samples of graphs data. The D3 [14] graphical library was used to create real graph drawings from the same dataset. D3 is being developed and improved by Mike Bostock and uses document driven approach to visualize any kind of information. Some graph drawings require additional data which had to be provided to allow test of all drawings. The last chapter 5 summarises the usability of selected visualization techniques.

2 Graph Drawing Theory

This section describes important therms from graph theory used for graph visualizations. Graph theory is an area of mathematical research with a large specialized vocabulary. Some authors use the same word with different meanings. Some authors use different words to mean the same thing. [1,7] A drawing of a graph or network diagram is a pictorial representation of the vertices and edges of a given graph. This drawing should not be confused with the graph itself, very different layouts can correspond to the same graph. In the abstract, all that matters is which pairs of vertices are connected by edges. In the concrete, however, the arrangement of these vertices and edges within a drawing affects its understandability, usability and aesthetics [1,3]. Some graphs even change over time by adding and deleting edges (dynamic graph drawing). Their goal is to preserve the user's mental map over time. This paper focus are solely static graph drawings

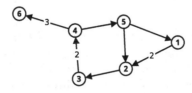

Fig. 1. Simple Graph Example

Figure 1 shows example graph with several vertices and weighted edges which is used further in this chapter to demonstrate some graph theory principles important for creating visualizations.

2.1 Adjacency and Degree

Degree or valency of a vertex is usually a measure of immediate adjacency. An edge connects two vertices. these two vertices are said to be incident to that edge, or, equivalently, that edge incident to those two vertices. All degree-related concepts have to do with adjacency or incidence. A vertex of degree 0 is an isolated vertex. A vertex of degree 1 is a leaf.

In the simple graph example, vertices 1 and 3 have a degree of 2, vertices 2, 4 and 5 have a degree of 3, and vertex 6 has a degree of 1. If number of edges is finite, then the total sum of vertex degrees is equal to twice the number of edges.

2.2 Weighted Graphs

A weighted graph associates a certain weight with every edge in the graph. Weights are usually real numbers. Weights may be restricted to rational numbers or integers. The weight of a path or the weight of a tree in a weighted graph is the sum of the weights of the selected edges. Sometimes a non-edge (a vertex pair with no connecting edge) is indicated by labeling it with a special weight representing infinity. Sometimes the word cost is used instead of weight. When stated without any qualification, a graph is always assumed to be unweighted.

In some writing on graph theory the term network is a synonym for a weighted graph. A network may be directed or undirected, it may contain special vertices (nodes), such as source or sink. The classical network problems include:

1. minimum cost spanning tree
2. shortest paths
3. maximal flow (and the max-flow min-cut theorem)

2.3 Direction

Directed edge is an ordered pair of vertices that can be represented graphically as an arrow drawn between these vertices. In such an ordered pair the first vertex is called the initial vertex or tail; the second one is called the terminal vertex or head (because it appears at the arrow head). An undirected edge disregards any sense of direction and treats vertices on both sides interchangeably. A loop in a digraph, however, keeps a sense of direction and treats both head and tail identically.

A set of arcs are multiple, or parallel, if they share the same head and the same tail. A pair of arcs are anti-parallel if one's head/tail is the other's tail/head. A digraph, or directed graph, or oriented graph, is analogous to an undirected graph except that it contains only arcs. A mixed graph may contain both directed and undirected edges; it generalizes both directed and undirected graphs. When stated without any qualification, a graph is almost always assumed to be undirected.

A digraph is called simple if it has no loops and at most one arc between any pair of vertices. When stated without any qualification, a digraph is usually assumed to be simple. A quiver is a directed graph which is specifically allowed, but not required, to have loops and more than one arc between any pair of vertices.

2.4   Quality Measures of Graph Drawings

Many different quality measures have been defined for graph drawings, in an attempt to find objective means of evaluating their aesthetics and usability [1,3,4]. In addition to guiding the choice between different layout methods for the same graph, some layout methods attempt to directly optimize these measures.

1. The crossing number of a drawing is the number of pairs of edges that cross each other. If the graph is planar, then it is often convenient to draw it without any edge intersections; that is, in this case, a graph drawing represents a graph embedding.
2. The area of a drawing is the size of its smallest bounding box, relative to the closest distance between any two vertices. Drawings with smaller area are generally preferable to those with larger area, because they allow the features of the drawing to be shown at greater size and therefore more legibly. The aspect ratio of the bounding box may also be important.
3. Symmetry display is the problem of finding symmetry groups within a given graph [4], and finding a drawing that displays as much of the symmetry as possible. Some layout methods automatically lead to symmetric drawings; alternatively, some drawing methods start by finding symmetries in the input graph and using them to construct a drawing.
4. It is important that edges have shapes that are as simple as possible, to make it easier for the eye to follow them. [1] In polyline drawings, the complexity of an edge may be measured by its number of bends, and many methods aim to provide drawings with few total bends or few bends per edge. Similarly for spline curves the complexity of an edge may be measured by the number of control points on the edge.
5. Angular resolution is a measure of the sharpest angles in a graph drawing. If a graph has vertices with high degree then it necessarily will have small angular resolution, but the angular resolution can be bounded below by a function of the degree.

3   Graph Visualization Layouts

This section describes traditional layouts strategies [7,10] as well as modern approaches [8,9,12,13] that are used to display complex graph data. Many of these layouts originate at different knowledge domains where they solve specific problems. These layouts can be used to display any kind of relational data from other domains as well.

3.1   Force Directed Graphs

Force-directed graph drawings are a family of algorithms that position the nodes of a graph in two-dimensional space so that all the edges are of more or less equal length and there are as few crossing edges as possible by assigning forces among the set of edges and the set of nodes based on their relative positions [7]. These forces are then used either to simulate the motion of the edges and nodes or to minimize their energy. Force-directed algorithms calculate the layout of a graph using only in formation contained within the structure of the graph itself, rather than relying on domain-specific knowledge.

Fig. 2. Force-directed Graph

3.2   Hive Plots

In hive plot vertices are mapped to and positioned on radially distributed linear axes [8]. This mapping is based on network structural properties. Edges are drawn as curved links. The purpose of the hive plot is to establish a new baseline for visualization of large networks. A method that is both general and tunable and useful as a starting point in visually exploring network structure.

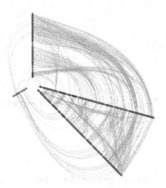

Fig. 3. Hive Plot

3.3   ARC Diagrams

An arc diagram uses a one-dimensional layout of vertices, with circular arcs to represent edges. Though an arc diagram may not convey the overall structure of the graph as effectively as a two-dimensional layout, with a good ordering of vertices it is easy to identify cliques and bridges. Further, as with the indented tree, multivariate data can easily be displayed alongside nodes.

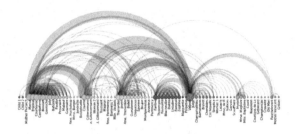

Fig. 4. Arc Diagram

3.4   Sankey Flow Diagrams

Sankey diagram is a specific type of flow diagram, where the width of the arrows is proportional to the flow quantity. [10] Sankey diagrams typically used to visualize energy or material or cost transfers between processes or the network flows [11].

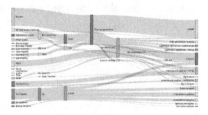

Fig. 5. Sankey Flow Diagram

3.5   Chord Diagrams

In chord diagram the vertices are arranged radially around a circle with the edges between them typically drawn as arcs. Chord diagrams are useful for exploring relationships between groups of entities. They have been heavily adopted by the biological scientific community for visualizing genomic data, but they have also been featured as part of infographics in numerous publications including Wired, New York Times, and American Scientist.

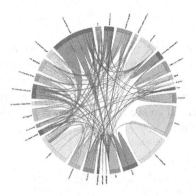

Fig. 6. Chord Diagram

3.6   Hierarchical Edge Bundles

While a small amount of data could be represented in a circular diagram using straight lines to show the interconnections, a diagram featuring numerous lines would quickly become illegible. To reduce the visual complexity, chord diagrams employ a technique called hierarchical edge bundling. There is a large class of data sets that contain both hierarchical components, i.e., parent-child relations between data items, as well as non-hierarchical components representing additional relations between data items. [9] If there is need for more insight in the hierarchical organization of graphed data it can be visualized by the hierarchical structure using any of the tree visualization methods. However, if we want to visualize additional adjacency edges on top of this by adding edges in a straightforward way, this generally leads to visual clutter. A possible way to alleviate this problem is to treat the tree and the adjacency graph as a single graph.

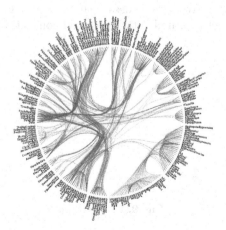

Fig. 7. Hierarchical Edge Bundle

3.7 Adjacency Matrices

An adjacency matrix represents a graph, where each value in row i and column j
in the matrix corresponds to the edge from vertex i to vertex j in the graph. [12]
Use of colors or saturation instead of text allows patterns to be perceived more
easily. The effectiveness of a matrix visualization is heavily dependent on the
order of rows and columns. If related vertices are placed closed to each other, it
is easier to identify clusters.

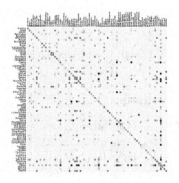

Fig. 8. Adjacency Matrix

3.8 Pivot Graphs

Pivot graph is technique for visualizing and analysing graph structures. [13] The
technique is designed specifically for graphs that are multivariate, where each
node is associated with several attributes. Unlike visualizations which emphasize
global graph topology, pivot graph uses a simple grid-based approach to focus
on the relationship between node attributes and connections.

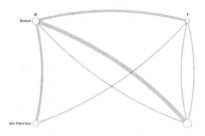

Fig. 9. Pivot Graph

4   Visualization Comparison

This section summarises some of the qualitative properties of different graph layouts. All parameters for evaluating graphs are in the following list.

Requires Domain Specific Data. This property states whether graph layout requires additional data to get displayed.

Maximum Number of Vertices. Maximum number of vertices where visualisation remains aesthetic and readable.

Maximum Number of Edges. Maximum number of displayed graph edges where visualisation remains aesthetic and readable.

Amount of Edge Crossings. How many edge crossings does the graph layout produce.

Multivariate Data Support. Does the graph layout support display of additional information.

Global Graph Topology. This property shows if drawing covers whole topology or just specific subgraphs.

All of these parameters were evaluated, tested agains a d3js graph implementation. The results of the tests are summarised in the table 1.

Table 1. Graph Drawing Layouts Comparison

Layout	Domain	Vertices	Edges	Crossings	Multivariete	Global topo.
Force-directed	no	1.000s	2.000s	high	no	yes
Hive plot	yes	2.000s+	5.000s+	medium	yes	yes
Adjacency matrix	no	100s	1.000s	no	yes	yes
Arc diagram	no	100s	1.000s	low	no	yes
Sankey diagram	no	100s	1.000s	low	yes	yes
Chord diagrams	yes	100s	1.000s	medium	yes	both
Pivot graph	yes	10s	10s	low	yes	no

Some values are estimates as the largest graph data set contained 2000 vertices and 5000 edges. Force directed graphs started to exhibit a hairball effect with growing number of vertices [8]. Other layouts did not loose understandability in such a rate, but other reasons prevented to get larger.

5   Conclusion

We have covered modern layout methods used for graph drawing in our paper. Each layout has it's area of use. Some visualisations like arc diagram and chord diagram display edge weights and directions better than others. Some visualisations like hive plots or hierarchical edge bundles require additional domain specific knowledge definitions that make them harder to maintain but provide more aesthetic and usable layouts for datasets where traditional layout approaches fail.

Graph layouts are located at http://lab.newt.cz/vis/relational/. The graphs are implemented in D3 javascript framework for data driven graphics.

References

1. Di Battista, G., Eades, P., Tamassia, R., Tollis, I.: Algorithms for Drawing Graphs: an Annotated Bibliography. ACM (1994). doi:10.1016/0925-7721(94)00014-x
2. Graph, H.: Graph visualization and navigation in information visualization: A survey. Visualization and Computer Graphics. IEEE doi:10.1109/2945.841119
3. Purchase, H., Carrington, D., Allder, J.-A.: Empirical Evaluation of Aesthetics-based Graph Layout. Empirical Software Engineering. Kluwer (2002)
4. McGrath, C., Blythe, J., Krackhardt, D.: The effect of spatial arrangement on judgements and errors in interpreting graphs. Social Networks **19**. Elsevier (1997)
5. Purchase, H.C., Hoggan, E., Görg, C.: How important is the mental map? – an empirical investigation of a dynamic graph layout algorithm. In: Kaufmann, M., Wagner, D. (eds.) GD 2006. LNCS, vol. 4372, pp. 184–195. Springer, Heidelberg (2007)
6. Di Battista, G., Gargb, A., Liottab, G., Tamassiab, R. ,Tassinaric, E. Vargiuc, F.: An experimental comparison of four graph drawing algorithms. Elsevier (1997). doi:10.1016/S0925-7721(96)00005-3
7. Kobourov, S.: Handbook of Graph Drawing and Visualization: Force-directed drawing algorithms (2013). http://cs.brown.edu/rt/gdhandbook/chapters/force-directed.pdf
8. Krzywinski, M.: Hive Plots - Linear Layout for Network Visualization - Visually Interpreting Network Structure and Content Made Possible (2011). http://www.hiveplot.net/
9. Holten, D.: Hierarchical, Edge Bundles: Visualization of Adjacency Relations in Hierarchical Data (2006)
10. Riehmann, P., Hanfler, M., Froehlich, B.: Interactive Sankey Diagrams (2005)
11. Holik, F., Horalek, J., Marik, O., Neradova, S., Zitta, S.: The methodology of measuring throughput of a wireless network. In: CINTI 2014 (2014)
12. Godsil, C., Royle, G.: Algebraic Graph Theory. Springer (2001). ISBN 0-387-95241-1
13. Wattenberg, M.: Visual Exploration of Multivariate Graphs. ACM (2006). ISBN 1-59593-372-7
14. Bostock, M.: D3 Data-Driven Documents Documentations (2015). https://github.com/mbostock/d3/wiki

A Proposal of a Big Web Data Application and Archive for the Distributed Data Processing with Apache Hadoop

Martin Lnenicka, Jan Hovad, and Jitka Komarkova[✉]

University of Pardubice, Faculty of Economics and Administration, Pardubice, Czech Republic
martin.lnenicka@gmail.com, {jan.hovad,jitka.komarkova}@upce.cz

Abstract. In recent years, research on big data, data storage and other topics that represent innovations in the analytics field has become very popular. This paper describes a proposal of a big web data application and archive for the distributed data processing with Apache Hadoop, including the framework with selected methods, which can be used with this platform. It proposes a workflow to create a web content mining application and a big data archive, which uses modern technologies like Python, PHP, JavaScript, MySQL and cloud services. It also shows the overview about the architecture, methods and data structures used in the context of web mining, distributed processing and big data analytics.

Keywords: Big web data · Big data analytics · Web content mining · Distributed data processing · Python · Apache Hadoop

1 Introduction

During the past few years there has been a growing interest in the term big data, fostered by the broad availability of various data acquisition, storing and processing platforms. The concepts of linked and open data have led to the necessity to process these large amounts of data very quickly to retrieve valuable information. Types of data frequently associated with the big data analytics include the web data (e.g. web log data, e-commerce logs and social network interaction data), the industry specific transaction data (e.g. telecommunications call data records, geolocation data and retail transaction data), the machine generated / sensor data and the text (e.g. archived documents, external content sources or customer and supplier data). These data are not only needed to be stored, but also be prepared to provide tools for the support of the decision-making process. Big data applications may change the way of mining, storing, processing and analyzing of websites content. Big data applications do not replace traditional applications, but complement them. Businesses as well as public sector can mine this big web data to provide insight to support future decisions.

Thus, authors focus on design and coding a web content mining application, which provides the basic layer to build a big web data archive and then propose a framework for the processing of these data with the Apache Hadoop cluster.

© Springer International Publishing Switzerland 2015
M. Núñez et al. (Eds.): ICCCI 2015, Part II, LNCS 9330, pp. 285–294, 2015.
DOI: 10.1007/978-3-319-24306-1_28

2   Related Work and Background

The growth of data sources and the ease of access that information technology affords also bring new challenges on data acquisition, storage, management and analysis. The most recent survey about the term big data is well conducted in [1], where authors present the general background of big data and review related technologies, such as distributed approach, Internet of Things, data centers, and Apache Hadoop. They also introduce the terms big data generation and acquisition, big data storage and big data applications including web data analysis, which are closely related to the topic of this paper. Tien [2] then compares major differences between the big data approach and the traditional data management approach with four components (acquisition, access, analytics, application) and three elements (focus, emphasis, scope).

The work by Dean and Ghemawat [3] describes the file system implemented by Google called the Google File System (GFS), which handles the big data operations behind the Google services and it is the main part of the MapReduce framework. Also the findings of Vilas [4] lend support to the claim that high performance computing platforms are required, which impose systematic designs to unleash the full power of big data. A comparison of approaches to large-scale data analysis can be found e.g. in Pavlo et al. [5] or also Chen et al. [1]. The evidence supporting the use of big data for analytics and the improvement of the decision-making process may lie in the findings of Power [6], who proposes the use cases and user examples related to analyzing large volumes of semistructured and unstructured data.

There are three types of mining: data mining, web mining and text mining. Web mining lies in between data mining and text mining, and copes mostly with semi-structured data and/or unstructured data. It can be distinguished in the three different types of categories: web content mining, web usage mining and web structure mining. Web content mining is performed by extracting useful information from the content of a web page. It includes extraction of structured data from web pages, identification, match, and integration of semantically similar data, opinion extraction from online sources, and concept hierarchy or knowledge integration [7].

3   Problem Statement and Used Tools

The primary objective of this paper is to build a big web data archive and introduce a framework for the distributed processing of these data with Apache Hadoop. For the purpose of this objective the web content mining application is proposed.

The partial objectives are intended to meet the following objectives. The first one is to compose a suitable architecture and choose the right web technologies. Save and clear strings of each source by regular expressions. Transform the HTML structure into the easy accessible objects. Implement the processing phase by Python and in the case of the big web data, by the distributed approach represented by Apache Hadoop. Save necessary history data in a simple MySQL database and utilize the replicated commercial cloud storage Dropbox. Create a framework for the analysis with Apache Hadoop over the obtained big web data.

These points form the outline of this paper and they are shown in the following scheme, which is represented by the Fig. 1.

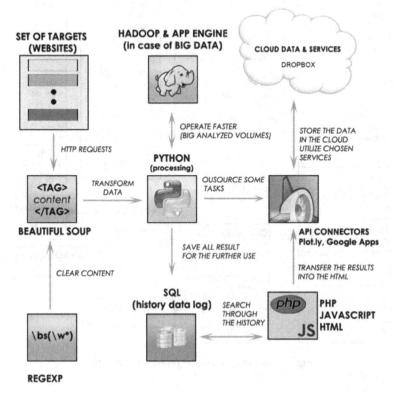

Fig. 1. The solution scheme of this paper

The main tool used is Python in the version 2.7, because the version 3.3 has limited compatibility with some web services e.g. Google. Processing the HTML data (website raw string) response to the sent request is done by parsing the output tree, which is described by Jackson [8]. Uniform data structure as a XML is replaced by a JSON, which is more suitable for a Python and it is also used by many API's as a data structure to transfer the optional adjustment arguments [9]. An API creation is covered in the work of Henning [10]. The MySQLdb connector is used to save the big web data during the script runs. The beautiful soup (BS) library is used for accessing the data. The processing tasks use the MapReduce approach [3], which is in this paper represented by its open-source implementation Apache Hadoop. Ubuntu Server 12.04, Java 7 and Hadoop 2.2.0 are used for the deployment of the Apache Hadoop cluster.

The code is written in objective way and it can be extended to cover new situations and handle number of different tasks such as crisis management, social media monitoring, fast financial market decisions, customer segmentation, crime area identification based on the location based services, etc. The class diagram of the proposed application and its connection to the tools used by authors can be seen from the Fig. 2.

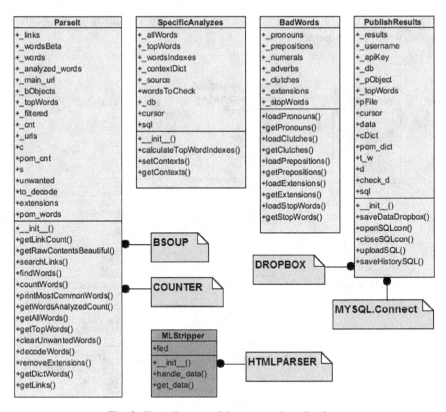

Fig. 2. Class diagram of the proposed application

4   A Description of the Big Web Data Application

4.1   Targets and the HTML Structure

Authors use the three biggest media servers in the Czech Republic. This fact is crucial, because the Czech language is characterized by hard inflection of individual words and their meanings. Targets are set to the www.novinky.cz, www.ct24.cz, www.m.idnes.cz. Desktop versions usually need more focus in the case of clearing the content, mobile versions are much clearer. The big web data mining application runs automatically in the specified time intervals on the server. This type of scan (over three servers with 20 000–30 000 words analysed) takes approx. 7 seconds.

The HTML request is made for each of these sources and the result is saved in the string variable. This variable contained everything, comments, JavaScript, tags and also a plain text of the articles. Then this text string is transformed by the BS library into the BS object (BO). Each source is then loaded by an object constructor and the title page is scanned for the relevant links that led to the individual articles. The only manual step before running the script is to explore the HTML structure for the identification of the targeted content.

4.2   Filtering and the Data Structure

Initially, the title page of the first mass media server is scanned for the all <h3> occurrences, because the new articles are characterized by this tag. All the content inside is scanned again for the <a> tag that contains links to the full text. Then, the content of this tag is returned and the string of the attribute "href" is saved into the associative array (dictionary).

Another iteration of this operation processes the founded links once again. They are converted into the BO one by one while the pure text of the article is extracted. Pure text is placed in the <div id="articleBody"> and this structure is global for the whole website. The content is returned and saved as a string ready for clearing. MLStripper class inherited from the HTMLParser and instantiated an object that cleares a string from the HTML tags and JavaScript. Punctuation is also removed by a simple regular expression. Words are stored in the JSON data structure, which is characterized as a key-value pair, where the value can be another associative array.

4.3   Removing Unwanted Content

Authors store all the words in the counter object. Short words (<=2 characters) are limited. Authors use a special class that operates with an access to the dictionaries of the unwanted words. These dictionaries are made during the tens of individual script runs and they are extended continuously.

In the Czech language, each word has many different shapes. For the purpose of this application, this problem is eliminated by using a JSON, where the key is the word in its default state and the value for this key is a list of alternatives.

4.4   Saving and Archiving of the Obtained Data

The obtained big web data are stored in the MySQL database on the daily basis since the deployment of the proposed application (January 2014) and then are saved to the Dropbox folder, as it is given in the following algorithm. Data from the MySQL database are then deleted daily. The plan is to build the big web data archive and save these big web data for the processing using the distributed approach and Apache Hadoop. Authors do not save or process these data in the real-time, so they did not use the technologies such as MySQL Binlog, Hadoop Applier, etc.

```
def saveHistorySQL(self):
    t_w = self._topWords
    d = self.openSQLcon()
    cursor = d.cursor()
    if (datetime.datetime.now().strftime("%H") == '00'):
        print("MIDNIGHT SCAN - CLEARING HISTORY DATA")
    cursor.execute("""TRUNCATE table realmining.HISTORY""")
        d.commit()
    else:
```

```
    print("ADDING HISTORY DATA")
    check_d = cursor.execute("""SELECT * FROM infor-
mation_schema.tables WHERE table_schema = 'realmining'
AND table_name = 'HISTORY' LIMIT 1;""")
    d.commit()
    if (check_d):
        print("History table is ready, inserting data")
    else:
        sql = """CREATE TABLE realmining.HISTORY
(WORD_NAME VARCHAR(100) NOT NULL,WORD_LENGTH
INT(3),WORD_COUNT INT(5),CONTEXT_STRING
VARCHAR(250),FULL_URL VARCHAR(250),START_TIME
TIME,START_DATE DATE) DEFAULT CHARACTER SET utf8 COLLATE
utf8_general_ci"""
        cursor.execute(sql)
        d.commit()
        print("History table doesn't exist")
    for top_w in self._topWords:
        wrdn = top_w[0].upper().encode("utf-8")
        wrdl = top_w[1]
        wrdc = top_w[2]
        cont = top_w[3]
        furl = top_w[4]
        cursor.execute("insert into realmining.HISTORY
(WORD_NAME,WORD_LENGTH,WORD_COUNT,CONTEXT_STRING,FULL_URL
,START_TIME,START_DATE) values
(%s,%s,%s,%s,%s,TIME(),DATE())",
(wrdn,wrdl,wrdc,cont,furl))
        d.commit()
    d.close()
def saveDataDropbox(self,path):
    pFile=open(path,"w+")
    for tpl in self._results:
pFile.write(str(tpl[0])+","+str(tpl[1])+","+str(tpl[2])+"
,"+str(tpl[3])+","+str(tpl[4])+"\n").encode("cp1250")
    pFile.close()
```

5   Distributed Approach and the Processing of Big Web Data

There are two main ways to process the big web data. The first one requires some hardware costs and investments or at least configuration of the computer cluster. This can be done by installing Apache Hadoop, an open-source platform written in Java for reliable, scalable and distributed computing, under the UNIX type operating system. The second option is to save all the hardware costs and use some of the commercial tools and sources (e.g. Google, Amazon, etc.). Both of these solutions have the similar base. The replicated File System (FS), which handles all the parallel and distributed processing operations automatically. It can be a job distribution, optimization, error handling, thread access and many more complicated tasks. In the case of the open-source solution, the file system is called the Hadoop File System (HDFS), in the case of the commercial solution, it is called the GFS. The descriptions of Master node and Slaves configuration together with the utilization of the commercial services are not included in this paper. Nevertheless, the topic of the big web archives and Apache Hadoop in practice has been clearly documented by Holmes [11].

5.1   A Big Web Data Archive and its Structure

Although there is a lot of technical information about Apache Hadoop, there is not much information about how to effectively structure data in a Hadoop environment. Even though the parallel processing systems provide an optimal environment for the processing big web data, the structure of the data itself plays a key role [12].

The MySQL relational data model of the proposed application is consisted of one database table, which is represented by the Fig. 3.

Fig. 3. Relational data model

Data stored in the database can be retrieved using Structured Query Language (SQL). On the daily basis, the data are moved from MySQL to the Dropbox folder as a single file. Through the run of the proposed application approx. 2–3 MB of the data were obtained daily. For the 15 months run it means 1.5 GB data file to process. This file was daily updated through incremental data transfer. The file was saved as a CSV formatted text file because in the CSV format a line break would start a new line. In order to use this type file, class CsvRecordInput in Apache Hadoop can be used.

5.2   Architecture of the Cluster for the Big Web Data Processing

Apache Hadoop is chosen especially for the advantages presented in Pavlo et al. [5] and Chen et al. [1], mostly because of much easier set up and use of this approach. Authors focus on a purely distributed solution to emulate a small-scale Apache Hadoop cluster for testing purposes with low costs and without utilization of a cloud service, not a large-scale production environment.

A simple architecture for the processing of the big web data is proposed. Only 5 PCs are used, one of them worked as a master, the others as slaves. The proposed cluster is based on Apache Hadoop 2.2.0 and Java 1.7. Ubuntu Server 12.04 is used as an operating system running on all PCs. All the computers are connected by 100 Mbps Ethernet to a dedicated Virtual Local Area Network (VLAN). A proposal for the speeding up the processing of the big web data by the proposed cluster and the distributed approach can be seen in the Fig. 4.

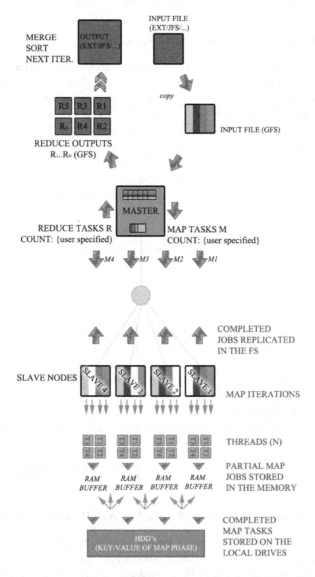

Fig. 4. Speeding up the processing of the big web data using the distributed approach

5.3   A framework for the Big Web Data Processing with Apache Hadoop

The main challenge for this type of the big web data is to compute statistics for all the records on various combinations of reference dimensions. Table 1 then shows some selected methods based on the web content mining techniques and also trend analysis, which can be used with Apache Hadoop.

Table 1. The framework of the selected methods for the big web data processing

The type of the analysis	Key feature with Apache Hadoop
Trend analysis of the targeted words through the time, e.g. the use of the words on the daily basis, which words are used more frequently than the others, the comparison of the word between the mass media servers, etc.	e.g. WordCount example reads text files and counts how often words occur. This script can be also easily modified based on the selected attributes (see the chapter: A big web data archive and its structure). The barrierless version of WordCount algorithm is presented by Verma et al. [13].
Association	e. g. finding the rules associated with frequently co-occurring words.
Attribute importance	e. g. ranks words according to strength of relationship with target word, extracting a vector of meaningful content words.
Distance matrix	e. g. the Euclidean distance calculation between the two words.
Correlation analysis	e. g. the most correlated words based on the different distances (1, 2, 3), correlation matrices.
Clustering	e. g. groups words together into clusters based on the time or distance.
Classification	e. g. the problem of identifying to which of a set of predefined categories (art, sport, etc.) a word belongs.
Web content mining techniques	A lot of scripts were already written for this area. More can be found e.g. in Holmes [11].
Business intelligence techniques	

6   Conclusions and Future Research

In this paper, authors offer an effective approach to analyze publicly available and valuable content on the Web with the primary aim of building the big data archive. The proposed big web data application demonstrates the basic principles, which can be further broadly expanded to mine other relevant information resources over the Web. The second part of this paper is focused on the importance of the distributed approach and Apache Hadoop in the processing of the big web data. The deployment of this solution can save the resources, especially costs, and get the results in the reasonable time, which is an important part in the decision-making process.

There are three phases of the big web data value chain covered by this paper: a) big web data generation; b) big web data acquisition; c) big web data storage. The last phase, which is focused on the big web data analytics, is only mentioned through the framework of the selected analyses. It will be content of the future work and it will be presented together with the comparison and processing of the obtained big web data, the performance analysis of the proposed cluster solution and the implementation of the selected methods in Python or Java for the use with Apache Hadoop.

Acknowledgements. This paper was supported by the SGSFES_2015001 fund.

References

1. Chen, M., Mao, S., Liu, Y.: Big Data: A Survey. Mobile Networks and Applications **19**(2), 171–209 (2014)
2. Tien, J.M.: Big Data: Unleashing Information. Journal of Systems Science and Systems Engineering **22**(2), 127–151 (2013)
3. Dean, J., Ghemawat, S.: MapReduce: simplified data processing on large clusters. Communications of the ACM **51**(1), 107–113 (2008)
4. Vilas, K.S.: Big Data Mining. International Journal of Computer Science and Management Research **1**(1), 12–17 (2012)
5. Pavlo, A., et al.: A comparison of approaches to large-scale data analysis. In: Proceedings of the 2009 ACM SIGMOD International Conference on Management of data, pp. 165–178. ACM (2009)
6. Power, D.: Using 'Big Data' for analytics and decision support. Journal of Decision Systems **23**(2), 222–228 (2014)
7. Zhang, Q., Segall, R.S.: Web mining: A survey of current research, techniques, and software. International Journal of Information Technology & Decision Making **7**(4), 683–720 (2008)
8. Jackson, Q.T.: Efficient formalism-only parsing of XML/HTML using the § -calculus. ACM SIGPLAN Notices **38**(2), 29–35 (2003)
9. Peng, D., Cao, L., Xu, W.: Using JSON for Data Exchanging in Web Service Applications. Journal of Computational Information Systems **7**(16), 5883–5890 (2011)
10. Henning, M.: API design matters. Queue **5**(4), 24–36 (2007)
11. Holmes, A.: Hadoop in Practice. Manning, Shelter Island (2012)
12. Bradley, C., et al.: Data Modeling Considerations in Hadoop and Hive. Technical paper (2013)
13. Verma, A., et al.: Breaking the MapReduce stage barrier. Cluster computing **16**(1), 191–206 (2013)

Centralized Data Flow Management System

Agáta Bodnárová(✉) and Michal Kökörčený

University of Hradec Králové, Faculty of Informatics and Management,
Hradec Králové, Czech Republic
agata.bodnarova@uhk.cz, michal@74.cz

Abstract. Ethernet is the most widely used technology on which the current local area networks are built up. The disability of loop detection belongs to the main disadvantages of this technology. The Ethernet technology contains no mechanisms to avoid loops; the frame header also contains no fields to allowing discard looped data frames. Loops occur due to redundant design of the network. The redundancy is necessary because eliminates single points of failure, where one failed device or design element brings down the service.

This article is focused on the mechanisms that prevent the creation of bridging loops while keeping the redundant topology. The currently used solutions are very inefficient; they do not reflect the needs of the modern communication techniques. The implementation of the newly developed solutions is extremely expensive. A new economic efficient solution is proposed, based on the requirements imposed on the design of the modern communication networks.

Keywords: Ethernet · Redundancy · Bridging loops prevention · L2

1 Introduction

The currently used Ethernet technology is wildly implemented for approximately 85 percent of the world's local area networks [4]. Switches are the most often used devices on such networks due their attractive features: easy implementation and management, low cost network solution, flexible topology design and fast data retransmission [10], [18].

Redundancy on these networks eliminates single points of failure, where one failed device or design element brings down service [5]. Redundancy increases the network availability by several hundred percent [2]. The side effect of the redundancy is that it evokes loops.

When the switch doesn't recognize destination MAC address (so it isn't in MAC address table), switch sends this frame to all interfaces, except one, which received that frame. This is called **unknown unicast flooding**. The same occurs if broadcast transmission is used. Every switch in topology acts the same way. In very short time the amount of repeating frames increases and the network is flooded. Broadcast storm occurs in the network.

© Springer International Publishing Switzerland 2015
M. Núñez et al. (Eds.): ICCCI 2015, Part II, LNCS 9330, pp. 295–305, 2015.
DOI: 10.1007/978-3-319-24306-1_29

Between the main disadvantages of the Ethernet standard belongs the disability of loop detection. The Ethernet technology does not contain mechanisms to avoid loops; the frame header also does not contain fields to allow discard looped data frames.

This problem can be prevented at the topology design level by interconnecting devices without any loops to tree-structure. But this topology cannot accomplish basic requirement on the network – link redundancy.

The usage of devices working on network layer of OSI model which have loops prevention is also a possible solution. However it requires changes in connected device setting with every move and it has slower data stream processing (because of decomposition to packet level). A router characteristic is also limited number of interfaces.

2   Existing solutions

The possible solutions are the followings:

2.1   The Family of the Spanning Tree Protocols

The currently used family of the Spanning Tree Protocols (STP) overcomes the possibility of bridging loops by keeping the redundant topology. In the case of redundant links, switches can be made aware that a link shut down for loop prevention [8]. The principles are based on graph theory. The network refers to a weighted graph and the algorithm creates a spanning tree on this graph. Between any two nodes exists a single path. This path may not be, and in many cases isn't the shortest. Links, they are not part of the spanning tree, are disconnected [5]. The convergence is very slow – up to 50 seconds.

STP has now been superseded by the RSTP (Rapid Spanning Tree Protocol) [9]. The RSTP is an IEEE standardized protocol that delivers a very fast convergence. But on a single network exists only one spanning tree, which evokes a very inefficient usage of the links and inefficient usage of the capacity of the links. There is also no possibility of multipath forwarding or achieve load balancing within the network.

The number of articles focused on the improvement of the family of spanning tree protocols is uncountable. In PVSTP (PerVLAN Spanning Tree Protocol) for every existing VLAN a new spanning tree is created. In the MSTP (Multiple Spanning Tree Protocol) it is possible to group VLANs and run one instance of spanning tree for a group of VLANs [3]. Articles [14] and [23] discuss about the mapping of a set of VLANs to a given number of spanning trees of the MSTP. Articles [15] and [17] are focused on solving the traffic engineering for MSTP by finding the set of spanning trees described by customer traffic demands and given network topology. Article [6] expands articles [15] and [17]. Its aim is to find solutions for large data center networks with hundreds of switches instead of small networks with dozen switches. The performance is evaluated on traffic matrices collected in data centers and compared to the traditional 802.1s MSTP with up to 16 VLANs. The contribution of these improvements is also disputable, because the basis of the solution remains the same.

For nowadays used virtualization techniques are also not usable because they need to run on one network or VLAN.

2.2   Routing on L2

One of the possible solutions is to replace the currently used switches by RBriges. RBridges are devices which uses network layer protocols at the data link layer. These network layer protocols are TRILL, Fabric Path or SPB. The term RBridges are used concretely for devices using TRILL, but for this work it can be generalized for Fabric Path or SPB too. This solution is very expensive; the forenamed protocols are implemented just in a few numbers of very expensive devices. Fabric Path is for example implemented on Cisco Systems Nexus 5500 switches or above. The price of the Nexus 5500 is 25 600 USD + 5 000 USD for license [1]. In the case of large network topology (the unofficial border of the enterprise networks is about 200 devices) it is an inconsiderable sum of money. Other costs are evoked by the edification of employers.

2.3   Spanning Tree Alternate Routing Bridge Protocol

STAR (The Spanning Tree Alternate Routing Bridge Protocol) attempts to find and forward frames over alternate paths to the STP paths. These STAR paths are shorter or the same than their corresponding STP paths [12]. The STAR bridge protocol is a patented protocol [11], founded by Mr. King-Shan and Mr. Chiu-Lee. It is backward compatible with the standard 802.1D STP, so it is possible to incrementally upgrade standard-based bridges to be STAR bridges. Both protocols also, the 802.1D STP and the STAR works seamlessly.

2.4   Other Related Solutions

Two more solutions of the overcome the possibilities of bridging loops by keeping the redundant topology are well known. Both they improve the performance of the network:

- The usage **VSS (Virtual Switching System)** or similar technology that pools two switches into one virtual switch increasing the performance of the system. These two switches operate as a single logical virtual switch [22]. The number of interconnected devices is limited to two. As it is a non-standardized technology, only devices of the same vendor can be interconnected. The network also loses on redundancy – the devices are dependent on each other; fault occurrence on one of the devices may cause failure of the whole system.
- The next possibility how to increase the performance of the network by increasing the horizontal throughput is to use the **EtherChannel** or other similar technology. EtherChannel is a link aggregation technology used for grouping of several physical Ethernet links between two neighboring nodes to create one logical link. The data flow is then balanced using a hash algorithm between these physical links [20]. This EtherChannel must be also configured on every pair of devices.

3   The Proposed Solution

3.1   Solution Requirements

Based on the study of the existing solutions, as well as the newly developed ones and also based on the needs on nowadays communication technologies, the requirements on the proposed solution are summarized. The proposed solution should fulfil the following requirements:

- Compatibility with currently used networking devices and the Ethernet technology
- Efficient usage of the links
- Efficient usage of the capacity of the links
- Efficiency on one network – without dividing it onto sub-networks
- Load balancing or forwarding data over multiple paths
- Fault tolerance
- Topology independent
- Speed of convergence
- Multicast, broadcast and unknown unicast forwarding
- Security
- Adaptability, QOS
- Follow the trends of virtualization
- Low cost of the solution
- Easy adoption

3.2   Centralized Management

To enforce the compatibility with currently used networking devices and the Ethernet technology is not an easy task. Many manufactures (for example Cisco Systems too) built their operation systems on their networking devices proprietarily. These operation systems are vendor specific and do not typically provide an open or programmable platform. The internal processes of these internetworking operation systems are hidden and differ from vendor to vendor.

If it is not possible to edit the internetworking operation systems processes; it is almost no practical way to experiment with newly developed protocols in real environment.

One of the possible solutions is to disable the decision making on these networking devices and to use static configuration to achieve the desired behavior. The decision making another, centralized device is provided.

3.3   The Proposed Solution - CEMA

New centralized L2 traffic engineering system is proposed in this section, named as CEMA (Centralized Data Flow Management System). CEMA replaces the spanning tree protocols; on a dedicated device data about the network are collected and analyzed. Afterwards the used algorithm calculates the forwarding paths between the devices. The results are distributed as configuration changes on networking devices. These devices do not need to be smart devices in a classical point of view. For the efficiency of the solution the centralized CEMA system is responsible. The solution may be self-adapting; it may recognize and reflect the actual needs of the network.

The proposed solution is similar to the idea of SDN (Software Defined Networking) [19]. In SDN architecture the control and the data planes are decoupled. The network intelligence and the state of the network are centralized while the underlying network structure is abstracted form the applications. SDN allows support of switching across multi-vendor hardware. OpenFlow is the first standard interface designed for SDN [19]. OpenFlow allows on OpenFlow capable devices easily deploy and run new routing and switching protocols [16]. Some commercial switches (or routers) are enhanced with the OpenFlow feature. Limited number of vendors in limited number of devices supports this solution. For example the list of OpenFlow supported HP switches is found in [7].

CEMA Processes
The CEMA protocol is based on the following processes:

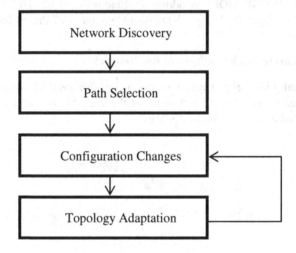

Fig. 1. The process diagram of CEMA

- **Network discovery process** – during this process the network devices and their interconnections as well as their characteristics are recognized.
- **Learning process** – is responsible for the first run of the used algorithm. The algorithm runs on the static characteristics of the network - on the network topology and on the available bandwidth.
- **Configuration changes** – this process is responsible for the application of the changes in the configuration of the devices. It must cover all the devices available on the network.

- **Topology adaptation** – reflects the actual needs of the network. The topology adaptation process is evoked by events (for example, congestion on the network) or it occurs when different data flow patterns are using the MF-ARTMAP fuzzy ANN recognized.

4   Experiments

Two aspects will be compared: The theoretical throughput of the network and the performance of the network – the actual throughput.

Used components:

- Switches: A, D, E: Cisco Catalyst 2960, Model: WS-C2960-24TT, IOS: C2960–Lanbasek9-M version 15.0 (2) SE, 26 ports; B, C: Cisco Catalyst 3560, Model: WS-C3560V2-24PS, IOS: C3560–Ipeservicesk9-M version 15.0 (2) SE, 26 ports
- Links: 100Mb/s ports are used to interconnect the devices with straight-through cat5e cable.
- Computers: 5x Dell Optiplex 745, Intel® Core™2 CPU, 6700 @ 2.66GHz 2.66GHz, 4 GB RAM, 232 GB HDD, Windows 7 Enterprise 64-bit, DELL Vostro 3460/ i3-2370M/ 4GB/ 320GB (7200)/ DVDRW/ 14"/ Windows 7 Pro 64-bit

4.1   The Theoretical Throughput of the Network

The Theoretical Throughput refers to the bandwidth available on the network. It can be calculated as the sum of bandwidths of the available or the non-blocked links. It is demonstrated on the following topology:

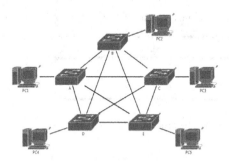

Fig. 2. Experiment 1 topology

Results of the STP: On the used switches PVSTP is the default STP. It is configured for one VLAN – the VLAN1. As it is possible to predict based on the MAC addresses of the switches, switch E has the lowest bridge ID and it is the root bridge for the used STP. The visualization of the created spanning tree is shown on the following figure:

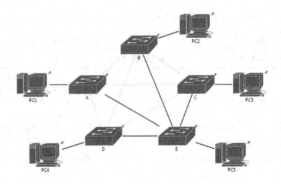

Fig. 3. The created spanning tree

Results of the CEMA Process: We are demonstrating the effectiveness of the Path Selection process. Therefor the used processes are reduced to Network Discovery, Path Selection and Configuration Changes.

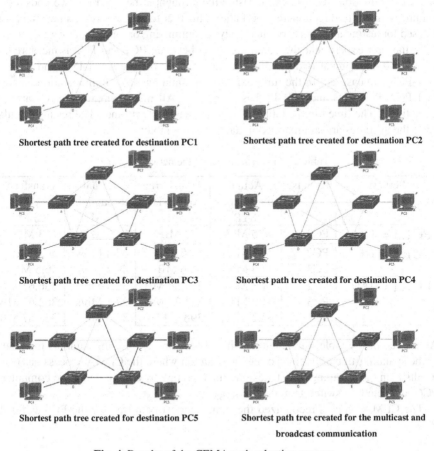

Fig. 4. Results of the CEMA path selection process

Table 1. The theoretical throughput of the network

	Blocked links	Available links
STP	A-B, B-C, C-D, A-C, A-D, B-D.	A-E, B-E, C-E, D-E
CEMA	-	A-B, B-C, C-D, A-C, A-D, B-D, A-E, B-E, C-E, D-E

As 100Mb/s ports are used to interconnect the devices, the theoretical throughput of the network on which the STP runs is in this case 4 * 100Mb/s = **400Mb/s.** The theoretical throughput of the network managed by the CEMA process is 10 * 100 Mb/s = **1 000 Mb/s.**

4.2   The Performance of the Network

The actual throughput using the **Iperf** network performance measurement tool is measured [13], [21]. Iperf measures the TCP Throughput of the network. It has a client and server functionality, and can measure the throughput between the two ends, either unidirectional or bi-directionally. Iperf allows the user to set various parameters that can be used for testing a network, or alternately for optimizing or tuning a network

In the first experiment the communication between PC1 and PC2 is measured. In the second experiment two parallel communications between PC1-PC2 and between PC1-PC3 is measured. In the third experiment simultaneous communication between PC1-PC2, PC1-PC3 and PC1-PC5 is monitored. All measurements are repeated for three times. The time t in the table refers to a measurement time. The results calculated as the averages are summarized in the following table:

Table 2. The performance of the network - part 1

Server:	Clients:	Actual throughput STP	Transferred data STP	Actual throughput CEMA	Transferred data CEMA
PC1, t = 10s	PC2	94,5 Mb/s	113 MB	94,5 Mb/s	113 MB
PC1, t = 10s	PC2	47,3 Mb/s	56,5 MB	47,4 Mb/s	56,6 MB
	PC3	47,4 Mb/s	56,6 MB	47,3 Mb/s	56,5 MB
PC1, t= 100s	PC2	31,7 Mb/s	377,34 MB	32,4 Mb/s	386,67 MB
	PC3	31,2 Mb/s	371,34 MB	32,1 Mb/s	382,67 MB
	PC5	33,2 Mb/s	396, 3 MB	31,6 Mb/s	376,67 MB

As we see in the table, the results of the CEMA process are the same as the results of the spanning-tree protocol. To see the situation where the CEMA process has better results than the spanning tree, let extend the topology by adding one more computer - PC6 connected to switch D in the topology:

The CEMA process recognized the same shortest path tree for the PC6 as for the PC4:

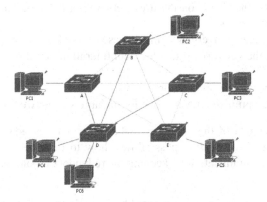

Fig. 5. Shortest path tree for PC6

In the same time communication between PC1 and PC4 and between PC6 and PC5 occurs. All measurements are three times repeated. The results calculated as the averages are summarized in the following table:

Table 3. The performance of the network - part 2

Server:	Clients:	Actual throughput STP	Transferred data STP	Actual throughput CEMA	Transferred data - CEMA
PC1, t = 100s	PC4	49,6 Mb/s	577,67 MB	94,6 Mb/s	1,1 GB
PC6, t = 100s	PC5	46,5 Mb/s	554,34 MB	94,6 Mb/s	1,1 GB

5   Conclusion

As it is shown, the theoretical throughput of the network is using the CEMA processes higher than the theoretical throughput of the network limited by the STP. The theoretical throughput of the topology created by the STP is 400Mb/s. The theoretical throughput of the network managed by the CEMA process is 10 * 100 Mb/s = 1 000 Mb/s.

The performance of the network managed by the CEMA has at less the same efficiency or it is more effective than the performance of the network managed by the spanning tree protocol. The improvements of the CEMA process are more significant in case of simultaneous communication between more end-devices. Thanks to the constructed per-destination (single source) shortest path trees the communication managed by STP on one link may be using the CEMA processes divided into several links.

CEMA is a modular centralized data flow management system. The further research can reach all of the parts of the proposed solution. The further main fields of interest are for example:

- Discovery of the needs set on the used networking devices. These devices do not need to make decisions,
- How to improve the convergence time in case of topology changes or congestion,
- How to adjust the networking topology configuration based on the actual needs of data flows,
- How to improve QOS policy enforcement,
- Investigate and optimize CEMA behavior on large size networks.

The field of interest of this article is highly attractive nowadays. There are many start-up companies investigating possibilities, how to improve the performance and efficiency of L2 communication by keeping the redundant topology.

Acknowledgement. This article was supported by the project of specific science "Application of Artificial Intelligence in Bioinformatics" (1/2015) at the University of Hradec Kralove, Faculty of Informatics and Management, Czech Republic.

References

1. Cisco price list. https://apps.cisco.com/WOC/WOConfigUI/pages/coreconfig/coreconfigui page.jsp#License@8
2. Clark, K., Hamilton, K.: Cisco LAN switching. Cisco Press (1999). ISBN-13: 978-1578700943
3. Donohue, D.: CCNP Switch. Quick Reference, 1st edn., pp. 642–813.Cisco Press, ASIN: B007BAOXI2 (2010)
4. Ethernet. http://www.cisco.com/en/US/tech/tk389/tk214/tsd_technology_support_protocol_ home.html
5. Froom, R., et al: Implementing Cisco IP Switched Networks (Switch) Foundation Learning Guide: Foundation learning for Switch, pp. 642–813, Cisco Press, Indianapolis, USA (2010). ISBN-13: 978-1-58705-884-4
6. Ho Trong, V., et al: Traffic engineering for multiple spanning tree protocol in large data centers. In: 23rd International Teletraffic Congress (ITC), pp. 23–30 (2011)
7. HP Switch Software, OpenFlow Supplement. http://h20000.www2.hp.com/bc/docs/ support/SupportManual/c03170243/c03170243.pdf
8. Hucaby, D.: CCNP Switch Official Certification Guide, 1st edn., pp. 642–813 Cisco Press (2010). ISBN-13: 978-1587202438
9. IEEE Standard for Local and metropolitan area networks - Media Access Control (MAC) Bridges, 802.1D. http://standards.ieee.org/getieee802/download/802.1D-2004.pdf
10. Industrial Ethernet: A Control Engineer's Guide. http://www.cisco.com/web/strategy/docs/ manufacturing/industrial_ethernet.pdf
11. King-Shan, L., Chiou, L.W.: Spanning tree alternate routing bridge protocol. U.S. Patent No. 7,027,453 (2006)
12. King-Shan, L., et al.: STAR: a transparent spanning tree bridge protocol with alternate routing, In: ACM SIGCOMM Computer Communication Review, vol. 32, no. 3, pp.33–46 (2002). ISSN: 0146-4833
13. Measuring end-to-end bandwidth with Iperf using Web100, http://www.csm.ornl.gov/~ dunigan/pam.pdf

14. Meddeb, A.: Multiple spanning tree generation and mapping algorithms for carrier class ethernet. In: Proceedings of the IEEE GLOBECOM Conference (2006)
15. Mirijalily, G., et al.: Best multiple spanning tree in metro ethernet networks. In: Proceedings of the Second International Conference on Computer and Electrical Engineering (2009)
16. OpenFlow: Enabling Innovation in Campus Networks. https://www.opennetworking.org/images/stories/downloads/white-papers/openflow-wp-latest.pdf
17. Padmaraj, M., et al.: Metro ethernet traffic engineering based on optimal multiple spanning trees. In: Wireless and Optical Communications Networks (2005)
18. RFC 5556 - Transparent Interconnection of Lots of Links (TRILL): Problem and Applicability Statement. http://datatracker.ietf.org/doc/rfc5556/?include_text=1
19. Software Defined Networking: The New Norm for Networks. http://www.bigswitch.com/sites/default/files/sdn_resources/onf-whitepaper.pdf
20. Understanding EtherChannel Load Balancing and Redundancy on Catalyst Switches. http://www.cisco.com/en/US/tech/tk389/tk213/technologies_tech_note09186a0080094714.shtml
21. Ventura, L.: Iperf on Windows. http://linhost.info/2010/02/iperf-on-windows
22. Virtual Switching System (VSS) Q&A. http://www.cisco.com/en/US/prod/collateral/switches/ps5718/ps9336/prod_qas0900aecd806ed74b.html
23. Xiaoming, H., et al.: Traffic engineering for metro ethernet based on multiple spanning trees. In: Proceedings of the International Conference ICNICONSMCL (2006)

Decision Support Smartphone Application
Based on Interval AHP Method

Richard Cimler[✉], Karel Mls, and Martin Gavalec

University of Hradec Kralove, Rokitanskeho 62 500 03, Hradec Kralove, Czech Republic
{richard.cimler,karel.mls,martin.gavalec}@uhk.cz,
http://www.uhk.cz

Abstract. Multicriteria decision making based on Analytic Hierarchy Process in the health care mobile application is introduced and studied. The application focuses on processing data from internal sensors of a smart phone as well as external sensors in order to monitor current state of a person. Measured data are partly evaluated in the device in order to identify critical situations such as fall of the person, and then are also sent to a server for the deeper analysis. While using AHP method, pairwise comparison matrices have to be created by experts - in our case doctors. Each expert can have different preferences and thus the resulting matrix, created based on the opinions of several experts, may be inconsistent. The method presented in this paper is based on interval judgments and shows how to merge inconsistent and uncertain preference matrices from several experts to deliver a robust and sensitive model for online machine decision making.

Keywords: Healthcare · AHP · Sensors · Interval judgments · Smartphone

1 Introduction

In a last few years many external sensors capable of monitoring different biometric data has been introduced. Number of different types of sensors and types of biometric data which can be measured grows steadily. Sensors are often combined in a various devices for an example in a shape of a wristband. Gathered data can be transmitted by a wireless technology to the computational unit which stores and presents the data. Some advanced devices are capable of the data evaluation itself and run the actions based on built-in algorithms. Most of the devices are able to communicate with the application for the smart phones. There are several health applications collecting, processing, and transforming healthcare data into information, knowledge, and action. [1] or [2].

Computational performance of the modern smartphones is comparable to the personal computers or laptops. Moreover, various sensors are embedded into smart phones. These sensors are also capable of monitoring numerous different biometric data, which makes smartphones, together with a smartphone computation performance, useful devices capable of monitoring and processing information about state and even health

© Springer International Publishing Switzerland 2015
M. Núñez et al. (Eds.): ICCCI 2015, Part II, LNCS 9330, pp. 306–315, 2015.
DOI: 10.1007/978-3-319-24306-1_30

status of the person. It is possible to monitor the position of the person, not only in the wide area using GPS but position of the body towards the earth surface as well. That enables to monitor the occurrence of critical situations such as a fall of the person to the ground [3]. Even health status can be monitored with a combination of appropriate sensors [4]. Using an accelerometer and a gyroscope it is possible, for example, to monitor the person's breathing. More information can be found at [5]. Other sensors can be connected to the smart phone by wireless connection such as the Bluetooth connection, which enables monitoring of further data.

There are also complex systems such as Framework of Ubiquitous Healthcare System Based on Cloud Computing for Elderly Living [6, 7 or 8], which are based not only on the smart phone capabilities. Several sensors located in the environment transmit data into the computational unit which processes data and executes actions on the actuators based on the built-in algorithms. Such complex systems can use information from smart phone sensors and combine it with information from sensors located in the environment. Computation resources are provided by the cloud based solution. Cloud solutions in the medical sector have been spreading rapidly in recent years [9].

Watch Dog is an application designed for Android mobile phones and its general purpose is to gather the information from internal and external sensors connected to the smart phone. The application was primarily focused on the monitoring of persons with serious health issues. On basis of information from the accelerometer and the gyroscope it was possible to monitor the respiratory arrest [5]. With the increasing number of external sensors and different devices capable to measure various biometric data, application was modified to be prepared to communicate with such sensors, process the data and be able to send the results to the appropriate destination – calling emergency or just notifying relatives about non-standard situation.

Internal sensors of the smart phone can be divided into three categories:

- Movement sensors – sensors capable of measuring the rotation force in three dimensional environment. Accelerometers, gravitation sensors and rotation sensors belong to this category.
- Environment sensors – barometers, photometers and thermometers are measuring environment conditions such as temperature, pressure or light intensity.
- Position sensors – physical position of the device is monitored by orientation sensors and magnetometers.

External sensors capable of measuring different biometric data can be connected by the wireless technology such as Bluetooth [2]. These sensors can measure different biometric data such as blood pressure, pulse, temperature, moisture or the position of the limbs. All the information from sensors is processed in the device or in a case of more complex computation on the server application. Aim of the system is to evaluate the actual state of the monitored person according to the measured values and to make appropriate actions. Some of the measured values can be normal, some of them can be suspicious and some can be strictly alerting. Suspicious values can signify danger only when are observed together with a suspicious values from some the other sensors. The importance of each sensor value can vary according to the context and various types of dangerous situations can be expressed by combination of measured values. That is why we are suggesting to use the Analytic Hierarchy Process in the decision making process. Based on data from particular sensors, the application is expected to make a proposal of the action – to call an emergency, to alert or to continue monitoring the situation. To meet these expectations, not only the data from sensors but expert knowledge is necessary as well. Multicriteria Analytic Hierarchy Process (AHP) [10] model is one of the perspective methods that can naturally merge these two requirements. The experts who will help to adjust the system are doctors – specialists on different medical sectors. There can be different opinions on the importance of the measured values and their combination. Each disease can have a different set of symptoms; moreover every specialist has his/her specific view on the importance of the measured data. Thus the evaluation matrices from the experts could be inconsistent and even contradicting. This situation has to be solved by suitable decision making approach.

2   Analytic Hierarchy Process: Basic Concepts

In AHP approach to decision making, priority scales are generally derived objectively after pair-wise subjective judgments are made. The construction and evaluation procedure of the AHP model can be summarized as follows [11]:

- Model the problem as a hierarchy containing the decision goal, the alternatives for reaching it, and the criteria for the alternatives' evaluation (Fig. 1).
- Establish priorities among the elements of the hierarchy by making a series of judgments based on pairwise comparisons of the elements.
- Synthesize these judgments to yield a set of overall priorities for the hierarchy.
- Check the consistency of the judgments.
- Reach the final decision based on the results of this process.

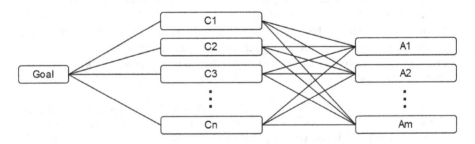

Fig. 1. General AHP model with *m* alternatives and *n* criteria.

In most practical cases, decision making is made by groups (two or more decision makers) rather than by individuals. To help decision makers identifying the most effective and acceptable decision style in the certain situation context, the so called contingency model were proposed [12]. The model, based mainly on observations and interpretations of past situations in companies attracted considerable attention of many practitioners, researchers and also critics [13, 14]. In [15], the new method of evaluation of decision makers' preferences in contingency model based on AHP was introduced.

There are various approaches to reach consensus or at least get the best from several and usually inconsistent experts' judgments in group decision making. Dong [16] proved that both consistency and consensus may not be perfect in general and so there is a need of some measure or degree for them. The consistency measure is then used to quantify the difference among individual decision makers and to help in finding acceptable solution of the consensus model.

Considering AHP as a platform for decision making support, opinions of individual experts can be merged in several ways. Aggregation of individual opinions can be done at the initial stage of the decision making process by searching consensus on the given judgments (aggregation of individual judgments – AIJ) or at the last stage by individually induced preferences with importance weights of the group members (aggregation of individual priorities - AIP) [16]. Between these stages, several other approaches, e.g. voting, can be applied to decomposed intermediate judgments aggregation [17].

3 AHP with Multiple Expert Evaluations

The search for consensus in group decision making is typical in many real world situations. Several methods and tools, based on AHP were suggested to build consensus [18]. In medical applications, the need for more cautious approach to uniting of individual expert opinions is vital [19]. To get the best evaluation of the model and to find the best alternative in the health monitoring application, not only consensus between experts, compromise or aggregation of individual judgments or priorities is sufficient. The individual's health is estimated from immediate and historical data provided by sensors, while these data can be (and usually are) interpreted in different ways by individual experts.

3.1 Measuring Inconsistency in Individual AHP Model

In AHP approach an $n \times n$ pair-wise comparison matrix A with positive elements a_{ij} is used to hold experts' judgments about alternatives. In this paper, for preferences of A_i with respect to A_j and for further calculations, original, multiplicative form will be considered.

$$A = \begin{pmatrix} 1 & a_{12} & \cdots & a_{1n} \\ 1/a_{12} & 1 & \cdots & a_{2n} \\ \cdots & \cdots & \cdots & \cdots \\ 1/a_{1n} & 1/a_{2n} & \cdots & 1 \end{pmatrix} \quad (1)$$

Normalized principal eigenvector w of the matrix A is computed to get priorities of respective alternatives.

$$Aw = \lambda_{max}w, \tag{2}$$

where λ_{max} is the principal eigenvalue of A, $\lambda_{max} \geq n$.

Another commonly used prioritization method is the row geometric mean method or the logarithmic least squares method:

$$w_i = \frac{\sqrt[n]{\prod_{j=1}^n a_{ij}}}{\sum_{i=1}^n \left(\sqrt[n]{\prod_{j=1}^n a_{ij}}\right)} . \tag{3}$$

We say that A is consistent if $a_{ij} \bullet a_{jk} = a_{ik}$ for each i, j, k. If $a_{ij} \bullet a_{jk} \neq a_{ik}$ for some i, j, k, than A is said to be inconsistent. The degree of inconsistency of A is measured by the consistency index CI $= (\lambda_{max} - n)/(n - 1)$. To provide a measure independent of the order of the matrix n, Saaty [20] proposed the consistency ratio CR=CI/CI$_{rand}$ $\lesssim 0.1$, where CI$_{rand}$ is an average consistency index over a large number of random reciprocal matrices of the order n.

3.2   Interval Comparisons in AHP

In general, precise values or comparisons of elements in multi criteria model cannot be obtained and only estimations based on Saaty's fundamental scale are available. This uncertainty in the information can be addressed by a decision maker through the use of an interval judgment rather than a single numerical value [21]. The width of the interval is then a measure of the uncertainty of the individual judgment and can be either minimized by obtaining more relevant information about the system, or it can be utilized for minimizing conflicts and different points of view in group decision making situations. By interval AHP, the priority weights of alternatives and importance weights of decision makers are obtained as interval from the given pairwise comparison matrices. The resulting interval priority weights reflect all possibilities in the given information and can help the group members to understand one another and reach the consensus [17]. Finally, some method of calculation the crisp values from aggregated interval weights can be provided.

Uncertainty of experts in formulating their pairwise comparisons is approximated by the interval $\frac{w_i}{w_j} = \left[\frac{\underline{w_i}}{\overline{w_j}}, \frac{\overline{w_i}}{\underline{w_j}}\right]$, where $\underline{w_k} > 0$ $\forall k$, then $a_{ij} \in \frac{w_i}{w_j} \forall(i,j)$, that is $\frac{\underline{w_i}}{\overline{w_j}} \leq a_{ij} \leq \frac{\overline{w_i}}{\underline{w_j}}$.

Decision makers don't need to express their intuitive subjective judgments as the only strict value, but, vice versa, they can correctly say "I don't know this or that for 100%" or „I am not sure". In this case, we get interval judgments in the comparison matrix in a form

$$A_{ij} = \left[\underline{a}_{ij}, \overline{a}_{ij}\right], \quad A_{ij} \in \frac{w_i}{w_j} \Leftrightarrow \frac{\underline{w_i}}{\overline{w_j}} \leq \underline{a}_{ij} \leq \overline{a}_{ij} \leq \frac{\overline{w_i}}{\underline{w_j}} \tag{4}$$

Uncertainty of the expert judgments can be measured by several methods, e.g. as a sum of all interval lengths in his/her judgments:

$$I = \Sigma_{i,j} \left| \overline{a}_{ij} - \underline{a}_{ij} \right| \tag{5}$$

Let suppose that decision makers use Saaty's fundamental scale, then for $\underline{a}_{ij_{min}} = 1$ and $\overline{a}_{ij_{max}} = 9$ we get $I_{min} = 0$ for no uncertainty and $I_{max} = 8 \cdot \frac{n^2-n}{2}$ for maximal uncertainty of the judgments of individual decision maker. Normalized uncertainty level of the decision maker can be utilized as a weight of the expert in the process of aggregation. Several normalization methods are discussed in more detail in [22].

3.3 Multiple Experts in Interval AHP

In group decision making, each participant can give different pairwise comparisons for the same set of alternatives based on his/her subjective knowledge. The comparison matrices of individual decision makers can be then aggregated by several methods. One of the simplest methods is based on maximum and minimum. The comparisons given by k decision makers are aggregated by taking their maximum and minimum:

$$A_{ij} = \left[\underline{a}_{ij}, \overline{a}_{ij} \right] = \left[\min_k a_{ijk}, \max_k a_{ijk} \right] \forall (i,j) , \tag{6}$$

Aggregated comparisons are interval and are included in the approximated interval priority weights.

Another method deals with direct aggregation of priority weights or weighted priority weights. If the quality of the decision makers cannot be quantified, taking the maximum and minimum of individual priority weights is used:

$$W_i = \left[\underline{w}_i, \overline{w}_i \right] = \left[\min_k w_{ik}, \max_k w_{ik} \right] \forall i , \tag{7}$$

When the importance or quality of decision makers is given by some authority, or estimated by some evaluation of relevant data (e.g. level of uncertainty of their judgments), weighted sum of priority weights can be used:

$$W_i = \left[\Sigma_k p_k \underline{w}_{ik}, \Sigma_k p_k \overline{w}_{ik} \right] \forall i , \tag{8}$$

where p_k stands for the chosen importance weight of the k-th decision maker.

4 Practical Implementation of the Method to WatchDog Application

In our model, several simplifying assumptions were determined. First, only seven sensors were selected to monitor health of the person. There are sensors directly embedded in the smartphone (gyroscope – S1, accelerometer – S2, magnetometer – S3 and proximity sensor – S4) and external wearable sensors for contact measurement of temperature – S5, blood pressure - S6 and heart rate – S7. Next, independence of the measured and preprocessed data from particular sensors is supposed. Corresponding AHP model of the application is shown on Figure 2, where the part of expert evaluation and the part of sensor data are marked.

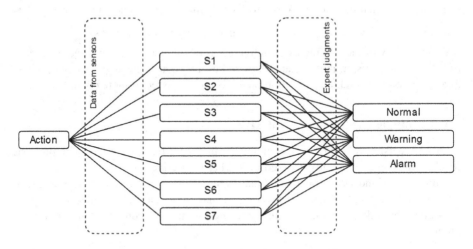

Fig. 2. AHP model for the WatchDog application.

4.1   Numerical Example

Before obtaining the expert opinion, eleven new criteria, based on selected sensors data were introduced in the model:

- Temperature normal
- Temperature high
- Temperature very high
- Heart rate normal
- Heart rate high
- Heart rate very high
- Activity – lying down
- Activity – sitting
- Activity – walking
- Activity – running
- Person out of reach

Six selected experts were asked to write his/her opinion on the preference of alternatives (Alarm, Warning and Normal) with respect to particular criteria. Obtained individual judgments were collected and processed in an Excel spreadsheet. Data from experts, with respect to the criterion *Temperature high*, are displayed in Table 1.

Table 1. Pairwise evaluation of the *Temperature high* component of the AHP model

		MIN			MAX		
		Alarm	Warning	Normal	Alarm	Warning	Normal
	Alarm	1	1	3	1	1	4
Expert1	Warning	1	1	5	1	1	6
	Normal	1/3	1/5	1	1/4	1/6	1
	Alarm	1	1/5	1	1	1	1
Expert2	Warning	5	1	5	1	1	1
	Normal	1	1/5	1	1	1	1
	Alarm	1	1/3	2	1	1/4	3
Expert3	Warning	3	1	4	4	1	5
	Normal	1/2	1/4	1	1/3	1/5	1
	Alarm	1	1/2	1	1	1/3	1
Expert4	Warning	2	1	2	3	1	3
	Normal	1	1/2	1	1	1/3	1
	Alarm	1	1/9	1	1	1	1
Expert5	Warning	9	1	9	1	1	1
	Normal	1	1/9	1	1	1	1
	Alarm	1	1/3	1/2	1	1	2
Expert6	Warning	3	1	1	1	1	1
	Normal	2	1	1	1/2	1	1

For the presented criterion Temperature high, partial uncertainty of the experts' judgments was calculated according (5) as:
$I_{E1} = 2.12$, $I_{E2} = 9.60$, $I_{E3} = 3.3$, $I_{E4} = 2.33$, $I_{E5} = 17.78$ and $I_{E6} = 5.67$, while maximal uncertainty in this case could be $I_{max} = 24$.

4.2 Multiple Experts in Interval AHP

Received individual judgment matrices were merged directly according to (6):

$$A_{ij} = \begin{bmatrix} 1,1 & \frac{1}{9},1 & \frac{1}{2},4 \\ 1,9 & 1,1 & 1,9 \\ \frac{1}{4},2 & \frac{1}{9},1 & 1,1 \end{bmatrix}$$

Partial preferences for the criterion *Temperature high* are:
$\underline{w}_{Alarm} = 0.400$, $\overline{w}_{Alarm} = 0.103$,
$\underline{w}_{Warning} = 0.524$, $\overline{w}_{Warning} = 0.559$,
$\underline{w}_{Normal} = 0.076$, $\overline{w}_{Normal} = 0.339$.

It can be seen, that for the selected criterion *Temperature high*, the *Warning* action has the highest priority weight among other alternatives, both for minimum and maximum expert judgments. Alternatively, individual judgment matrices can be employed to get individual priority weights, and formulas (7) or (8) provided the output of the model.

5   Conclusions and Future Work

The study of several methods of dealing with uncertainty and inconsistency in individual expert's and group of experts' evaluation of the multicriteria AHP model were conducted. The model is expected to become a part of the monitoring and alerting health care application Watch Dog, intended for non-stop online surveillance of elderly or people with health problems. The proposed method is flexible and can be easily extended, depending on new sensor types or demands for different behaviors.

Future study, based on practical results of the application in real tests will be focused on improvements of the model. Analytic network process (ANP) method which can also reflect internal relations within the model, implementation of limiting rules and especially comparison of extremal algebraic methods in computation of the overall model results will be considered as well.

Acknowledgement. The support of Czech Science Foundation GACR project #14-02424S and SPEV project #2106 is gratefully acknowledged.

References

1. Demirkan, H.: A Smart Healthcare Systems Framework. IT Professional **15**(5), 38–45 (2013)
2. Nandkishor, B., Shinde, A. Malathi, P.: Android smartphone based body area network for monitoring and evaluation of medical parameters. In: First International Conference on Networks & Soft Computing (ICNSC 2014), Guntur, pp. 284–288 (2014). doi:10.1109/CNSC.2014.6906663
3. Abbate, S., Avvenuti, M., Bonatesta, F., Cola, G., Corsini, P., Vechio, A.: A Smartphone-Based Fall Detection System. Pervasive and Mobile Computing **8**(6), 883–899 (2012)
4. Pantelopoulos, A., Bourbakis, N.: A Survey on Wearable Sensor-Based Systems for Health Monitoring and Prognosis. IEEE Transactions on Systems, Man, and Cybernetics, Part C (Applications and Reviews) **40**(1), 1–12 (2010)
5. Suba, P., Tucnik, P.: Mobile monitoring system for elder people healthcare and AAL. In: Conference on Intelligent Environments, Athens, vol. 17, pp. 403–414 (2013)
6. Ou, Y.Y., Shih, P.Y., Chin, Y.H., Kuan, T.W., Wang, J.F., Shih, S.H.: Framework of ubiquitous healthcare system based on cloud computing for elderly living. In: Asia-Pacific Signal and Information Processing Association Annual Summit and Conference (APSIPA), Kaohsiung (2013). doi:10.1109/APSIPA.2013.6694298
7. Cimler, R., Matyska, J., Sobeslav, V.: Cloud based solution for mobile healthcare application. In: Proceedings of the 18th International Database Engineering & Applications Symposium on, IDEAS 2014, pp. 298–301 (2014)
8. Suryadevara, N., Mukhopadhyay, S., Rayudu, R., Huang, Y.: Sensor data fusion to determine wellness of an elderly in intelligent home monitoring environment. In: Proceedings of the 2012 IEEE International Instrumentation and Measurement Technology Conference, Graz, pp. 947–952 (2012)

9. Dolezal, R., Sobeslav, V., Hornig, O., Balik, L., Korabecny, J., Kuca, K.: HPC cloud technologies for virtual screening in drug discovery. In: Nguyen, N.T., Trawiński, B., Kosala, R. (eds.) ACIIDS 2015, Part II. LNCS, vol. 9092, pp. 440–449. Springer, Heidelberg (2015)
10. Saaty, T.L.: Decision Making for Leaders: The Analytic Hierarchy Process. RWS Publications, Pittsburgh (2008)
11. Saaty, T.L.: Theory and Applications of the Analytic Network Process: Decision Making with Benefits, Opportunities, Costs, and Risks. RWS Publications, Pittsburgh (2009)
12. Jago, A.G., Vroom, V.H.: Decision Making as a Social Process. Decision Sciences **5**, 743–755 (1973)
13. Brown, F.W., Finstuen, K.: The Use of Participation in Decision Making: a Consideration of the Vroom-Yetton and Vroom-Jago Normative Models. Journal of Behavioral Decision Making **6**(3), 207–219 (1993)
14. Field, R.G.: A Test of the Vroom-Yetton Normative Model of Leadership. Journal of Applied Psychology **67**(5), 523–532 (1982)
15. Mls, K., Otčenášková, T.: Analysis of complex decisional situations in companies with the support of AHP extension of vroom-yetton contingency model. In: Int. conference, IFAC MIM 2013, Part 1, Saint Petersburg, Russia, vol 7, pp. 549–554 (2013)
16. Dong, Y., Zhang, G., Hong, W.-C., Xu, Y.: Consensus Models for AHP Group Decision Making Under Row Geometric Mean Prioritization Method. Decision Support Systems **49**(3), 281–289 (2010)
17. Entani, T.: Interval AHP for a group of decision makers. In: Proceedings of IFSA/EUSFLAT 2009 Congress, pp. 155–160. Calouste Gulbenkian Foundation, Lisbon, Portugal (2009)
18. Moreno-Jiménez, J.M., Aguarón, J., Escobar, M.T.: The Core of Consistency in AHP-Group Decision Making. Group Decision and Negotiation **17**(3), 249–265 (2008)
19. Pecchia, L., Martin, J.L., Ragoyyino, A., Vanyanella, C., Scognamiglio, A., Mirarchi, L., Morgan, S.P.: user needs elicitation via analytic hierarchy process (AHP). A Case Study on a Computed Tomography (CT) Scanner. BMC Medical Informatics and Decision Making 2013 **13**(2) (2013). doi:10.1186/1472-6947-13-2
20. Saaty, T.L., Vargas, J.G.: Diagnosis with Dependent Symptoms: Bayes Theorem and the Analytic Hierarchy Process. Operations Research **46**(4), 491–502 (1998)
21. Moreno-Jimenez, J.M., Vargas, L.G.: A Probabilistic Study of Preference Structures in the Analytic Hierarchy Process with Interval Judgments. Mathematical and Computer Modelling **17**, 73–81 (1993)
22. Pavlačka, O.: On Various Approaches to Normalization of Interval and Fuzzy Weights. Fuzzy Sets and Systems **243**, 110–130 (2014)

Impact of Weather Conditions on Fingerprinting Localization Based on IEEE 802.11a

Peter Brida[✉] and Juraj Machaj

FEE, Department of Telecommunications and Multimedia,
University of Zilina, Univerzitna 1 010 08, Zilina, Slovakia
{peter.brida,juraj.machaj}@fel.uniza.sk

Abstract. In this paper we deal with implementation of outdoor positioning system based on WiFi network works in 5 GHz frequency band (IEEE802.11a). Positioning solution based on 5 GHz WiFi seems to be interesting, because the interference in this bandwidth with other networks is not so critical in comparison with 2.4 GHz WiFi. Positioning was based on fingerprinting method utilizing received signal strength information. The goal of the paper is to investigate an impact of different weather conditions during positioning process on positioning accuracy. Performance of the implemented positioning method was tested in two basic weather conditions, i.e. bad condition: raining and snowing, good condition: sunny. The experimental scenarios were implemented in the outdoor environment.

Keywords: Fingerprinting · IEEE802.11a · Mobile localization · Weather condition · Wifi

1 Introduction

In mobile services the position of mobile device became very important information. Over the last location based services (LBSs) become very popular mainly due to rich offer of smart devices with GNSS (Global Navigation Satellite System) support. In the last years, other commercial positioning solution emerged and can be used to estimate position of users in order to provide LBS services. Growing LBS market introduced new potential, stimulates development of new systems and technologies.

Basic problem for a service provider is to decide, which is the most appropriate positioning system for the deployment of the given LBS. Such decision should be based on the requirements of the given LBS, e.g. required positioning accuracy, mobile device parameters etc. It is important to note that implementation and operating costs may also play very critical role [1] - [4].

Currently positioning systems are commonly implemented in many areas of our life e.g. intelligent transport systems, commercial sector, industrial military and etc. [5, 6]. These systems are most frequently be used for navigation purposes, tracking systems and searching systems.

Each of many existing positioning systems has its own advantages and disadvantages. Most commonly used GNSSs offer world-wide coverage with satisfactory

M. Núñez et al. (Eds.): ICCCI 2015, Part II, LNCS 9330, pp. 316–325, 2015.
DOI: 10.1007/978-3-319-24306-1_31

accuracy in the outdoor environment. On the contrary, performance of these systems significantly decreases in dense urban areas due to poor signal quality. It is mainly caused by obstacles in line of sight and multipath propagation. In environments like this, positioning systems based on the radio communication networks seem to be more suitable [7] - [10].

Positioning systems used in the outdoor environment are generally based on observations of various signal parameters. The most common positioning methods are based on measurements such as Cell Identification (Cell ID), Time of Arrival (ToA), Received Signal Strength (RSS), or Angle of Arrival (AoA) [1], [3], [7], [9], [11] - [14]. Most of the measurements utilized by positioning methods are sensitive to multipath propagation. Consequently, these methods are not suitable for use in the dense urban environments.

Time based methods have potential to be the most accurate, however they are too expensive to implement in current networks. On the opposite side, basic methods (Cell ID and RSS) do not have high demands on implementation costs, but their accuracy is lower. Therefore it is necessary to utilize positioning method, which can combine advantages of both systems e.g. is accurate and does not require high implementation costs.

Research in the area of positioning has been mainly focused on the utilization of the received signal strength information [15] - [18]. This might be based on the fact that signal strength information is measured in almost all mobile devices. Therefore, utilization of this measurement does not require any modification to the mobile device, thus reducing the implementation costs. There are many various ways how to process RSS information in order to estimate position of the device. The most popular methods based on RSS are trilateration and fingerprinting method. Fingerprinting method seems to be more suitable for Non-Line-of-Sight (NLoS) environment, due to its immunity to multipath environment. Based on this analysis we decided to use Fingerprinting method in this work.

In our work we use fingerprinting method that utilizes RSS measurements. Generally, there is the need to perform calibration phase. This can be performed either by measurements in a real environment or by prediction as described in [16]. The first case can be quite time consuming, however measurements are more precise since real RSS information are collected. On the other hand, prediction of RSS is faster and more comfortable. In such case data used for the position estimation are highly dependent on the quality of map model of the given environment and on propagation model used during the simulation. There is a compromise between the demanding effort and accuracy as introduced in [16].

Large number of positioning systems based on fingerprinting is implemented in IEEE802.11 networks. These systems are mainly developed for the indoor environment, but can also be successfully implemented in outdoor. In this paper, we have implemented positioning system based on IEEE802.11a in the outdoor environment. This type of wireless network is not implemented as often as standards, which operate in 2.4 GHz frequency band, since it is mainly used in the outdoor environment.

This paper investigates the accuracy of the implemented positioning system based on the IEEE802.11a network and utilizing fingerprinting positioning method. We focus on an impact of different weather conditions and number of moving reflectors in the area on the accuracy of implemented system. It is assumed that weather conditions should play important role in outdoor environment in the observed frequency band (5 GHz). The positioning accuracy is evaluated by mean, median and 95% error.

The paper is structured as follows; Section 2 deals with the theory and related work in area of fingerprinting localization. Test scenarios are described in the Section 3. In the Section 4 the achieved results are shown and discussed. The Section 5 concludes the paper.

2   Fingerprinting Positioning Method

From the mathematical point of view fingerprinting algorithms can be implemented in various ways. The most commonly fingerprinting algorithms are divided into two main frameworks - deterministic and probabilistic. In this paper fingerprinting positioning based on the deterministic algorithm Nearest Neighbor (NN) is implemented and tested. This algorithm should achieve similar accuracy compared to more advanced algorithms, if density of reference points in radio map is high enough [19].

It can be stated that accuracy of positioning system based on deterministic fingerprinting algorithms is affected by two basic assumptions. The first assumption is that signal properties vary a lot at relatively small area. For instance, signal from an AP (Access Point) can get attenuated, get lost or be replaced with a stronger one in the relatively short distance. The second is that these signals are assumed to be relatively constant in time. Based on these it is possible to use date stored in the radio map to estimate position of mobile device.

Main drawback of the fingerprinting positioning is high sensitivity for changes in the area of positioning such as movement of objects (e.g. people, cars), since these can alter signal properties.

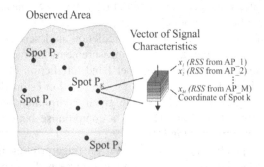

Fig. 1. Radio map for fingerprinting using RSS

Function of all fingerprinting methods can be divided to two phases. During the first phase, radio map must be created for an area where localization service will be provided (see Fig. 1). This radio map is commonly stored in the database at the localization server. Radio map is in fact database of reference points with known position (coordinates) coupled with measured radio signal properties, e.g. RSS, AoA, or signal propagation time. This phase is frequently called off-line phase or calibration phase.

When the radio map database is created and stored in the localization server, the second phase can be performed. This stage is widely named on-line phase or positioning phase. Required signal properties are measured by MS at unknown position and sent to localization server. Afterwards, the algorithm implemented at the server is used to find the best match from reference points stored in the radio map. The match is actually the reference point with the highest similarity and its coordinates are considered as position estimate (for NN method).

2.1   Localization Algorithm

Euclidean algorithm is the most common method to find the most similar reference point from the radio map database and actually estimate position. Let's assume a radio signal fingerprint stored in the radio map and characterized by vector **P**:

$$\mathbf{P} = [x_j] = [x_1, \ldots, x_M], \tag{1}$$

where x_j defines the reference point, i.e. measured values of the given signal property, in our case RSS. M represents number of APs detected during the radio map construction. In general, let's consider that the radio map consists of fingerprints collected at N reference points.

$$\mathbf{P}_i = [x_{ij}] = [x_{i1}, \ldots, x_{iM}], \ i = 1, \ldots, N. \tag{2}$$

Together with measured signal properties unique identifiers of APs (commonly MAC address and SSID) as well as coordinates of reference points are stored in radio map. These are coupled with x_i. In order to keep the model simple we decided not to show them here. The whole radio map contains fingerprints from all reference points $\mathbf{P}_i$ and creates the set **S** defined as

$$\mathbf{S} = \{\mathbf{P}_i : i = 1 \ldots N\}. \tag{3}$$

During the localization process, the MS measure desired signal properties at unknown position and new fingerprint **Q** is obtained

$$\mathbf{Q} = [y_j], \ j = 1, \ldots, M. \tag{4}$$

After the fingerprint **Q** is received at the localization server the algorithm calculate Euclidean distance d_k between vectors $\mathbf{P}_k$ and **Q**. The Euclidean distance is defined as:

$$d_k = |\mathbf{P}_k - \mathbf{Q}| = \sqrt{\sum_{j=1}^{M} (x_{kj} - y_j)^2}. \tag{5}$$

When Euclidean distance is calculated for all fingerprints stored in the radio map, the vector of distances **D** is obtained. The vector **D**, contains all distances between vectors $\mathbf{P}_i$ stored in the radio map and vector **Q**, and can be calculated as

$$\mathbf{D} = [d_i] = \left[|\mathbf{P}_i - \mathbf{Q}| \right] = \left[\sqrt{\sum_{j=1}^{M} (x_{ij} - y_j)^2} \right], \; i = 1,...,N. \tag{6}$$

The position estimate is defined by the element of vector $\mathbf{D}$ with minimum value. Based on the position of this element within the radio map, it is possible to find coordinates of the reference point that are considered as the position estimate.

3   Experimental Scenario

We proposed the experimental setup in order to investigate the impact of weather conditions and amount of surrounding movable reflectors. The environment can be characterized as sub-urban/rural outdoor environment with buildings, roads, trees and etc. Experimental setup was implemented at the campus of the University of Zilina. The observed area is shown in the Fig. 2. Size of the area where positioning system was deployed is 450x100 meters. Trajectory of the mobile terminal was the same during both phases and is highlighted by red line.

The experiments were carried out using positioning system WifiLOC. The system is in development at the University of Zilina as a research tool. WifiLOC utilizes RSS information from surrounding Wi-Fi APs to estimate position of a mobile device. The system is based on the above mentioned fingerprinting positioning algorithm and RSS information. Detailed architecture and functions of WifiLOC are described in [22]. The positioning system is able to utilize any communication platform implemented at the mobile device. In this paper we focus on the IEEE802.11a technology, i.e. 5 GHz ISM (Industrial, Scientific and Medical) frequency band.

Six APs (Nanostation M5 Loco - Ubiquiti Networks™) were distributed around the area on the roofs of the surrounding building. Each AP worked in different frequency channel, i.e. channels were not interfering themselves. All used APs were in the communication range over the whole observed area and Line of Sight (LoS) signal propagation was dominant. Measurements were performed using mobile terminal *Lenovo ThinkPad T431s* equipped with reference external GPS device *GeoExplorer GeoXH* for accurate position determination. Since the mobile terminal was not equipped with omnidirectional WiFi antenna, it is necessary to take into account its orientation during the measurements. Hence, the antenna orientation was same during the all experiments.

The signal power in the used frequency band can be influenced not only by moving reflectors (pedestrians, cars) in the observed area, but also by multipath propagation and fading. In order to eliminate the problem of stochastic signal power fluctuations, an estimation of local average power was used. Local average power was calculated from the measured data as

$$\overline{RSS} = \frac{1}{N_s} \sum_{i=1}^{N_s} RSS_i, \tag{7}$$

where N_s is the number of samples, in the measurements we used $N_s = 10$.

Fig. 2. Experimental area – University of Zilina campus

During the off-line phase, fingerprints were created at the reference points (at red line). Local average powers from all APs were calculated and sent to the localization server as a fingerprint. At the server the fingerprints were stored in the radio map database. Accurate coordinates of the reference points were estimated using GPS module. During the both phases, the speed of mobile terminal was app. 3 km/h. The distance between particular reference points was about 3 m. During the on-line phase, 100 independent position estimations per scenario were performed to investigate the impact of various conditions on the positioning system.

The measurements were performed under various conditions. In principle, we investigate two factors: weather conditions during positioning process and amount of movable reflectors in the observed area with interaction with the first factor. The mentioned factors were monitored and distinguished in both phases.

Firstly, we tried to investigate an impact of various weather conditions (snowing, raining and sunny) during the positioning process on the accuracy. Therefore, the snowing and raining is denoted as bad weather conditions in these experiments. On the other hand, the good weather condition is identified as sunny weather. Based on previous experiments impact of amount of movable reflectors on positioning accuracy has to be taken into account. The impact of movable reflectors on positioning accuracy was previously investigated in [20, 22]. Therefore, following conditions were defined:

1. Condition 1: the bad weather conditions during the experiment.
2. Condition 2: the good weather conditions during the experiment.
3. Condition 3: many of movable reflectors were situated in the area.
4. Condition 4: no movable reflectors were situated in the area.

According to goals of the experiment, the mentioned conditions were purposely combined into experimental scenarios which are shown in Table 1. Particular conditions for each scenario were valid for both phases, i.e. off-line and on-line.

Table 1. Experimental scenarios – 1st experiment

	Conditions	
Scenario 1	1	3
Scenario 2	1	4
Scenario 3	2	3
Scenario 4	2	4

The next experiment was focused on investigation of different impact of the above mentioned factors in on-line and off-line phase. For example whether the bad conditions during the on-line phase have greater impact compared to the off-line phase. Therefore we defined two more scenarios in Table 2.

Table 2. Experimental scenarios – 2nd experiment

	Off-line phase		On-line phase	
	Conditions		Conditions	
Scenario 5	1	4	2	4
Scenario 6	2	4	1	4

The situation with influence of reflectors on fingerprinting positioning accuracy is rather clear. The static reflectors should not have significant negative impact, e.g. building, trees, parking cars. Some of the results were introduced in [22]. The achieved results are analyzed in the following section.

4   Experimental Results

This section describes results obtained by WifiLOC positioning system based on IEEE802.11a. The performance of the system is evaluated by standard statistical characteristics.

Impact of weather conditions on the performance of WifiLOC system were investigated in the first experiment. The obtained results are depicted in the following Table and Fig. 3.

Table 3. Experimental results – 1st experiment

	Positioning error [m]		
	Median	Mean	95%
Scenario 1	12.53	20.12	43.02
Scenario 2	9.33	11.95	29.15
Scenario 3	12.15	19.37	42.74
Scenario 4	9.14	11.69	28.52

On the basis of the obtained results, it can be concluded the weather conditions have not significant impact on the positioning accuracy. The results are almost same for Scenarios 1 and 3 and Scenarios 2 and 4. The impact of bad weather conditions was not significant in comparison to general error of fingerprinting positioning method. It means that the weather effect is minimal. Quantity of movable reflectors in the area influenced the positioning accuracy more significantly. The error was higher by approximately 3 m (median value) in case when higher number of reflectors was present in the area. This caused higher instability of received signal strength. This problem can be partially solved by implementation of RBF (Rank Based Fingerprinting) positioning method into WifiLOC [21].

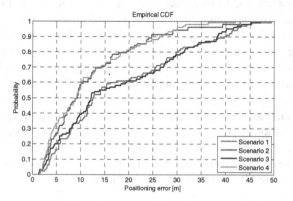

Fig. 3. Experimental area – University of Zilina campus

The second experiment was proposed to determine dependency between different weather conditions during on-line and off-line phase and positioning accuracy. Achieved result can be seen in Table 4.

Table 4. Experimental results – 2nd experiment

	Positioning error [m]		
	Median	Mean	95%
Scenario 5	9.19	11.55	27.22
Scenario 6	9.47	11.96	28.42

The obtained results confirmed the first experiment results that the impact of weather conditions is not significant. The difference is minimal, but the bad weather conditions during the on-line phase have affected the positioning accuracy more negatively than bad weather conditions in the off-line phase.

Based on the results it can be seen that in all combinations of weather conditions the achieved accuracy is very similar. The difference between best and worst case is only 0.4 m, which in comparison to the error of the positioning method (in this case approximately 9 m) is not significant.

5   Conclusion

We investigated the impact of weather conditions on outdoor positioning system based on fingerprinting positioning method. The implemented positioning system was based on IEEE802.11a platform, i.e. 5 GHz in ISM band. The advantage of this platform is smaller interference compared with mostly used 2.4 GHz band.

On the basis of achieved results, it can be concluded that IEEE802.11a is suitable for mobile positioning in outdoor environment as backup solution when GNSSs cannot be used. Based on the presented results it can be concluded that the change of weather conditions does not have negative effect on the positioning accuracy. This fact is very important due to radio map creation issue. The system providers do not need more variants of radio maps for different weather conditions.

According to obtained results, quantity of movable reflectors in the area where positioning service is provided has more significant influence on the positioning accuracy compared with weather conditions. In such case, higher number of signal reflections is presented in the area, causing higher variations of the RSS.

Acknowledgement. This work was supported by the Slovak VEGA grant agency, Project No. 1/0394/13 and by EUREKA project no. E! 6752 – DETECTGAME: R&D for Integrated Artificial Intelligent System for Detecting the Wildlife Migration and by Centre of excellence for systems and services of intelligent transport, ITMS 26220120028 supported by the Research & Development Operational Programme funded by the ERDF.

References

1. Wu, C., Yang, Z., Liu, Y., Xi, W.: WILL: wireless indoor localization without site survey. In: INFOCOM 2012, pp. 64–72 (2012)
2. Ibrahim, M., Youssef, M.: Cell Sense: An Accurate Energy-Efficient GSM Positioning System. IEEE Trans. Vehicular Technology **61**, 286–296 (2012)
3. Tokarcikova, E., Kucharcikova, A.: Diffusion of Innovation: The Case of the Slovak Mobile Communication Market. International Journal of Innovation and Learning **17**(3), 359–370 (2015)
4. Cerny, M., Penhaker, M.: Wireless Body Sensor Network in Health Maintenance Systems. Journal Electronics and Electrical Engineering **115**(9), 113–116 (2011)
5. Alzantot, M., Youssef, M.: UPTIME: ubiquitous pedestrian tracking using mobile phones. In: Wireless Communications and Networking Conference 2012 (WCNC), pp. 3204–3209 (2012)
6. Krejcar, O., Jirka, J., Janckulik, D.: Use of Mobile Phone as Intelligent Sensor for Sound Input Analysis and Sleep State Detection. Sensors **11**(6), 6037–6055 (2011)
7. Bahl, P., Padmanabhan, V.N.: RADAR: an in–building RF–based user location and tracking system. In: IEEE Infocom 2000, vol. 2, pp. 775–784 (2000)
8. Krishnan, P., Krishnakumar, A.S., Wen-Hua, J., Mallows, C.: A aystem for LEASE: location estimation assisted by stationary emitters for indoor RF wireless networks. In: IEEE Infocom 2004, vol. 2. pp. 1181–1190 (2004)

9. Cipov, V., Dobos, L., Papaj, J.: Cooperative Trilateration-Based Positioning Algorithm for WLAN Nodes Using Round Trip Time Estimation. Journal of Electrical and Electronics Engineering. **4**, 29–34 (2011)
10. Matic, A., Papliatseyeu, A., Osmani, V., Mayora-Ibarra, O.: Tuning to your position: FM-radio based indoor localization with spontaneous recalibration. In: IEEE International Conference on Pervasive Computing and Communications (PerCom), pp. 153–161 (2010)
11. Wang, X., Wang, Z., O'Dea, B.: A TOA-Based Location Algorithm Reducing the Errors Due to Non-line-of-sight (NLOS) Propagation. IEEE Tran. Vehicular Technology **52**(2003), 112–116 (2003)
12. Haibo, Y., Tao, G., Xiaorui, Z., Jingwei, X., Xianping, T., Jian, L., Ning, J.: ftrack: infrastructure-free floor localization via mobile phone sensing. In: IEEE International Conference on Pervasive Computing and Communications (PerCom 2012), pp. 2–10 (2012)
13. Abdellatif, M., Mtibaa, A., Harras, K., Youssefy, M.: Greenloc: an energy efficient wifi localization on mobile phones. In: IEEE International Conference on ICC 2013 - Selected Areas in Communications Symposium, pp. 4425–4430 (2013)
14. Sen, S., Radunovic, B., Choudhury, R.R., Minka, T.: You are facing the mona lisa: spot localization using PHY layer information. In: 10th international conference on Mobile systems, applications, and services 2012, pp. 183–196 (2012)
15. Moghtadaiee, V., Dempster, A.: Indoor Location Fingerprinting Using FM Radio Signals. IEEE Trans. on Broadcasting **60**(2), 336–346 (2014)
16. Kurner, T., Meier, A.: Prediction of Outdoor and Outdoor-to-Indoor Coverage in Urban Areas at 1.8 GHz. IEEE Journal on selected areas in communications **20**(3), 496–506 (2002)
17. Laoudias, C., Piché, R., Panayiotou, C.G.: Device Self-Calibration in Location Systems Using Signal Strength Histograms. Journal of Location Based Services **7**(3), 165–181 (2012)
18. Nerguizian, C., Despins, C., Affès, S.: Indoor geolocation with received signal strength fingerprinting technique and neural networks. In: de Souza, J.N., Dini, P., Lorenz, P. (eds.) ICT 2004. LNCS, vol. 3124, pp. 866–875. Springer, Heidelberg (2004)
19. Tsung-Nan, L., Po-Chiang, L.: Performance Comparison of Indoor Positioning Techniques Based on Location Fingerprinting in Wireless Networks. In International Conf. on Wireless Networks, Communications and Mobile. Computing **2**, 1569–1574 (2005)
20. Kyklos, P.: Implementation of Localization System WifiLOC Based on IEEE802.11a for Outdoor Positioning. Diploma thesis, University of Zilina, p. 63 (2013)
21. Machaj, J., Brida, P., Piché, R.: Rank based fingerprinting algorithm for indoor positioning. In International Conference on Indoor Positioning and Indoor Navigation (IPIN), pp. 1–6 (2011)
22. Brida, P., Machaj, J.: Mobile positioning solution suitable for intelligent transportation system based on IEEE 802.11a. In: New Trends in Software Methodologies, Tools and Techniques SOMET_2014, Langkawi, Malajzia, pp. 327–336 (2014)

Enterprise ICT Transformation
to Agile Environment

Ales Komarek, Vladimir Sobeslav$^{(\boxtimes)}$, and Jakub Pavlik

Faculty of Informatics and Management, University of Hradec Kralove,
Rokitanskeho 62, Hradec Kralove, Czech Republic
{ales.komarek,jakub.pavlik.7,vladimir.sobeslav}@uhk.cz
http://uhk.cz/

Abstract. This paper describes how modern configuration management tools change the rate of change of enterprise processes. Enterprise rely on standard frameworks like ITIL, TOGAF or COBIT to cope with change. These frameworks define process flows how to model, deploy and run complex systems in predictable way. Using these frameworks enterprise can archive predictable outcomes in given time. The speed of required changes has grown and traditional configuration tools like IBM Tivoli Configuration and Change Management Database and IBM SmartCloud Control Desk soft cannot cope.

This papers shows how new generation of CM tools like Puppet, Salt with cloud computing resources can speed up the process of bringing the system architecture models to reality deployments. This process is realised by several automated steps and draws all needed information from CMDB and transformation process can be governed by ITIL framework.

Keywords: Network · Graph · Hierarchy · Flow

1 Introduction

This paper explains how adopting configuration management [7–9] and cloud technologies within enterprise can rapidly speed up service transition processes within standard enterprise architecture frameworks [1,4]. This leads to increased efficiency in resource usage or management and gives valuable competitive advantage.

The enterprise environment usually consists of many interconnected systems that support or enable vast majority of company processes. Some of the system may be even named as legacy systems. This pejorative term, referencing a system as "legacy", often implies that the system functions are out of date or in need of replacement. The configuration management with cloud services can provide vital option in choosing the right replacement for handling or replacing legacy systems.

The section introduces frameworks that govern modern enterprise architectures [1,2] within large scale organisations. It brings up the the most serious

© Springer International Publishing Switzerland 2015
M. Núñez et al. (Eds.): ICCCI 2015, Part II, LNCS 9330, pp. 326–335, 2015.
DOI: 10.1007/978-3-319-24306-1_32

problems that organisations are facing today. The problem in not in the frameworks but more in tools used to enforce it. Rigid monolithic solutions are not capable to cope with changes modern technologies require.

The next section covers new options in handling service transitions. The rise of cloud computing and open-source configuration management tools allowed alignment of service developers and service operators to form a new, more flexible ways to deliver designed services into actual operations.

The section compares transition metrics of various hardware providers as well as deployment speeds of configuration management. The test scenario was clean installation and update of sample service stack. The first stack is standalone application server with database server simulating common staging environment. The second stack is the same application in cluster with database backup server on standby simulating service production environment. The hardware models were created for cloud provider services (VMWare, Amazon and OpenStack) and software models for configuration management tools (Puppet, Salt, Ansible).

2 Standard IT Governance Problems

There are several frameworks today that govern IT services alignment with the real business needs. The most renown are ITIL and TOGAF. Information Technology Infrastructure Library (ITIL) is a set of practices for IT service management (ITSM) that focuses on aligning IT services with the needs of business [4]. In its current form ITIL is published as a series of five core volumes, each of which covers a different ITSM lifecycle stage. The Open Group Architecture Framework (TOGAF) is a framework for enterprise architecture which provides an approach for designing, planning, implementing, and governing an enterprise information technology architecture [1].

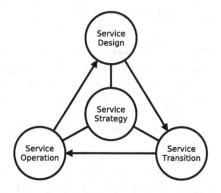

Fig. 1. ITIL Process Cycle

These frameworks deal with complete lifecycle of services. As shown in figure 1 the process starts with service design, where necessary service models are created, then follows the transition phase which is responsible to create real system according to the designed model. [4] The last part cover actual service operation which eventually leads to start of another cycle.

2.1   Problems Statement

This subsection describes most common problems of enterprise architectures that traditional companies are facing today.

Heterogeneous Infrastructure. In many cases no only single systems are legacy but entire infrastructure is shattered into many departments, hardware platforms and corresponding management tools. The one of the reasons for this state where organisation needs many domain experts covering each of the systems. The heterogenity starts by operating system in use, AIX, RHEL, Solaris, OracleLinux running different hardware platforms for example. Each department usually deals with subset of the systems and chooses or creates its management tool to ease the deployment tasks on both hardware and software level.

Simple Compute Virtualization. The increasing power of hardware cannot not be efficiently utilised by single system which is reason for the rapid spread of virtualization techniques. Virtualization allows the creation of virtual resources, in this case mostly computers, within physical servers. The differences in virtualization approaches is beyond scope of this paper, but the important thing is that this allowed better and faster allocation of resources to projects. Simple virtualization did not change the process of creating and managing entire systems (which often consist of many computers, networks, storage disks, external services, etc) but eased one step in the process. The problem is that not every system supports virtualization so this bonus leads to further fragmentation of infrastructure from management point of view.

Systems Deployments. Different software or hardware technologies need IT specialist for long-term substitutability. This leads to gradual separation of teams to very domain specific units, which in term rapidly increase cost and decrease efficiency and agility of the organization as whole.

"Legacy world lives with its makers"

Manual and temporarily installations of complex systems are very often closely associated with their makers. In this context maker stands for IT specialist who deploys systems into production. Although is usually some documentation guide about installation and troubleshooting available, it is not usually accurate and repeatable without further knowledge of the system. The system description is usually uncomplete and after some time no one really knows how to build or migrate it. This creates need for larger operational teams that eventually leads to a formula **"system = IT specialist (or a team)"**.

The fragmentation of the infrastructure and it's management by the different departmets (Storage dpt, Network dpt, Compute dpt, Development dpts) leads to the processes that spawn accross several departments and people. The process flow suffers from the information transfer delays as the work orders are passed to the people in the chain. This is rather rigid process can take several days as not all orders are processed in timely manner.

2.2 The Goals

This section shows possible solutions to address some of the problems.

Speed up Resource Life-Cycles. The best case scenario has fully automated processes for entire life-cycles of the systems. For hardware stack this means all the time from the purchase, installation, operation and finally decommission. For software this means dealing with several operation environments, usually production, staging, testing and multiple development environments. Systems are usually combination of hardware and software resources, where applications run on dedicated servers with dedicated storage drivers, with the possibility of having multiple instances of the system serving operation environments for software development. This can lead to real nightmare, when for example having 700 systems which are crucial and need further development, where at least 3 operations are needed leeds easily to more than 2000 system instances that need to be handled.

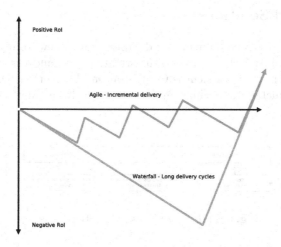

Fig. 2. Value of Agile Release Cycles

New Product Development. The inability to change the infrastructure fast enough, allocate hardware resources for new systems or environments can cripple the business as the new products do not reach the market in time ahead of competition. Developers need get quickly to point where the colleagues have stopped and make their additions without worrying too much how this point was archived. This can be solved by DevOpsteams where members from traditional operation deparments are placed to each development team to support their needs [5]. As we have seen this often lead to even further fragmention in tooling where each operator gets separated and heavily influenced by the assigned development team.

Hardware Resources Life-Cycles. Software relies entirely on the hardware stack, where compute resources are easy to virtualise, the storage and networks virtualization is a little different story. Each vendor provide its tools to manage it's hardware so when a company owns storages from sereral vendors it needs several times more experts to support each vendor and it's tool. The same applies for the network where many companies use Cisco devices and not having experts for other vendors forces companies to use it. The cloud technologies have answers to many of the problems posed by hardware management, abstracting different vendors under unified API for virtual compute, storage and network resources and hide it's entire life-cycle behind several API calls. Hardware compute systems that do not support virtualization can be managed by LCM tools in very similar manner as cloud compute resources, this is covered further in the text.

3   Proposed Solutions

This section proposes components and groups for a new configuration management approach, which includes transformation way and target cloud-ready infrastructure. The basic element is Configuration Management Database which is standardized metadata storing way for either Hardware and Software components.

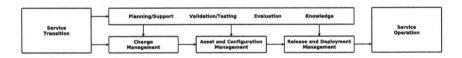

Fig. 3. ITIL Service Transformation Phase

3.1   Configuration Management Database

As already mentioned Configuration Management Database is cornerstone for whole solution. It is formalized metadata configuration in hierarchical database for each specific entity in infrastructure like network, storage, compute, application, database, etc. The advantages of hierarchical databases are in direct integration with specific configuration management. This enables immediately and dynamically run infrastructure orchestration, changes, deployments and it is not only statical description for current state [10].

There representation of a new dynamic hierarchical CMDB are tools like Foreman, Hiera/Puppet, Reclass, etc. The Section 4 introduces several tools and applies them to sample service application stack.

Our transformation solution divides enterprise IT infrastructure into to domains:

- Hardware Configuration Management - infrastructure as group of resources managed in a standardized way.
- Software Configuration Management - application and middle-ware components runs on top of hardware resources.

3.2 Hardware Configuration Management

Hardware or resource configuration management includes environments like Virtualization, Cloud Computing, Infrastructure as a Service, Platform as a Service, etc. In general, CM looks at these components like resources with appropriate relationships between them.

Hardware Configuration Management works with following components:

- Hardware devices - Routers, Switches, Wireless devices, ...
- Virtualization - VMware vSphere, Microsoft Hyper-V, Linux KVM/QEMU, XEN, Containers, PowerVM ...
- Infrastructure as a Service - VMware vCloud Director, Amazon AWS, OpenStack, CloudStack, OpenNebula, ...
- Platfrom as a Service - Heroku, RedHat OpenShift, Cloud Foundry ...

From here on all infrastructure is defined as precise models and governed by the configuration management tool. Transformation of platform infrastructure is steady process after services are processed into configuration management infrastructure. Transformation is done system after system and rate depends on quality of existing documentation, models and standards.

3.3 Software Configuration Management

All major systems must be identified and decomposed to separate atomic services. This is rather difficult task as all of the specific domain experts must cooperate in creation of suitable service models and automation of installation and deployment processes.

- Operating Systems - AIX, RHEL, Solaris, OracleLinux, Windows, ..
- Database Systems - OracleDb, MS SQL, MySQL, ...
- Application Platforms - Tomcat, JBoss, Glassfish, ...
- Support systems - Monitoring, Firewalls, Access, ...

The goal is to identify possible problems in incorporating specific services into configuration management infrastructure.

3.4   Target Cloud-Ready Infrastructure

Target Cloud-ready Infrastructure exposes all required interfaces for configuration management infrastructure covered further in the text. Not the entire infrastructure is cloud-driven but majority of virtualization resources should be reasonable well ready already for operation in cloud environments.

It actually consist of only 2 phases. In the 1st the CM orchestration client creates or prepare the necessary HW/cloud resources like servers, networks, load balancers, public ip addresses, routers, etc. After all resources have been successfully deployed the 2nd phase is stated. It orchestrates services (Software Configuration Management) across the application stack in the right order, ie. the database services are installed before the application services that require them.

It maps previous CM systems in following way:

- Hardware Configuration Management - 1st phase creation and deployment of infrastructure resources - Hardware Resource Orchestration
- Software Configuration Management - 2nd phase orchestration of application services - Software Orchestration

The following schema illustrates how this deployment process works with the IaaS platform.

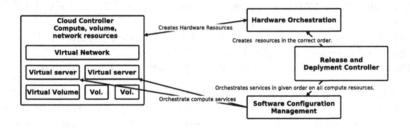

Fig. 4. General Virtual Resource Orchestration

Except simple resource management it can offer auto-scaling features. This is integration with telemetry, which can create alarms based on instance CPU usage and associate actions like spinning up or terminating instances based on CPU load. For example if target application enables adding arbitrary number of front ends, this feature can automatically launch and terminate instances based on load.

Target Cloud-ready enables rapidly decrease time for application life-cycles development from weeks to hours, dynamically responds on a new business requirements and reduces manual errors.

Following Section 4 takes sample application clusters and analyses and measure whole processes by different tools for hardware and software configuration management.

4   Service Transition Tools Comparison

This section provides comparison of service transition speeds of several different resource orchestrators and configuration management tools during deployment of sample enterprise application stack. The compared metric is a transition speed, time for resource allocation and time for software deployment.

Sample Enterprise Application Stack. The tested application stack was business intelligence software CloverETL which is a rapid, end-to-end data integration solution. Measurements and comparison was done in two separate scenarios:

- Single node scenarion - 1x application server (java, RedHat JBOS, CloverETL), 1x database server instance (Oracle DB)
- Cluster scenario - 1x load balancer, 3x application server (CloverETL in clustered mode), 2x database servers in active-standby mode (Oracle DB in High Availability)

Unfortunatelly, we cannot explain all aspects of CM tools, because the space is limited in this paper. Therefore only time for transition deployment is compared.

4.1   Hardware Resource Management

Virtual hardware resource management tools orchestrate first phase of deployment process. Several different platforms that provide required functions to provision virtual and physical resources exist today. The three most reknown solution from VMware, Amazon and open source community OpenStack were analyzed and tested.

- **VMware vRealize Automation** is an enterprise tool that gives end users the ability to consume your organization's cloud while giving cloud administrators the tools to curtail the consumption [14].
- **OpenStack Heat** is the main project of the OpenStack orchestration program. It allows users to describe deployments of complex cloud applications in declarative text files called templates. These templates are then parsed and executed by the Heat engine and projected into virtual resources [12].
- **CloudFormation** is easy way to create and manage a collection of related AWS resources, provisioning and updating them in an orderly and predictable fashion [13]. Uses similar templates as Openstack Heat.

Table 1 shows time how long takes provisioning of virtual resources for CloverETL in sigle and cluster scenario. Virtual resources for cluster scenario are provisioned up to 30 minutes. Nowadays enterprise companies with legacy systems spends more than two weeks with manual provisioning through different IT departments.

Table 1. Speed of service deployment comparison

CM tool	Single node scenario	Cluster cluster
VMware vRealize Automation	10-15 min	25-30 min
OpenStack Heat	2-4 min	5-10 min
AWS CloudFormation	5-10 min	10-15 min

4.2   Software Orchestration

Following tools were chosen as configuration management tools for software configuration.

- Puppet - Is ruby based configuration management tool which started in 2005. Uses HTTPS or AMQ communication (mCollective) to communication, Agent based, modules.
- SaltStack - Python based, Started 2011, now version 2014.7, Standalone, 0MQ (SSH) communication, Agent based, states, formulae, grains
- Ansible - Python based, Started 2012, now version 0.17.2, SSH (0MQ) communication, Agentless, commands, playbooks

Table 2 shows service transition speed comparison of speed deployment.

Table 2. Speed of deployment comparison

CM tool	Single node scenario	Cluster scenario
Puppet	45-50 min	50-60 min
SaltStack	20-25 min	30-40 min
Ansible	20-25 min	30-40 min

5   Conclusion

The analysis introduced problems that traditional enterprise is facing today. Growing complexity and pressure on rapid changes pose sometimes unbeatable challenges. The goals was to introduce solution for ICT transformation from legacy to new Cloud-ready architecture with legacy fallbacks. Proposed solutions define ways for Hardware and Software Configuration Management. The key component of the solution is Configuration Management which provides single unified approach for standardization, configuration and change management.

Our comparison was proven in real IT telco enterprise, where we finally applied proposed CMDB solution based on Heat Orchestration and SaltStack CM. The company is now able to clone production environment in minutes instead of month.

The outcomes of the solution are better hardware utilization, scalability and faster lifecycle management. On application side the solution allows much quicker development cycles which will eventually allow products to reach the markets in time to get the right competitive advantage.

References

1. Van, H.: TOGAF Version 9.1. Van Haren Publishing (2011). ISBN: 9087536798
2. Buckl, S., Ernst, A.M., Matthes, F., Ramacher, R.: Using enterprise architecture management patterns to complement TOGAF. In: IEEE International Enterprise Distributed Object Computing Conference, EDOC 2009 (2009). ISBN: 978-0-7695-3785-6
3. Blevin, T., Spencer, J.: TOGAF ADM and MDA. OMG (2005)
4. Shirley, L., Ivor, M.: Service Transition Book. The Stationery Office (2007). ISBN: 011331048X
5. Klosterboer, L.: Implementing ITIL Change and Release Management. IMB Press (2008). ISBN: 0138150419
6. Puppet Labs. Creating Hierarchies. http://docs.puppetlabs.com/hiera/1/hierarchy.html
7. Myers, C.: Learning Saltstack. Packt Publishing (2015). ISBN: 1784394602
8. Mohaan, M.: Learning Ansbible. Packt Publishing (2014). ISBN: 1783550635
9. Franceschi, A.: Extending Puppet. Packt Publishing (2014). ISBN: 178398144X
10. Holik, F., Horalek, J., Marik, O., Neradova, S., Zitta, S.: Effective penetration testing with metasploit framework and methodologies. In: CINTI 2014 (2014)
11. The Foreman. The Manual: Compute Resources. http://theforeman.org/manuals/1.4/index.html
12. John, R., de Jan, C., Novak, F.; Openstack Cloud Computing. ISBN: 0956355684
13. van, Jurg, V., Paganelli, F., Geurtsen, J.: Resilience And Reliability On AWS. Sebastopol, CA: O'Reilly (2013). ISBN: 1449339190
14. Daniel, L.: Vmware Vrealize Orchestrator Cookbook. ISBN: 1784392243

Analysis of the Use of Cloud Services and Their Effects on the Efficient Functioning of a Company

Josef Horalek, Simeon Karamazov, Filip Holik[(⊠)], and Tomas Svoboda

Faculty of Electrical Engineering and Informatics,
University of Pardubice, Pardubice, Czech Republic
{josef.horalek,simeon.karamazov}@upce.cz,
{filip.holik,tomas.svoboda5}@student.upce.cz

Abstract. The aim of this article is to introduce a long-running research in the Czech Republic focused on utilizing cloud technologies and services and their effect on efficient functioning of a company. The results of the research reflect more than 80 representatives of national and international companies in the environment of the Czech Republic, and government organizations classified into autonomous territorial units.

Keywords: Cloud computing · Cloud principles · Research cloud · Efficiency cloud · Corporate environments

1 Introduction

Cloud computing is a concept that originated in 1960, but its significant development was seen between the years 2007 and 2008, when the term *cloud* was first put into use [1]. Cloud computing, according to [2], should meet the following criteria: on-demand self-service, broad network access, resource pooling, rapid elasticity, and measured service. Some main advantages of utilizing a cloud are high efficiency, great availability, elastic scalability, and fast deployment. Nowadays, more and more mobile devices are being connected to the cloud and thus the development of Mobile Cloud Computing (MCC) is being developed [3]. MCC tries to overcome the major flaws, such as limited performance and capacity of storage, of mobile devices. Such flaws manifest themselves mainly while running demanding services like streaming mobile video, surveillance, playing games etc. [4]. Besides that, a cloud is utilized in a new branch of Internet computing technologies called Web 2.0, where services like social computing and collective intelligence dominate [5]. Implementing a cloud solution is advantageous for both sides: to the company it brings the option to pay only for services they really use, and it enables the cloud owner to maximize the use of their computer resources while minimizing costs for maintenance [6]. But with the gradual expansion of clouds into individual companies, the question of energy consumption of such solutions is more and more topical, and any effort to minimize the impact of the energy footprint could bring significant lowering of costs. One way to lower the energy intensity of clouds is an intelligent allocation scheme based on a performance

© Springer International Publishing Switzerland 2015
M. Núñez et al. (Eds.): ICCCI 2015, Part II, LNCS 9330, pp. 336–345, 2015.
DOI: 10.1007/978-3-319-24306-1_33

analysis described in [7], or a dynamic resource management scheme, which can lower the consumption of the cloud solution by up to 50,3% [8]. One chapter on its own, is security of a cloud solution provided by a cloud provider. Some of the biggest security risks are the unavailability of data to legitimate users and breach of privacy. Cloud security is fully in the hands of a provider, to whom the users trust with their data. Users often do not have the option to ascertain what security measures the provider has implemented to ensure security and whether they abide by the agreed SLAs [9]. For these reasons common providers deem themselves trustworthy: we believe that data is secured against outside attacks, but not from the inside [10].

2   Cloud Computing, Its Definition and Principles

The term cloud computing can nowadays be found almost everywhere and for that reason there is a wide scale of used definitions that are, however, not unified and ambiguous. For this article, the definition by the National Institute of Standards and Technology (NIST) is the most suitable. According to it, a cloud provides an instant, easy, and on-demand network access to shared configurable computer resources (networks, servers, data storage, applications, and services) that can be provided or vacated for minimal administrative costs and without coordination with the service provider. This cloud model consists of five basic characteristics, three models of service providing, and four models of implementation. Another viewpoint comes from Gartner, Inc. [11], who defined cloud computing as scalable and elastic resources provided as a service to external customers with utilization of internet technologies. Just like with the definition of cloud computing, there are many views on characteristic features. For maximum relevance of the results introduced below, the most significant features are the ones identical to what Mell and Grance [2] discuss. **On-demand self-service** is where the customer themselves can choose, when they will use the given service and what resources they will use, without having to interact with the service provider. The customer can then flexibly react to their needs in a short period of time. **Broad network access** are services available through a computer network, which uses standardized mechanisms that support thin or thick clients (mobile phones, tablets, workstations, laptops). **Resource pooling** is where computer resources of the owner are provided to individual customers who share them, but who are shielded from one another just like their data. But the customer has no option to ascertain where these resources are physically present, which is actually one of the main problems concerning the security of using cloud services. The customer should have the option to get information about the locations of their data, at least on a state level, or at the data centre level. **Rapid elasticity** is a characteristic important from both viewpoints. The service provider can dynamically change computer resources assigned to individual customers, while keeping the interaction with them to a minimum. The customer can then see the resources as unlimited and can be assigned the resources at any amount and at any given time. Cloud systems enable the owner to automatically check the quality of the provided services and to optimize utilization of resources.

This option is called **measured service.** Using such an option of monitoring services provides a transparent outlook for both the provider and the customer, who then pays only for services they actually use.

3   Methodology of Research and Used Methods

Questionnaire research using a standardized questionnaire was used as a research method, which is actually a form of a controlled written interview. Questions, that are written, unlike in an actual interview, are answered also in a written form. For reasons of maintaining logical and stylistic correctness of individual questions, a pilot research, to which participants of lectures about the issue of cloud computing and it deployment were invited, was conducted. The questionnaire was distributed as purely anonymous, which led to maximal objectivity and honesty of all participating subjects. Another significant element of the questionnaire research was personal contact with persons responsible for the given issue in a particular company, such as IT managers, directors of computer system management, etc. The issue of cloud computing was discussed with all respondents during a motivation interview, the aim of which was to motivate the respondents to provide their objective answers. To evaluate the questionnaire, quantitative methods, enabling fast and straightforward data collection and relatively easy generalized results, were used. Maximal elimination of dependency on a particular researcher had a significant effect on using quantitative methods of evaluating the research. Selected questions aimed mainly at the characteristics of the researched subject were secondarily verified from answers collected during the motivating interview. The presented questionnaire contained 30 closed questions with the option of selecting an answer, in which the respondent could have select the option 'other answer' and provide their own answer in the edit field. Questions were individually selected based on responses in previous questions. There were questions aimed at the characteristics of the organization; at using or not using cloud services and their reasons for doing so; at deploying cloud services; and the last set questions were aimed at the security issue of cloud services.

4   Results and Evaluation of Obtained Data

The first result, obtained during a research conducted from February to November 2014 in the Czech Republic, was realization that only 43% of companies promised to participate in our research about using cloud services. Within our research and results introduced below, we obtained information from 87 organizations. 73% of these organizations use cloud services, and the remaining 27% do not. This is caused by the fact that the organizations that refused to share information for the research do not have cloud services implemented. Counting of organizations that refused to take part in the research, and which we presume do not use cloud services, we can assume that 35% of approached organizations do use cloud services.

4.1   Structure of Organizations Using/Not Using Cloud Services

Fig. 1 depicts the percentage of organizations already using cloud services. It is not surprising that 40% of organizations working in IT use cloud services in their operation. But an unexpected result was that 33% of organizations operating in the field of power engineering, in both production and management, use cloud services. This result can be attributed to dynamic development of deploying smart networks and smart metering, which is supported in the Czech Republic not only by company managements, but also by government grant policy. Overall, a small percentage of 14% of industrial companies (oriented on production in the area of engineering, construction, and glazier production), that use cloud services, can be explained by the relatively high costs for innovating current infrastructures.

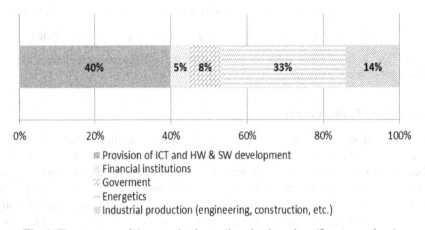

Fig. 1. The structure of the organizations using cloud services (Source: authors)

4.2   The Type of Deployed Cloud Solution Depending on the Type of Organization

Another interesting result coming from our research is the type of the deployed cloud solution depending on the type of organization that deployed it. The first interesting result depicted in Fig. 2 is that there are only three types of cloud solutions used - a hybrid cloud, a hosted private cloud, and a custom private cloud. It is not surprising that the custom private cloud is used by government organizations, for which using some other solutions is, more or less, impossible for legal reasons, requiring high level of security for data stored in the cloud environment. Management and security of organization data is still the most significant element that affects the deployment of cloud services. This fact is apparent from the usage of the hosted private cloud and the hybrid cloud solutions. Both of these alternatives give the user the option to store and secure their data in compliance with requirements set by the law on cyber security, law on the protection of personal data of both employees and clients, and in compliance with internal materials of the company (defined by internal policies of companies in compliance with ISO 27000).

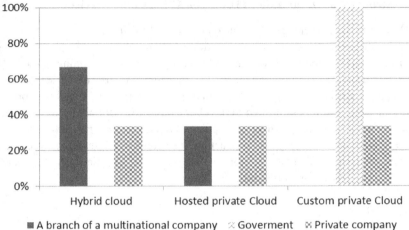

Fig. 2. Type deployed cloud solutions depending on the type organization structure of the organizations (Source: authors)

4.3   Cloud Platform Used

The following results, depicted on Fig. 3. follow the results introduced before. Those results are about utilizing technological platforms that the users of cloud services use. Obtained results can be deemed surprising, mostly because there were only four used platforms among the collected answers. More than 40% of organizations using cloud services use them in the form of a private cloud. This technological solution is provided by various distributors. For our research it was not important to know what technology was used. Furthermore, from the results, it is apparent that the solution by the strong international corporation, Microsoft is prefered in the Czech Republic. This solution is called Microsoft Azure and enables utilization of all characteristics of IaaS and PaaS models. Microsoft Azure supports runtime operating system Windows Azure and the provider thus gets a complex package of cloud solutions with support of all most used technologies, such as Live Services, SQL Azure, AppFabric, Dynamics CRM Services, and SharePoint Services. Sharepoint platform was then presented as an independent cloud service by 8% respondents. The last interesting result from this phase of research is the 18% percentage of utilized cloud services based on the OpenStack platform, which is implemented by *Technologicke centrum Pisek* in the Czech Republic.

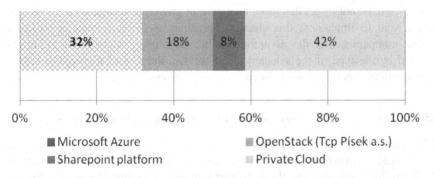

Fig. 3. Used cloud platform (Source: authors)

4.4 Analysis of Time Needed for Implementing Cloud Services

In Fig. 4, the results of research in the area of the time needed for implementing cloud services into the structure of organization are depicted. The time needed to implement new technologies is a very important factor that affects the choice of whether to deploy new technologies into the current and fuctioning infrastructure alone. In our research, the respondents were asked about the time needed to fully integrate selected cloud services into the organization. In this period, the time needed for testing and individualization of the given service for the organization needs is included. From the graph below it is apparent that the shortest implementation times were noted in companies that are subsidiaries of a international company. These results are affected by several factors, like wide base of IT specialists, experienced international partners, and high work efficiency. The result, that more than 60% of state organizations that have already implemented cloud services in the period of six months, can be deemed positive.

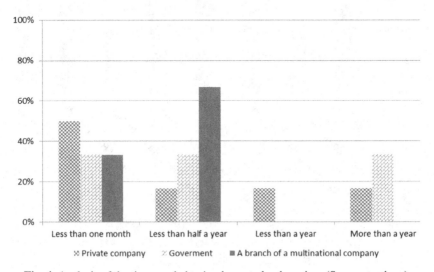

Fig. 4. Analysis of the time needed to implement cloud services (Source: authors)

But we must not forget that more than 30% of government organizations took more than a year to implement this change. The time needed for implementation with private companies met the assumptions that their implementation is significantly affected by the type of the selected cloud service, the size of the organization, and their different experience of IT specialists and their accessibility.

4.5   Evaluation of Favourableness of Implementing Cloud Services

When choosing to implement cloud services into their infrastructure, the organization has to keep in mind both the negatives and the positives of such implementation. Realizing such reasons then plays significant part in deciding whether to take such a step. For this reason the organizations that had decided to implement cloud services and had done it, were asked about the reasons reflected in this decision. Individual questions can be divided into two areas, one of which is oriented on technical benefits and the second on benefits from the viewpoint of business and organization management. From the data collected it is apparent that the technical advantages of implementing cloud services (e.g. better option of mobile access, compatibility, scalability and security, and data protection) affected individual organizations only by 65% on average. On the other hand, managerial-economic benefits (e.g. increased productivity at work, reducing the cost of running IT infrastructure, optimizing

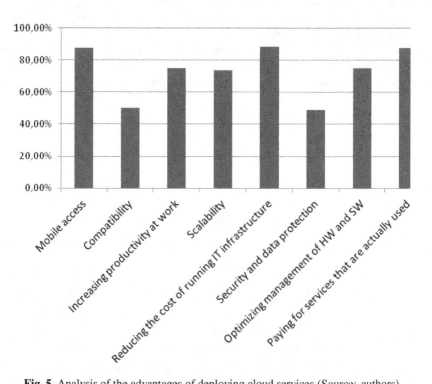

Fig. 5. Analysis of the advantages of deploying cloud services (Source: authors)

management of HW and SW, and paying for services that are actually used) were taken into consideration in 80% of cases. From this data we can deduce that implementing cloud services is, according to our research, beneficial not only from the technical viewpoint, but also from the viewpoint of the organization. Major fiscal benefits are not only the level of HW and SW cost optimization, but also in the area of efficiency and work productivity.

4.6   Reasons Affecting Utilization of Cloud Services

The last results introduced in this article are the reasons necessary for cloud implementation, which cause issues for organizations that had not yet implemented cloud services. Independent checking of cloud service providers by an external subject seems to be not such a significant factor. This viewpoint, logically concerning public and hybrid cloud services, takes up nearly 25% of organizations. Another factor that can be deemed as limiting for not implementing cloud services is the inaccessibility of fast connection for end users. Making it more accessible is a deciding aspect for implementing cloud services for more than 49% of organizations. This result is important for corporate organizations that provide both the connectivity and cloud services. Over 50% of organizations then deem insufficient, the amount of provided information about the location, security, and manipulation of data stored in the cloud. This result fully corresponds with the interpretation of previous parts of the research, in which organizations are inclined to implementat a private or a hybrid cloud solution mostly because of data security and control over data. The most important aspect according to the respondents is then, insufficient or complicated legal scope defining relationships between the provider and the client of the cloud, and concerning legal responsibility in the case of an outage, abuse, or loss of data stored in the cloud.

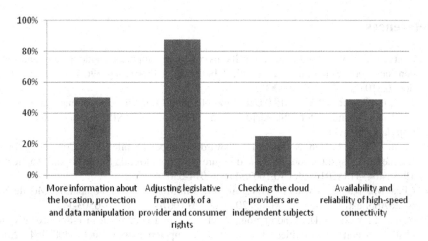

Fig. 6. Analysis of the reasons affecting the use of cloud služeb (Source: authors)

344 J. Horalek et al.

5 Conclusion

Cloud computing brings, from the viewpoint of potential users and organizations who already use it, many advantages as it is apparent from the results of the conducted research. The strong side is mostly the lowering of costs for purchasing and operating an infrastructure. The customer does not have to care about purchasing servers and creating a network infrastructure. That is all in the hands of the provider. Also, significant savings on salaries for IT employees occur, as the employees no longer have to maintain hardware and thus their number can be reduced in some cases. Payments for used services or infrastructure is usually paid by the company in the form of a subscription, where the provider can estimate their future profits based on previous payments. Scalability is for many organizations a feature of cloud computing that can be a part of financial savings. One weakness can be the accessibility of providers' servers, on which user data is stored. Such servers are usually scattered all over the world. Lastly, it is necessary to mention potentional threats, one of which is data security and the risk of its abuse by a third party. Potentially, the provider can go out of business and the customer is then facing the risk of losing their data, which was also mentioned by the respondents in our research. We can say that the implementation of cloud services in the Czech Republic is about average, but we can positively mention that organizations that use cloud are fully aware of its possibilities and positive sides and they are aware of the possible risks and threats connected with it. Better support of using cloud services from the legislature and corporations providing such services could lead to better quality of data connections and their utilization.

Acknowledgment. This work and contribution is supported by the project of the student grant competition at the University of Pardubice, Faculty of Electrical Engineering and Informatics Security smart grid networks and cloud computing No. SGSFEI 2015002.

References

1. Arutyunov, V.V.: Cloud computing: Its history of development, modern state, and future considerations. Scientific and Technical Information Processing **39**(3), 173–178 (2012). doi:10.3103/S0147688212030082
2. Mell, P., Grance, T.: The NIST Definition of Cloud Computing, Recommendations of the National Institute of Standards and Technology (NIST), September 2011. http://goo.gl/3nGfE0
3. Shon, T., Cho, J., Han, K., Choi, H.: Toward Advanced Mobile Cloud Computing for the Internet of Things: Current Issues and Future Direction. Mobile Networks and Application **19**(3), 404–413 (2014). doi:10.1007/s11036-014-0509-8
4. Chen, M., Wu, Y., Vasilakos, A.V.: Advances in Mobile Cloud Computing. Mobile Network Application, 131–132, March 2014. doi:10.1007/s11036-014-0503-1
5. Yang, X., Wallom, D., Waddington, S., et al.: Cloud computing in e-Science: research challenges and opportunities. The Journal of Supercomputing **70**(1), 408–464 (2014). doi:10.1007/s11227-014-1251-5

6. Di Modica, G., Tomarchio, O.: Matching the business perspectives of providers and customers in future cloud markets, Cluster Computing, March 2014. doi:10.1007/s10586-014-0364-1

7. Lee, H., Jeong, Y., Jang, H.: Performance analysis based resource allocation for green cloud computing. The Journal of Supercomputing **69**(3), 1013–1026 (2014). doi:10.1007/s11227-013-1020-x

8. Gao, Y., Guan, H., Qi, Z., Song, T., Huan, F., Liu, L.: Service level agreement based energy-efficient resource management in cloud data centers. Computers and Electrical Engineering **40**(5), 1621–1633 (2014). doi:10.1016/j.compeleceng.2013.11.001

9. Hussain, M., Abdulsalam, H.M.: Software quality in the clouds: a cloud-based solution. Cluster Computing **17**(2), 389–402 (2014). doi:10.1007/s10586-012-0233-8

10. Fabian, B., Ermakova, T., Junghanns, P.: Collaborative and secure sharing of healthcare data in multi-clouds. Information Systems **48**, 132–150 (2015). doi:10.1016/j.is.2014.05.004

11. About Gartner. Gartner, Inc. Gartner [online]. [cit. 2014-08-28] (2014). http://www.gartner.com/technology/about.jsp

Proposal to Centralize Operational Data Outputs of Airport Facilities

Josef Horalek, Sona Neradova, Simeon Karamazov,
Filip Holik$^{(\boxtimes)}$, Ondrej Marik, and Stanislav Zitta

Faculty of Electrical Engineering and Informatics,
University of Pardubice, Pardubice, Czech Republic
{josef.horalek,sona.naradova,simeon.karamazov}@upce.cz,
{filip.holik,ondrej.marik,stanislav.zitta}@student.upce.cz

Abstract. This article presents analysis and a proposal to centralize operational data outputs of airport facilities for the Armed Forces of the Czech Republic. In the first part, the current state of infrastructure and individual systems necessary for controlling flight operations of the Armed Forces of the Czech Republic is presented. Within the analysis, a complex packet capturing was conducted, from which the list of most used protocols was chosen. The second part is focused on proposing a monitoring system with emphasis on its expansibility and security whilst minimizing changes to the current solution, which would mean the necessity of getting new system certifications from designated offices.

Keywords: Monitoring systems · Category monitoring systems · Proactive solution · Principles monitoring systems · Monitoring active systems

1 Introduction

Monitoring network infrastructure is a key feature in production networks, and its main function is to identify patterns and trends of network devices and operation of the network itself in order to monitor its current state [1]. Some of these trends are current workload of network and end devices, workload of network connections, or inspection of individual data streams and their communication as described in [2]. The key feature of network monitoring is summarizing individual information in order to make it clear and easy to interpret [3]. Such information can then be utilized in the whole lifecycle of the network [1]. Another important area, that network monitoring solves, is network security. Monitoring provides higher security against various security threats and can implement functionality of IPS (Intrusion Prevention System) or IDS (Intrusion Detection System). But in order for the system to efficiently prevent various types of attacks, such as DoS (Denial of Service), Probing, and Worm infestations, it must be a fairly complex system and not just an average firewall [4]. Such systems may implement network forensics, which is a modern discipline focused on detecting malicious activities and saving proofs by capturing packets and then analyzing them [5]. Another problem when using IPS is the performance needed for wire-speed packet capturing, in the high speeds of the network that IPSs require. One possible

© Springer International Publishing Switzerland 2015
M. Núñez et al. (Eds.): ICCCI 2015, Part II, LNCS 9330, pp. 346–354, 2015.
DOI: 10.1007/978-3-319-24306-1_34

solution could be using the force of parallel processing by multicore processors [6]. The problem with traffic monitoring is, however, what has to be monitored and which packets should be captured. In a production network, in which there is a transmission of a huge amount of various types of flows, there is no possibility to use a tool summarizing these flows, such as NetFlow. The load of network devices and the amount of aggregated information could be disproportionately high [7]. One possible solution is to use a sampling technique, which searches for a compromise between the accuracy of monitoring and network load, just like in [8]. Another question is whether to use active or passive monitoring and it is also necessary to choose between periodic and event-triggered polling when using protocols like SNMP (Simple Network Management Protocol) [1]. This article describes a proposal of an NMS (Network Management System) to centralize operational data outputs of AB (Airbase) of AFCR. This system implements hierarchic division according to AB, without which the data analysis would be very difficult due to the huge amount of it, which could lead to overlooking a problem and cause consequent downtime of the whole network [10]. A distributed solution even raises the reliability in the case of malfunction of connection in isolation of individual AB, or in case of malfunction of the central monitoring element [11]. This distributed solution does not bring too many bandwidth requirements [12], the capacity of which, in the case of AFCR (Armed Forces of the Czech Republic), is severely limited.

2   Analysis of the Current State

Sources for analysis of the current state come from personal consultations at selected workplaces of AFCR airforce bases, documentation and specifications of individual subsystems, and testing of current systems at key suppliers. The project of the centralized monitoring system is primarily planned to be implemented on stations securing flight operations of AFCR. So currently, it is the Air Traffic Control (ATC), Airbases 01 to 04, based in towns of Čáslav, Praha-Kbely, Náměšť nad Oslavou, and Pardubice. Stations are interconnected via a permanent command connecting network and airforce control, which provides redundant and constant connection between the stations. This network is not owned by individual stations securing flight operations, and therefore monitoring it is not being currently considered. From the viewpoint of the monitoring system and also the end user, the network is accessed as a closed entity that guarantees its functionality. Individual stations are connected to this network via a shared line with guaranteed minimal speed of 4 Mbps. This network connection, depicted in Fig.1, is implemented thanks to a guaranteed bandwith, i.e. every application, that uses this line, cannot use its full capacity, but only its part, which will be assigned to it by an administrator. This method of assigning connectivity is used exactly because of mutual isolation and guarantee of the line for individual systems. This factor is critical for future decisions about potential centralization of operating information from individual stations. In case this scenario was required, it would be necessary to restrict the data flow in order for the data to be delivered in time for monitoring in real time.

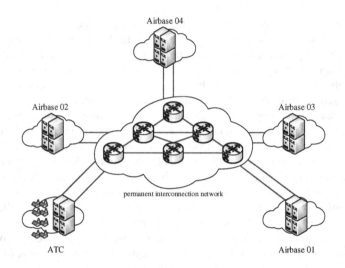

Fig. 1. Logical diagram of connection of workplaces for the air traffic of Army of the Czech Republic (Source: authors)

2.1 Currently Used Technologies

In Fig. 2, there is the logical topology and mutual connection between systems on AB, which were analyzed and observed and which are imperative to guarantee all provided services. In the schema, the systems are divided into three basic groups. **DANESE** – the main purpose of which is to connect into the permanent connecting network. From the viewpoint of the monitoring system and AB, the main focus point is the Cisco 3950 border router (behind this router it is not possible to monitor the network, because it is not in ownership of the individual AB, but the general staff). **Closely specialized systems** – currently not creating individual autonomous units and directly dependant on cooperation, or control from other superior systems (depicted in grey). And **critical systems** – for securing flight services, which can in many cases exist independently (depicted in white). Such systems are given a lot of attention, because they are the main objects of the planned monitoring system.

Segment LAN is the first element for which attention needs to be paid. It is based mostly on products by Cisco and together they ensure the division of AB from the permanent interconnection network. In the infrastructure there is a switch that ensures main connectivity for support of external systems, and currently this segment is used by a system called Zabbix, which uses the SNMP protocol to monitor itself (it displays limited statistical information about active network elements). The **AMS segment** has the duty to control and signal the status of light-signaling equipment of AB and to monitor and control other AB's operation systems. Among the systems, that the AMS system services and manages, belong light-signaling and security devices of runways and taxiways, radionavigation devices and systems, ATIS system, electroenergetic systems, meteorology and fire devices, etc. The AMS systems in their current version include internal monitoring application, which is capable of presenting the user with some basic statuses, information about the whole system, and

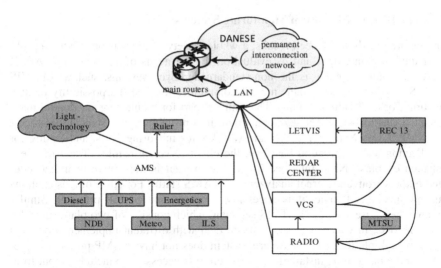

Fig. 2. Logical diagram of interconnection systems (Source: authors)

communication of elements via a logical scheme and color-coded individual system parts. However, it is not capable of providing such information for further processing outside the AMS network.

A **LETVIS segment** is used for radar and procedural management of flight operations including planning and managing the airspace. It can be configured as a source of radar data and information for other systems within operating and monitoring flight operations or for external users of the LETVIS system. The system consists of assemblage of servers and workstations, which are located on TS. Currently the LETVIS system is partially modernized and it is thus built on two platforms - x86-64 using OS Suse 11 and SPARC using Solaris 9. A special part is the SUP subsystem, which serves as an internal monitoring system for the LETVIS system and which provides technicians with basic information about the system and its state. It is an independent workstation, which takes care of data collection and its analysis from all workstations and servers, and also provides tools for basic diagnosis and configuration of systems (ping, ftp, telnet, etc.).

Segment Voice Communication System (VCS) is used for integrating voice and data communication via radios, phones, and intercom systems, and it uses its modular and scalable architecture, which is possible to access as a 'normal' computer system that uses standardized communication elements and tools. For managing and configuring the system, the Frequentis VCS has its own tool TMCS based on a client-server architecture, in which the server stores all configuration entities and databases and the client serves as an interface for the user and managing the system. Connection between them is realized via LAN CFI.

2.2   The Issue of Disunity of Monitoring Systems

From the previously introduced results, a whole variety of impacts on solving a problem of unified data output and its evaluation for the needs of airforce bases AFCR emerges. A major problem is the non-standard deployed systems, such as LETVIS and VCS, and not securing their mutual communication and the possibility of data collection from individual systems. The simplest part for monitoring and central management seems to be a LAN segment, where it is possible to use standard protocols like SNMP, CDP, NetFlow, SPAN and RSPAN for mirroring ports, and Syslog for log collection and archivation based on the client-server principles. The AMS segment then enables SNMP interface usage, however, it is imperative to utilize integrated outputs from an internal analyzer of the AMS status. For some of these outputs requiring specialized logic, it is necessary to implement SNMP protocol. Similar problems then arise with the LETVIS segment, which consists of two platforms x86-64 and SPARC, in which it is necessary to deal with different implementations and outputs of both platforms as the Solaris system does not have SNMP protocol implemented in the basis of its installation. Even here it is necessary to include output from LETVIS SUP into the resulting communication. Even the lastly introduced segment, VCS, has its specifics mostly for ensuring the reliability of radio communication reserves, in which the proposed system must be able to inform users about potential outages or faults of these specialized devices. Before introducing the proposal itself, it is imperative to mention that all systems deployed in AFCR bases must be approved beforehand by appropriate government departments, which is not only time-consuming, but also technologically and economically demanding and it is therefore necessary to change the current subsystems only minimally, and thus maximizing their standardized outputs.

3   Proposal of a Complex System for Monitoring

During the analysis, with focus on security and individual subsystem coordination, there was a long-term operation observation conducted. The original idea was to create a classic federated architecture with central monitoring system on the ATC station, where it would be possible to monitor information from all systems in real-time. Based on testing and the analysis of available communication lines, this solution is not feasible as the transfer capacities between AB and the permanent interconnection network are severely limited. For these reasons a hybrid solution was proposed in Fig.3. Instead of a full federated architecture it was decided to not transfer all data to the central system in real time, but to limit the central system only to the reporting system and collection of statistics (RSC) where the standard local surveillance system (LSS) would transfer collected statistics about the monitored systems at the end of each day. RSC would also be accessible for the needs of General Staff of the Army and Ministry of Defence, where it would be possible for responsible office clerks to get overviews and statistics for the purpose of controlling life cycles of the systems at AB. It is natural that another request to the proposed system is the option of implementation of a fully federated architecture resulting in possibility of future technical changes and needs for a monitoring system.

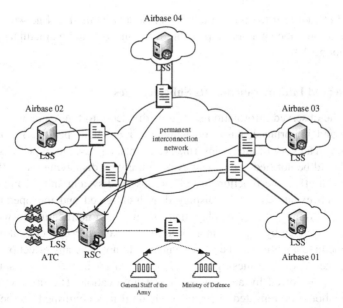

Fig. 3. The proposed architecture of communications surveillance systems (Source:authors)

3.1   System of Data Collection

By implementing a local monitoring system, that would be completely independent of the "outer" environment and other systems, on every AB we get an advantage of independence in case of an outage of connection (for example, natural disaster) into the permanent interconnection network. The proposed primary communication protocol for data collection is SNMP, and other protocols will be additional for the needs of specific systems. Considering the amount of collected data, the best way to collect data seems to be periodic collection by requesting, while SNMP traps would be used in systems only for critical events that should be signaled immediately. Signalling should enable not only to control the load of individual systems that would be caused by the monitoring system, but also to select priority elements, the status of which can be checked more frequently. The implemented system should then enable change of the interval for query of each monitored device/group separately. The second area, which needs attention, was the source of data collection. The monitoring system which collects data by itself directly from hardware, was first taken into consideration, during the analysis, testing, and proposal creation. Later, when discussing the tests with suppliers of systems in question, this way seemed plausible, but not really ideal because many systems already have some sort of an internal monitoring/diagnosis and the central monitoring system would only duplicate their function, which would mean double load and configuration problems (two communities). The best alternative seems to be utilizing already present monitoring systems and data sources, provided that each supplier of a newly supplied system would have to prepare an SNMP interface for the central monitoring system. This solution also allows the user to avoid bigger intervention into certified and tested systems. The only directly observed elements of the infrastructure then should be the LAN network, which is mostly shared; and the supplier responsible for managing and maintaining these elements cannot be defined.

3.2   Observed Information and Its Subcategories

Generally, the observed information should be divided into four categories according to severity and importance, based on which their detailed specification is created. **Information** - category of purely informative events, e.g. user login, CPU status, etc. The user should be notified about such events purely by displaying them. **Warning** - events that can affect the functionality of the observed system and require operator's attention. Such events should be displayed as highlighted and the operator should confirm them. Along with the display, there should be a notification forwarded to a responsible technician by e.g. an email or a text message. **Error** - critical events that require immediate attention of the technician, but do not have any effect on providing services to the ATC. The message should be displayed in red with a sound being played with it, followed by an email or a text notification. The display and sound notification should be repeated at set intervals until it is confirmed by the operator. **Critical error** - an error that affects providing services for ATC. In case such a scenario occurs, there should follow notifications to other affected stations, besides standard treatment, as their operation can be affected by such an error (i.e. outage, or problems with LETVIS should be reported to higher places).

3.3   Data Saving

Besides presenting and monitoring the communication environment on AB, the task of the central monitoring system will also be analyzed according to the load and lifespan of individual systems. It is then desirable for the system to store the biggest possible amount of data, which will affect the overall size of the database that the system generates and therefore it would be appropriate to establish a strategy for data storage. The main task of LSS should be local controlling and monitoring of the infrastructure on AB. Analysis of statistics and examining systems from the viewpoint of historical data, will be secondary on AB and so a good strategy seems to be to store only a part of data on LSS and mirror the rest on the RSC. RSC will then serve only for analytical purposes of planning further development and controlling the system's lifespan. From the viewpoint of time period, it is recommended to store data on LSS for one year. But, on RSC all available data should be collected for maximum of 7 years.

3.4   Options of Integrating and Modifying the System

One of the very important areas and characteristics of the planned system should be its expandability, modifiability, and options to integrate it into other systems solved via optional creation of own data connectors, which would enable reading data via specialized protocols. From the medium-term point of view, expansion of the central monitoring system by a portal of work processes control, and document circulation for individual stations are taken into account. Another significant functionality, which is supposed in the future, is the system of controlling remote access of external service suppliers. The main idea lies in creating such a mechanism that would enable the service suppliers to conduct small modifications on a system that does not require physical presence. The aim is to create a VPN solution enabling the opening a dynamic window for specific suppliers only on demand and via a two-factor authentication while monitoring changes to the system. Another reason for openness and modifiability of the system can then be integration with other information systems on AFCR, such as registry of entry permits on AB, camera and security systems, fire systems, inventory systems, localization system via wireless networks [13] or via optical flow methods [14] etc. The central monitoring system then should be open not only inwards (expandability), but also outwards.

4   Conclusion

The aim of this article was to introduce the current system of communication of AB operated by AFCR and to propose a complex system of central data collection and monitoring of infrastructure. From the conducted analysis, long-term operation monitoring, and simulation of the environment and system in laboratory conditions, a complex solution including the specification of the AFCR environment and current system was proposed. From the future point of view it is then necessary to keep in mind that the system, that will newly be implemented on AB within expansion and modernization, should meet the standards for integration into the monitoring system. The main requirement and standardized protocol on AB should be the SNMP protocol, version 2 or 3. Systems should be able to provide possible maximum meaningful data via this protocol. In the case of operating systems, it is possible to use standardized protocols to monitor specific operating systems. Minimal data, that the systems should provide, belong to CPU load, RAM load, parameters and use of HDD, RAID status, load of network connection, information about running daemons and services, status of system logic units (inner subsystems) and the relations between them, status and configuration of specialized interfaces, system events and changes (change in configuration, user login, etc). All future systems and applications also must support user identification and user roles with support of a current authenticating mechanism used on AB. Based on the above introduced proposals, tender documentation to realize the complex monitoring system in the environment on AFCR AB is now being prepared.

Acknowledgment. This work and contribution is supported by the project of the student grant competition at the University of Pardubice, Faculty of Electrical Engineering and Informatics Security smart grid networks and cloud computing No. SGSFEI 2015002.

References

1. Lee, S., Levanti, K., Kim, H.S.: Network monitoring: Present and future. Computer Networks **65**, 84–98 (2014). doi:10.1016/j.comnet.2014.03.007
2. Braun, L., Volke, M., Schlamp, J., von Bodisco, A., Carle, G.: Flow-inspector: a framework for visualizing network flow data using current web technologies. Computing **96**(1), 15–26 (2014). doi:10.1007/s00607-013-0286-4
3. Hoplaros, D., Tari, Z., Khalil, I.: Data summarization for network traffic monitoring. Journal of network and computer applications **37**, 194–205 (2014). doi:10.1016/j.jnca.2013.02.021
4. Wattanapongsakorn, N., Charnsripinyo, C.: Web-based monitoring approach for network-based intrusion detection and prevention. Multimedia Tools and Applications, 1–21, May 25, 2014. doi:10.1007/s11042-014-2097-9
5. Chen, S., Zeng, K., Mohapatra, P.: Efficient Data Capturing for Network Forensics in Cognitive Radio Networks. IEEE/ACM Transactions on Networking **22**(6), 1988–2000 (2014). doi:10.1109/TNET.2013.2291832
6. Dashtbozorgi, M., Azgomi, M.A.: A high-performance and scalable multi-core aware software solution for network monitoring. The Journal of Supercomputing **59**(2), 720–743 (2012). doi:10.1007/s11227-010-0469-0
7. Keys, K., Moore, D., Estan, C.: A robust system for accurate real-time summaries of internet traffic. In: SIGMETRICS 2005, pp. 85–96, June 2005
8. Hernandez, E.A., Chidester, M.C., George, A.D.: Adaptive Sampling for Network Management. Journal of Network and Systems Management **9**(4), 409–434 (2001). doi:10.1023/A:1012980307500
9. Case, J., Fedor, M., Schoffstall, M., Davin, J.: A Simple Network Management Protocol (SNMP) [Online], RFC 1157. http://www.ietf.org/rfc/rfc1157.txt [cit. 2015-02-02]
10. Frye, L., Liang, Z., Cheng, L.: Performance Analysis and Evaluation of an Ontology-Based Heterogeneous Multi-tier Network Management System. Journal of Network and Systems Management **22**(4), 629–657 (2014). doi:10.1007/s10922-013-9272-6
11. Jung, J., Chang, Y., Yu, Z.: Advances on intelligent network management. Telecommunication Systems **52**(2), 633–634 (2013). doi:10.1007/s11235-011-9507-3
12. Gaglio, S., Gatani, L., Lo Re, G., Urso, A.: A Logical Architecture for Active Network Management. Journal of Network and Systems Management **14**(1), 127–146 (2006). doi:10.1007/s10922-005-9012-7
13. Nemec, Z., Dolecek, R., Silar, Z.: The model of communication channel in the 802.11b standard wireless network. In: Advances in Elecrical and Electronic Engineering, vol. 7 (2008). ISSN: 1336-1376
14. Silar, Z., Dobrovolny, M.: Comparison of two optical flow estimation methods using Matlab. Applied Electronics, pp. 1–4, September 2011. ISSN: 1803-7232

IED Remote Data Acquisition and Processing in Power Distribution

Ladislav Balik, Ondrej Hornig, Josef Horalek[✉], and Vladimir Sobeslav

Department of Information Technologies, Faculty of Informatics and Management,
University of Hradec Kralove, Hradec Kralove, Czech Republic
{ladislav.balik,ondrej.hornig,josef.horalek,
vladimir.sobeslav}@uhk.cz

Abstract. The article is directed toward to remote data acquisition and processing in power distribution with a focus on specific communication aspects of Intelligent Electronic Devices. The paper is divided in two main parts. First part focuses on a description of the main principles and control of remote data acquisition and processing. Second part of this paper presents specific requirements for utilization of data gathered from the Intelligent Electronic Devices and other key elements of substation automation system. This paper is a result of cooperation with the biggest national power distributor in Czech Republic – CEZ, a.s.

Keywords: IED · Remote management · Data processing · Architecture for data collection · IEC 61850 · IED data classification

1 Introduction

Modern automation systems of electric power substation are built upon the requirement of complex IED (Intelligent Electronic Device) data deployment [1] and monitoring as stated in [2]. The primary purpose of this requirement is based on the need to ensure the fully automated substation operation. This means mainly the requirement of a reliable and functional system for remote control and monitoring of the dispatching center, network operation and operational commands [3]. The most important elements in substation automation are mainly the protection misuse, prevention of malfunctions, which could lead to injury or property losses and prevent the erroneous action of protection devices leading to unnecessary interruption of electrical power in distribution system. With the development of information technologies, the new possibilities and challenges in substation automation systems appear. It is particularly the IED data utilization to increase the operational efficiency of power distribution networks [4], which can ensure the efficient use of network capacity, quick identification of the fault location and remote switching protection settings for regime change in distribution network. Another, yet important fact of the efficient IED data utilization is the maintenance and improvement of primary components and elements of substation, which is based on statistical evaluation of traffic in distribution networks and switching elements and their comparison with the guaranteed-priced parameters.

© Springer International Publishing Switzerland 2015
M. Núñez et al. (Eds.): ICCCI 2015, Part II, LNCS 9330, pp. 355–364, 2015.
DOI: 10.1007/978-3-319-24306-1_35

Finally and also beneficial option is a comprehensive remote management IED and it's monitoring as is, for example, proposed in [5, 6]. Currently deployed concept of substation automation systems is based on the international standard IEC 61850 (Fig. 1). The main part of the substation system consist of IEC 61850 communication bus and Intelligent Electronic Devices. This devices are involved to ensure connection between the dispatching centers or business information systems and stations for data transmission paths and corporate information systems [4, 7, and 8]. The goal of communication gateway is to collect data on the substation level and send them to other simultaneously connected systems. Furthermore, it is used to receive managerial commands which are used to control the distribution from dispatching center. The substation level is consequently used for the emergency or control functions, for example in the case of the planned shutdowns or within the substation revisions.

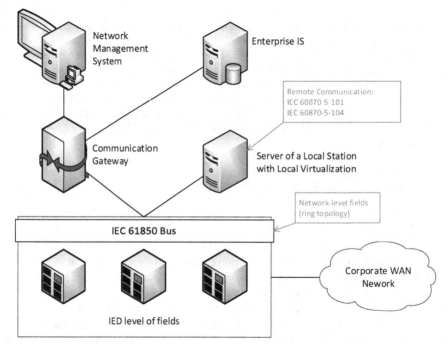

Fig. 1. Principle of the modern system of substation automation

2   IED Data Utilization

The development of modern technologies increases the amount of data and level of quality for the description of the state and treatment of energy distribution networks [9]. Furthermore this approach, used by energy companies for operating these networks, allows a better and more comprehensive use of gathered information [10]. Information derived from IED (resp. electrical protection units) were historically utilized only to protect the substation equipment or certain parts of the network. This information was presented on the one hand from contact output of protection relays, which were used in the case of substation failure, and on the other hand by setting individual protection modules. This setups were tested and recorded in test reports. The only memory functions, in this case, were mechanical protection indicators by which the operator was able to determine which protective functions impressed. Gradually, this information was further processed with the emphasis on eliminating inaccuracies and errors caused of transients fault in electric power. This states are very short (sometimes only a few milliseconds). It was very difficult to determine the exact cause of these processes. These meter readings contained inaccuracy in the form of random errors caused by inaccurate data reading and non-synchronized data records. The implementation of IED (Fig. 2.) devices allows automatic data logging and information retrieval from selected elements of the substation. The most important element was the introduction of recorded information in electronic form on the storage media. Thus, the data can be electronically archived on central repositories and processed via various software tools and make them available to every relevant department or employee of power plant.

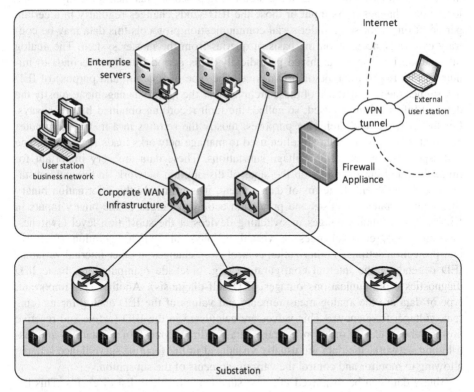

Fig. 2. The principle of the power substation involvement in industry/business networks and public internet network

2.1   IED Data Classification

Data obtained from the IED can be categorized according to the type of value (analog, digital), according to the reason formation (operational failure) or by the place of origin (physical information directed from the substation, virtual data gathered from the further IED processing). From the perspective of this article, the data processing at substation level is very important, however, this task has to be distinguished by the way how the information get into the system and is further processed. The IED data are send out spontaneously and thus based on events in the monitoring or in a regular cycle of time, or on data provided at the request of the operator or master system. In other words, this information and data can be divided into "On-line" and "Off-line" data categories.

Spontaneous Data (On-line)
As spontaneous data can be considered as those the IED spontaneously sends to the master system based on events in the monitoring process. This means mainly the analog values changes at its input or those the IED sends changes regularly in a certain period of time. For some older serial communication protocols this data may be contrary cyclically accused on the basis of queries from the master system. The analog information is typically archived periodically. This period last from seconds to minutes according to the type of IED device. The type of function and purpose of IED determines the intended use of these archives. For the failure management, mostly the detailed information are used, so called, the fault recording obtained by other ways. For the network load archiving purposes, mostly the settings in a matter of minutes are used. Spontaneous data are often used to manage networks loads for high voltage and especially in very high voltage substations. These data are very important for immediate and current view on the state of distribution network, individual substations and equipment. In terms of data content, it is primarily the information binary (bit) events, states, impulses and protection actions, changes of the binary inputs in IED and RTU units, changes in switching devices at the substation level (switches, disconnections, etc.); changes in virtual or physical switch operation (fault repair/protection settings/ authorization control etc.), changes in other internal variables IED depend on the internal configuration (e.g. blockade manipulation), basic IED diagnostics (communications outages, IED self-diagnosis). Another, yet important, type of data are the analog measurements and values of the IED analog inputs (current, voltage, frequency). This values are calculated in the IED (active and reactive power, labor, etc.). In order to leverage data for direct control of substations and distribution network, the data are usually visualized in the form of surveillance scheme allowing to monitor and control the various elements of the substation.

Binary data can be displayed in a text statement so called, list of events. This list contains all the changes that have occurred in the reporting process. Each change is fitted with a time stamp identifying the object resp. equipment and information about the status of this change. Selected information can also be used for automated alerts generation ("alarm"). Moreover it is used to alert staff about the malfunction or unexpected event in a substation process.

In some cases an archive of analog dumping values is evaluated for dispatching survey. These surveys are used for short overview of actual network load or partial network utilization withdrawing. These values are presented in operation table or a simple graph.

On Demand Data ("Off-line")
These data are not transmitted immediately when they origin. They might be transmitted automatically and periodically, but usually are transmitted and request based. This are the data stored in the registry of IED or in other parts of registers in substation automation systems. Mostly, these records are the history of past failures, thus are saved in a fault recorder (waveforms values measured before and at fault). Even more, these fault locators record maximum values of analog measurements at the time of failure.

The further information processing is performed within an interval after a breakdown or other events that were triggered. These data are not needed for immediate service response, therefore the periodical recoding of IED information is unnecessary (it is limited by the amount of IED memory). Consequently the reading of these kind of data is traditionally requested by the operator or technical staff in the following manner (e.g. once a day at specified time) or based on events in the IED, where the level of memory utilization is beyond a certain capacity of IED (typically 70%), or after a start-up/ shutdown delay caused due to the IED protection function.

After the reading from IED, data is stored in a file in standardized formats, mostly COMTRADE - Common Format for Transient Data Exchange for Power Systems ("a common format for data processing in power transmission"). Data files may be stored in a place with its file names, as of individual staff choice. This data can then be further shared.

The goal of IED "off-line" data utilization lays mainly in the analysis of faults and events. Those skilled operators or IED suppliers of substation automation systems, use this data mainly to detect failures and for advanced analysis of distribution networks (medium voltage, high voltage and very high voltage). Further use of these data might be, maintenance and replacement of primary elements planning (especially switches and transformers). From the recorded data a numbers of switching connections and the amount of energy that has been switched off can be derived. For the transformers, it is possible to determine the performance and power transmission limit. These data have a major impact on the life of the primary device. In addition to information about failures and transformers, IED contain also its internal function settings – mainly the electrical protection. This information is stored in IED memory. Their activation and setup is performed by the responsible personnel outside the IED with a specialized software (according to manufacturer IED). The pre-configured settings are uploaded to IED, if necessary, their communication settings is backuped at the same time by the grid operator or power distribution network outside the substation automation systems.

3   Utilization of IED Data in Power Distribution Company

Data and information obtained from the IED and the overall substation automation systems can be widely used in many departments and business processes in power distribution company, smart grid networks respectively. These data are utilized not only within the directly related processes, but also for the transmission and distribution of electricity, as for example in the processes of managing the company assets. To understand in detail the functionality and needs of the energy business model, the description of the most important processes and sub-processes along with an outline of the main activities and their relation to data and information derived from IED units is needed.

The main operating processes of a power distribution network model can be divided into five separate parts, processes management, network management, network recovery management, last but not least is the finance and property management.

3.1   Distribution Company Management

These are the highest representatives of the company, who have responsibility for the company's results and strategic development in the medium and long term. There are two departments, for which the gathered IED data are essential for its efficient work and contribute to the development and evolution. Firstly, the department of strategic development, whose task is the company development planning, including strategic acquisitions or distribution network development based on needs and government energy distribution policy. Secondly, the department of quality management, which monitors the quality of the delivery and processing of strategic plans regarding to the distribution network maintenance and development. The department of strategic IED data development is then responsible for long term mapping of historical production and procurement of process date, which are used for prediction of electricity consumption in the region. The department of quality management is consequently responsible for long-term historical data utilization in production environment, purchasing and delivering energy, thus the IEDs are used to predict potential weak points in the network.

3.2   Power Distribution Network Management

This process is highly influenced of the data and information obtained from the IED. It ensures the operation functionality of distribution networks, substations systems and dispatching control. It also has direct responsibility for the operation and maintenance of primary (main power devices at substation) and secondary parts (equipment, including individual substations station control systems and IED). Furthermore this process is touching the installed components to optimize their operations in terms of economic and quality of electricity supply.

The first sub-process of the main process is the preparation of substation engagement, which it plans the temporary shutdown of distribution network and substation according to the planned upgrades, revisions or an unplanned repairs. It use medium term historical data to estimate the consumption and utilization of network with a respect to supply behavior in each node.

Another sub-process is the operation management of distribution network, which guarantees the maintenance and operation of substations and power lines, conducts the regular inspections and maintenance of substation equipment and carry out the repairs in case of failures. The IEDs utilize all of the spontaneous data traffic for network monitoring system for the online status checking of all primary and secondary elements and is also used for transmission failure functions. Its task is also to process the long-term data and parameters about switching elements and transferred energy for following scheduling of revisions and exchanges.

The optimization of operation level is another important sub-process which belongs to the main process of network control. The aim of this sub-process is to improve the economics asset of distribution network traffic, increase the transmission capacity and supply reliability. Even more, it handles historical data about the limit states in the network (planned overload), losses on individual network segments (plan-effective load of network elements) and information about failures (reliability optimization). It also uses the spontaneous data traffic for the case of implementation and monitoring of the optimization measures results.

Controlling systems process utilize a wide range of available data from IED devices. This process is also responsible for the operation, maintenance and failure repair of the substation control systems parts, such as the master control systems and their repair in case of failures. Furthermore, the evaluation records the protections in case of failure on the primary part of the network. The IED uses spontaneous data concerning the operation of control systems (communications outages, internal diagnostics systems and IED), even all of the data provided on IED request (records of failures in the network for their evaluation to access the settings of protection functions via IED remote administration).

The last sub-process process is then the dispatching management, which is responsible for the direct management of the distribution network, substations and dispatching control system. For its work it use the spontaneous data regarding the primary elements of substations along with a short history of events at the substation.

3.3   Power Distribution Network Recovery

This process is touching the construction and reconstruction of electro-energy networks and their components, such as lines and lines, substations and switchyards components (fields, transformers, IED, parts of controlling system).

Restoration, construction and reconstruction is based on periodic service life planning according to the individual actual state of individual components or on the basis of the company development plans. The main type of necessary data and information needed are as follows, historical data about the development and transfer of energy. This is very important for detailed distribution network planning and also to gain more information about the failures, component life cycle, and substation development, thus for the planned replacement of these components in relevant life cycles.

The first sub-process of the main process is the development of networks, which is in charge of planning, construction and expansion of the existing network according to specification resulting from the process of strategic development. It uses some historical data about the distribution network utilization (load). The second, yet important process, utilizes the data from IED, therefore it is a process of technical preparation that specifies the technical basis for the preparation of construction, covers the project documentation, works on vendor selection and is in charge of the technical scrutineering of realized buildings. For his activities some historical data on the reliability and operational parameters are used, this is mainly the individual parts of networks and substations. These results are further utilized for tender parameters or corporate standards settings/establishment.

3.4   Finance and Property Management

This process provides the enterprise activities related to company finance and investment management, financial resources and capital equipment. From the perspective of this article, this is an interesting sub-process of poverty management. This sub-process is responsible for security and fire protection of buildings and structures in power substations. Due to the historic practices, various IED provides common reporting that often involve information subsystems such as an electronic fire alarm system, electronic security system assets and buildings and operating systems in buildings. Spontaneous data from these subsystems can be used for building management planning of maintenance or to intervene in terms of the safety of persons or property.

4   Conclusion

The area of substation automation and especially the medium, high and extra high voltage is very topical and specialized part of information technologies and systems. Together with the areas of management, operation and development of power distribution networks, we can see an extensive development over the last decade. This development happened mainly due to two main factory, namely:

- Technological development of information and electronic systems.
- General acceptation and implementation of IEC (International Electrotechnical Commission) standards in industry.

The basic stone of the substation automation are the devices - IED (Intelligent Electronic Device). These are a special microprocessor devices equipped with their own memory, operating system and application program. They are also equipped with various communication interfaces such as digital and analog input/output signals connection. This device integrates several functions, which were used historically for self-isolation. These function include:

- Protection of electrical substation equipment or parts of distribution networks.
- Collecting data and information from the substation (binary and analog values).
- Enable remote control of active elements switchgear (switches, connectors).

- Records acquisition and retention (various kinds of transients or faults in the substation or distribution network.
- Ensure the interlocking within all or part of the substation system.

Data obtained from IED devices (and others) are used in the substation automation systems form high to very high voltage, thus it can be widely used not only in the management of distribution networks, but also in their operation and development. Data and information are available as:

- Current analog values and binary states.
- Archives of analog values and archives binary states.
- Data available on demand for managing and setting functions IED or to evaluate transformation and disturbances in substations and power networks.

Information from the substation automation systems cover a large range of monitored devices. Basic monitored devices are:

- Primary substation equipment of electric power and distribution networks.
- Secondary substation equipment (including IED and station control systems).
- Other operating data (fire and security systems, buildings etc.).

The development of information technology has also led to the development of data itself. Following this trend, more accurate and faster algorithms in various expert software systems, which enable the prediction of the behavior of a large range of energy networks, are used. This is based on the existence and analysis of long-term archive data gathered from the IED. Furthermore, we can also see the rapid increase of usable capacity, reliable storage for data backup and the ability to share the saved data across large numbers of users and systems. This development leads naturally to the extension of possibility for using information systems, substation automation of various processes in enterprise distribution networks. These information and data are no longer used for simple control and monitoring of substations, but even for the development planning of entire networks that enables the modeling of different situations and predicting the network behavior with regard to possible or expected failures. The development of communication and networking options in IED communication elements generally allows far better and more comprehensive remote management of IED. Nowadays, the authorized personnel with the appropriate access rights to edit protecting features or evaluate recorded IED events is able to perform this action from any place through the public Internet. The ability to simultaneously manage all the IEDs deployed on any geographical area (e.g. the office) is becoming common, but on the other hand, the utilization of traditional computer networks, opens new challenges in security.

Acknowledgement. The authors would like to thank the Faculty of Informatics and Management of University Hradec Kralove for the continuous cooperation and the support over the past years.

References

1. Hadbah, A., Ustun, T.S., Kalam, A.: Using IEDScout software for managing multivendor IEC61850 IEDs in substation automation systems. In: 2014 IEEE International Conference on Smart Grid Communications, SmartGridComm 2014, art. no. 7007624, pp. 67–72 (2015)
2. Blair, S.M., Booth, C.D.: A Practical and Open Source Implementation of IEC 61850-7-2 for IED Monitoring Applications. IEEE Transactions on Smart Grid, Article in Press (2015)
3. Clavel, F., Savary, E., Angays, P., Vieux-Melchior, A.: Integration of a new standard: A network simulator of IEC 61850 architectures for electrical substations. IEEE Industry Applications Magazine 21(1), 41–48 (2015). art. no. 6942162
4. Xiong, H., Wang, X., Zhu, Y., Zhang, C., Wang, D.: Modeling and verification of real-time interaction for IEC 61850 intelligent electronic device. Dianli Xitong Zidong-hua/Automation of Electric Power Systems 38(19), 90–95 and 121 (2014)
5. Duan, B., Tan, C., Liu, Y., Zhang, C., Yuan, H.: An online tracking system of protection settings based on IEC 61850 and multi-agent technology. Dianli Xitong Zidong-hua/Automation of Electric Power Systems 38(7), 70–76 (2014)
6. Wang, D.W., Liu, R.K.: An improvement of IED configuration tool and modeling for transformer condition monitoring IED. Applied Mechanics and Materials 519–520, 1307–1310 (2014)
7. Xia, Y., Cheng, Y., Huang, S., Yang, N., Chang, W., Bi, J.: The research on transformer IED excitation source. In: Proceedings of POWERCON 2014 - 2014 International Conference on Power System Technology: Towards Green, Efficient and Smart Power System, art. no. 6993641, pp. 1306–1313 (2014)
8. Sidhu, T.S., Yin, Y.: Modelling and Simulation for Performance Evaluation of IEC61850-Based Substation Communication Systems. IEEE Transactions on Power Delivery 22(3), 1482–1489 (2007). doi:10.1109/TPWRD.2006.886788. [online] [cit. 2015-04-08]
9. Piringer, G., Steinberg, L.J.: Reevaluation of Energy Use in Wheat Production in the United States. Journal of Industrial Ecology 10(1–2), 149–167 (2006). doi:10.1162/108819806775545420. [online] [cit. 2015-04-08]
10. Pullen, S.F., Steinberg, L.J.: Energy used in the Construction and Operation of Houses. Architectural Science Review 43(2), 87–94 (2000). doi:10.1080/00038628.2000.9697439. [online] [cit. 2015-04-08]

Social Networks and Their Potential for Education

Petra Poulova[(⊠)] and Blanka Klimova

University of Hradec Králové, Rokitanského 62, Hradec Králové, Czech Republic
{petra.poulova,blanka.klimova}@uhk.cz

Abstract. The purpose of this article is to explore whether there is a potential in using social networks among first-year undergraduates at the Faculty of Informatics and Management of the University of Hradec Kralove, Czech Republic. The authors want to discover which social networks are the most popular among these students so that the most frequently used could be implemented into teaching and learning processes.

Keywords: Cognitive aspects · Education · E-learning · Network

1 Introduction

At present information and communication technologies (ICT) are inseparable part of everyday activities. Hardly anyone could imagine this 20 years ago. And the same is now becoming true for the use of social networks. Originally, social networks were created for the purpose of entertainment, but gradually became used in all spheres of human activities, including education thanks to their abilities to assist in creating, collaborating and sharing content (cf. [1]; [2]). Boyd & Ellison add that social network sites (SNSs) are increasingly attracting the attention of academic and industry researchers who are intrigued by their affordance and reach [3]. They define SNSs as web-based services that allow individuals to construct a public or semi-public profile within a bounded system; articulate a list of other users with whom they share a connection; and view and traverse their list of connections and those by other within the system. They also put the main emphasis on their social aspect since these social network sites enable users to articulate and make visible their social networks.

SNSs are now on their rise, particularly in the institutions of higher education because most of their users, the so-called digital natives, study at these institutions. Generally, there is a trend worldwide to adopt this opportunity of using SNSs in order to support their academic activity [4]. However, the use of SNSs in education must be carefully planned and the right teaching and learning techniques and assessments must be employed. Thus, one should pay close attention to the following factors when implementing SNSs into a learning process (cf. [4]): the background and behavior of user; the university policy on the Internet access; the behavior of university communication; role and rule of social network in daily communication; and the attitude of user. In addition, Ratnesway & Rasiah claim that technology alone cannot bring

© Springer International Publishing Switzerland 2015
M. Núñez et al. (Eds.): ICCCI 2015, Part II, LNCS 9330, pp. 365–374, 2015.
DOI: 10.1007/978-3-319-24306-1_36

transformation because more important is the all-compassing pedagogy without which, no amount of technology can transform a student [5].

Many studies in the field of SNSs now confirm that SNSs can be very effective learning platforms that enhance students' engagement and learning experience, transforming them into active learners with an increased motivation to learn while fostering big quality exchange of ideas and knowledge among participants (cf. [6]; [5]; [7] or [8]).

Thus, the purpose of this article is to explore whether there is a potential in using social networks among first-year undergraduates at the Faculty of Informatics and Management of the University of Hradec Kralove, Czech Republic. Furthermore, the authors of this article want to discover which social networks are the most popular among these students so that the most frequently used could be implemented into teaching and learning processes.

2   Survey

The authors conducted the field research to discover whether young people, in this case Czech university students, are aware of social networks and their potential for education. The survey itself and its results are described below.

2.1   Survey Aim, Methods and Sample

The aim of the survey was to discover students' awareness of social networks and their potential use in education. This was done on the basis of online questionnaires which were distributed among the FIM students. These questionnaires were then analyzed, evaluated and the findings were compared with those of other researchers dealing with these issues.

Altogether 322 respondents participated in the survey. They were all enrolled in the first year of study. Out of these respondents 159 (49%) were females, 160 (50%) were males and three students (1%) did not answer this question. See Fig. 1 below for an illustration. The biggest age group of the respondents (196/61%) was between 20-29 years, which was then followed by students under 20 years (82 respondents/25%). See Fig. 2 for more information.

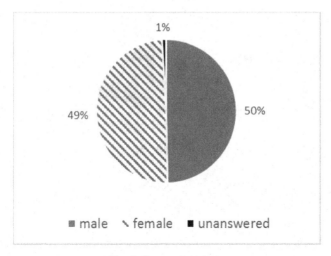

Fig. 1. Respondents' sex

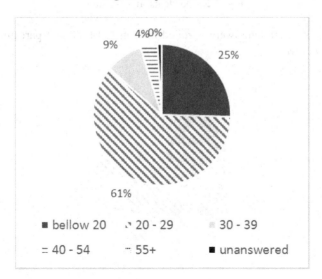

Fig. 2. Respondents' age

In addition, the respondents were relatively evenly spread across the main fields of study which are taught at FIM, such as Applied Informatics (68 respondents/21%); Information management (95 respondents/30%); Financial Management (65 respondents/20%); or Management of Tourism (88 respondents/27%). See Fig. 3 below.

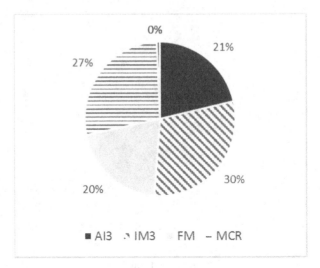

Fig. 3. Respondents' field of study

Moreover, these students were enrolled both in full-time and part-time studies as Fig. 4 shows below.

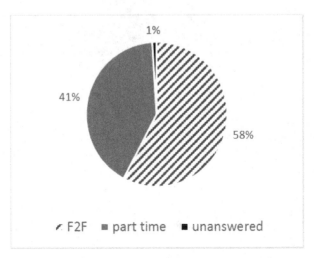

Fig. 4. Respondents' form of study

2.2   Findings

Within a larger survey focused on mobile devices and social networks, the respondents were asked seven questions which concerned their attitude to social networks.

Question 1: Do you have a social network account?
As Fig. 5 indicates, almost all respondents (95%) have the social network account. Only 3% of the respondents do not have any. And almost half of them possess two or three accounts.

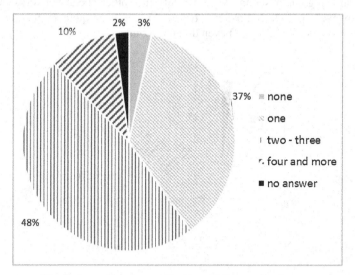

Fig. 5. A number of respondent's social network accounts

Question 2: If your answer to the previous question was positive, which of the following social networks do you use?
Fig. 6 below demonstrates that 298 respondents (93%) mostly use Facebook, which is then followed by Google+ (122 respondents/38%), Schoolmates (63 respondents/20%) and Twitter (63 respondents/20%), while only 37 students (11%) exploits LinkedIn.

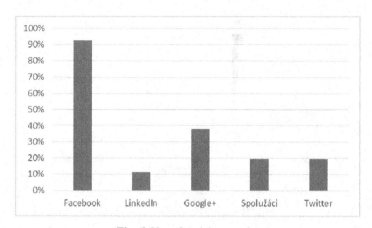

Fig. 6. Use of social networks

370 P. Poulova and B. Klimova

Question 3: How often do you login into Facebook?

The authors of this article were also interested how often students login into these social networks and since Facebook was the most popular, they analyze here just the frequency of this network because, for example, the respondents do not login into Google+ on a regular basis and very scarcely into other social networks mentioned above. However, as Fig. 7 illustrates below, Facebook is quite frequently accessed and some respondents (112/35%) even do not logout.

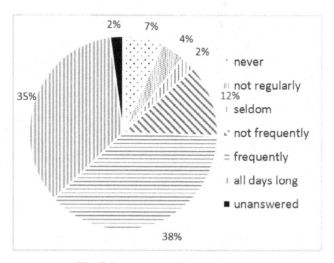

Fig. 7. Respondent's Facebook access

Question 4: Do you consider social networks useful?

The majority of the respondents find the social networks useful or very useful (269 students/83%). Only 27 respondents consider them useless or totally useless. See Fig. 8 below.

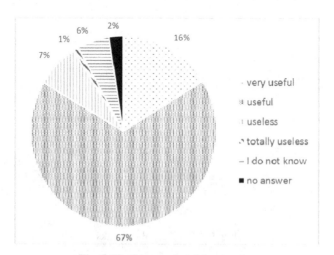

Fig. 8. Usefulness of social networks

Question 5: What do you use social networks most often for?
The authors of this article were curious whether students also use the social networks for study purposes and they were quite nicely surprised because besides communicating with friends and planning events, study purposes came third on the list. See Fig. 9 below for more information.

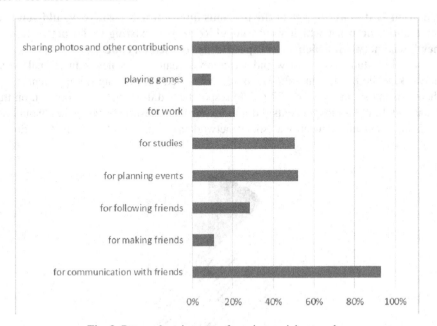

Fig. 9. Respondents' reasons for using social networks

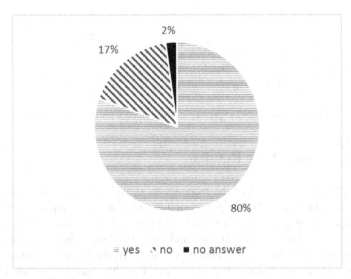

Fig. 10. Possibility of an educational social network

Question 6: Would you welcome an opportunity to have a social network connected with education?
Most of the respondents (259 students/80%) would welcome social networks connected with education, while only 56 respondents (17%) would not (Fig.10).

Question 7: If your answer to the previous question was positive, would you prefer a social network which would be added to you existing social network, or a new social network which would specifically focused on education?
Almost 158 students (49%) would welcome a chance of a new educational social network, while 85 students (26%) would be satisfied with having it implemented into their existing social network. 37 (11%) respondents did not have any opinion on this issue and 42 (13%) respondents did not answer this question because they would not probably want any educational social network as it has resulted from Question 6. (Fig. 11).

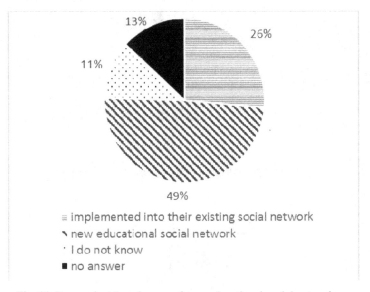

Fig. 11. Respondents' preferences for an educational social network

3   Discussion of the Results

As the findings show, most of the respondents (95%), mainly between 20 - 29 of an age across all the fields of study (i.e. even those who are not computer specialists such as students of Management of Tourism), use the social networks and they find them useful. However, the most dominating and commonly used social network nowadays is undoubtedly Facebook [9], including the users in the Czech Republic [10], [11], [12]. [13]. At present Facebook in the Czech Republic has 3.8 million of its users, which makes 56% of the whole population. And as this survey proved, also 93% of the FIM students used Facebook. In addition, besides using it mainly as a means of communication with their friends (93%), planning events (52%), they also use Facebook for their studies (51%). Therefore, teachers should consider implementing them into their teaching. There have already existed several empirical studies ([5], [6], [7], [8] or [14]) describing the use of social networks, particularly Facebook in teaching and learning processes. For example, Falahah & Rosmala's research conducted among 300 Indonesian university students confirmed that the most dominant social network was Facebook [4]. These students use it at school mainly for resource material sharing and searching and task assignment. Furthermore, Ratneswary & Rasiah claim in their study that Facebook creates a more positive and less threating learning environment which can enhance students learning and rapport between them and their teaching [5]. In addition, it can expand team-based learning, including interpersonal and intrapersonal skills, team-work skills, time management skills, discipline-specific knowledge and digital literacy.

The survey also revealed that there was an interest among the respondents (80%) in the introduction of the social network which would concentrate on education. One of the examples of such network which already exists and it is on its rise is the web portal called Piazza which was founded by an Indian girl Pooja Sankar. Its name comes from the Italian word for plaza--a common city square where people can come together to share knowledge and ideas. Piazza is a free platform for instructors to efficiently manage class questions and answers. Students can post questions and collaborate to edit responses to these questions. Instructors can also answer questions, endorse student answers, and edit or delete any posted content. Piazza is designed to simulate real class discussion [15].

4   Conclusion

This survey proved that there was a big potential for using social networks, particularly Facebook, in education since they attract thousands of users every day all over the world, including the Czech users. Facebook has many applications that support teaching and learning. Moreover, it was confirmed that it could cultivate positive learning experiences as well as to enhance the rapport between the educators and their students [16].

Acknowledgment. This paper is supported by the SPEV Project N. 2108.

References

1. Bicen, H., Cavus, N.: Social network sites usage habits of undergraduate students: case study of Facebook. Procedia - Social and Behavioral Sciences **28**(1), 943–947 (2011)
2. Uzumboylu, H., Bicen, H., Cavus, N.: The efficient virtual learning environment: A case study of web 2.0 tools and Windows Live Spaces. Computers & Education **56**(3), 720–726 (2011)
3. Boyd, D.M., Ellison, N.B.: Social network sites: Definition, history and scholarship. Journal of Computer-Mediated Communication **13**(1), 210–230 (2008)

 4. Falahah, Rosmala, D.: Study of social networking usage in higher education environment. Procedia - Social and Behavioral Sciences **67**(1), 156–166 (2012)
 5. Ratneswary, R., Rasiah, V.: Transformative higher education teaching and learning: Using social media in a team-based learning environment. Procedia - Social and Behavioral Sciences **123**(1), 369–379 (2014)
 6. Bosch, T.E., Preez, A., Michell, L.: Using online social networking for teaching and learning: Facebook use at the University of Cape Town. South African Journal for Communication Theory and Research **35**(2), 185–200 (2009)
 7. Selwyn, N.: Faceworking: Exploring students' education-related use of Facebook. Learning, Media and Technology **34**(2), 157–174 (2009)
 8. Wang, Q., Woo, H.L., Quek, C.L., Yang, Y., Liu, M.: Using the Facebook group as a learning management system: An exploratory study. British Journal of Educational Technology **43**(3), 428–438 (2012)
 9. Facebook Statistics (2014). http://www.statisticbrain.com/facebook-statistics
10. Cerna, M., Poulova, P.: Social software applications and their role in the process of education from the perspective of university students. In: Proceedings of the 11th European Conference on e-Learning (ECEL 2012), Groningen, pp. 87–96 (2012)
11. Poulova, P., Cerna, M.: Role of social media in academic setting awareness, utilization and willingness. In: Szakal, A. (ed.) 11th IEEE International Conference on Emerging eLearning Technologies and Applications (ICETA 2013), pp. 59–62. IEEE, Stara Lesna (2013)
12. Poulova, P., Cerna, M.: Social applications in engineering education. In: IEEE Global Engineering Education Conference. EDUCON, pp. 206–209. IEEE, Istanbul (2014)
13. Social Network Statistics (2014). http://www.statisticbrain.com/social-networking-statistics/
14. Roblyer, M.D., McDaniel, M., Webb, M., Herman, J., Witty, J.V.: Findings on Facebook in higher education: A comparison of college faculty and student uses and perceptions of social networking sites. The Internet and Higher Education **13**(3), 134–140 (2010)
15. Sankar, P.: Our story (2013). https://piazza.com/about/story (retrieved September 2, 2014)
16. Mazer, J.P., Murphy, R.E., Simonds, C.J.: I'll see you on "Facebook": The effects of computer-mediated teacher self-disclosure on student motivation, affective learning, and classroom climate. Communication Education **56**(1), 1–17 (2007)

ITiB: Special Session
on IT in Biomedicine

Development of Information and Management System for Laboratory Based on Open Source Licensed Software

Pavel Blazek[1,2(✉)], Kamil Kuca[1,3], Daniel Jun[2], and Ondrej Krejcar[1]

[1] Faculty of Informatics and Management, Center for Basic and Applied Research,
University of Hradec Kralove, Hradec Králové, Czech Republic
ondrej.krejcar@uhk.cz
[2] Faculty of Military Health Sciences, University of Defence,
Hradec Králové, Czech Republic
{pavel.blazek,daniel.jun}@unob.cz
[3] Biomedical Research Centre, University Hospital Hradec Kralove,
Hradec Králové, Czech Republic
kamil.kuca@uhk.cz

Abstract. The value of data is hidden in their organization, accessibility and usability. During a research the data are collected from external sources and created too. If they are not stored in a central database, it is difficult to reuse them and it could lead to inefficiency in a form of duplicated information and time wasting for finding the same information in public sources. An output of research can also be a long time expected medication as a biohazard substance. Both ideas have a common problem of sensitivity and security. It is necessary to protect these data either for secreting against competitors or hiding information from terrorists. For the research team, the Biomedical Database System offers besides security also improvement of data flow in an organization, clear arrangement and efficiency.

Keywords: Biomedical research · Information system · Data collecting · Data processing · Security

1 Introduction

There are three collaborating subjects in the campus of Hradec Kralove which are focused on medicine and pharmacy. They are interconnected to other subjects and cooperate with them on various types of research.

The Biomedical research centre (BRC) belonging to the University hospital [1] provides comprehensive services and offers cooperation in several areas of basic and applied research. One of the main BRC aims is a preparation of new drugs and molecules which can be used for diagnostics of selected diseases. The BRC works closely with both clinical and interdisciplinary workplaces in the University hospital [2].

The Faculty of Military Health Sciences, University of Defence in Hradec Kralove [3] is a centre of medical education and research of the Czech Army and belongs among key elements of military university education. The Department of Toxicology

© Springer International Publishing Switzerland 2015
M. Núñez et al. (Eds.): ICCCI 2015, Part II, LNCS 9330, pp. 377–387, 2015.
DOI: 10.1007/978-3-319-24306-1_37

and Military Pharmacy (DTMP) is the unique research facility and is focused on diagnostics, prophylactics, treatment and decontamination of intoxication caused by highly toxic compounds including chemical warfare agents.

The University of Hradec Kralove [4] develops scientific disciplines within the research and teaching activities of individual faculties. The Faculty of Informatics and Management (FIM) has a study program - Applied Informatics which provides the grounding necessary for careers in computer systems including both hardware and software aspects. This BASE gives a great potential for successful research [5]. On the other hand, this group produces and collects enormous amount of data which has to be saved, organized, secured and ready to use for research.

2   Problem Definition

To support standardized procedures, generate valid results of above mentioned subjects and last but not least to run labs effectively, it was necessary to implement a unified information system in order to increase their competitive ability. The assignment came out from the analysis of the present status at a management level, lab equipment and potential possibilities of the background and from the study of already applied projects [6], [7], [8]. The use of Open source platform was primarily discussed in the course of its development with the emphasis on data security particularly at the application level. Simultaneously, the system required to be modular i.e. open to potential subsequent functions, and modules could be mutually linked, which is actually the trend of all LIMS [9], [10]. The processes described in the following chapters in more detail were defined after a deeper analysis.

3   Related Works

Deployment of Laboratory information management system (LIMS) followed from the requirement to enhance a quality, efficiency and data security in campus labs [11]-[13]. A developing system design has been under discussions for a long time, however, they have not been finished with the final version documentation. Recently, the issue of an effective environment for theoretical calculations of molecular bonds and anticipated properties has been solved. This project was described in [14] and comprises a reference on the Cloud environment at the University of Hradec Kralove. A following logical step appeared to be the effective usage of this environment for facilitating work in labs, it means to introduce LIMS. In compliance with the general definition, Management information system should represent the system of leading processes either in a laboratory or business which should improve data acquisition of output processes and according to described LIMS evolution [10], the study of potential solutions was focused to satisfy current trends [6], [9]. As a base for our project were chosen Open Source activities. Due to of this, it was necessary to take into account possible security risks [15].

The selection of a proper candidate was based on both publicly available information from the Internet [7] and information from science sources [10], [13] The research was carried out in two lines. The first one was targeted at IMS in general (e.g. ILIAS, Open-KM, Alfresco ...). Shortly after the beginning of reconnaissance we realized that it brings more complexity without assets expected. These systems are mostly focused to business and corporation sphere and solve information and document flow generally so we narrowly focused to exploration of LIMS (e.g. AgiLAB, BIKA LIMS, Open-LIMS, LABkey, LISA ...). The commercial systems were also studied for inspiration of possible functionalities [8], [16]-[19]. The client-server applications ported to the Private Cloud environment were preferred, when the client was replaced by the web interface which enabled access from an independent HW platform.

Based on assignment we aimed farther than implement simple LIMS [20]. Deployment of our own IS concentrates more functionalities than available products. Process of similar module integration is also discussed in [21], [22] but without connection to private knowledgebase. Simple integration of collected open source applications should leads to unpredictable problems. There could be collection of several different databases for each implemented module and connect them each to other could be impossible so it was necessary find one central module or make decision to create our own one. It was chosen to create LIMS tailored to environment defined above.

There is one exception what was made during deployment. Due to security reason is prohibited to use commercial datastores (e.g. OneDrive, Dropbox, Google drive, Rapidshare ...) to store laboratory data and share them. More secure way is to use ownCloud [23], [24] service provided by CESNET. That means store and share data over association founded by all universities of the Czech Republic for supporting their noncommercial activities.

4 Laboratory Information System – LabIS

The private cloud environment and support of the UHK allows us to test and implement all modules of LabIS. The system development was divided into several logical parts. Finally, it consists of four main modules:

— Security ... authentication authority
— Groupware ... cooperating tool
— eLearning ... electronic learning materials
— LabApp ... laboratory application fully deployed

The main page of the LabIS offers basic information about cooperating subjects and contains links to the modules mentioned above but only authenticated users are allowed to open these links. Security module is hidden behind all activities and is connected to all modules and only authenticated users can access modules of LabIS.

Groupware implementation was done for better communication, elastic cooperation inside the teams. It allows sharing unfinished documents and other digital materials in the sub-module called simply the Library. Now, it is not necessary to send links and files via emails. The System automatically sends only a notification about group activity. It is possible to extract all files shared this way to put keywords into the metadata and to add our own. It is allowed to use versioning of text documents. The System collects information about users´ activities in group communication [25, 29-31]. The staff members can prepare courses for students or import there free available ones into the eLearning module.

4.1   LabApp Module

The module is built on the idea of the standard cycle when the study of documents and patents has firstly the theoretical idea then it is applied in an experimental environment and consequently analyzed. Data linked to the whole process are processed and subsequently published. (Fig.1)

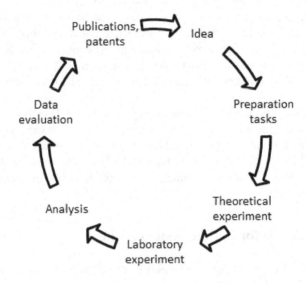

Fig. 1. General lab process circle

The application itself is physically based on the database and designated for labs that study substances synthesis. The environment of the University of Hradec Králové campus primarily consists of two workplaces which use comparable work procedures. In the time of the module proposal, it was possible to compare their specificity and to suggest the information system with respect to both sides' requirements frequently based on standards [11].

Future users' responses revealed they want to build the system to which they will not have to adjust their routine activities, and which will not require any investments or only minimal. The aim of the application is to cover all phases of participating teams' activities in the lab and to transfer paper and autonomous digital records into the central system. The study of follow-up activities revealed a possibility to offer new functionalities to the users which could contribute to higher efficiency. Mapping of logical links of the lab environment became the basis for a proposal of the future information system.

4.2   Organizational Structure

For comprehension of the roles, it is necessary to describe the operated structure. We can divide it into five groups:

— Management – supervisors, who lead activities in the lab and assign authorizations and tasks for processing
— *In silico* team – the group of computer chemistry collects information and data to the assigned task. Their members design molecule models and study their bonds. With the use of computer simulation, they model and analyze behavior and forces of mutual bonds. The management uses the outputs to decide which compound will be physically synthetized.
— Synthetic team prepares compounds according to obtained theoretical materials Biochemical, toxicological and pharmacokinetic team – groups which carry out thorough basic tests of the synthetized sample in vivo and in vitro independently on each other, if needed pharmacokinetic tests. The result reveals an effect of the tested substance on living organisms in the respect of verification of biological effect for which the substance was proposed, its toxicity or possibly a capability of living organisms to degrade the substance.

The whole process is conducted by the management group and is shown in (Fig. 2). This group assigns tasks to all teams. The first task is mostly directed to the *In silico* team for provision of theoretical tests. The preparation lies, besides the theoretical plane, in building a local "ligand database" and a biological target" model. It is not complicated to acquire data for the actual database owing to a possibility to download different molecular "subsets" (e.g. lead-like, fragment-like, drug-like)" from the huge on-line "drug" databases (e.g. zinc.docking.org).

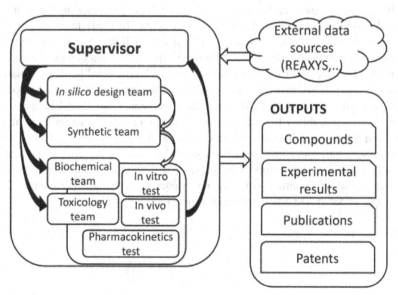

Fig. 2. Management flow in a lab

For further data processing in the "docking" program, it is necessary to carry out a format conversion and preliminary chemical calculation. A similar situation occurs in proteins, the structures of which are available in on-line databases (i.e. www.pdb.org) already "adjusted for molecular docking".

The configurative file comprises data which define parameters of modelling program cycle (e.g. AutoDock Vina). The processing requires a sufficient computing power which may offer the environment dedicated in the cloud [14]. Potential biological and physical properties of the studied substances are the results of the mentioned above procedure. The team uses either data stored in the database itself or in an external source to design the model. By chemical structures modelling, the team will get a group of potential substances which correspond with the assigned task. Selecting these substances, the team will get the formula of one or more theoretically significant substances which is then handed over to the Supervisor for assessment. The next procedure depends on the Supervisor's decision, e.g. which substance or if the substance will be synthetized. He will subsequently assign the formula with running comments to a particular member in the group of organic chemistry. The designed substance is handed over to the Supervisor for evaluation of biological activities and toxicological properties. Then, the Supervisor will give the tested substance to a person responsible for a particular test and the person will provide testing and elaborate required reports. When the experiment is generally successful, pharmacological tests follow, in which a capability of the living organism to keep and eliminate a potential medicament is tested.

4.3   Function Proposal

Now, the procedure from the point of the system proposal is concerned. As it was already mentioned, the data from the *In silico* team are presented in the form of the chemical formula and comments, it means metadata. Just this team participates in creation of metadata most, particularly with inserting links to external information sources. Those can have links to more than one experiment, that is why it was decided the modules LabApp and Library would be linked with the bond N:M. Information on chemical properties and procedures received from the external sources is stored in the system of digital Library. Consequently, their references are stored in the LabApp database with a possibility to write personal notes to a particular experiment, it means 1:N, which are inserted into the metadata table with a characteristic of the given experiment. Simultaneously, the identification of the person who inserted the data into the system is automatically stored here together with the timestamp which defines the relevance in case of a conflict with data in subsequent verifying experiments.

Example of database table is shown in (Fig. 3).

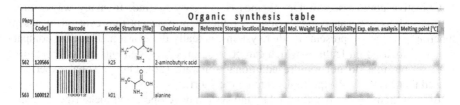

Fig. 3. Sample of Synthetics table

The record prepared this way facilitates processing of citations sources into reports, scientific articles or for patent procedure. (Fig. 4)

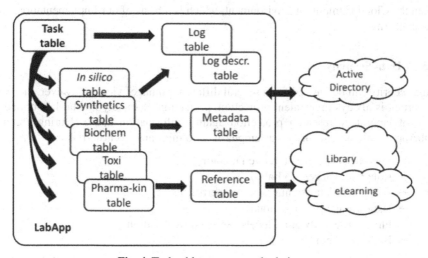

Fig. 4. Task table structure and relations

4.4   Database Engine

To create a simple database for the given purpose, which would comprise above described procedure, is quite a simple task. It can be solved on a personal computer for example by means of MS Access out of the office tools MS Office Professional package. However, a few requirements which negate this option emerged from the assignment:

    — Prerequisite for regressive digitalization of records from the lab notebooks
    — Simultaneous data processing and the database access
    — Data security at all levels of processing
    — Integration of other LabIS parts (Authentication, Library, eLearning)
    — Access from different devices (PC, NB, Tab, CellPhone
    — Unified interface
    — Possibility to search in stored data according to selected parameters

Objective comparison of DB engines is not easy to find. In many ways, information is biased because of developers' reluctance to accept that evasion of absent functions, which are present in other systems, is really the evasion. Thus, the comparison was focused on features which can be used in SQL statements or self-contained SQL scripts that do not require additional software [26]. Our personal experience caused that MySQL, PostgreSQL were on the short-list.

Both of them are frequently used by the system developers and make right decision during selection was theirs goal.We have had experience with MySQL from other especially web applications so that it was our choice. Nevertheless, from some load tests arise that PostgresSQL would be more powerful under a higher load. MySQL offers broader range of Storage engines. Our chosen InnoDB fulfils demands for a support of required system properties (e.g. transactions, row locking, foreign key…). Then the Cloud Computing Environment, which is a basis of our implementation, was our concern.

4.5   Data Security

Data security can be viewed from several different points of view. However, the most important is always investment protection, protection of society mental labour and in case of laboratory research protection against information abuse. During creating LabApp mainly data security was emphasized to minimize security risks.

— User authentication (Active Directory)
— Role based access to LabIS
— Encrypted data communication (certificate)
— Data storage (Private cloud)
— Physical security at all levels (Servers, workstations)
— Backup schema

The security module is one of the main parts of LabIS and unites users' accounts of all people in the central database. Its centre is current MS Windows Server in the role of a domain controller. It was integrated in its structure within building LabIS. Each individual module, the LabApp as well, uses Active Directory for users' accounts verification.

Security requirements, i.e. data theft protection, damage or loss due to HW failure, were taken into account. A part of that is provided by a sub-model, through a graphic interface of which an authorized person is allowed to set rights to individual users according to their roles in the organization, which simplifies the supervisor's administration of the system in assigning partial tasks.

Due to the danger of information leak, e.g. to a competitive group, it was necessary to secure it thoroughly. After analyzing the information flow and observed data it was found out that partial data (a chemical formula, the melting point, the boiling point, solubility…) themselves do not provide any relevant information on a biological activity of the tested substance. Analogically, it is not possible to acquire the substance chemical formula from the data describing the biological activity in vitro or in vivo again. Therefore, the proposed model for the application interface do not allow to display the whole group of data to anybody else than to the Supervisor role. The subsequent logic of delivering the data within the task follows.

4.6   Communication and GUI

The graphic user's interface is unified for the whole LabIS system. As it is in principle a SaaS implementation of the solution which has its infrastructure out of the laboratory and to access to it connectivity to the Internet is necessary, it was agreed that the web interface based on HTML5 standard with mobile device support will be used primarily [27]. The communication goes through potentially dangerous environment, therefore SSL certificate implementation was used as a common commercial security standard. It was necessary to enable safe data inserting into the LabApp database and into the Library according to the task from any place with the connectivity and it was reached. The reason for this requirement is the fact that some experiments take a lot of hours of machine time, their completion is not possible to estimate and they can continue at weekends as well. The data from the devices, which are used for the experiments, can be read through the web interface, which is also accessible from home.

4.7   Infrastructure and Back-Up

Whole infrastructure is situated in the environment of the cooperating university private Cloud in whose gestion is not only physical security but also the care for backing-up data within the system administration [28]. It means, this activity is not provided by the people whose responsibility does not include IT so they can pay their full attention to their work.

5   Conclusion

The described LabApp has accomplished the requirements given by the task and has also extended them with other functionalities. The security policy was fully set and it eliminated possibilities of sensitive information leak and at the same time it made the task process more effective. Concurrently, the control of the laboratory processes, laboratory staff activity and systematic processing of the data gained in laboratory measurements became better. The central system of information sources register made the work with information more effective. Apart from other things, the system introduction also meant elimination of mistakes connected with inaccurate reading of the records from laboratory notebooks caused by handwriting. Also, the additional replenishment of a module for observing inserted source materials and metadata into the Library system according to people can be considered as a positive side effect. Through a periodic evaluation it is possible to view activities of particular team members from the supervisor's position. Nowadays, we have already had more suggestions to extend the current LabApp functionalities. Solve connection between compound database and IBM SPSS Statistic software for semi-automated processing of statistical data output. Make search routines faster and more scalable without necessity of SQL language knowledge for operators. Improve sample signing process based on barcode or replace it with much more suitable QR. Utilization of mobile devices in the laboratory environment to insert data to the database and deploy secure wireless network for them.

Acknowledgement. This work and the contribution were also supported by project "Smart Solutions for Ubiquitous Computing Environments" FIM, University of Hradec Kralove, Czech Republic (under ID: UHK-FIM-SP-2015-2103).

References

1. Biomedical Research Center (2011). http://eng.fnhk.cz/cbv/introduction (cit. March 11, 2015)
2. Blazek, P., Krenek, J., Kuca, K., Jun, D., Krejcar, O.: The system of instant access to the life biomedical data. Computational Intelligence and Informatics, pp. 261–265 (2014). doi:10.1109/CINTI.2014.7028686
3. University of Defence. Faculty of Military Health Sciences (2014). http://www.unob.cz/en/fmhs/Pages/default.aspx (cit. March 21, 2015)
4. University of Hradec Kralove (2015). https://www.uhk.cz/en-GB/UHK (cit. March 10, 2015)
5. Blazek, P., Krenek, J., Kuca, K., Jun, D., Krejcar, O.: The Biomedical Data Collecting System, Microwave and Radio Electronics Week, pp. 1–4 (2015)
6. Çağındı, Ö., Ötleş, S.: Importance of laboratory information management systems (LIMS) software for food processing factories. Journal of Food Engineering **65**(4), 565–568 (2004). doi:10.1016/j.jfoodeng.2004.02.021. ISSN 0260-8774
7. LIMSwiki (2015). http://www.limswiki.org (cit. March 20, 2015)
8. Quo, C.F., Wu, B., Wang, M.D.: Development of a laboratory information system for cancer collaboration projects. In: IEEE Engineering in Medicine and Biology 27th Annual Conference (2005). doi:10.1109/IEMBS.2005.1617070
9. Kammergruber, R., Robold, S., Karlic, J., Durner, J.: The future of the laboratory information system – what are the requirements for a powerful system for a laboratory data management? Clinical Chemistry and Laboratory Medicine (CCLM) **52**(11) (2014). doi:10.1515/cclm-2014-0276
10. Prasad, P.J., Bodhe, G.L.: Trends in laboratory information management system. Chemometrics and Intelligent Laboratory Systems **118**, 187–192 (2012). doi:10.1016/j.chemolab.2012.07.001
11. ISO/IEC 17025 - General requirements for the competence of testing and calibration laboratories
12. Ekins, S.: Computer Applications in Pharmaceutical research and Development, pp. 57–61. Sean Ekins, Wiley (2006). ISBN 0-471-73779-8
13. Hu, Y.: Development of information management system used in laboratory. Advanced materials research. In: 27th Annual International Conference of the Engineering in Medicine and Biology Society. IEEE-EMBS 2005, Shanghai, China, pp. 2859–2862 (2012). doi:10.4028/www.scientific.net/AMR.605-607.2518
14. Dolezal, R., Sobeslav, V., Hornig, O., Balik, L., Korabecny, J., Kuca, K.: HPC cloud technologies for virtual screening in drug discovery. In: Nguyen, N.T., Trawiński, B., Kosala, R. (eds.) ACIIDS 2015. LNCS, vol. 9012, pp. 440–449. Springer, Heidelberg (2015)
15. Hansen, M., Köhntopp, K., Pfitzmann, A.: The Open Source approach - opportunities and limitations with respect to security and privacy. Computers & Security **21**(5), 461–471 (2002). doi:10.1016/S0167-4048(02)00516-3
16. LABWARE Result Count (2015). http://www.labware.com/ (cit. February 28, 2015)
17. Lab Collector (2015). http://labcollector.com/ (cit. February 28, 2015)

18. Qi Analytica: LISA.lims (2014). http://lisa.lims.cz/ (cit. February 28, 2015)
19. Delaney, N.F., Echenique, J.I.R., Marx, C.J.: Clarity: An Open-Source Manager for Laboratory Automation. Journal of Laboratory Automation **18**(2), 171–177 (2012). doi:10.1177/2211068212460237
20. Stephan, Ch., Kohl, M., Turewicz, M., Podwojski, K., Meyer, H.E., Eisenacher, M.: Using Laboratory Information Management Systems as central part of a proteomics data workflow. Proteomics **10**(6), 1230–1249 (2010). doi:10.1002/pmic.200900420
21. Machina, H.K., Wild, D.J.: Laboratory Informatics Tools Integration Strategies for Drug Discovery: Integration of LIMS, ELN, CDS, and SDMS. Journal of Laboratory Automation **18**(2), 126–136 (2012). doi:10.1177/2211068212454852
22. Sahiti, M., Vimla, L.P.: Organization of biomedical data for collaborative scientific research: A research information management system. International Journal of Information Management **30**(3), 256–264 (2010). doi:10.1016/j.ijinfomgt.2009.09.005. ISSN 0268-4012
23. CESNET: Data Care (2015). https://du.cesnet.cz/en/start (cit. March 20, 2015)
24. ownCloud (2015). https://owncloud.org/ (cit. March 27, 2015)
25. Kasik, V., Penhaker, M., Novák, V., Bridzik, R., Krawiec, J.: User interactive biomedical data web services application. In: Yonazi, J.J., Sedoyeka, E., Ariwa, E., El-Qawasmeh, E. (eds.) ICeND 2011. CCIS, vol. 171, pp. 223–237. Springer, Heidelberg (2011)
26. SQLworkbench (2015). (cit. February 24, 2015)
27. http://www.sql-workbench.net/dbms_comparison.html
28. Behan, M., Krejcar, O.: Adaptive graphical user interface solution for modern user devices. In: Pan, J.-S., Chen, S.-M., Nguyen, N.T. (eds.) ACIIDS 2012, Part II. LNCS, vol. 7197, pp. 411–420. Springer, Heidelberg (2012)
29. Pavlik, J., Komarek, A., Sobeslav, V.: Security information and event management in the cloud computing infrastructure. Computational Intelligence and Informatics, pp. 209–214 (2014). doi:10.1109/CINTI.2014.7028677
30. Cerny, M., Penhaker, M.: Wireless body sensor network in Health Maintenance systems. Elektronika Ir Elektrotechnika **115**(9), 113–116 (2011). doi:10.5755/j01.eee.115.9.762
31. Penhaker, M., Cerny, M., Rosulek, M.: Sensitivity analysis and application of transducers. Paper presented at the 5th International Workshop on Wearable and Implantable Body Sensor Networks, BSN 2008, in conjunction with the 5th International Summer School and Symposium on Medical Devices and Biosensors. ISSS-MDBS 2008, Hong Kong (2008)

Determining of Blood Artefacts
in Endoscopic Images Using a Software Analysis

Lukas Sulik[1], Ondrej Krejcar[1(✉)], Ali Selamat[1,2],
Reza Mashinchi[2], and Kamil Kuca[1,3]

[1] Faculty of Informatics and Management, Center for Basic and Applied Research,
University of Hradec Kralove, Rokitanskeho 62, 500 03 Hradec Králové, Czech Republic
{Lukas.Sulik,kamil.kuca}@uhk.cz, Ondrej@Krejcar.org
[2] Faculty of Computing, Universiti Teknologi Malaysia, 81310 Johor Baharu, Johor, Malaysia
aselamat@utm.my, r_mashinchi@yahoo.com
[3] Biomedical Research Center, University Hospital Hradec Kralove,
Hradec Králové, Czech Republic
kamil.kuca@uhk.cz

Abstract. Wireless capsular endoscopy is a novel method of gastroenterology investigation during last years. Most important as well as the most difficult part of such investigation is the blood artefact determination. These artefacts are only be localised by medical experts trained in such field of gastroenterology medicine. This article describe the process of development a software solution which can be used to help localise some specific artefacts using developed algorithms. Such algorithms are firstly developed by Matlab solution while they are consequently transformed to developed software solution. Our solution was already preliminary tested by one of specific artefacts – blood artefacts, while the results have been found as sufficient to several use of this software. Future research will be focussed on other specific artefact description and algorithm development for detection.

Keywords: Capsular endoscopy · Computer vision · Feature detection · OpenCV

1 Introduction

The article focuses on the analysis of biomedical data obtained using the capsular endoscopy technique. The data, which are received from cameras, are further processed with the help of computer vision. The suspicious anomalies are searched which could at the end result in an illness occurrence or bleeding.

The small intestine is the most complicated part of the intestine to inspect. This is the case because of the distance from the mouth and from the anus, as well as because of its complex shape which is representative by multiple loops. The conventional endoscopic techniques, for example enteroscopy or colonoscopy, are also limited by the length of the intestine (3.35–7.85 m) [1].

The capsular endoscopy consists of a small camera which has the size of a bigger pill that also possesses its own source of light. The camera scans the image of the intestine

© Springer International Publishing Switzerland 2015
M. Núñez et al. (Eds.): ICCCI 2015, Part II, LNCS 9330, pp. 388–397, 2015.
DOI: 10.1007/978-3-319-24306-1_38

and sends it wirelessly to the receiver which is situated outside of the patient's body. The capsule's battery lasts for 7-8 hours. This time is sufficient enough in most of the cases for exploring the whole small intestine [2]. After the whole process is finished, the camera is removed from the body through alimentary tract.

It is necessary to analyse the obtained data. This can be conducted by the person with accurate knowledge about the topic. This can take a long time due to the number of the medical examinations and the length of the recording.

There is also the option to evaluate the data using special software which are able to detect potential illnesses and shorten the process that is necessary for identifying the illness.

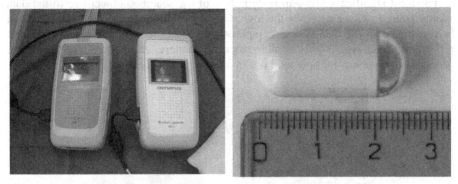

Fig. 1. The wireless capsule endoscope is a disposable capsule with an outer diameter of 11 mm, length of 26 mm and weighing 3.8 g. It moves through the small bowel, propelled by peristalsis and transmits data to a portable data recorder. [3]

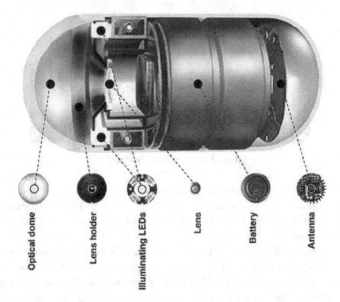

Fig. 2. Structure of camera which provides direct colour video images of the small intestinal mucosa and of the intestinal content at a rate of 2 images per second (for approx. 8 hours). [3]

2   Problem Definition

The developed application should be able to evaluate the record entry and detect the possible bleeding in intestines. The data from the camera are extracted into the AVI video format. However, rather than video, it is a sequence of subsequent pictures. The problem arises during the identification of whether the intestine contains blood. It is necessary to describe or identify the searched object in order to be able to accurately find it using the image processing, with the least number of mistakes. Therefore, it is needed to find EER (acceptable number of mistakes) between FAR (objects which are described as acceptable) and FRR (objects which are not recognised, but they are positive) [4].

The most significant problem is definition of the searched object due to the data quality being low as a result of the low camera resolution. Numerous anomalies can be also found in the data. These complicate the recognition of the object (different reflections, food remainders, similar colourful matters, etc.). Therefore, it will be necessary to combine multiple methods in order to obtain the best result.

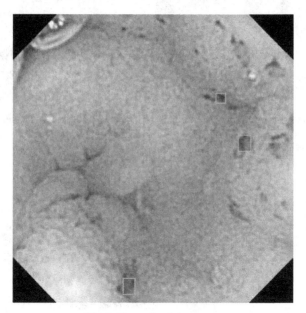

Fig. 3. Example of the entry

In the article called "Detection of bleeding in wireless capsule endoscopy images using range ratio color" [5], the authors discuss the blood detection option using the colour range in RGB model. However, this approach has many disadvantages, such as in the real environment it is possible to find a pixel of the size of blood on almost each image. A group of authors of the article called "A technique for blood detection in wireless capsule endoscopy images" [6] use more sophisticated approach. Firstly, they filter the image and dispose in this way of any unwanted objects. Secondly, they distinguish whether the found object is blood or not, using the combination of the anomaly detection and colour filter.

3   New Solution

The new algorithm is based on the combination of more techniques. Two independent operations take place here which are at the end compared and the possible blood can be discovered.

In the first part of the algorithm, the edges are detected which show possible objects that are different from the background. Canny edge detection [7-11] algorithm was used for this detection. This algorithm is considered as the optimal solution. After the edge detection, the approximation of found objects into a simple geometric shape (ellipse or circle) is performed in order to be able to further work with those objects. Subsequently, the large approximated objects are taken from the selection. These are the mistakes which originate from edge detection (e.g. the edges of the video). After the removal of the unwanted objects, the rest of the selection is drawn onto an instrumental image which is used for the end of the algorithm.

In the second part of the algorithm, the ROI is searched (the area of interest – possible area where the blood could be found. This part consists of more subparts which are independent. Firstly, the conversion of the image colours from RGB into HSV is performed as this model is more suitable for filtration according to colours. The colours from the given spectrum are filtered out from HSV image. Moreover, the all of the outlines which were created after the filtration are found. The selection among these outlines is made and too small outlines are removed. This eliminates various colour anomalies. The morphologic operation of dilatation [8] is applied on the outlines. This ensures the connection of small parts into larger ones. These larger parts will be iterated and for each of them the specification will be made. This specification is based on the next colour filtration according to stricter criteria for each area of the part and after filtration each part will be saved into an instrumental image.

After the first two stages of the algorithm, the instrumental images are compared. If an object from the first stage is discovered in the second stage, the final image shows an area where the possible blood can be found.

One part of the new solution will be the evaluation software that will be able to analyse larger amount of data and show results. The software will be designed in a way that it will be easy to operate and enlargeable for further algorithms for searching other anomalies. The parameters of each algorithm will be possible to modify without the need to change the programme. It will be possible to use the same algorithm multiple times with different setting [12].

4   Implementation

The software is designed using several standard technologies like JavaSE 8, JavaFX 8, WeldSE – CDI for JavaSE, OpenCV, Apache, Maven, Git, [12]. The implementation of the software is complex. Therefore, in this part, only the basic parts will be described.

4.1   Modules Package

This package contains the programme's modules which will process the video. These are the classes with specific algorithm. Each module must be implemented in a Module range or must divide the AbstractModule class which has a defined basic behaviour. The module itself is configured using the external file where all its characteristics are given. It is possible to have more configuration files and run only one module with different characteristics.

4.2   Core Package

The main parts and the business logic of this programme are found in this part. Figure (Fig. 4) shows the simplified model of the classes.

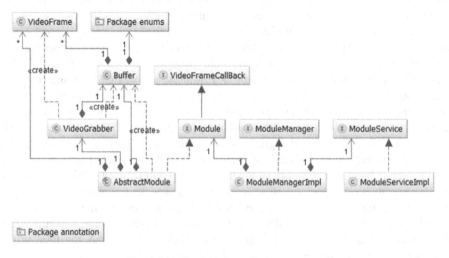

Fig. 4. Model of the three Core packages

During the initiation of the programme, the first method used is the AllModules() initiation method from the ModuleManager class. The aim of this method is to load the configuration files of modules and create module objects. All takes place using reflection and annotation.

The abstract class called AbstractModule is the basic implementation of the Module range. This class is designed in a way that modules which benefit from her need to focus only on their own algorithm. The class obtains VideoGrabber which loads the video from the disc and saves is into the storage that is represented by Buffer class. The abstract module has also its own storage, but it is only for the purpose of development. It is not necessary to keep all data in the production version. The VideoGrabber class is supplied to the module during the analysis of the video and can serve also to other modules. The loading of the video from the disk is realised using the inbuilt solution in OpenCV. Each modul receives its own fibre. This enables multiple modules to run at the same time and process the same file without the need to wait for the end of previous operations. It is possible to process even more files at once.

4.3   Commons Package

This package is identical for each parts of the programme and contains the supporting or sharing classes. Among the main classes belong the classes for exceptions maintaining in the programme and the utilisation of the class for maintaining the localisation, buffer and reflection. Furthermore, there is also the aspect for measuring the time of method duration which is solved using the interceptor, offered by Weld. Then the interceptor is used for improving of the method optimisation and area finding which can be written more effectively. The method that is to be measured, must contain annotation PerformanceLog.

4.4   Application Package

The package contains application logic of the application. The user's interface and controllers are defined here. These work further with the Core Package. Moreover, the exceptions are captured here and the information about the mistake is displayed to the user. The whole application is localised into the variable language version. It is only necessary to translate the localised files.

Since there are no existant developing tools for Java which would be created for OpenCV, it was necessary to build a new interface that would simplify the algorithm's development. This is established using the reflection and for each module it is possible to create an interface in which the module's configuration in real time can be changed. Furthermore, it is possible to follow individual steps of the algorithm that are saved in the buffer.

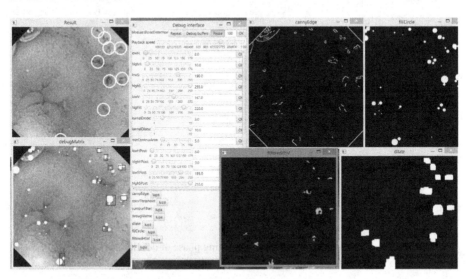

Fig. 5. Example of the algorithm's steps during development process

4.5   Implication of the Algorithm for Blood Detection

In the first part of the algorithm, the search of edges is conducted. Since the implementation of Canny Edge algorithm in OpenCV does not contain the whole process of the algorithm, but only its main part, it is necessary, due to different entry data, to firstly process the data in order to fit the implementation of OpenCV. This is realised during the conversion of RGB model into the grey scale using the cvtColor method. For better results of the detector, it is suitable to firstly blur the image, precisely the Gaussov's filter is used which is implemented in the blur method. After the data processing, the Canny edge detector can be used (Fig. 5) [12].

5   GUI of Developed Software – "BioMedical Analyser"

During the development process the GUI of application for image processing was developed (Fig. 6). The goal was to develop a smart and intuitive GUI with a strong demand on speed of image analysis processing and easy and usable outputs in the form of reports (in Portable Document Format - PDF and Microsoft XLSX format).

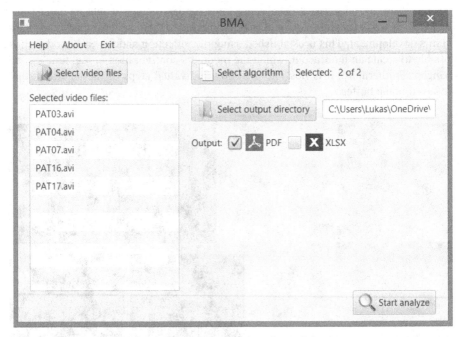

Fig. 6. Steps of the algorithm during testing phase of development process

6   Testing of Developed Application

The testing was conducted on the sample of 31 video files where each video has 201 frames. On those videos, the expert defined artefact which should be found [3]. The programme should closely resemble the expert's finding while we used standard testing methodology used during other research projects [13-22].

The testing process was run empirically. On one chosen frame, where the problem was displayed in the most suitable fashion, the values of the algorithm were set in order to obtain the most accurate result. According to these settings, all of the videos were analysed. On the basis of the videos' analysis, the algorithm was repeatedly adapted, until the required result was reached [17-20].

Table 1. Testing results of 31 video files processed

Video file	True acceptance rate	False acceptance rate
NOR1	-	23,38%
NOR7	-	1,00%
PAT2	-	1,49%
PAT7	92,00%	-
PAT14	10,00%	-
PAT15	10,00%	-
PAT16	25,00%	-
PAT17	10,00%	-

In the table 1, it is possible to see the values of the algorithm's success. Only the data where there was a detection of an object were recorded here. In the case of NOR1, where the largest number of mistakes was found, it was not possible to remove the mistakes due to the colourful spectrum which closely depicted the colour of blood. In the other cases, the errors of the solution can be neglected. During the attempt to dispose of these errors, the algorithm's success was rapidly decreased.

On the PAT7 video, the blood was correctly detected with the success of 92%. The other 8% were not very well recognisable due to small difference from the surroundings and due to very small size. This was the only sample where blood was present in reality.

The PAT14, 15, 16, 17 show no traces of blood, but red flare was detected by the application, as well as by the expert who identified it as an illness. Therefore, it is possible to consider this as an additional advantage of the algorithm. During the adjustment of the algorithm to the values which reflect the red marks, the success rate of 80% was reached for the PAT16 video in regards to the detected marks.

It is very complicated to compare the solutions from other authors as the same sample of data or at least the data from the same camera would need to be supplied.

7   Conclusions

As it is obvious from the results, the blood on the sample was detected with almost 100% success rate. For better checking of the given solution, it would be useful to obtain a larger sample of data for testing or to implement the software in production environment.

The usability of the solution is seen mainly in the simplifying of people's work that includes analysing the videos, one by one, which can be very time-consuming. The people's time which could be saved by this solution could be used in areas, where automated tools are not possible to establish.

However, the problem is whether it is possible to completely replace this process as the software will not be able to replace all the experts' knowledge. In the case of incorrect diagnosis, it could lead to fatal results.

Acknowledgement. This work and the contribution were supported by project "SP-103-2015 - Smart Solutions for Ubiquitous Computing Environments" Faculty of Informatics and Management, University of Hradec Kralove, Czech Republic. The Universiti Teknologi Malaysia (UTM) and Ministry of Education Malaysia under Research University grants 02G31, and 4F550 are hereby acknowledged for some of the facilities that were utilized during the course of this research work.

References

1. Appleyard, M., Fireman, Z., Glukhovsky, A., Jacob, H., Shreiver, R., Kadirkamanathan, S., Lavy, A., Lewkowicz, S., Scapa, E., Shofti, R., Swain, P., Zaretsky, A.: A randomized trial comparing wireless capsule endoscopy with push enteroscopy for the detection of small-bowel lesions. Gastroenterology **119**(6), 1431–1438 (2000)
2. Fireman, Z., Mahajna, E., Broide, E., Shapiro, M., Fich, L., Sternberg, A., Kopelman, Y., Scapa, E.: Diagnosing small bowel Crohn's disease with wireless capsule endoscopy. Gut **52**(3), 390–392 (2003)
3. Kunes, M., Kvetina, J., Tacheci, I., Kopacova, M., Bures, J., Nobilis, M., Krejcar, O., Kuca, K.: Imaging and evaluating method as part of endoscopical diagnostic approaches. In: Nguyen, N.T., Attachoo, B., Trawiński, B., Somboonviwat, K. (eds.) ACIIDS 2014, Part II. LNCS, vol. 8398, pp. 605–614. Springer, Heidelberg (2014)
4. Alpar, O.: Intelligent biometric pattern password authentication systems for touchscreens. Expert Systems with Applications **42**(17–18), 6286–6294 (2015)
5. Penhaker, M., Krejcar, O., Kasik, V., Snášel, V.: Cloud computing environments for biomedical data services. In: Yin, H., Costa, J.A., Barreto, G. (eds.) IDEAL 2012. LNCS, vol. 7435, pp. 336–343. Springer, Heidelberg (2012)
6. Barbara, P., Tammam, T., Marco, G., Enrico, M., Gabriella, O.: A technique for blood detection in wireless capsule endoscopy images. In: EUSIPCO, vol. 2009
7. Slaby, R., Hercik, R., Machacek, Z.: Compression methods for image processing implementation into the low capacity device. Technical Gazette **20**(6), 1087–1090 (2013)
8. Machacek, Z., Hercik, R., Slaby, R.: Smart user adaptive system for intelligent object recognizing. In: Nguyen, N.T., Trawiński, B., Jung, J.J. (eds.) New Challenges for Intelligent Information and Database Systems. SCI, vol. 351, pp. 197–206. Springer, Heidelberg (2011)
9. Machacek, Z., Slaby, R., Hercik, R., Koziorek, J.: Advanced system for consumption meters with recognition of video camera signal. Elektronika ir Elektrotechnika **18**(10), 57–60 (2012)
10. Haska, J., Pies, M., Machacek, Z.: Image signal processing, analysis and detection for robotic system. In: 14th International Conference on Control, Automation and Systems, (ICCAS 2014), pp. 1060–1063 (2014)

11. Hercik, R., Slaby, R., Machacek, Z.: Image processing for low-power microcontroller application. Przeglad elektrotechniczny **88**(12A), 343–346 (2012)
12. Sulik, L., Krejcar, O.: Blood Detection in Biomedical Images using a Software Analysis. In: 13th Conference on Programmable Devices and Embedded Systems. IFAC paper online (2015)
13. Penhaker, M., Cerny, M., Rosulek, M.: Sensitivity analysis and application of transducers. In: 5th International Summer School and Symposium on Medical Devices and Biosensors. ISSS-MDBS 2008, pp. 103–106 (2008)
14. Krejcar, O., Jirka, J., Janckulik, D.: Use of Mobile Phone as Intelligent Sensor for Sound Input Analysis and Sleep State Detection. Sensors **11**(6), 6037–6055 (2011)
15. Horák, J., Růžička, J., Novák, J., Ardielli, J., Szturcová, D.: Influence of the number and pattern of geometrical entities in the image upon PNG format image size. In: Pan, J.-S., Chen, S.-M., Nguyen, N.T. (eds.) ACIIDS 2012, Part II. LNCS, vol. 7197, pp. 448–457. Springer, Heidelberg (2012)
16. Kasik, V., Penhaker, M., Novák, V., Bridzik, R., Krawiec, J.: User interactive biomedical data web services application. In: Yonazi, J.J., Sedoyeka, E., Ariwa, E., El-Qawasmeh, E. (eds.) ICeND 2011. CCIS, vol. 171, pp. 223–237. Springer, Heidelberg (2011)
17. Kasik, V., Cerny, M., Penhaker, M., Snášel, Václav, Novak, Vilem, Pustkova, Radka: Advanced CT and MR image processing with FPGA. In: Yin, Hujun, Costa, José A.F., Barreto, Guilherme (eds.) IDEAL 2012. LNCS, vol. 7435, pp. 787–793. Springer, Heidelberg (2012)
18. Krejcar, O., Janckulik, D., Motalova, L.: Complex biomedical system with mobile clients. In: IFMBE Proceedings of the World Congress on Medical Physics and Biomedical Engineering, Munich, Germany, vol. 25/5, pp. 141–144, September 07–12, 2009. Springer (2009)
19. Krejcar, O., Janckulik, D., Motalova, L., Musil, K.: Real time processing of ECG signal on mobile embedded monitoring stations. In: Proceedings of Second International Conference on Computer Engineering and Applications, ICCEA, Bali Island, Indonesia, vol. 2, pp. 107–111, March 19–21, 2010
20. Behan, M., Krejcar, O.: Modern Smart Device-Based Concept of Sensoric Networks. EURASIP Journal on Wireless Communications and Networking **2013**(1), No. 155 (2013). doi:10.1186/1687-1499-2013-155. ISSN 1687-1499
21. Krejcar, Ondrej: Threading possibilities of smart devices platforms for future user adaptive systems. In: Pan, Jeng-Shyang, Chen, Shyi-Ming, Nguyen, Ngoc Thanh (eds.) ACIIDS 2012, Part II. LNCS, vol. 7197, pp. 458–467. Springer, Heidelberg (2012)
22. Behan, Miroslav, Krejcar, Ondrej: Adaptive graphical user interface solution for modern user devices. In: Pan, Jeng-Shyang, Chen, Shyi-Ming, Nguyen, Ngoc Thanh (eds.) ACIIDS 2012, Part II. LNCS, vol. 7197, pp. 411–420. Springer, Heidelberg (2012)
23. Behan, M., Krejcar, O.: Modern Smart Device-Based Concept of Sensoric Networks. EURASIP Journal on Wireless Communications and Networking **2013**(1), No. 155 (June 2013). doi:10.1186/1687-1499-2013-155

Combating Infectious Diseases
with Computational Immunology

Martina Husáková[(✉)]

Faculty of Informatics and Management, University of Hradec Králové,
Hradec Králové, Czech Republic
martina.husakova.2@uhk.cz

Abstract. Computational immunology aims at investigation of the immunity with computer science, mathematics, physics or statistics. Immune simulator is one of the outputs of research in computational immunology. Development of immune simulator is derived from the analysis of application domain and design of simulator with particular conceptual modelling language. Conceptualization is then used for simulator programming. The paper reviews and compares the most cited approaches of conceptual modelling and computational models building.

Keywords: Biological system · Computational immunology · Conceptual modelling · Computational model

1 Introduction

Artificial immune system (AIS) can be perceived in two different points of view. Immunological computation (IC) uses the AIS as a computational system based on the immunity-based algorithms that are aimed at solving different types of problems occurring e. g. in computational security, scheduling, optimization, pattern recognition, machine learning or bioinformatics. Immunity-based algorithms use the immunity as an inspiration source for problem solving. Computational immunology (CI) uses approaches of computer science, mathematics, physics or statistics for studying processes occurring in the natural immune system (NIS). Various conceptual or computational models are the output of the research in the CI. These models are focused on understanding of immune processes which are hardly investigated in vivo, in vitro, or behaviour prediction of this amazing complex system. Main aim of the paper is to review the conceptual and computational approaches used in the CI. The paper is organized as follows. The section 2 presents the computational immunology in historical point of view. The section 3 introduces means used for conceptualization of immune processes. Computational approaches are mentioned in section 4. Section 5 concludes the paper where the most important differences between conceptual and computational approaches are mentioned.

M. Núñez et al. (Eds.): ICCCI 2015, Part II, LNCS 9330, pp. 398–407, 2015.
DOI: 10.1007/978-3-319-24306-1_39

2   Computational Immunology

The CI is the subarea of computational biology trying to investigate processes occurring in the natural immune system (NIS). It uses approaches of artificial intelligence, computational intelligence, statistics, mathematics or physics for answering questions that are hardly received by in vivo or in vitro approaches. The NIS is explored with in silico approaches in the CI. Computational models try to predict behaviour of the NIS under different conditions and offer strategies how to influence its behaviour. Beginning of the CI is dated on 1974 when N. K. Jerne (Nobel Price Winner in Physiology or Medicine) proposed the idiotypic network theory. The theory claims that the NIS can be stimulated by the presence of B-cells (white blood cells) with their surface structures which can be detected by the others B-cells. Philosophical aspects of the theory are mentioned in [1]. Formalization of dynamics between lymphocytes and antibodies was presented in the same year in [2]. Shape space (SS) concept was proposed for quantitative representation of properties and interactions between entities of the AIS in 1979 [3]. Each entity of the SS is represented as a point in the coordinate system where each one has the radius a_r (affinity threshold) representing the ability to recognize others entities. Higher value of the a_r indicates higher probability to interact with another entity. Research of J. D. Farmer, N. H. Packard and A. S. Perelson is crucial in the IC and CI. They used differential equations for modelling dynamics of the immune cells, antibodies and antigens, similarly as N. K. Jerne in [2]. Binary strings (0, 1) were used for representation amino acids which are part of the binding sites of antibodies [4]. Differences between immune cells were calculated according to various distance measures, e. g. Hamming distance measure. Researchers investigated the NIS as the computational system and connected the theoretical immunology with the computational intelligence in [4].

3   Conceptualization of Immune Processes

An immune simulator is one of the outputs in the CI. Development requires investigation of application domain for which the simulator is built. Facts of application domain are "transferred" into the semi-formal shape including maps, graphs or diagrams. Graphical representations of the most important concepts and relations between them play the role of building blocks that are rewritten into the programming code. Conceptualization of particular immune entities and processes can be done with a wide variety of approaches. These approaches are briefly described and compared in the Tab. 1 (see Appendix A).

A concept map (CM) is an information and knowledge-based structure for representation of concepts and relations between them in graph-based form. The approach was primarily aimed at understanding how the knowledge of child is changed during study [5]. Concept mapping is powerful tool for generation or sharing of new ideas, memorizing of new facts, capturing tacit knowledge of experts or in decision making. Concept maps are suitable for informal representation of crucial facts about biological systems and in medical education [6], [7].

Entity-relationship diagram (ERD) is graphical representation of an entity-relationship model designed in early stages of databases development. Non-traditional usage of the ERD is mentioned e. g. in [8] where relations between enzymes, their reactions and properties of proteins and biopolymers are modelled.

An ontology is a tree or graph-based structure for information and knowledge representation and sharing [9]. Bio-ontologies are specific ontologies playing the role of common vocabulary of bio-medical terms and relations between them for support of knowledge sharing, communication between interested parties, annotation of web sources or building knowledge bases of biomedicine systems [10]. Gene ontology (GO) is one of the best known bio-medical ontology. It is a part of the Gene Ontology project aiming at the standardization of genes representation and gene products of various organisms. GO contains vocabulary of terms for annotation of gene products of various animals. Cellular components, molecular functions or biological processes are represented by the GO [11]. Various bio-medical ontologies are available thanks to the OBO Foundry initiative providing space for sharing of bio-ontologies [12].

A statechart is different approach in comparison to the concept maps, ERD and ontologies. Statecharts are focused on modelling behaviour of particular entity in the view of states and transitions between them during its life. This formalism was proposed as the extension of state diagrams that were not efficient for modelling complex phenomena [13]. Statecharts has been already used in conceptualization of processes occurring in the lymph node e. g. in [14]. Inner processes of immune cells (B-cells, T-cells and FDC (Follicular Dendritic Cells)) are represented by the statecharts and the following immune processes: movement of cells, proliferation of cells, differentiation of B-cells into plasmatic and memory form and antibodies secretion. Similar research study is published in [15]. It is aimed at modelling humoral immune response in the lymph node. Various immune cells (Th-cells, antibodies, antigens and B-cells) and immune processes (e. g. migration of immune cells, cloning of immune cells, immune cells differentiation and antibodies generation) are taken into account. Research is aimed at the statistical analysis of the occupancy of the immune cells in various segments of the lymph node. Statecharts are extended by the Biochart conceptual language used for conceptual modelling of biological systems. Biochart language is applied e. g. in modelling of chemotactic movement of the bacteria E.Coli in [16].

The UML (Unified Modelling Language) is standardized language used for graphical representation of object-oriented systems during analysis and design. Several research studies advocate the usefulness of the UML for modelling biological systems. Collection of UML class diagrams is applied for representation of protein structures in [8]. Components of eukaryotic genomes and relations between them are modelled also with the UML class diagram in [17]. UML-based diagrams are used also in modelling of metabolic pathways (especially glycolytic pathway) as a demonstration of the design methodology CellAK (Cell Assembly Kit) used for building cellular models [18]. The UML-based domain model of the experimental autoimmune encephalomyelitis (EAE) is mentioned in [19]. EAE is complex autoimmune disease occurring in mice. The disease plays the role of a murine model for the multiple sclerosis in humans. The most important entities responsible for the EAE are represented by the UML class diagram. The UML activity diagram is applied for modelling dynamic inter-cellular processes leading to the EAE. The UML state machine diagrams represent behaviour of biological entities which are important in the EAE (Th1 CD4 cells,

Tc CD8 cells, Treg CD4 cells, Treq CD8 cells or dendritic cells). Experimental Visceral Leishmaniasis (EVL) is an infectious disease caused by the intra-cellular parasite of the Leishmania genus. If the disease is untreated, the result is almost always the death of the host. The EVL is often related with the granuloma formation. The granuloma formation occurring in the liver is modelled with the UML diagrams in [20]. The UML class diagram represents biological entities occurring during granuloma formation. The UML state diagram models behaviour of T-cells, innate liver cells or macrophages and the UML sequence diagram represents granuloma formation during time.

Approaches presented above are mainly used for conceptualization of the investigated domain of interest. Conceptualization is also related with markup languages, more based on formal semantics.

SBML (Systems Biology Markup Language) is the XML-based declarative language used for modelling biological systems and processes occurring inside them [21]. SBML is not intended to read by humans or written by hand. It should be used by computers as a machine-readable format for exchanging information about investigated biological system. SBML is useful for modelling cell signalling pathways, metabolic pathways, biochemical reactions or gene regulatory networks, for more details see e. g. [22].

The CellML is also XML-based declarative language. It is primarily used for description of cellular biological systems with the mathematical expressions, i. e. ordinary differential equations, differential equations or simple linear algebra [27]. The CellML is mainly applied for development, storage and sharing mathematical models of cellular biological systems with other developers.

SBGN (Systems Biology Graphical Notation) is a graphical notation transforming biological pieces of knowledge into more formal shape [23]. The aim is to offer standardized and unambiguous notation for depicting complex processes occurring in various biological systems. SBGN-based notation consists of three languages. The Process Description language (PD) aims at modelling particular molecular processes changing states of biological entities [24]. Temporal characteristics are taken into account in the PD-based models in comparison to the next language of the SBGN - the Entity Relationships language (ER). The ER is focused on modelling interactions between biological entities and impacts of these interactions on the behaviour of other biological entities. The ER is suitable for modelling of signalling pathways [25]. The Activity Flow language (AF) models activities of biological entities and information flow between them. It is mainly useful for description of signalling pathways or gene regulatory networks [26].

4   Computational Models of Immunity

Conceptual models can be used for implementation of programming code intended computational model. Top-down and bottom-up approaches used for modelling biological systems are distinguished in [28]. Top-down approaches investigate a biological system in a macroscopic point of view where aggregated characteristics are in the centre of the interest. Differential equations (DE) or Petri nets (PN) are typical examples of top-down approaches. Bottom-up approaches are focused on modelling micro-scopic levels of the investigated biological system where characteristics and behaviour of particular entities are in the centre of the interest. Cellular automata (CA) or multi-agent systems (MAS) are main representatives of the bottom-up approaches. These approaches are briefly described and compared in the Tab. 2 (see Appendix B).

The DE are mainly used for studying changes in concentrations of components in the biological system per time unit. Simulation with DE means to solve equations. Complexity of the system decides how many equations are necessary. DE were successfully applied for formalization of the first formal immune network model proposed by N. K. Jerne [2]. DE are used also in modelling avascular tumor growth [29], angiogenesis [30] or signalling pathways [31]. Control mechanism of the JAK-STAT signal transduction pathway, responsible for the cell growth or apoptosis regulation, is also modelled by the DE [32]. Petri nets are mathematical models mainly focused on modelling distributed systems including biological systems. Petri nets are useful for modelling metabolic signalling pathways [35], signal transduction pathways [36] or gene regulatory networks [33], [34].

The CA is discrete system consisting of artificial cells using very simple rules for decision making. These rules are based on the state of surrounding artificial cells. CA are useful for discovering emergent characteristics of investigated biological systems, but they are often criticized for too many simplifications during building CA-based models. Foundations of CA-based models were specified by F. Celada and P. E. Seiden in [37]. They proposed the IMMSIM simulator - 2D-CA-based computational model used for investigation processes occurring during humoral immune response. Components of the NIS are represented as bit-string data structures. Programming code of the IMMSIM was published in 1992 in [38]. IMMSIM3 version is extended by the cellular immunity. SIMMUNE simulator is based on the 3D-CA, including interactions between cells and viruses [38]. Infection caused by the HIV virus is modelled by the CA e. g. in [40] or [41]. Advanced version of the IMMSIM called C-IMMSIM was used for simulation infection caused by the Epstein-Barr virus in [42].

The MAS are similar to the CA, but they model an investigated system in more details in comparison to the CA. MAS are mainly used for modelling processes occurring in cellular or tissue scale of the biological system. They can explore emergent phenomena in biological systems similarly as CA-based systems. Production of antibodies caused by viral infection is simulated with MAS in the Breve simulation environment [43]. Granuloma formation in Visceral Leishmaniasis is modelled with MAS and published in [20], see section 3. Agent-based approach is also applied for investigation serious disease called Experimental Autoimmune Encephalomyelitis [19], see section 3. Formation of functional epithelium is modelled as MAS and published in [44]. Formation of the epithelium is an emergent property of the biological system. Main aim of the study is to model normal behaviour of the epithelium structure and to investigate anomalies in behaviour possibly leading to the pathologies.

5   Conclusion

The paper presents various conceptual approaches used in the CI. Each conceptual approach is intended for specific purpose and it is difficult to say which one is the best. If "a modeler" needs to have efficient and fast approach for finding out whether his/her understanding of more or less complex immune process is correct, an informal concept map is the most suitable strategy for conceptualization, because a concept map is easily created, extended and compared with the knowledge of expert. SBGN is a formalism also visualizing "concepts", but these are depicted with more colourful and precise notation in comparison to the concept maps. If the intention is to develop a (bio)medical database, it is not surprising that analytical activities are focused on modelling of ER diagrams. UML and statecharts are verified solutions for analysis and design of immune simulators, but they can be also used for visualization of static or dynamic aspects of biological entities without the intention to develop an immune simulator. This last view is closer to concept mapping, but the notation of the UML (statecharts) offers more elements in comparison to concept maps. The UML represents behavior and properties of biological systems more precisely than concept maps. If a conceptual model should be machine-processable, an ontology, SBML or CellML should be used for modelling of biological systems. Ontologies need not be inevitable machine-processable. They can play the same role as concept maps and additionally model semantics of concepts in more detail in comparison to the concept maps.

The paper presents various computational approaches used for simulation of biological phenomena. If "a modeler" deals with the large amount of entities, differential equations or Petri nets are more promising approaches for simulation in comparison to cellular automata and multi-agent systems, because they are not focused on modelling individual entitites, but on their aggregated characteristics. If a "modeler" needs to represent interactions between particular entities, to model living conditions or spatial characteristics of the environment where entities should exist, cellular automata and multi-agent systems are the most suitable.

The paper does not introduce all results in the CI. Multi-scale modelling is one of the research directions focusing on modelling biological systems in all known scales - molecular, cellular, tissue and organ. The actual research in the CI can be described as a tendency to find effective and computationally undemanding solutions for building efficient and useful computational models. Combination of various approaches of computational models can bring added value for better understanding of processes occurring in the NIS.

Acknowledgements. This article was supported by the project No. CZ.1.07/2.3.00/30.0052 - Development of Research at the University of Hradec Králové with the participation of post-docs, financed from EU (ESF project) and the project of specific science "Application of Artificial Intelligence in Bioinformatics" (1/2015) at the University of Hradec Kralove, Faculty of Informatics and Management, Czech Republic.

Appendix A: Review of Conceptual Approaches

Table 1. Conceptual approaches used in computational immunology

Attribute	Conceptual approach			
	CM	**ERD**	**Statecharts**	**UML**
Origin	1972	1976	1987	1996 (ver. 0.9)
Author	J. D. Novak	P. Chen	D. Harel	G. Booch I. Jacobson J. Rumbaugh
Elements	concept linking-word cross-link	entity type relation attribute	state transition action	It depends on a type of a diagram.
Purpose	visualization of information and knowledge	conceptualization of data for database systems development	modelling behaviour of reactive systems	analysis and design of software systems with the UML-based diagrams
Software support	CmapTools VUE Visio	LucidChart Visio SmartDraw	Visio SmartDraw AnyLogic	EA StarUML ArgoUML
Attribute	Conceptual approach			
	Ontology	**SBML**	**CellML**	**SBGN**
Origin	1980	2001 (Level 1)	2001 (ver. 1.0)	2005
Author	J. McCarthy	J. Doyle H. Kitano M. Hucka, et al.	D. Bullivant W. Hedley P. Nielsen, et al.	H. Kitano (initiator)
Elements	class property individual	XML-based structure	XML-based structure	It depends on the language dialect: PD, ER, AF.
Purpose	knowledge representation for communication between humans, machines or humans and machines	specification of machine-readable biological models	description of cellular biological systems with the usage of mathematical expressions	unambiguous graphical notation for modelling biological systems
Software support	Protégé CmapTools Ontopia	Cytoscape Biographer CellDesigner	VirtualCell JSim OpenCell	CellDesigner Athena Arcadia

Appendix B: Review of Computational Approaches

Table 2. Top-down and bottom-up approaches used in computational immunology

Characteristic	Computational approach			
	DE	**PN**	**CA**	**MAS**
Type	Top-down	Top-down	Bottom-up	Bottom-up
Components	equation	place transition arc	1D/2D/3D grid cell	agent
Scale of biological system (the most cited)	molecular cellular tissue	molecular	cellular tissue	cellular tissue (molecular)
Modelling of hetero-geneous entities	No	No	No	Yes
Mathematical theory necessary	Yes	Yes	No	No
Modelling emergent phenomena	No	No	Yes	Yes
Modelling interac-tions between biolog-ical entities	No	No	Yes	Yes
Modelling spatial characteristics	No	No	Yes (3D CA)	Yes

References

1. Jerne, N.K.: Towards a Network Theory of the Immune System. Annales d'immunologie **125C**(1–2), 373–389 (1974)
2. Jerne, N.K.: Clonal selection in a lymphocyte network. In: Edelman, G.M. (ed.) Cellular Selection and Regulation in the Immune Response, p. 39. Raven Press, New York (1974)
3. Perelson, A.S., Oster, G.F.: Theoretical studies of clonal selection: Minimal antibody repertoire size and reliability of self-non-self discrimination. Journal of Theoretical Biology **81**(4), 645–670 (1979)
4. Farmer, J.D., Packard, N.H., Perelson, A.S.: The immune system, adaptation and machine learning. Physica D, pp. 187–204. The Elservier Science Publishers, Amsterdam (1986)
5. Novak, J.D.: Concept maps and vee diagrams: Two metacognitive tools for science and mathematics education. Instructional Science **19**, 29–52 (1990)
6. Bentley, F.J.B., Kennedy, S., Semsar, K.: How Not To Lose Your Students with Concept Maps. Journal of College Science Teaching **41**(1), 61–68 (2011)
7. Gomes, A.P., et al.: The Role of Concept Maps in the Medical Education. Revista Brasileira de Educação Médica **35**(2), 275–282 (2011)
8. Bornberg-Bauer, E., Paton, N.W.: Conceptual data modelling for bioinformatics. Briefings in bioinformatics **3**(2), 166–180 (2002)

9. Gruber, T.R.: A Translation Approach to Portable Ontology Specifications. Knowledge Acquisition **5**(2), 199–220 (1993)

10. Robinson, P.N., Bauer, S.: Introduction to Bio-Ontologies, 1st edn, p. 517. Chapman and Hall/CRC Mathematical and Computational Biology (Book 41) (2011)

11. Ashburner, M., et al.: Gene Ontology: tool for the unification of biology. Nat. Genet. **25**(1), 25–29 (2000)

12. Smith, B., et al.: The OBO Foundry: coordinated evolution of ontologies to support biomedical data integration. Nature Biotechnology **25**, 1251–1255 (2007)

13. Harel, D.: Statecharts: A visual formalism for complex systems. Science of Computer Programming **8**, 231–274 (1987)

14. Swerdlin, N., Cohen, I.R., Harel, D.: The Lymph Node B Cell Immune Response: Dynamic Analysis In-Silico. Proceedings of the IEEE **96**(8), 1421–1442 (2008)

15. Belkacem, K., Foudil, Ch.: An anylogic agent based model for the lymph node lymphocytes first humoral immune response. In: Proceedings of the International Conference on Bioinformatics and Computational Biology (ICBCB 2012), vol. 34, pp. 163–169. IACSIT Press, Singapore (2012)

16. Kugler, H., Larjo, A., Harel, D.: Biocharts - a visual formalistm for modelling biological systems. J. R. Soc. Interface **2010**(7), 1015–1024 (2010)

17. Paton, N., Khan, S., Hayes, A., et al.: Conceptual modelling of genomic information. Bioinformatics **16**(6), 548 557 (2000)

18. Webb, K., White, T.: UML as a Cell and Biochemistry Modeling Lanugage. Carleton University Cognitive Science Technical Report 2003–05. Biosystems **80**(3), 283–302 (2005)

19. Read, M., Timmis, J., Andrews, P.S., Kumar, V.: A domain model of experimental autoimmune encephalomyelitis. In: Proceedings of the CoSMos 2009, pp. 9–44. Luniver Press (2009)

20. Flugge, A., et al.: Modelling and simulation of granuloma formation in visceral leishmaniasis. In: Proceedings of the Congress on Evolutionary Computation (CEC), pp. 3052–3059. IEEE Press (2009)

21. Hucka, M., et al.: The systems biology markup language (SBML): a medium for representation and exchange of biochemical network models. Bioinformatics **19**, 524–531 (2003)

22. Finney, A., Hucka, M.: Systems Biology Markup Language: Level 2 and Beyond. Biochemical Society Transactions **31**(6), 1472–1473 (2003)

23. Le Novère, N., et al.: The Systems Biology Graphical Notation. Nature Biotechnology **27**(8), 735–741 (2009)

24. Moodie, S, et al.: Systems Biology Graphical Notation: Process Description language Level 1. Nature Precedings (2011). doi:10.1038/npre.2011.3721.4

25. Le Novere, N., et al.: Systems Biology Graphical Notation: Entity Relationship language Level 1 (Version 1.2). Nature Precedings (2011). doi:10.1038/npre.2011.5902.1

26. Huaiyu, M., et al.: Systems Biology Graphical Notation: Activity Flow language Level 1. Nature Precedings (2009). doi:10.1038/npre.2009.3724.1

27. Lloyd, C.M., Halstead, M.D.B., Nielsen, P.F.: CellMl: its future, present and past. Progress in Biophysics & Molecular Biology **85**, 433–450 (2004)

28. Xiang-Hua, L., et al.: Modelling Immune System: Principles, Models, Analysis and Perspectives. Journal of Bionic Engineering **6**(1), 77–85 (2009)

29. Roose, T., et al.: Mathematical Models of Avascular Tumor Growth. Siam Review **49**(2), 179–208 (2007)

30. Chaplain, M.A.I., et al.: Mathematical Modeling of Tumor-Induced Angiogenesis. Annual Review of Biomedical Engineering **8**, 233–257 (2006)

31. Randamani, P., Iyengar, R.: Modelling cellular signalling systems. Essays Biochem. **45**, 83–94 (2012)
32. Yamada, S., et al.: Control mechanism of JAK/STAT signal transduction pathway. FEBS Letters **534**(16), 190–196 (2003)
33. Hardy, S., Robillard, P.N.: Modeling and simulation of molecular biology systems using Petri nets: modelling goals of various approaches. Journal of Bioinformatics and Computational Biology **2**, 595–613 (2004)
34. Pinney, J.W.: Petri Net representation in systems biology. Biochemical Society Transactions **31**, 1513–1515 (2003)
35. Reddy, V.N., et al.: Petri net representations in metabolic pathways. Proceedings of International Conference on Intelligent Systems for Molecular Biology **1**, 328–336 (1993)
36. Sackmann, A., et al.: Application of Petri net based analysis techniques to signal transduction pathways. BMC Bioinformatics **7** (2006)
37. Celada, F., Seiden, P.E.: A Computer Model of Cellular Interaction in the Immune System. Immunology Today **13**, 56–62 (1992)
38. Cellada, F., Seiden, P.: A model for simulating cognate recognition and response in the immune system. Journal of Theoretical Biology **158**(3), 329–357 (1992)
39. Meier-Schellersheim, M., Mack, G.: SIMMUNE, a tool for simulating and analyzing Immune System behavior, p. 23 (1999). http://arxiv.org/abs/cs/9903017
40. Zorzenon dos Santos, R.M., Countinho, S.: Dynamics of HIV Infection: A Cellular Automata Approach. Physical Review Letters **87**(16) (2001)
41. Grilo, A., Caetano, A., Rosa, A.: Immune System Simulation through a Complex Adaptive System Model. Soft Computing and Industry 675–698 (2002)
42. Castiglione, F., et al.: Simulating Epstein-Barr virus infection with C-ImmSim. Bioinformatics **23**(11), 1371–1377 (2007)
43. Jacob, C., Litorco, J., Lee, L.: Immunity through swarms: agent-based simulations of the human immune system. In: Nicosia, G., Cutello, V., Bentley, P.J., Timmis, J. (eds.) ICARIS 2004. LNCS, vol. 3239, pp. 400–412. Springer, Heidelberg (2004)
44. Walker, D.C., et al.: The Epitheliome: Agent-Based Modelling of the Social Behaviour of Cells. BioSystems **76**, 89–100 (2004)

Mobile Cloud Computing in Healthcare System

Hanen Jemal[✉], Zied Kechaou, Mounir Ben Ayed, and Adel M. Alimi

REGIM: REsearch Groups in Intelligent Machines, National School of Engineers (ENIS),
University of Sfax, 1173, Sfax 3038, Tunisia
{hanen.jemal,zied.kechaou,mounir.benayed,adel.alimi}@ieee.org

Abstract. Mobile Cloud Computing (MCC) is a potential technology for mobile web services. Accordingly, we assume that MCC is likely to be of great avail to healthcare domain. MCC offers new kinds of services and facilities for patients and caregivers. In this regard, we have tried to propose a new mobile medical web service system. The proposed system called Medical Cloud Multi Agent System is a complex system which integrates MCC and Multi Agent System in healthcare with view to improving healthcare system.

Keywords: Mobile Cloud Computing · Mobile web services healthcare · Cloud Computing · Medical Cloud Multi Agent System · Multi Agent System

1 Introduction

Mobile Cloud Computing (MCC) is an integration of Cloud Computing (CC) into the applications of mobile devices. New advances in CC and mobile technology have inspired different patterns of cloud health care services and devices. In the cloud system, health-data can be stored and transmitted to medical caregivers from everywhere and response can be returned to patients through web service network. In this article, we present a MCC solution through a healthcare web service.

The motivations for the use of MCC in healthcare consist in several benefits that can be derived from the combinations of mobile and CC: ABI Research study showed that more than 240 million businesses will use cloud mobile services by 2015. "Traction will push the revenue of mobile cloud computing to $5.2 billion" [1].

In our study MCC offers significant benefits to our healthcare solution such as:

- *Collaboration:* MCC technology maintains collaboration and team care delivery.
- *Performance:* MCC model can improve rapid access to computing, share information more easily, large storage of big data (cloud-based medical records) and reduce costs.
- *Modernization:* MCC will lower the barriers for modernization and innovation of healthcare applications.
- *Scalability:* several patients utilizing the healthcare applications.
- *Portability:* the ability to remotely access applications and data and provide functionality for managing information in distributed and ubiquitous applications.

M. Núñez et al. (Eds.): ICCCI 2015, Part II, LNCS 9330, pp. 408–417, 2015.
DOI: 10.1007/978-3-319-24306-1_40

Our research defines hybrid system that combines Muli Agent System (MAS), Web Service and MCC. In this paper we will only detail the MCC and our healthcare web services. This paper is organized as follows: At first we started with an introduction, subsequently section 2 describes the existing cloud healthcare solutions and a state of the art of MCC. Section 3 presents our cloud hospital architecture designed to mobile healthcare applications. Finally, we conclude with a summary of the article with further suggestions for additional advancement.

2 Cloud Computing and Mobile Cloud Computing in Healthcare

2.1 Cloud Computing in Healthcare

CC is simply an architectural model that employs many of the same components used in datacenters around the world today in a more flexible, responsive, and efficient way [2].

According to NIST (National Institute of Standards and Technology, USA), "Cloud Computing is a model for enabling convenient, on-demand work access to a shared pool of configurable resources (e.g. networks, servers, storage, applications and services) that can rapidly be provisioned and released with minimal management effort or service provider interaction"[3].

CC consists of hardware and software resources made available on the internet as managed third-party services. These services typically provide access to advanced software applications. Cloud technology providers deliver applications via internet, which are accessed from a web browser, while the business software and data are stored on servers at a remote location.

In cloud technology the information is shared from clients to the organization through the virtual data centers. The cloud technology includes three models: SaaS (Software as a service), PaaS (Platform as a service) and IaaS (Infrastructure as a service).

Cloud in healthcare information is speedily becoming the most important trend for the development of healthcare information systems [4]. CC can improve medical services and benefit biomedical research providing Centralization, Collaboration and Virtualization:

- Centralization: by saving time and cost through providing an easier access to and retrieval of data.
- Collaboration: doctors can collaborate together on cases, researches, through sharing resources, information and files.
- Virtualization: it is the core axis of IaaS (Infrastructure-as-a-Service).

Recent studies indicate that CC can facilitate the biomedical informatics research communities. Reports show that as many as 30% of healthcare organizations are either implementing or operating cloud-based solutions[1] .

---

[1] CDW 2011 Cloud Computing Tracking Poll http://www.cdw.com/cloudtrackingpoll

JiunnWoei and his colleague [5] identify a key factor for hospitals to make decision through Cloud Computing technology. The purpose of this study is to examine the factors that will change the decision to adopt CC technology in Taiwan's hospital. Pankaj & Inderveer in 2013 [6] presents a Cloud Based Intelligent Health Care Service (CBIHCS) that performs real time examination of patient health data for diagnosis of chronic illness, like diabetes, collected from different wireless sensory medical equipments. The authors utilized Principal Component Analysis for attribute selection and k-nearest neighbor (KNN) and Naïve Bayes for classification of patient health status.In [7] authors utilizes Aneka [8] framework to create an autonomic cloud environment for hosting ECG (Electrocardiography) data analysis services.

Kuo in 2011 [9] designate that Cloud Computing can change the execution and adoption of medical information technology, in particular for the development of EHR (Electronic Health Records). In [10] authors use high performance computing of Amazon Web Service to facilitate genomic findings.

In 2011 Piette and his colleagues [11] present a CC based Voice over IP (VoIP) service for diabetic patients, the patients subscribing the cloud service received VoIP calls with prerecorded voice messages as self-care reminders. In this study, the advantages of this cloud computing based healthcare service are cost effective, and it can be extended globally easily. In practice, the Fujitsu in Japan [12] propose a cloud computing solution for hospitals. In addition, Microsoft also defines a cloud computing technology in order to improve the quality of care. This solution reduced costs in Italy Pediatric research [13]. In its healthcare industry project [14], IBM proposed the use of CC in the United States hospitals: patients' data are stored in a cloud database, patients can be monitored at home via this cloud service and doctor can use the cloud platform to make a diagnosis at home.

2.2   Mobile Cloud Computing in Healthcare

The purpose of Mobile healthcare (m-healthcare) is to provide mobile healthcare users easy and quick access to the resources (e.g., PHR patient health records) and offers a variety of distributed services. The aim of applying MCC in healthcare applications is to reduce the limits of traditional medical applications (e.g., security, small storage, and medical errors [15,16]).

The Mobile Cloud Computing Forum considers MCC as "an infrastructure where both the data storage and the data processing happen outside of the mobile device" [17]. In [18] MCC is defined as a new model for mobile applications: "it will be transferred to a centralized and powerful computing platform in the cloud".

There are several advantages of MCC such as [19]:

- Extending battery lifetime: the technique of computation offloading is proposed in order to migrate the complex treatment from limited devices (mobile devices).
- Improving data storage capacity and processing power: MCC is developed to enable mobile users to store/access cloud data.
- Improving reliability: the storage of data on a number of computers in the clouds improves reliability.

- Dynamic on demand provisioning of resources and scalability: It is a flexible way for s running the applications without advanced reservation of resources and adding services.
- Multi-tenancy: providers of service can share the resources in order to support a variety of applications and large number of users.
- Ease of Integration: several services from diverse service providers can be integrated simply through the cloud.

In the literature there are only few works about MCC applications in healthcare such as, Upkar Varshney [20] who presents five key of mobile applications in the pervasive healthcare environment:

1. Comprehensive health monitoring services enable patients to be monitored at anytime and anywhere.
2. Intelligent emergency management system can manage the large call volume received from accidents or incidents.
3. Health-aware mobile devices which detect blood pressure, pulse-rate and level of alcohol.
4. Pervasive access to healthcare information allows caregivers and patients to access medical data.
5. Pervasive lifestyle incentive management can be used for paying healthcare expenses and other healthcare charge.

In the same way, [21] proposed a prototype of m-healthcare information management system called @HealthCloud and based on MCC and Android operating system.

In practice, Tang and his colleagues implement a telemedicine homecare management system [22] in Taiwan to monitor patients with hypertension and diabetes. The system examines 300 patients and stores more than 4736 records of sugar and blood data on the cloud.

Then, [23,24] present a solution to protect the patient's health information: [23] utilizes P2P paradigm to ensure security and data in the clouds. In [24] authors present security as a service to defend mobile applications.

This solutions present very talented research in the MCC applications as it is the result of the collective profits of mobile and cloud technologies, but these solutions presents several disadvantage listed in table1 and there are no complete systems which offer complete solutions for healthcare. In section 3 we present our MCMAS solution that combines MAS and MCC in healthcare domain.

3   Proposed Architecture

Our research defines a hybrid System that combines Multi Agent System, Web Service and Mobile Cloud Computing.

In this paper we will only detail the MCC and our healthcare web services.

This architecture (Fig. 1) is related to the previous works in the healthcare domain when authors introduced MAS for healthcare and present the interaction in an agent-based architecture for healthcare [25,26]. This architecture is composed of a set of autonomous agents adapted to the interaction.

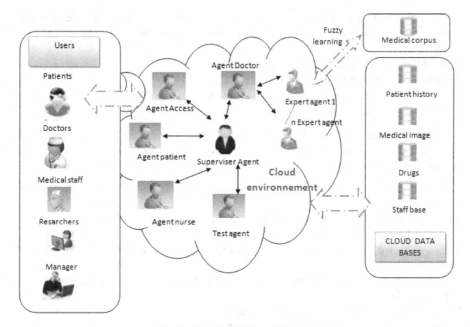

Fig. 1. Our MCMAS architecture

The agents of our hospital can be classified in two groups or layers [25]:

- Intelligent agents or super agent (Agent doctor, Agent patient, Agents nurses....)
- Swarm layer inspired from the Swarm Intelligence fields (applies the collective behavior of groups to resolve a problem.) such as office, medical materials ...

This architecture includes the two previous layers of intelligent agents, representing a medical organization with different roles and communication patterns, and facilitating interoperability, and the accessibility to information.

The purpose of our study consists in the development of a medical framework able to solve a large variety of medical problems. Consequently, we designed cloud architecture consisting of multiple distributed agents placed in the cloud environment.

The main functionality of the prototype is to provide users with a mobile interface in order to manage healthcare information, the applications platform is a tool for several users:

1. Hospital staff: it represents all the agents that are concerned in providing care to patients in the healthcare system. This community should be able to know and confirm the abstract representation of the theoretical model of the hospital as well as manipulate and utilize the results of the system. The main categories of this category are:

- Medical personnel: physician, surgeon, radiologist, anesthetist, etc.
- Nursery personnel;
- Technicians: such as laboratory personnel

- Admission and discharge personnel;
- Medical support personnel: security, archive, supplying, cleaning, etc.

2. Team study: such as researchers, engineers.
3. Non hospitalized patients represent the main users of the healthcare system and the core of the care system.

Our aim is that agent's work in the background to provide ambient intelligence to the users (Doctors, patient's, nurses...) who reside in the cloud environment. In other word the agents communicate with each other, acquire their behavior and receive information through medical cloud data.

4   Implementation of a Prototype

A new medical system called MCMAS (Medical Cloud Multi-Agent System) is proposed. This is a complex system which integrates hybrid solutions, i.e., MCC and Multi agent System in healthcare, in order to make care as efficient as possible. The web services and service-oriented architecture (SOA) technologies have been used to our prototype. This architecture offers a model combining the benefits of both Mobile technology and CC (Fig. 2).

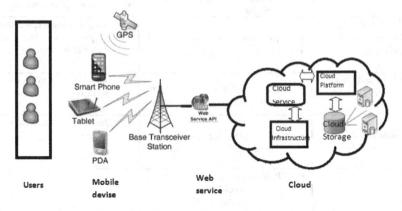

Fig. 2. Mobile Cloud Computing Architecture

We define several services such as:

- Patient appointment: patient can choose a date for remote consultation
- Remote consultation: patient can make consultation in real times by sending and receiving message from doctors.
- Resource allocation: doctors and patients can remote allocate resources (IRM, scanner...).

- Connection to CC storage (e.g. patient health records): The system application allows caregivers to save and upload distributed medical data.
- Image viewing by supporting the DICOM image.
- Patient registration: patients can make registration remotely as well as choose the medical center and the suitable medical unit.
- Medical analysis results viewing: the content of the test resides remotely into the cloud storage.

Table 1 lists the most positive points, the disadvantages of existing works and the positive features of our Medical Cloud Multi Agent System (MCMAS). In fact there are no complete systems which offer complete solutions for healthcare.

Table 1. Advantages and disadvantages of existing works in healthcare and the position of our system

Existing works	Advantages	Disadvantages
Doukas *et al,* [21]: @HealthCloud	• Seamless connection to CC storage (e.g., medical images, patient health records) utilizing web services Patient health record management system. • Image viewing and annotation. • Data encryption and user authentication.	• Absence of remote consultations. • No cloud appointment. • Just a prototype. • Focus only on the medical imaging. • Absence of security.
Tang *et al,* [22]: homecare management system	• Telemedicine system. • Homecare management.	• Specific patient (only monitor patients with hypertension and diabetes).
Hoang and Chen, [23]: MoCAsH	• Utilizes P2P paradigm to ensure security and data in the clouds.	• Just a solution to ensure security.
Nkosi and F. Mekuria [24]	• Present security as a service to defend mobile applications.	• Just a solution to ensure security.
Our MCMAS	• Patient appointment. • Remote consultation. • Resources allocations. • Connection to CC storage (e.g. patient health records). • Image viewing. • Patient registration. • Medical analysis results viewing	

To implement our proposed application we use JADE (Java Agent Development Framework) as a software framework for developing multi-Agent systems. JADE provides a set of interfaces for the design of agents implemented in Java [27].

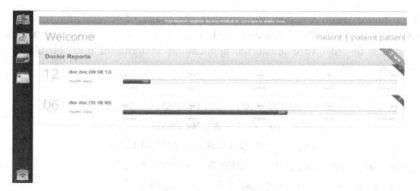

Fig. 3. Consultation report

The main purpose of this article is to define a prototype of mobile internet application in the cloud environment for healthcare system, in which patients are able to get immediate aid ((Fig. 3) interface doctor reports present the heath status of patient) and make an appointment (Fig. 4) form the corresponding doctors.

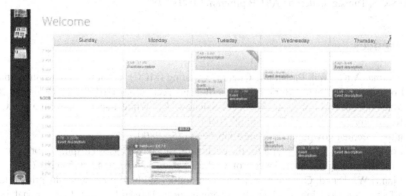

Fig. 4. Patient appointment

5   Conclusion

The presented work is an effort towards the integration of MCC in the health delivery process. Advances in Information Technologies such as mobile computing and CC are creating new opportunities to improve the health care ecosystem. Thus Mobile and CC provide the essential functionalities that make them suitable to be used in the remote healthcare domain.

The proposed architecture in this article will be appreciated in the future, as it is the consequence of the combined profits of both mobile and cloud technologies.

MCC is related to cloud computing although both are at their early stage of development. Besides, CC has five significant challenges such as availability, performance, security, interoperability and privacy. Those basic challenges are shown in (Fig. 5).

As a future work, we plan to introduce the usage of several improvements or possible extensions to the architecture and deploy our prototype in the real healthcare environment in order to evaluate the applications in terms of user tolerability and performance.

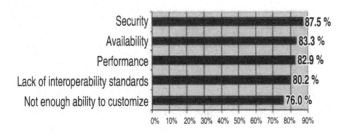

Fig. 5. Challenges in Cloud Computing [28]

Acknowledgements. The author would like to acknowledge the financial support of this work by grants from the General Direction of Scientific Research and Technological Renovation (DGRSRT), Tunisia, under the ARUB program LR11ES48.

References

1. Sharma, A.K., Soni, P.: Mobile Cloud Computing (MCC): Open Research Issues. International Journal of Innovations in Engineering and Technology (IJIET) 2(1) (February 2013)
2. MacVittie, L., Murphy, A., Silva, P., Salchow, K.: White Paper Controlling the Cloud: Requirements for Cloud Computing (2010)
3. http://csrc.nist.gov/publications/nistpubs/800-145/SP800-145.pdf
4. Low, C., Chen, Y.H.: Criteria for the evaluation of a cloud based Hospital information system outsourcing provider. Journal of Medical Systems 36(6), 3543–3553 (2012)
5. Lian, J.W., David, C., Wang, Y.T.: An exploratory study to understand the critical factors affecting the decision to adopt cloud computing in Taiwan hospital. International Journal of Information Management (2013). Elsevier
6. Kaur, P.D., Chana, I.: Cloud based intelligent system for delivering health care as a service. Computer methods and programs in biomedicine (2013). Elsevier
7. Pandey, S., Voorsluys, W., Niu, S., Khandoker, A., Buyya, R.: An autonomic cloud environment for hosting ECG data analysis services. Future Generation Computer Systems 147–154 (2012)
8. Vecchiola, C., Chu, X., Buyya, R., Aneka, A.: Software platform for.NET based cloud computing. In: IOS 2009, pp. 267–295 (2009)
9. Kuo, A.M.H.: Opportunities and challenges of cloud computing to improve health care services. Journal of Medical Internet Research (2011)
10. Fusaro, V.A., Patil, P., Gafni, E.D., Wall, P., Tonellato, P.J.M.: Biomedical cloud computing with Amazon web services. PLoS Comput Biol (2011)
11. Piette, J.D., Maendoza-Avelares, M.O., Gancer, M., Mohamed, M., Marinec, N., Krishnan, S.: Preliminary study of a cloud-computing model for chronic illness self-care support in an underdeveloped country 40(6), 629–632 (2011)

12. Japan, N.E.C.: Fujitsu roll out cloud computing services for hospitals. In: Telecom paper Africa/Asia (2012). http://www.accessmylibrary.com/article1G1290052791/japannecfujitsuroll.html%20Accessed%2020.02.13

13. Lisa, B.A.: Hospital uses cloud computing to improve patient care and reduce costs. http://www.microsoft.eu/cloudcomputing/casestudies/hospitalusescloudcomputingtoimprovepatientcareandreducecosts.aspx (accessed February 20, 2013)

14. Ostrovsky, G.: IBM's cloud computing coming to a hospital near you (2010). http://medgadget.com/2010/08/ibmscloudcomputingcomingtoahospitalnearyou.html (accessed February 20, 2013)

15. Kohn, L.T., Corrigan, J.M., Donaldson, S.: To Err is human: building a safer health system. National Academy Press, Washington (1999)

16. Kopec, D., Kabir, M.H., Reinharth, D., Rothschild, O., Castiglione, J.A.: Human errors in medical practice: systematic classification and reduction with automated information systems. Journal of Medical Systems 27(4), 297–313 (2003)

17. Ali, M.: Green cloud on the horizon. In: Jaatun, M.G., Zhao, G., Rong, C. (eds.) Cloud Computing. LNCS, vol. 5931, pp. 451–459. Springer, Heidelberg (2009)

18. White Paper: Mobile Cloud Computing solution brief. AEPONA (2010)

19. Sailaja, R., Sheela Devi, M.V.: Mobile Cloud Computing: a lookup. In: International Conference on Computing and Control Engineering (ICCCE) (2012)

20. Varshney, U.: Pervasive healthcare and wireless health monitoring. Journal on Mobile Networks and Applications 12(2–3), 113–127 (2007)

21. Doukas, C., Pliakas, T., Maglogiannis, I.: Mobile healthcare information management unitizing cloud computing and android OS. In: Annual International Conference of the IEEE on Engineering in Medicine and Biology Society (EMBC), pp. 1037–1040, October (2010)

22. Tang, W-T., Hu, C-M., Hsu, C-Y.: A mobile phone based homecare management system on the cloud. In: Proceedings of the 3rd International Conference on Biomedical and Informatics (BMEI), vol. 6, pp. 2442 (November 2010)

23. Hoang, D.B., Chen, L.: Mobile Cloud for Assistive Healthcare (MoCAsH). In: Proceedings of the 2010 IEEE Asia-Pacific Service Computing Conference (APSCC), p. 325, February 2011

24. Nkosi, M.T., Mekuria, F.: Cloud computing for enhanced mobile health applications. In: Proceedings of the 2nd IEEE International Conference on Cloud Computing Technology and Science, p. 629, February (2011)

25. Jemal, H., Kechaou, Z., Ben Ayed, M.: Swarm intelligence and multi agent system in healthcare. In: 6th International Conference of Soft Computing and Pattern Recognition (SoCPaR) (2014)

26. Jemal, H., Kechaou, Z., Ben Ayed, M., Alimi, A.M.: A Multi agent system for hospital organization. International Journal of Machine Learning and Computing 5(1), 51–56 (2015)

27. JADE Home Page: Java Agent Development Framework (JADE) (2004). http://jade.tilab.com/

28. Shen, Z., Tong, Q.: The security of cloud computing system enabled by trusted computing technology. In: Proceedings of International Conference on Signal Processing Systems (ICSPS) (2010)

Parallel Flexible Molecular Docking in Computational Chemistry on High Performance Computing Clusters

Rafael Dolezal[1,4(✉)], Teodorico C. Ramalho[1,2],
Tanos C.C. França[1,3], and Kamil Kuca[1,4]

[1] Faculty of Informatics and Management, Center for Basic and Applied Research,
University of Hradec Králové, Rokitanskeho 62 50003, Hradec Králové, Czech Republic
{rafael.dolezal,teodorico.ramalho,tanosfranca,kucakam}@gmail.com
[2] Department of Chemistry, Federal University of Lavras, Lavras, Brazil
[3] Department of Chemistry, Military Institute of Engineering, Rio de Janeiro, Brazil
[4] Biomedical Research Center, Sokolska 581 50005, Hradec Králové, Czech Republic

Abstract. The main objective in pharmaceutical research is development of novel drugs with improved biological effect in specifically afflicted organisms. A common practice in drug design focuses on systematic organic derivatization of chemical structures exhibiting certain biological activity and subsequent biological *in vitro* evaluation of the resulted benefits. However, this classical approach can be more or less classified as a chance drug discovery, being very arduous, expensive and time consuming. Nowadays, a lot of enthusiasm is given to rationally oriented drug research techniques like computer-aided drug design, virtual screening, bioinformatics, chemometrics, quantitative structure-activity relationships, etc. In the present article, we deal with designing a high performance computing (HPC) support for flexible molecular docking (FMD) which can be beneficially utilized in structure-based virtual screening (SBVS). The principles of FMD are briefly introduced and a solution combining message passing interface (MPI) with multithreading is proposed. The merits (e.g. availability, scalability, performance) of MPI-HPC enhanced SBVS/FMD are compared with other HPC techniques utilized for novel lead structures discovery in medicinal chemistry.

Keywords: Molecular docking · Virtual screening · HPC · AutoDock vina

1 Introduction

Molecular docking can be characterized as a computational task which aims at finding the optimal geometrical arrangement of a ligand molecule with the respect to the position and conformation of a receptor molecule. The ligand-receptor complex is described by a number of molecular mechanics equations which use atomic Cartesian coordinates and a set of force field parameters to express the potential energy of the system. In simple terms, the total energy of the ligand-receptor complex is a distance dependent quantity the global minimum of which is sought by a molecular docking program [1]. Depending on the accuracy and precision of the calculations,

© Springer International Publishing Switzerland 2015
M. Núñez et al. (Eds.): ICCCI 2015, Part II, LNCS 9330, pp. 418–427, 2015.
DOI: 10.1007/978-3-319-24306-1_41

the predicted binding of ligands in the receptor molecule can substitute, to certain extent, for X-ray analysis of the ligand-receptor complex and biological activities determined *in vitro*.

If 3D molecular model of the receptor is available, many ligands can be tested by molecular docking like a set of keys to fit a lock. This extended mode of molecular docking represents an *in silico* surrogate for classical high throughput screening (HTS) used in biochemical drug discovery [2]. However, the present state-of-art in *in silico* drug design does not limit itself to rigid molecular docking as implied by the key-lock model, but takes into account the molecular flexibility, thermodynamics and the presence of ions and solvents. Accordingly, the flexible molecular docking (FMD) is preferred in this area to find appropriate settlement of ligands in the active site of the receptor.

Currently, many computational methods have been applied in so called computer-aided drug design (CADD). The main approaches of CADD can be classified into three groups: i) ligand-based virtual screening (LBVS) [3]; ii) structure-based virtual screening (SBVS) [4]; iii) quantitative structure-activity relationships (QSAR) [5]. In the present article we focus particularly on SBVS coupled with FMD, which can be considered as the most convenient method within *in silico* lead discovery. Provided suitable software and hardware is available, this type of computational tool can serve to scout out promising potential lead structure from extent virtual ligand databases containing even millions of ligands. This way, considerable costs and time can be saved comparing to ordinary drug research. Nonetheless, in a large scale SBVS/FMD, high performance computing (HPC) clusters or supercomputers have to be employed to complete the tasks in a reasonable time span.

In several last decades, a number of molecular docking programs has been developed, such as Dock, Glide, Gold, Icm, Molegro and FlexX. One of the most favorite program for FMD is AutoDock Vina (abbreviated as Vina) which can be also utilized as flexible molecular docking engine in SBVS [6]. However, for deployment of millions of docking jobs special adjustments of the program have to be rendered in order to reach sufficient scalability. Although Vina program natively implements multithreading, further improvements like message passing interface (MPI) is necessary to employ in order to evenly exploit the power of computer clusters or supercomputers while not overloading the I/O system. In this article, flexible molecular docking, virtual screening, and MPI technologies are analyzed to support SBVS/FMD based drug discovery in medicinal chemistry. Eventually, the performance and scalability of SBVS/FMD is compared for the case of classical batch distribution method and the MPI enhanced approach.

2   Flexible Molecular Docking in AutoDock Vina

As already mentioned, molecular docking is a calculation process ranking among molecular mechanics modeling which tries to fit a small molecule (e.g. ligand) into a larger molecule of receptor or enzyme (Fig. 1). The aim of this calculation is to find such a mutual geometrical position of ligand and receptor that exhibits the lowest potential energy. Prediction of the binding Gibbs free energy associated with preferred binding

mode can serve as an indicator of the interaction strength and thus as a measure of biological activity of the ligand. In addition, molecular docking can be used not only for
modeling of non-covalent interactions between one ligand and one receptor but also for
large scale screening of virtual ligand libraries to discover novel lead structures.

Basically, molecular docking is principally dependent on the 3D model of the target protein (i.e. enzyme or receptor). Although the protein of interest can be prepared
for instance by homology modeling and molecular dynamics, X-ray crystallography
or NMR spectrometry is widely preferred in order to base the molecular docking on
experimentally determined protein structure. In contrast, small ligand molecules can
be modeled completely *ab initio*. For the purposes of molecular docking, thousands of
3D enzyme models are on-line available in www.pdb.org database and millions of
various ligands can be downloaded from zinc.docking.org.

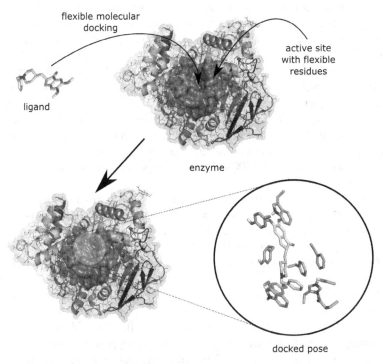

Fig. 1. Flexible molecular docking of donepezil (ligand, yellow) in human recombinant acetyl-
cholinesterase (*h*AChE) (pdb id: 4EY7) by AutoDock Vina. Flexible aminoacid residues are
rendered in blue.

In general, molecular docking problem is solved as an optimization task in which
the global minimum in the potential energy landscape is searched for. The potential
energy within molecular docking is commonly simplified by a scoring function. Additionally in Vina, not only the minima but also the shape of the energy profile and the
temperature is taken into consideration to estimate the free binding energy. The global
scoring function consists of several components responsible for: i) bond stretching

(E_s); ii) bond bending (E_b); iii) dihedral torsion (E_t); iv) van der Waals interactions (E_{vdW}); v) hydrogen bonding (E_{hb}); vi) electrostatic interactions (E_e) (Eq. 1) [7].

$$E_s = \sum_{i=1}^{n} K_s (r_i^a - r_i^0)^2; \; E_b = \sum_{i=1}^{n} K_b (\theta_i^a - \theta_i^0)^2; \; E_t = \sum_{i=1}^{n} \frac{V_n}{2} (1 + \cos(n\varphi_i^a - \varphi_i^0));$$

$$E_{vdW} = \sum_{i,j \in vdW}^{n} \left[\frac{A_{ij}}{R_{ij}^{12}} - \frac{B_{ij}}{R_{ij}^{6}} \right]; \; E_{hb} = \sum_{i,j \in hb}^{n} \left[\frac{C_{ij}}{R_{ij}^{12}} - \frac{D_{ij}}{R_{ij}^{10}} \right]; \; E_e = \sum_{i,j \in e}^{n} \left[\frac{q_i q_j}{\varepsilon R_{ij}} \right]$$

$$\tag{1}$$

Regarding the components, the scoring function can be expressed as a summation of two contributions: intramolecular $(E_s, E_b, E_t, E_{vdW}, E_{hb}, E_e)$ and intermolecular (only E_{vdW}, E_{hb}, E_e). In Vina program, the scoring function c is given as follows (Eq. 2.) [8]:

$$c = \sum_{i<j} f_{t_i t_j}(r_{ij}) = c_{inter} + c_{intra} \tag{2}$$

Here, $f_{t_i t_j}$ denotes a symmetric interaction function for atoms i and j assigned a type t_i and a type t_j, respectively, separated by r_{ij} distance. The symmetric interaction function f includes all 6 energy contributors outlined in Eq. 1. which depend particularly on atomic coordinates. As the rigid part of enzyme cannot change its conformation the task reduces to finding the optimal position and conformation of the ligand and the residues set as flexible. On the other hand, introduction of the ligand and the receptor flexibility significantly improves the ability of molecular docking in finding realistic binding modes. It is obvious that such minimization of the scoring function is not a trivial task and has to be solved numerically. Vina program employs for the global minimization a type of memetic algorithm that interleaves global and local searches by stochastic and deterministic techniques [9]. The open source code of this algorithm was thoroughly investigated by Handoko et al. who expressed it in the following pseudocode [10]:

```
Initialize conformation vector x, best
For step := 1 to max_step
  mutate (x);
  local_search (x);
  if (metropolis_accept (x) and score (x) < score (best))
    local_search (x);
    if (score (x) < score (best))
      best :=x;
    endif
  endif
endfor
```

The global optimizing algorithm implemented in Vina is related to Markov chain Monte Carlo algorithm with restart supplemented by random mutating the current solution. The result of local search is accepted if it fits the Metropolis criterion and is of a descending trend. After a sufficient number of iterations, the algorithm converges to a local minimum. Noteworthy, a series of randomly initialized *runs* can be deployed in parallel utilizing the benefits of multicore machines. In the local search, Vina utilizes Broyden-Fletcher-Goldfard-Shanno (BFGS) algorithm aimed at nonlinear unconstrained optimization. As a quasi-Newton method, it calculates the first and second order derivatives of the scoring function with respect to all involved variables (i.e. position and conformation of the ligand and the flexible parts of the receptors). By means of the gradient of the scoring function, a direction to the local minimum is estimated. When the gradient becomes close to zero, the local search is terminated. The BFGS algorithm is described by the bellow pseudocode [10]:

```
function local_search (vector x) // x describes the con-
formation
  vector g:=derivative (x);
  initialize H; // H is a Hessian approximate
  for step:=1 to max_step
    find a direction p by solving Hp=-g;
    do line search along p, update x and g;
    if (|g| -> 0)
      break;
    endif
    update H;
  endfor
endfunction
```

In Vina, completing all *runs* within the optimization is enhanced by multithreading, taking advantage of the shared-memory hardware parallelism. The optimization algorithm can split the task into several *runs* and perform them concurrently. Finally, the results of separate *runs* are clustered and analyzed, picking up the top scoring solution. The number of *steps* (i.e. the max_step parameter) within a *run* is determined heuristically depending on the conformational complexity of the ligand and the flexible receptor parts. However, the number of *runs* which are processed in parallel is set by the user through exhaustiveness parameter. Accordingly, to utilize all the CPUs available by Vina exhaustiveness parameter should be set greater than the number of CPUs reserved for the job. Increasing exhaustiveness proportionally prolongs the time needed for completing the task whilst, on the other hand, it decreases the probability of not finding the minimum exponentially. Vina is a robust molecular docking program that employs a sophisticated gradient method for localization of the global minimum of the scoring function. It is a free available software (http://vina.scripps.edu). Thanks to multithreading it can take advantage of multiple CPU machines to speed up the calculations. This feature is especially appreciated in SBVS where high throughput is demanded. Nonetheless, further support of Vina functions might be of a great convenience specifically in large scale SBVS on supercomputers.

3   Structure-Based Virtual Screening

Similarly to FMD, SBVS is an alternative knowledge-based technology to biochemical research attempting to estimate the binding affinity of various ligands towards a known target protein by computational chemistry methods. To certain extent,

SBVS/FMD can serve as a computational substitute for experimental screening (HTS) to evaluate biological properties of many thousands of potential drugs. Contrary to ligand-based virtual screening (LBVS) which does not utilize the structure of biological targets, SBVS can explicitly model the intermolecular interactions in the ligand-receptor complex predicting the preferred binding mode and the affinity for each database compound [11]. Thus, when the structure of the target protein (i.e. receptor or enzyme) is available the binding energies of all compounds in a virtual ligand library can be calculated and ranked suggesting which the best candidates for the lead structure in drug discovery process are.

SBVS/FMD involves several steps like target and ligand database preparation, designing the size and position of the binding site, selection of the flexible residues, deploying the calculation, docking, analysis of the outputs and finally prioritization of compounds for biological testing (Fig. 2.).

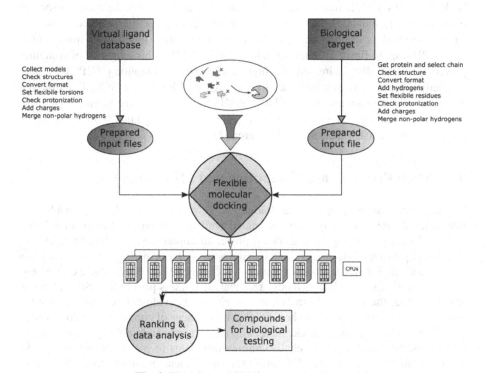

Fig. 2. Workflow of SBVS employing FMD.

A starting point in SVBS/FMD is the preparation of the virtual ligand library and the target protein. The ligand models are often downloaded from free online databases (e.g. zinc.docking.org) collecting all physically available compounds from chemical vendors but they can be prepared also as models of novel chemical entities (NCE). Accordingly, virtual ligand libraries may be prepared by *in silico* derivatization (e.g. scaffold hopping) which enables to develop completely new potential drugs. Nonetheless, application of sophisticated physico-chemical filters, QSAR models, or

LBVS is necessary in order to reject from the virtual ligand library non-synthesizable substances or chemical structures with undesirable biological properties. 3D ligand models can be generated in many chemical software like HyperChem, Spartan, Maestro, etc. Final preparation of the ligand files for SBVS/FMD in Vina including addition of charges, merging non-polar hydrogens into heavy atoms, defining the rotable torsions can be done by MGL Tools. On condition that the 3D structure of the target protein is available (e.g. in www.pdb.org databank), the preparation of the active site starts by localizing the co-crystalized ligand in the model. According to the ligand position, a gridbox center and size as well as the number of flexible aminoacid residues are configured to perform FMD in an adequate spatial extent. Various modifications of the protein model (i.e. adding charges, removing water molecules and superfluous protein chains, relaxing the structure, adding hydrogens, etc.) and splitting into the rigid and the flexible parts can be done, for example, in Maestro and MGL Tools, respectively. Within SBVS/FMD many thousands of docking tasks have to be accomplished in order to find promising ligands with significant affinity for the selected biological target. Nonetheless, proper handling the ligand and protein flexibility requires sufficient CPU time to finish all intended jobs in SBVS/FMD. This exigency of computer power is classically addressed by high performance computing (HPC) clusters coupled with a distributing interface like TORQUE, PBS Pro or SGE. Even though there are many applications of such HPC in SBVS described in the literature [12, 13], novel approaches for utilization of clouds, graphic processing units (GPU) and supercomputers are actually analyzed in these area [2].

4   High Performance Computing and MPI Support

Fast development of computer technologies in several past decades has enabled remarkable advances in many sciences related with demanding mathematical calculations and data-processing. The growing power of supercomputers has brought also significant prospects into drug research and discovery by providing a breakthrough *in silico* platform for simulations of variety of chemical and biological experiments. Amongst plethora of CADD methods, a special attention is paid to SBVS/FMD that is able to predict the results of biochemical *in vitro* tests evaluating experimentally the potency of drug candidates. The beneficial consequences of computerized drug design have been highly praised in academia as well as in the research, although considerable effort must be made to satisfy challenging requirements of pharmacology and medicinal chemistry. Commonly, HPC provides an efficient approach to deploy many thousands of SBVS/FMD jobs utilizing either ordinary personal computers integrated into clusters or specialized supercomputers. However, the profit of modern HPC technologies depends on the proper use of parallel processing of the submitted jobs. Currently, top supercomputers can function above 10^{15} floating-point operation per second (i.e petaflop), which requires sophisticated program approaches to exploit efficiently, quickly and reliably the available power. One of the HPC approaches to parallel processing is launching the jobs through a dispatching system such as portable batch system (PBS). The task scheduler manages all submitted jobs and deploys them over available computer nodes. The architecture of HPC clusters usually consists of a front-server for task scheduling and a powerful compute node domain interconnected by high-speed network like Infiniband or Myrinet. The file storage is often implemented by high performance parallel file systems such as Lustre. However, the major limit of this approach arises when the communication between individual jobs uses intermediate files, which burdens the I/O transfer, slows down the performance and restricts the number of CPUs applicable in parallel. Other approach to HPC is to employ computer clouds services. According to the online published report (www.schrodinger.com), Schrödinger carried out extensive virtual screening of 21 million compounds utilizing the Amazon cloud platform. The ligand database with the size about of 10 gigabytes was deployed and analyzed by a SBVS method within outstanding 3h. Although this internet-based solution, known as cloud computing, is very attractive due to its flexibility and availability on demand, it can be complicated to reach optimal scalability, especially at the beginning of the calculations. An interesting hardware solution for HPC has been found in utilization of graphic processing unit (GPU) in GPU or hybridized GPU/CPU clusters. Several molecular docking programs (e.g.Dock6 Amber, Plants) have been developed to increase the computing performance by implementing the NVIDIA CUDA platform [14, 15]. The current molecular docking programs employing single GPU can speed up the calculations approximately by factor of 10 in comparison to single CPU. However, development of efficient programs for large scale SBVS/FMD is still associated with a number of challenges regarding the distribution process, job scheduling, and resource control. The last but not least approach in HPC that we would like to highlight in the present article is implementation of a message passing interface (MPI) to parallelize SBVS/FMD applications in standard CPU clusters. Despite other promising approaches, MPI-HPC still remains the mainstream method in effective parallel calculations. Specifically, a MPI enhanced version of Vina program developed in Lawrence Livermore National Laboratory (VinaLC) is reviewed in this article with respect to classical employment of Vina [16]. The original C++ code of Vina was extended by implementation of MPI to enable distributing the individual FMD jobs across different nodes. In order to utilize evenly the computer power and reduce the idling time, VinaLC adopts a master-slave MPI scheme which consists of three "for loops" for the master process and of one infinite "while loop" for each slave process. Briefly, the first master for loop imports every combination of the receptor and ligand, and distributes the FMD jobs across different nodes through two pairs of MPI send/recv calls to the slave process. After receiving the data, the slave processes start the FMD jobs. The second master loop performs the output data collecting and assigns the slave process another job. The third master loop is responsible for sending a finished job flag to free the slave processes, which finishes the MPI run. Within this master-slave MPI scheme, the master part takes care of job dispatching, whilst the slave processes control FMD tasks until they are completed. Zhang et al. have demonstrated that the VinaLC program is a portable application able to run on various types of supercomputers (e.g. Linux clusters of Intel Xeon processors, AMD Opteron processors, IBM 20-petaflop supercomputer with BG/Q architecture, etc.) [16]. The scalability of MPI enhanced VinaLC program was evaluated in SBVS/FMD project employing 600 – 15 408 CPU cores on a Linux cluster of Xeon processors (Intel Xeon 5660 dual-socket 6-core nodes 2.8GHz, 24GB RAM) and the SLURM job scheduler. Thus, they used 1284 nodes with the total of 15 408 CPUs, launching 12 threads for each docking calculation. The `exhaustiveness` parameter was set to 12. In the project, around 100 000 compound were docked into the active site of *Thermus thermophylus* gyrase B ATP-binding domain. It was observed that MPI wall time for the docking jobs dramatically dropped as the CPU cores increased. The CPU time necessary for completing one docking job changed from 82.98s to 89.69s as the number of allocated CPUs increased from 600 to 15 408. The factor of speed up was found nearly ideal with the respect to the number of employed CPUs, although increasing the CPUs over 6 000 slightly decreased the speed up factor. Completing 1 million FMD tasks for the same receptor took 1.4h on 15 408 CPUs.

As the source code of VinaLC is free available, we compiled the program and evaluated its performance on the Anselm cluster (94.5 Tflop) of the Czech National Supercomputing Center IT4Innovations (www.it4i.cz). So far, we employed only 8 nodes (2 x Intel Xeon E5-2665, 8-core, 2.4GHz) to compare the calculation of 200 same FMD jobs (a tacrine derivative in *h*AChE, pdb id: 4EY7) by VinaLC, using "mpirun", and Vina, using "qsub" of PBS Pro 12 scheduler. Utilizing 128 CPUs in parallel, VinaLC completed all 200 FMD jobs in **5m16s**. The same batch of 200 FMD jobs was completed by Vina on the same 8 nodes in **13m13s**. It is evident that VinaLC was able to better utilize the available amount of CPUs than a batch distribution of separated Vina tasks per one node. Basing on these results, significant improvements of SBVS/FMD are expected after providing the VinaLC program with sufficient computer resources.

5   Conclusion

Thanks to advanced HPC approaches it is currently possible to carry out computerized virtual screening of millions potential drug candidates without the necessity to synthesize and test the chemical compounds. SBVS/FMD can find a promising chemical structure only by calculations, which eliminates considerable time and cost spent on experiments with biologically inactive substances. However, sufficient computer resources and sophisticated programs have to be applied in order to take advantage of these *in silico* drug design approaches. In the article, we demonstrated in simple terms the fundamentals of FMD and SBVS. However, performing large scale SBVS/FMD demands for suitable program like VinaLC. We presented in the article that this free program outperforms previous Vina program in utilization of computer resources, especially by implementing MPI. It is evident that a HPC-MPI support for SBVS/FMD remains relevant approach amongst cloud and GPU computing.

Acknowledgements. The paper is supported by the project of specific science "Application of Artificial Intelligence in Bioinformatics" at the Faculty of Informatics and Management, University of Hradec Kralove, Czech Republic, by IT4Inovations projects no.: CZ.1.05/1.1.00/ 02.0070, LM2011033, and the long term development plan of FNHK.

References

1. Kuntz, I.D., Blaney, J.M., Oatley, S.J., Langridge, R., Ferrin, T.E.: A geometric approach to macromolecule-ligand interactions. J. Mol. Biol. **161**, 269–288 (1982)
2. Dolezal, R., Sobeslav, V., Hornig, O., Balik, L., Korabecny, J., Kuca, K.: HPC cloud technologies for virtual screening in drug discovery. In: Nguyen, N.T., Trawiński, B., Kosala, R. (eds.) ACIIDS 2015. LNCS, vol. 9012, pp. 440–449. Springer, Heidelberg (2015)
3. Lavecchia, A., Di Giovanni, C.: Virtual screening strategies in drug discovery: a critical review. Curr. Med. Chem. **20**, 2839–2860 (2013)
4. Horvath, D.: A virtual screening approach applied to the search for trypanothione reductase inhibitors. J. Med. Chem. **40**, 2412–2423 (1997)
5. Gramatica, P.: Principles of QSAR models validation: internal and external. QSAR Comb. Sci. **26**, 694–701 (2007)
6. Trott, O., Olson, A.J.: Software News and Update AutoDock Vina: Improving the Speed and Accuracy of Docking with a New Scoring Function, Efficient Optimization, and Multithreading. J. Comput. Chem. **31**, 455–461 (2010)
7. Kuczera, K.: Molecular Modeling of Peptides. Comp. Pept., pp. 15–41. Springer (2015)
8. Trott, O., Olson, A.J.: AutoDock Vina: improving the speed and accuracy of docking with a new scoring function, efficient optimization, and multithreading. J. Comput. Chem **31**, 455–461 (2010)
9. Ong, Y.S., Keane, A.J.: Meta-Lamarckian learning in memetic algorithms. IEEE T. Evolut. Comput. **8**, 99–110 (2004)
10. Handoko, S.D., Ouyang, X., Su, C.T.T., Kwoh, C.K., Ong, Y.S.: QuickVina: accelerating AutoDock Vina using gradient-based heuristics for global optimization. IEEE ACM T. Comput. Bi. **9**, 1266–1272 (2012)
11. Lyne, P.D.: Structure-based virtual screening: an overview. Drug Discov. Today **7**, 1047–1055 (2002)
12. Peréz-Sánchez, H., Fassihi, A., Cecilia, J.M., Ali, H.H., Cannataro, M.: Applications of high performance computing in bioinformatics, computational biology and computational chemistry. In: Ortuño, F., Rojas, I. (eds.) IWBBIO 2015, Part II. LNCS, vol. 9044, pp. 527–541. Springer, Heidelberg (2015)
13. Imbernón, B., Llanes, A., Peña-García, J., Abellán, J.L., Pérez-Sánchez, H., Cecilia, J.M.: Enhancing the parallelization of non-bonded interactions kernel for virtual screening on GPUs. In: Ortuño, F., Rojas, I. (eds.) IWBBIO 2015, Part II. LNCS, vol. 9044, pp. 620–626. Springer, Heidelberg (2015)
14. Korb, O., Stützle, T., Exner, T.E.: Accelerating molecular docking calculations using graphics processing units. J. Chem. Inf. Model. **51**, 865–876 (2011)
15. Wang, J., Wolf, R.M., Caldwell, J.W., Kollman, P.A., Case, D.A.: Development and testing of a general amber force field. J. Comput. Chem. **25**, 1157–1174 (2004)
16. Zhang, X., Wong, S.E., Lightstone, F.C.: Message passing interface and multithreading hybrid for parallel molecular docking of large databases on petascale high performance computing machines. J. Comput. Chem. **34**, 915–927 (2013)

ColEdu: Special Session on Collective Computational Intelligence in Educational Context

Meta-Learning Based Framework for Helping Non-expert Miners to Choice a Suitable Classification Algorithm: An Application for the Educational Field

Marta Zorrilla[✉] and Diego García-Saiz

Department of Computer Science and Electronics, University of Cantabria,
Avenida de los Castros s/n, 39005 Santander, Spain
{marta.zorrilla,diego.garcias}@unican.es

Abstract. One of the most challenging tasks in the knowledge discovery process is the selection of the best classification algorithm for a data set at hand. Thus, tools which help practitioners to choose the best classifier along with its parameter setting are highly demanded. These will not only be useful for trainees but also for the automation of the data mining process. Our approach is based on meta-learning, which relies on the application of learning algorithms on meta-data extracted from data mining experiments in order to better understand how these algorithms can become flexible in solving different kinds of learning problems. This paper presents a framework which allows novices to create and feed their own experiment database and later, analyse and select the best technique for their target data set. As case study, we evaluate different sets of meta-features on educational data sets and discuss which ones are more suitable for predicting student performance.

Keywords: Meta-learning · Regression · Student performance

1 Introduction

Currently, the possibility of automatising the Knowledge Discovery Process (KDD) is still an open problem. As it is well-known, the KDD process [4] consists in several phases (preprocessing, modeling, mining and testing) and each one, in turn, includes a large number of tasks which should be performed. As every complex problem, a way to deal with it is to follow a divide and rule strategy. That is why this paper is focused on the mining phase, that means, the step in which the data miners have to choose the best algorithm for their data set at hand.

Rice [15] was the one who first formulated this issue and since then different approaches have been proposed, for instance: a) a traditional approach based on a costly trial-and-error procedure; b) an approach based on meta-learning, able to automatically provide guidance on the best alternative from a set of meta-features; or c) the use of ensemble methods to obtain better predictive performance.

© Springer International Publishing Switzerland 2015
M. Núñez et al. (Eds.): ICCCI 2015, Part II, LNCS 9330, pp. 431–440, 2015.
DOI: 10.1007/978-3-319-24306-1_42

As derived from the *no-free-lunch theorem*, no learning algorithm can be specified as outperforming on the set of all real-world problems [20], we therefore search a mechanism which allows us to characterise the algorithms from the meta-features of the data sets which they classify with better accuracy and use these to build an algorithm recommender, that means, we rely on meta-learning.

Meta-learning is a subfield of machine learning that aims at applying learning algorithms on meta-features extracted from machine learning experiments in order to better understand how these algorithms can become flexible in solving different kinds of learning problems, hence to improve the performance of existing learning algorithms [19] or to assist the user to determine the most suitable learning algorithm(s) for a problem at hand [7], among others.

In particular, we provide a framework which recommends practitioners the set of algorithms which should be applied to a concrete data set according to its characteristics and show its feasibility carrying out an experiment with data sets from the educational arena. Concretely, we address one of the oldest and best-known problems in educational data mining [16] that is predicting students performance. Furthermore, we discuss which meta-features, according to our experimentation, are more suitable for this challenging problem. We utilise regression techniques in this case study, unlike this paper [21] in which classification algorithms were applied with the aim of selecting the best classifier.

This paper is organised as follows: Section 2 briefly describes the different elements which comprise our framework, introducing previously the meta-learning field. Section 3 describes the methodology used in our case study and the setting of our experiment. Section 4 presents and discusses the results obtained showing the feasibility of our proposal. Finally, conclusions and future works are outlined in Section 5.

2 Background and an Overview of our Framework

Our aim, as previously mentioned, is to provide novice miners with a tool which help them to analyze the behaviour of different machine learners on different data sets and enable them to choose the more suitable algorithm for their data set at hand.

Meta-learning aims at learning the relationship between the meta-features extracted from the data sets and the algorithms performance applied on them. Thus, a meta-learning system consists of two main stages: a training phase and a prediction phase. In the training stage, data sets are first characterized by a set of measurable characteristics and next, a set of algorithms are executed on these data sets and their performance evaluations such as accuracy, f-measure, error rate, etc. are linked to the characteristics of the involved data set. Later, a learning algorithm is trained on the collected meta-data, which will yield a model which will be used to predict which the best algorithm to be applied on a new data set is. In other occasions, instead of selecting an algorithm, a ranking of algorithms is provided [2]. Different approaches for building the predictor are found, mainly based on classification [10,13,21] and regression [8,14]. Recently, a new approach called meta-learning template [9] has arisen with the aim of recommending a hierarchical combination of algorithms.

Regarding the kind of meta-features that these systems generally use, these can be classified in:

- Simple or general features, such as the number of attributes, the number of instances, the type of attributes (numerical, categorical or mixed), the number of values of the target attribute and dimensionality of the data set, i.e., the ratio between the number of attributes and the number of instances.
- Statistical features, like skew, kurtosis among others which measure the distribution of attributes and their correlation [17,19].
- Information theoretic features used for characterising data sets containing categorical attributes such as class entropy or noise to signal ratio [5].
- Model-based meta-features, which collect the structural shape and size of a decision tree trained on the data sets [11].
- Landmarkers, which are meta-features calculated as the performance measures achieved by using simple classifiers [12].
- Complexity features, that characterize the apparent complexity of data sets for supervised learning [6]. These are provided by DCoL (data complexity library) [1,21].
- Contextual features, i.e., characteristics related to data set domain [21].

Figure 1 depicts our proposal graphically. As can be observed, the framework basically makes use of four workflows, one for extracting meta-features of the data sets, another one for loading the descriptive information of each experiment performed on each data set into the database; the third one, responsible for building a regressor for each type of algorithm used in the training phase; and the last one, responsible for carrying out the predictive phase, that means, reading a new data set, extracting its meta-features, applying this meta-data set to the regressors previously built and showing the algorithms ranked according to the value of accuracy predicted by themselves. Accuracy is directly calculated by running each regressor.

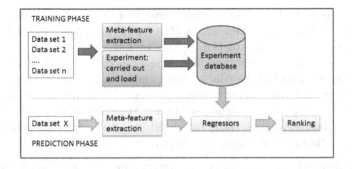

Fig. 1. Overview of our framework

The database schema used to gather the experiments is shown in Figure 2. We designed this based on the one proposed in [18] which gathers machine learning experiments. But this had to be extended to collect meta-features of each data set and the set of meta-features which comprise each training data set. A complete description of this database model is found in [3].

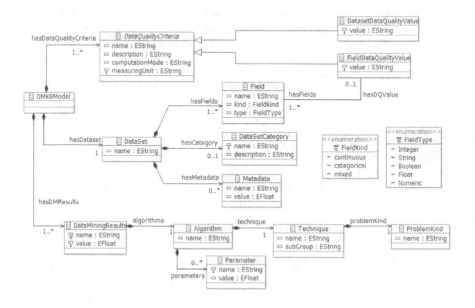

Fig. 2. Database model

3   Experiment Design

Next, we explain how the experiment has been carried out to show the feasibility of our approach.

The data sets used came from educational arena, in particular, they gather the activity performed by students in thirty different e-learning and blended courses hosted in a Moodle platform. This activity is measured by means of several metrics such as the total number of sessions open by each student in the course and in each tool of the course (tests, contents, forum,...), the number of self-tests performed, the number of messages posted and answered in the forum, among others. All attributes are numeric except the class attribute which collects if the learner failed (positive class) or passed (negative class) the course. Next, we extracted their meta-features. Concretely, we used the following ones:

- Simple meta features: number of instances, number of attributes, number of instances of the positive class (fail) and number of instances of the negative class (pass).

- Statistical measures: min, max and average value of the skewness and kurtosis of all the attributes of the data set calculated by means of the MATH3-apache Java library.
- Landmarkers. In this case we included as meta-features the accuracy achieved by the following weak classifiers: LinearDiscriminant (LD), BestNode with gain-ratio criterion (BN), RandomNode (RN), NaïveBayes (NB) and 1-N, all available in Weka or RapidMiner.
- Complexity meta features: we used the fourteen features offered by DCoL software that are the maximum Fisher's discriminant ratio (F1), the directional-vector maximum Fisher's discriminant ratio (F1v), the overlap of the per-class bounding boxes (F2), the maximum (individual) feature efficiency (F3), the collective feature efficiency (sum of each feature efficiency)(F4), the fraction of points on the class boundary (N1), the ratio of average intra/inter class nearest neighbor distance (N2), the training error of a linear classifier (N3), the fraction of maximum covering spheres (T1), the average number of points per dimension (ratio of the number of examples in the data set to the number of attributes)(T2), the leave-one-out error rate of the one-nearest neighbor classifier (L1), the minimized sum of the error distance of a linear classifier (L2) and the nonlinearity of a linear classifier (L3).

As our data sets only have numeric features, no information-theory measures were used. In a future work, we will include these meta-features by using multiclass data sets along with nominal predictor attributes.

As can be observed in Tables 1,2 and 3 meta-features extracted from training data sets take a wide range of values.

Table 1. Range of values of the complexity meta-features

F1	F1v	F2	F3	F4	L1	L2
0.04-29.46	0.03-370.82	0-0.2	0.02-0.88	0.05-1	0.27-0.92	0.09-0.45

L3	N1	N2	N3	N4	T1	T2
0.07-0.5	0.05-0.93	0.25-1.23	0-0.67	0-0.49	0.6-1	0.82-33.62

Table 2. Range of values of the simple and statistical meta-features

#N Ins.	#N Att.	#N Fail	#N Pass		
13-504	3-28	5-220	3-433		

Max. Skew.	Min Skew.	Avg. Skew.	Max. Kurt	Min. Kurt	Avg. Kurt
8.02-500.12	(-)1.51-15.22	2.99-131.43	1.43-22.33	(-)13.48-3.23	(-)2.3-10-36

Next, we run five classifiers on these thirty data sets using their default setting and 10-fold cross-validation. These were C4.5 (J48 version from Weka),

Table 3. Range of values of the landmarkers

Acc. NB	Acc. LD	Acc. BN	Acc. RN	Acc. 1NN
23.33-95.35	35-97.5	23.08-100	31.33-100	40-100

RandomForest, RIPPER (JRip version from Weka), k-NearestNeighbourds with auto-selection of the k number o neighbours (iBk in Weka), and LogisticRegression. As can be observed, each one follows a different learning paradigm. The accuracy achieved by each classifier on each data set during the training was stored in the data base.

Then, we generated five meta-data sets, one for each classifier. Each data set contained the meta-features of the training data sets along with the accuracy achieved by that specific classifier. For the sake of studying the behaviour of each group of meta-features, we built different linear regression models by using different combinations of meta-features:

1. Using all the meta features available.
2. Using only the meta-features which belong to each group (simple, statistical, complexity or landmarkers) separately.
3. Using only the most relevant meta-features chosen by a feature-selection algorithm. For this purpose, we used the ClassifierSubSet algorithm, offered by Weka, with the BestFirst algorithm as search method and linear regression as base classifier. The leave-one-out method was used for its evaluation. Tree thresholds were used, 10%, 40% and 70% to choose features according to their relevance.

All the linear regression models were generated using leave-one-out strategy as evaluation process. The RMSE of each regression model computed was also stored in the database.

Finally, we used two new data sets for testing our approach. We compared the predicted accuracy achieved by our regressors (recommenders) with the real accuracy achieved by the same classifiers. This allows us to measure at what extent our proposal actually shows a reliable ranking.

4   Results and Discussion

First of all, we show a summary of the results achieved in the regressors building phase. Table 4 displays, in the first column, "Avg. Acc.", the RMSE which results as a consequence of averaging the accuracy achieved for each classifier on all data sets. This will be used as base result in our experiment. The rest of the columns gather the RMSE obtained in the building of each linear regression model by using all meta-features ("All"), only the meta-features of a concrete group, or by applying a leave-one-out feature selection (FS) with a threshold of 10%, 40% and 70% ("FS 10%","FS 40%", "FS 70%").

Reading this table, one notices that using all the meta-features to build the regressors does not lead to a better result than using the average accuracy

Table 4. RMSE of the regressors built

	Avg. Acc.	Simple	Statistic	Complexity	Landmark	All	FS 10%	FS 40%	FS 70%
RF	0.108	0.118	0.110	0.274	0.056	0.153	0.153	0.090	0.060
Jrip	0.105	0.113	0.104	0.333	0.063	0.261	0.099	0.050	0.056
J48	0.098	0.118	0.110	0.336	0.052	0.321	0.091	0.067	0.051
LR	0.122	0.135	0.139	0.252	0.078	0.435	0.296	0.122	0.061
iBk	0.141	0.126	0.121	0.265	0.066	0.279	0.074	0.067	0.062
Avg.	0.115	0.122	0.117	0.292	0.063	0.289	0.143	0.079	0.058

directly. This is mainly due to the fact that there are several meta-features that hinder the building of a good model. This is clearly endorsed by the results obtained when the meta-features with a relevance lower than 10% are removed from the meta-data set. In this case the regression models based on iBk, J48 and Jrip performed better than the base case. The results improve further when more meta-features are dropped. In fact, when the meta-data set only include the features with a relevance higher than 70%, all regressors performed better than the base case.

Regarding using only a meta-features group to generate the regression models, the results show that the landmakers are good predictors since the RMSE achieved is close to the one obtained with "FS 70%". In fact, the model built using RamdomForest with only landmakers has the lowest RMSE. The use of the remaining groups of meta-features individually does not seem to get more accurate models. The results thus yield that the best way to achieve a good ranking system based on meta-learning is applying a feature-selection algorithm with a high removal threshold on the meta-features data set.

Table 6 displays the number of times that each meta-feature selected by "FS 70%" was used in the building of our five regressors (min of 1, max of 5).

Table 5. Times that the meta-features are selected with "FS 70%"

F1	F1v	F2	F3	F4	T1	T2	skMin	kurtMin	1NN	BN	RN	LD	NB
5	2	1	2	2	1	1	1	2	5	5	5	3	1

The landmarkers calculated as the accuracy achieved by the 1-NN, BN and LD algorithms were always selected. This fact is aligned with our first conclusion which stated that the landmarkers are the best meta-features for our purpose. Nevertheless, there are also other meta-features which show a high relevance. That is the case of the complexity measure labelled as F1, which was chosen 5 times too. Furthermore, as a result of the fact that regressors built from the "FS 70%" meta-data sets performed better in 4 out of 5 techniques, we can say that the most suitable strategy to build an algorithm ranking system is to calculate all meta-features and use the most relevant. Although using only landmarkers would also lead to good results, as was concluded after carrying out a paired t-test with a significance level of 0.01 between the case base regressor and landmakers-based regressors.

Regarding the predictive phase, we tested our proposal with two data sets from two different courses, with the aim of discovering the ranking of techniques which suggests us for each one. We used the "FS 70%" regression model of each one of the algorithms. The real (R.Acc.) and predicted (Pr.Acc.) accuracy for each data set and classifier are shown in Table 6, as well as the ranking of the classifiers (1 to 5 from better to worse performance).

Table 6. Output of our recommender system for the two test data sets

	Dataset1		Dataset2	
	Pr.Acc.(rank)	R.Acc.(Rank)	Pr.Acc.(rank)	R.Acc.(Rank)
RF	0.81 (3)	0.9(1)	0.67 (3)	0.64 (3)
Jrip	0.82 (2)	0.87(3)	0.75 (1)	0.70 (1)
J48	0.94 (1)	0.88(2)	0.65 (5)	0.64 (3)
LR	0.81 (3)	0.85 (4)	0.73 (2)	0.70 (1)
iBk	0.80 (5)	0.78 (5)	0.66 (4)	0.67 (2)

As can be observed, the recommendation for the first data set would be to use the J48 algorithm, since it has the higher predicted accuracy. This classifier achieved the second real higher accuracy, so that although our recommender did not recommend the best classifier, its accuracy is very close to it. Regarding the second data set, our proposal recommended the two better classifiers, Jrip and Logistic Regression. So, we can conclude that our approach works fine and can be a useful tool for helping novice data miners to decide which algorithms to utilise for their data set at hand.

5   Conclusions

This paper describes a framework which allows practitioners to discover what algorithm is more suitable for applying on certain data set. In fact, it offers a ranking of algorithms, thus the data miners can also evaluate what other techniques could well be used if their difference in accuracy with respect to the first one is not very large. Our proposal relies on meta-learning, that means, the use of a database which collects the results of different learning experiments along with the meta-features of the data sets involved with the aim of building a recommender which informs us about the better technique for a problem at hand.

The experimentation carried out shows that the use of linear regression for building this recommender works suitably and that the most significative meta-features for our purpose are landmakers, although, as demonstrated, the most effective recommender was built by applying a feature selection algorithm on all meta-features establishing a high threshold.

This work has a few limitations that will be addressed in a near future. On the one hand, we should extend our experimentation by including multiclass data sets and data sets with both numeric and nominal attributes. Furthermore, we should feed our database with more experiments applying more mining algorithms and registering different performance measures. Finally, we should build recommenders with other regression techniques and evaluate which one is the most suitable.

References

1. Cavalcanti, G., Ren, T., Vale, B.: Data complexity measures and nearest neighbor classifiers: a practical analysis for meta-learning. In: 2012 IEEE 24th International Conference on Tools with Artificial Intelligence (ICTAI), vol. 1, pp. 1065–1069, November 2012

2. Romero, C., Olmo, J.L., Ventura, S.: A meta-learning approach for recommending a subset of white-box classification algorithms for Moodle datasets. In: Proc. 6th Int. Conference on Educational Data Mining, pp. 268–271 (2013)

3. Espinosa, R., García-Saiz, D., Zorrilla, M.E., Zubcoff, J.J., Mazón, J.: Development of a knowledge base for enabling non-expert users to apply data mining algorithms. In: Accorsi, R., Ceravolo, P., Cudré-Mauroux, P. (eds.) Proceedings of the 3rd International Symposium on Data-driven Process Discovery and Analysis, Riva del Garda, Italy, August 30, 2013. CEUR Workshop Proceedings, vol. 1027, pp. 46–61. CEUR-WS.org (2013). http://ceur-ws.org/Vol-1027/paper4.pdf

4. Fayyad, U., Piatetsky-Shapiro, G., Smyth, P.: The kdd process for extracting useful knowledge from volumes of data. Commun. ACM **39**(11), 27–34 (1996)

5. Hilario, M., Kalousis, A.: Building algorithm profiles for prior model selection in knowledge discovery systems. Engineering Intelligent Systems **8**, 956–961 (2002)

6. Ho, T.K.: Geometrical complexity of classification problems (2004). CoRR cs.CV/0402020

7. Kalousis, A., Hilario, M.: Model selection via meta-learning: a comparative study. In: Proc. 12th IEEE International Conference on Tools with Artificial Intelligence, pp. 406–413 (2000)

8. Köpf, C., Taylor, C., Keller, J.: Meta-analysis: from data characterisation for meta-learning to meta-regression. In: Proceedings of the PKDD-00 Workshop on Data Mining, Decision Support, Meta-Learning and ILP (2000)

9. Kordík, P., Cerný, J.: On performance of meta-learning templates on different datasets. In: IJCNN, pp. 1–7. IEEE (2012)

10. Molina, M.M., Luna, J.M., Romero, C., Ventura, S.: Meta-learning approach for automatic parameter tuning: a case study with educational datasets. In: Proc. 5th International Conference on Educational Data Mining, pp. 180–183 (2012)

11. Peng, Y.H., Flach, P.A., Soares, C., Brazdil, P.B.: Improved dataset characterisation for meta-learning. In: Lange, S., Satoh, K., Smith, C.H. (eds.) DS 2002. LNCS, vol. 2534, pp. 141–152. Springer, Heidelberg (2002)

12. Pfahringer, B., Bensusan, H., Giraud-carrier, C.: Meta-learning by landmarking various learning algorithms. In: Proceedings of the 17th International Conference on Machine Learning, pp. 743–750. Morgan Kaufmann (2000)

13. Reif, M., Leveringhaus, A., Shafait, F., Dengel, A.: Predicting classifier combinations. In: Proceedings of the 2nd International Conference on Pattern Recognition Applications and Methods. INSTICC, SciTePress (2013)

14. Reif, M., Shafait, F., Goldstein, M., Breuel, T., Dengel, A.: Automatic classifier selection for non-experts. Pattern Analysis and Applications **17**(1), 83–96 (2014). http://dx.doi.org/10.1007/s10044-012-0280-z
15. Rice, J.: The algorithm selection problem. Adv. Comput. **15**, 65–118 (1976)
16. Romero, C., Ventura, S.: Data mining in education. Wiley Interdisciplinary Reviews: Data Mining and Knowledge Discovery **3**(1), 12–27 (2013)
17. Segrera, S., Pinho, J., Moreno, M.N.: Information-theoretic measures for meta-learning. In: Corchado, E., Abraham, A., Pedrycz, W. (eds.) HAIS 2008. LNCS (LNAI), vol. 5271, pp. 458–465. Springer, Heidelberg (2008)
18. Vanschoren, J., Blockeel, H., Pfahringer, B., Holmes, G.: Experiment databases. Machine Learning **87**(2), 127–158 (2012). http://dx.doi.org/10.1007/s10994-011-5277-0
19. Vilalta, R., Drissi, Y.: A perspective view and survey of meta-learning. Artificial Intelligence Review **18**, 77–95 (2002)
20. Wolpert, D.H.: The supervised learning no-free-lunch theorems. In: Proc. 6th Online World Conference on Soft Computing in Industrial Applications, pp. 25–42 (2001)
21. Zorrilla, M., García-Saiz, D.: Meta-learning: can it be suitable to automatise the KDD process for the educational domain? In: Kryszkiewicz, M., Cornelis, C., Ciucci, D., Medina-Moreno, J., Motoda, H., Raś, Z.W. (eds.) RSEISP 2014. LNCS, vol. 8537, pp. 285–292. Springer, Heidelberg (2014). http://dx.doi.org/10.1007/978-3-319-08729-0_28

Fuzzy Logic Based Modeling for Building Contextual Student Group Recommendations

Krzysztof Myszkorowski and Danuta Zakrzewska[✉]

Institute of Information Technology Lodz University of Technology, Lodz, Poland
{kamysz,dzakrz}@ics.p.lodz.pl

Abstract. Group learning plays an important role in Web based educational process. Groups of students who are to learn together should be characterized by similar features. However course needs may differ depending on the context of the system usage. Each new student, who intends to join the community, should obtain context-aware recommendation of the group of colleagues matching his preferences. In the paper, using fuzzy logic for modeling students and groups is considered. We propose to describe student characteristics by means of fuzzy sets and to use the possibility-based representation of each group. We assume that context is represented by a vector of weights. Then recommendations for new students are determined by applying pattern matching technique including respective context vector. We examine the presented approach by taking into account learning style dimensions as attributes which characterize student preferences. The method is evaluated on the basis of experimental results obtained for real student data.

Keywords: Recommender systems · Fuzzy logic · Context awareness · Group modeling

1 Introduction

Nowadays, group learning becomes an important part of e-learning process. However its performance depends on effective group formation. Students who are to share the same information content, should have similar profiles, what is necessary to adjust learning resources appropriately. But different student traits may be important taking into account course requirements as well as the context of the system use. Context aware differentiation of teaching materials becomes an important feature of an e-learning system [1]. Accordingly, group recommendations should enable each new student to join the suitable context dependent group of colleagues while enrolling on different courses.

New learner assignment should guarantee his similarity to all the group members. However if our knowledge of the students' characteristics is imperfect one has to apply tools for describing uncertain or imprecise information. As the most consistent with human being decision making process Shakouri and Tavassoli [2] mentioned fuzzy logic methods, which are based on the fuzzy set theory [3].

© Springer International Publishing Switzerland 2015
M. Núñez et al. (Eds.): ICCCI 2015, Part II, LNCS 9330, pp. 441–450, 2015.
DOI: 10.1007/978-3-319-24306-1_43

A concept of fuzzy set has been used in [4] for group recommending in e-learning systems. A fuzzy group representation has been defined with the use of linguistic terms corresponding to attribute values. The recommendation process was based on the cardinality of the defined fuzzy sets. A possibility-based group representation has been proposed in [5]. Attribute values were represented by means of possibility distributions. Recommendations for new students were determined taking into account a degree of possibility of matching together with the respective context parameters. In both papers students' characteristics were given precisely. The presented paper extends the idea described in [5] by using fuzzy values as attribute values in the description of students. Thus, students' characteristics are modeled by means of fuzzy sets which represent appropriate linguistic terms. Recommendations are made by aggregation via arithmetic mean. We assume that context is modeled by a vector of weights. In the recommendation process we apply pattern matching technique [6]. To check the quality of recommendation we introduce fuzzy recommendation error which takes into account fuzziness in student models. The proposed method is examined for student traits based on their learning style dimensions. It is validated, on the basis of experiments, done for real students' groups.

The remainder of the paper is organized as follows. The relevant research is described in the next section. Then the methodology for building recommendations, including student and group modeling as well as context usage are depicted. In Section 4 the case study of students described by dominant learning style dimensions is considered. In the following section results of experiments carried out on real students' data are presented and discussed. Finally, concluding remarks and future research are outlined.

2   Related Work

Many authors considered recommendations as an important tool for supporting personalization of activities in e-learning environments (see [7,8]). However the researchers emphasized an importance of a context in a personalization process (see [9,10] for example). The broad review of context parameters as well as context aware e-learning systems was presented in [11]. Incorporating of contextual information into the recommendation process was considered in [12]. Special attention was paid to using recommendations for group learning. Muehlenbrock [13] discussed both: learner profile and context information for group formation from the perspective of ubiquitous computing and ambient intelligence. He proposed probabilistic approach for automatic learning of individual characteristics. The problem of assigning learners into groups was examined in [14]. Wang and Zhao [15] considered a fuzzy Bayesian network for building a user group recommendations based on context of scenario analysis and inference. Masthoff [16] described using of group modeling and group recommendation techniques for recommending to individual users. Zheng et al. [17] considered differential context weighting approach, in which the contribution of each contextual variable is weighted.

Using fuzzy logic for student modeling was examined by several researchers. Most of them considered applications in e-learning systems. Student fuzzy models applied for evaluation of intelligent learning systems were described in [18]. Hogo proposed applying fuzzy clustering models for evaluation of e-learners behavior [19]. In [20] authors defined two metrics based on fuzzy logic for evaluation of different personalization strategies. Lu [21] proposed a fuzzy matching method to find suitable learning materials. Using fuzzy logic for evaluation and classification of student performance was presented in [22].

In several works researchers considered fuzzy approach to adapting and personalizing e-learning environments. Almohammadi and Hagras presented fuzzy logic based system, which learns student preferred knowledge delivery to create a personalised learning environments [23]. Chrysafiadi and Virvou combined fuzzy theory with user stereotypes and the overlay model to user modeling [24]. Accordingly, they build the module of an e-learning environment for computer programming, that aimed at identifying alterations on the state of student's knowledge level.

3   Contextual Student Group Recommendations

For ordinary sets membership of elements is defined with the use of two-valued logic. Each element u of the universe of discourse U has for the set F an associated value $I_F(u)$ belonging to $\{0,1\}$. For $u \in F$ $I_F(u) = 1$, otherwise $I_F(u) = 0$. The concept of fuzzy set allows for partial membership by replacing $\{0,1\}$ with $[0,1]$. The grade of membership in a fuzzy set F is defined by a membership function $\mu_F(u)$ which associates with each element $u \in U$ a number from $[0,1]$. For full membership $\mu_F(x) = 1$, for non-membership $\mu_F(u) = 0$. The set of elements $u \in U$ such that $\mu_F(u) > 0$ is called the support of F, denoted by $\mathrm{supp}(F)$. A fuzzy set F is called normal if there exists $u \in U$ such that $\mu_F(u) = 1$. Based on the concept of fuzzy sets there has been introduced the concept of possibility distributions [25]. Each element $u \in U$ is assigned with a number $\pi_X(u) \in [0,1]$ which expresses the possibility degree of $X = u$, where X denotes a variable on U. A fuzzy value expressed by a fuzzy set F with the membership function $\mu_F(u)$ is represented by the possibility distribution such that $\mu_F(u) = \pi_X(u)$.

We will assume that students are described by N nominal attributes A_i, $i = 1,..., N$, which represent their traits such as cognitive styles, usability preferences or characteristics of their historical behaviors. A tuple ST representing a student is of the form:

$$ST = (st_1, st_2, ..., st_N), \quad st_i \in DOM(A_i) \ , \tag{1}$$

where $DOM(A_i)$ stands for the domain of A_i. We will also assume that there exist NG groups of students GS_k, $k = 1, 2, ..., NG$ of similar features. Let us denote by $c_{k,i}$ cardinality of the most frequent value of the attribute A_i in the group GS_k:

$$c_{k,i} = \max_j card\left\{ST \in GS_k : ST(A_i) = a_{i,j},\ a_{i,j} \in DOM(A_i)\right\} \ . \tag{2}$$

As the representation of the attribute A_i for the group GS_k we will consider a fuzzy set $FGS_{k,i}$ with the following membership function:

$$\mu_{FGS_{k,i}}(a_{i,j}) = card\{ST \in GS_k : ST(A_i) = a_{i,j}\}/c_{k,i} . \qquad (3)$$

Let a group GS_k be represented by a tuple $(gs_{k,1}, gs_{k,2}, ..., gs_{k,n})$. The membership function $\mu_{FGS_{k,i}}(u)$ may be interpreted as a measure of possibility that objects from GS_k are characterized by a certain attribute value [26]. Thus each value $gs_{k,i}$ is represented by a normal possibility distribution:

$$gs_{k,i} = \{\pi_{k,i}(a_{i,j})/a_{i,j} : a_{i,j} \in DOM(A_i),\} . \qquad (4)$$

The biggest values of $\pi_{k,i}(a_{i,j})$ indicate dominant attribute values in groups. Possibilistic form of the group representation allows to determine matching degrees of attributes of new students and classify them to appropriate groups.

Let a tuple $NST = (ns_1, ns_2, ..., ns_N)$ represent a new student. Let us assume that ns_i is modeled by means of a fuzzy set FNS_i over $DOM(A_i)$:

$$FNS_i = \{< x, \mu_{FNS_i}(x) >: x \in DOM(A_i), \ \mu_{FS_i}(x) : DOM(A_i) \to [0,1]\} , \qquad (5)$$

where $\mu_{FNS_i(x)}$ denotes a membership function of FNS_i. The set FNS_i might be labeled with a certain linguistic term. Let us consider a group of students GS_k represented by a tuple $(gs_{k,1}, gs_{k,2}, ..., gs_{k,n})$. The compatibility degree between NST and GS_k can be estimated with the use of possibility measure, denoted by $Pos(NST, GS_k)$. This measure expresses the extent to which the considered values satisfy a comparison relation.

The problem of fuzzy pattern matching has been discussed in [6]. According to [6] a degree of possibility of matching NST to GS_k for the attribute A_i equals:

$$Pos(NST(A_i), GS_k(A_i)) = \sup_{x \in DOM(A_i)} \min(\pi_{k,i}(x), \mu_{FNS_i}(x)) . \qquad (6)$$

The total matching degree $Pos(NST, GS_k)$ for the group GS_k is expressed by the aggregation via arithmetic mean:

$$Pos(NST, GS_k) = \sum_{i=1}^{N} Pos(NST(A_i), GS_k(A_i))/N . \qquad (7)$$

Maximal value of $Pos(NST, GS_k)$ indicates the group that should be recommended for students.

The described way of recommendation assumes that all the attributes are of equal importance. However, if certain attribute A_i is less important for the choice of the group and the matching degree $Pos(NST(A_i), GS_k(A_i))$ is low, then the group may be rejected regardless of matching degrees of other attributes. The problem may be resolved by introduction of weights. Let $w_i \in [0,1]$ denote the grade of importance of A_i and let $\sum w_i = 1$. For attributes which are not considered during the recommendation process $w_i = 0$. A degree of possibility of matching NST to GS_k for A_i equals:

$$Pos(NST(A_i), GS_k(A_i)) = \sum_{i=1}^{N} w_i * Pos(NST(A_i), GS_k(A_i)) . \qquad (8)$$

Let us assume, that there are NG student groups, then the whole process of recommendation building will take place in the following way:

```
[Input]:A set of NG groups GS_k, containing students
        described by N nominal attributes;
        a tuple NST representing a new student;
        a set of weights assigned with attributes;
Step 1: For each group GS_k, k = 1, 2, ..., NG find its
        possibility-based representation according to (4);
Step 2: For the student NST find the group GL_rec
        with the maximal value of the matching degree (8);
Step 3: Recommend GL_rec to the student.
```

The quality of recommendation depends on values of $\pi_{k,i}(x)$ for $x \in \text{supp}(FNS_i)$. Let CS_i denote cardinality of $\text{supp}(FNS_i)$: $CS_i = card(\text{supp}(FNS_i))$. Thus a recommendation error Err_k is defined as follows:

$$Err_k = \sum_{i=1}^{N} w_i * err_{k,i} \ , \tag{9}$$

where

$$err_{k,i} = \sum_{x \in \text{supp}(FNS_i)} ((1 - \pi_{k,i}(x)) * \mu_{FNS_i}(x))/CS_i \ . \tag{10}$$

4   Student Models Based on Learning Styles

To validate the proposed methodology, we will consider student models based on dominant learning styles. Das et al. [11] mentioned learning styles as context parameters corresponding to media used by the learner. We will apply Felder & Silverman [27] model, where learning styles are described by means of 4 attributes which indicate preferences for 4 dimensions from among excluding pairs: active vs. reflective (L_1), sensing vs. intuitive (L_2), visual vs. verbal (L_3), and sequential vs. global (L_4) or balanced if the student has no dominant preferences. Attribute values belong to the set of odd integers from the interval [-11, 11]. These numbers describe scores for features represented by respective attributes. They are determined on the base of the questionnaires which are filled by students. Each student, who filled ILS questionnaire [28], can be modeled by a vector SL of 4 integer attributes:

$$SL = (sl_1, sl_2, sl_3, sl_4), \ \ sl_i \in \{-11, -9, -7, -5, -3, -1, 1, 3, 5, 7, 9, 11\} \ . \tag{11}$$

Negative values of sl_1, sl_2, sl_3, sl_4 mean scoring for active, sensing, visual or sequential learning styles, respectively. Positive values of them indicate scoring for reflective, intuitive, verbal or global learning styles. Values -5,-7 or 5,7 mean that a student learns more easily in a learning environment which favors the considered dimension; values -9,-11 or 9,11 mean that learner has a very strong preference for one dimension of the scale and may have real difficulty learning in an environment which does not support that preference.

As not always students' characteristics can be precisely determined, in further considerations we will assume that components of SL can be modeled by means of integer fuzzy numbers. For creating of the fuzzy representation of learning style dimensions we will define in the interval [-11, 11] fuzzy sets FL_j with the following membership functions:

$$\mu_{FL_1}(x) = \begin{cases} 1, & \text{if } x = -11 \\ 1/2, & \text{if } x = -9 \end{cases}, \tag{12}$$

$$\mu_{FL_{12}}(x) = \begin{cases} 1, & \text{if } x = 11 \\ 1/2, & \text{if } x = 9 \end{cases} \tag{13}$$

and for j = 2, 3, ... , 11

$$\mu_{FL_j}(x) = \begin{cases} 1, & \text{if } x = 2j - 13 \\ 1/2, & \text{if } x \in \{2j - 15, 2j - 11\} \end{cases}. \tag{14}$$

Each set FL_j contains exactly one value, denoted by lm_j, which fully belongs to it. The membership grades of neighboring values are equal to 1/2. Fuzzy sets FL_j represent linguistic terms l_j corresponding to attribute values. Let a tuple $NST = (l_{j_1}, l_{j_2}, l_{j_3}, l_{j_4})$, represent a new student. A degree of possibility of matching NST to GS_k for the attribute L_i equals:
for $j_i = 1$:

$$Pos(NST(L_i), GS_k(L_i)) = \max(\pi_{k,i}(-11), \min(\pi_{k,i}(-9), 0.5)) , \tag{15}$$

for $j_i = 12$:

$$Pos(NST(L_i), GS_k(L_i)) = \max(\pi_{k,i}(11), \min(\pi_{k,i}(9), 0.5)) , \tag{16}$$

for $j_i = 2,3,...,11$:

$$Pos(NST(L_i), GS_k(L_i)) =$$
$$\max(\pi_{k,i}(lm_{j_i}), \min(\pi_{k,i}(lm_{j_i} - 2), 0.5), \min(\pi_{k,i}(lm_{j_i} + 2), 0.5)) . \tag{17}$$

The quality of recommendation depends on values $\pi_{k,i}(lm_{j_i})$, $\pi_{k,i}(lm_{j_i} - 2)$ and $\pi_{k,i}(lm_{j_i} + 2)$. A recommendation error Err_k is defined as follows:

$$Err_k = \sum_{i=1}^{4} w_i * err_{k,i} , \tag{18}$$

where for $j_i = 2,3,...,11$

$$err_{k,i} = (1 - \pi_{k,i}(lm_{j_i}) + 1/2 * (2 - \pi_{k,i}(lm_{j_i} - 2) - \pi_{k,i}(lm_{j_i} + 2))/3 , \tag{19}$$

for $j_i = 1$

$$err_{k,i} = (1 - \pi_{k,i}(-11) + 1/2 * (1 - \pi_{k,i}(-9))/2 \qquad (20)$$

and for $j_i = 12$

$$err_{k,i} = (1 - \pi_{k,i}(11) + 1/2 * (1 - \pi_{k,i}(9))/2 \ . \qquad (21)$$

5 Experiments

The experiments aimed at evaluating the performance of the proposed recommendation technique taking into account different context parameters. The evaluation was done by comparison of recommendation results obtained by the system, with the ones which match students the best according to the recommendation error defined by (18)-(21) taking into account error tolerance of 10%. The tests were carried out in the context of different courses represented by the set of weight parameters.

The experiments were done for two different sets of real students' data representing their dominant learning styles as was presented in SL model (see (11)). Data was collected from students, who filled self-scoring Index of Learning Styles (ILS) questionnaire [28] to obtain scores for the respective learning style dimensions. The first set contains data of 194 Computer Science students from different levels and years of studies, including part-time and evening courses. This data was used for building student groups to learn together. The second dataset, used for recommendation purpose, consists of 31 data of students studying the same master's course of Information Systems in Management.

The groups were created by application of unsupervised classification. There were considered three well known algorithms: partitioning - K-means, statistical - EM and hierarchical Farthest First Traversal (FFT). Clusters were created by using Open Source Weka software [29]. In the case of K-means algorithm 2 different distance functions: Manhattan and Euclidean were taken into account. Such approach allows to check the proposed technique for groups of different structures as well as different similarity degrees. During experiments, sets of student groups were created by each of the clustering algorithms taking into account 3,4,5,6 and 7 clusters to enable comparison of the recommendation technique performance depending on the number of groups and their sizes. Recommendations were built for five different courses, with context represented by vectors: $w1 = (1/4, 1/4, 1/4, 1/4); w2 = (1/2, 0, 1/2, 0); w3 = (0, 1/2, 1/2, 0); w4 = (0, 1/2, 0, 1/2); w5 = (1/3, 1/3, 1/3, 0)$.

Quantitative analysis of the results showed that the majority of the students obtained the best recommendations. No dependency between clustering schema and the percentage of properly assigned recommendations has been notified. The detailed results of quantitative analysis are presented in Table 1. The first column shows a group schema information including clustering technique and the number of clusters, where K1M and K2M denote K-means algorithm with Euclidean and Manhattan distance functions respectively. Next columns show percentage of students who obtained the best recommendations for different weight vectors.

Qualitative analysis aimed at checking the cases of the worst performance of the considered method (when percentage of correct recommendations were lower than 60%). Such situations take place for context represented by w2 end group schemas EM6 and FFT3; for context w3 and schemas K1M7 and K2M7; for context w5 and schemas K2M6 and K2M7. In all these cases wrong recommendations concern visual students. In contexts w2, w3 and w5 emphasis were done on the third learning style dimension. At the same time, EM6, K1M6, K2M6 and K2M7 schemas contain more than one group with the majority of visual students. In the case of FFT3 schema and the context w2, most of the wrong recommendations concern the biggest group of 142 instances. The analysis showed the big influence of the group sizes as well as weight parameters on recommendations. When the differences between cluster sizes were big, usually the larger group was suggested. In most of the cases, the recommendation errors took on the smallest values when all student attributes were of the same importance. The biggest error values can be notified when there are students, whose profiles do not match any of existing groups. In such a situation student should be recommended to contact a tutor.

Table 1. Quantitative analysis depending on group schemas

Group schema		w1	w2	w3	w4	w5
EM	3	100%	77.42%	90.32%	93.55%	87.10%
	4	96.77%	74.19%	64.52%	70.97%	64.52 %
	5	87.10%	77.42%	70.97%	64.52%	70.97%
	6	83.87%	54.84%	80.65%	70.97%	61.29%
	7	83.87%	64.52%	77.42%	67.74%	64.52%
FFT	3	100%	58.06%	90.32%	100%	87.10%
	4	96.77%	77.42%	90.32 %	96.77%	77.42%
	5	96.77%	80.65%	80.65%	90.32%	70.97%
	6	90.32%	70.97%	64.52%	90.32%	67.74%
	7	87.10%	70.97%	64.52%	67.74%	70.97%
K1M	3	87.10%	74.19%	83.87%	74.19%	77.42%
	4	96.77%	77.42%	83.87%	83.87%	90.32%
	5	87.10%	90.32%	80.65%	70.97%	67.74%
	6	90.32%	74.42%	74.19%	74.19%	74.19%
	7	96.77%	67.74%	45.16%	80.65%	67.74%
K2M	3	87.10%	77.42%	70.97%	67.74%	70.97%
	4	90.32%	77.42%	83.87%	87.10%	83.87%
	5	87.10%	87.10%	77.42%	67.74%	61.29%
	6	93.55%	80.65%	70.97%	74.19%	54.84%
	7	90.32%	77.42%	54.84%	74.19%	41.94%

6 Concluding Remarks

In the paper, building student group recommendations in the context of different courses is considered. Unlike the precise models considered in [4,5] we take into account the case where characteristics of the new student are not determined precisely which is the most common situation. In the current research application of fuzzy sets for representation of ill-known values is examined. Characteristics of the new student are represented by means of integer fuzzy numbers. For building recommendations a possibilistic representation of groups is proposed. The considered technique was examined in the case of students described by dominant learning styles and contexts described by respective weight vectors. Recommendation accuracy was measured by fuzzy logic based recommendation error. Experiments done for datasets of real students and sets of groups of different structures and sizes showed that for the majority of the students best recommendations were provided. However there have been noticed big influence of group sizes and weight parameters on the proposed suggestions. Quantitative and qualitative analysis showed that using of fuzzy logic for student and group modeling guaranteed good quality of recommendations. Such approach can be applied by educators during the process of group learning or course management.

Future research will consist in further development and implementation of the recommendation tool as well as examination of attributes of different types.

References

1. Schmidt, A., Winterhalter, C.: User Context Aware Delivery of E-learning Material: Approach and Architecture. J. Univers. Comput. Sci. **10**, 38–46 (2004)
2. Shakouri, H.G., Tavassoli, Y.N.: Implementation of a Hybrid Fuzzy System as a Decision Support Process: A FAHP-FMCDM-FIS Composition. Expert Syst. Appl. **39**, 3682–3691 (2012)
3. Zadeh, L.A.: Fuzzy Sets. Inform. Control **8**, 338–353 (1965)
4. Myszkorowski, K., Zakrzewska, D.: Using fuzzy logic for recommending groups in E-learning systems. In: Bădică, C., Nguyen, N.T., Brezovan, M. (eds.) ICCCI 2013. LNCS, vol. 8083, pp. 671–680. Springer, Heidelberg (2013)
5. Myszkorowski, K., Zakrzewska, D.: Building contextual student group recommendations with fuzzy logic. In: Cornelis, C., Kryszkiewicz, M., Ślęzak, D., Ruiz, E.M., Bello, R., Shang, L. (eds.) RSCTC 2014. LNCS, vol. 8536, pp. 358–365. Springer, Heidelberg (2014)
6. Dubois, D., Prade, H., Testemale, C.: Weighted Fuzzy Pattern Matching. Fuzzy Set Syst. **28**, 313–331 (1988)
7. Bobadilla, J., Serradilla, F., Hernando, A.: Collaborative Filtering Adapted to Recommender Systems of E-learning. Knowl-Based Syst. **22**, 261–265 (2009)
8. Severac, Z., Devedzic, V., Jovanovic, J.: Adaptive Neuro-Fuzzy Pedagogical Recommender. Expert Syst. Appl. **39**, 9797–9806 (2012)
9. Jovanowic, J., Gasewic, D., Knight, C., Richards, G.: Ontologies for Effective Use of Context in E-learning Settings. Educ. Technol. Soc. **10**, 47–59 (2007)
10. Yang, S.J.H.: Context Aware Ubiquitous Learning Environments for Peer-to-Peer Collaborative Learning. Educ. Technol. Soc. **9**, 188–201 (2006)

11. Das, M.M., Chithralekha, T., SivaSathya, S.: Static Context Model for Context Aware E-learning. Int. J. Eng. Sci. Technol. **2**, 2337–2346 (2010)
12. Adomavicius, G., Tuzhilin, A.: Context-aware recommender systems. In: Ricci, F., et al. (eds.) Recommender Systems Handbook, pp. 217–253. Springer Science+Business Media (2011)
13. Muehlenbrock, M.: Formation of learning groups by using learner profiles and context information. In: 12th International Conference AIED 2005, pp. 507–514 (2005)
14. Christodoulopoulos, C.E., Papanikolaou, K.A.: A group formation tool in an E-learning context. In: 19th IEEE ICTAI 2007, vol. 2, pp. 117–123 (2007)
15. Wang, J., Li, H., Zhao, H.: The Contextual group recommendation. In: 5th International Conference on Intelligent Networking and Collaborative Systems, pp. 127–131 (2013)
16. Masthoff, J.: Group recommender systems: combining individual models. In: Ricci, F., et al. (eds.) Recommender Systems Handbook, pp. 677–702. Springer Science+Business Media (2011)
17. Zheng, Y., Burke, R., Mobasher, B.: Recommendation with differential context weighting. In: Carberry, S., Weibelzahl, S., Micarelli, A., Semeraro, G. (eds.) UMAP 2013. LNCS, vol. 7899, pp. 152–164. Springer, Heidelberg (2013)
18. de Arriaga, F., El Alami, M., Arriaga, A.: Evaluation of fuzzy intelligent learning systems. In: Mendez-Vilas, A., et al. (eds.) Recent Research Developments in Learning Technologies. Formatex, Badajoz (2005)
19. Hogo, M.: Evaluation of E-Learners Behavior Using Different Fuzzy Clustering Models: A Comparative Study. International Journal of Computer Science and Information Security **7**, 131–140 (2010)
20. Essalmi, F., Ayed, L., Jemni, M., Kinshuk, Graf, S.: Evaluation of personalization strategies based on fuzzy logic. In: 11th IEEE International Conference on Advanced Learning Technologies, pp. 254–256 (2011)
21. Lu, J.: A personalized e-learning material recommender system. In: the 2nd International Conference on Information Technology for Application, pp. 374–379 (2004)
22. Vrettaros, J., Vouros, G.A., Drigas, A.S.: Development of an intelligent assessment system for solo taxonomies using fuzzy logic. In: Mellouli, K. (ed.) ECSQARU 2007. LNCS (LNAI), vol. 4724, pp. 901–911. Springer, Heidelberg (2007)
23. Almohammadi, K., Hagras H.: An adaptive fuzzy logic based system for improved knowledge delivery within intelligent E-learning platforms. In: 2013 IEEE International Conference on Fuzzy Systems (2013)
24. Chrysafiadi, K., Virvou, M.: Fuzzy Logic for Adaptive Instruction in an E-learning Environment for Computer Programming. IEEE T. Fuzzy Syst. **23**, 164–177 (2015)
25. Zadeh, L.A.: Fuzzy Sets as a Basis for a Theory of Possibility. Fuzzy Set Syst. **1**, 3–28 (1978)
26. Dubois, D., Prade, H.: The Three Semantics of Fuzzy Sets. Fuzzy Set Syst. **90**, 141–150 (1997)
27. Felder, R.M., Silverman, L.K.: Learning and Teaching Styles in Engineering Education. Eng. Educ. **78**, 674–681 (1988)
28. Index of Learning Style Questionnaire. http://www.engr.ncsu.edu/learningstyles/ilsweb.html
29. Witten, I.H., Frank, E.: Data Mining: Practical machine learning tools and techniques, 2nd edn. Morgan Kaufmann Publishers, San Francisco (2005)

Mining Social Behavior in the Classroom

Roberto Araya[1(✉)], Ragnar Behncke[1], Amitai Linker[1], and Johan van der Molen[1,2]

[1] Centro de Investigación Avanzada en Educación, Universidad de Chile,
Periodista Carrasco 75, Santiago, Chile
roberto.araya.schulz@gmail.com
[2] School of Mathematics, The University of Edinburgh, Edinburgh, UK

Abstract. Classrooms are very well suited for research on social interactions in the wild. Hundreds of hours of student interactions are rapidly accumulated. In this paper we analyze two months of video recordings from a fourth grade class, where the teacher and a sample of 3 students selected each day wore a mini video camera mounted on eyeglasses. The data reveals different gaze patterns between groups according to gender, subject, student grade point average, sociometric scale and time of day. The patterns that were found demonstrate the promising power of first-person video recordings for understanding social interaction in the classroom.

Keywords: Mining social behavior · Gaze patterns · Classroom practices · Video analysis · Educational process mining

1 Introduction

School plays a central role in our society. Students spend a large proportion of their time at school. In particular, the classroom is preponderant for understanding the learning process since it is where formal education takes place. Students spend thousands of hours inside the classroom every year. The classroom is therefore the natural laboratory for studying situated learning. Moreover, it is a very controlled environment that lends itself to studying and identifying recurring patterns of social interactions while learning, as well as the impact of teaching strategies.

When studying learning in the classroom, it is very important to identify patterns of social interaction between the teacher and students, as well as between students themselves [1],[5],[9],[17]. Both types of interaction have an effect on the classroom environment and on student learning. Most studies of classroom interaction are ethnographic studies that are manually recorded on paper by experienced teachers, principals or expert analysts during classroom observations. Some studies use video recordings of classes, with trained coders performing a detailed analysis of the video recording [1],[10],[11],[18]. These are third-person videos recorded with one or two cameras fixed to the wall or ceiling, or set up on tripods. As a process, analyzing the videos is labor-intensive, slow, tedious, and prone to mistakes. It is therefore not used very often. The problem with third-person videos is that they do not easily capture social interactions. It is difficult to work out who is looking at whom from these videos. Furthermore, it is

M. Núñez et al. (Eds.): ICCCI 2015, Part II, LNCS 9330, pp. 451–460, 2015.
DOI: 10.1007/978-3-319-24306-1_44

difficult to identify the emotional state of an individual from third-person videos. In order to identify such emotions, it is ideal to have a front-on view of the subject's face. In addition to this, it is also very difficult to capture dialogue using third-person cameras [2].

Given the limitations of traditional classroom observations and third-person videos, it is important to find alternatives that facilitate the detection of social interaction patterns. Furthermore, it is also important to explore the possibility of automating the detection process. In this paper, we review our experience of using first-person videos. These are videos obtained from mini video cameras mounted on eyeglass frames. While similar to head-mounted cameras [13], they are less distracting for fourth graders. With first-person videos it is much easier to identify who is looking at whom at any given moment. Third-person videos require several cameras in order to solve complex situations involving obstacles and people hidden from view. With first-person videos, on the other hand, the problem is much easier and only requires the camera of the observer. It is also simpler with first-person videos to identify long-distance gazes towards other individuals. Such gazes are normally difficult to identify with third-person videos due to the fact that accuracy decreases as the distance increases. With first-person videos it is also much simpler to identify the main person that is interacting with the subject.

There are also other first-person methods for studying social patterns. One very powerful method is the Experience Sampling Method [8]. With this method, a beeper or smartphone periodically asks the subject to complete a mini survey on their current subjective experience and to state with whom they are interacting. This is performed several times a day for each subject and continues for several weeks. Interesting personal perspectives are obtained with this instrument. However, the method can be a major distraction for students; particularly those below sixth grade. Furthermore, although the method records important personal and social information, it does not capture fast, automatic and unconscious social interactions. These unconscious interactions can reveal very important social dynamics.

2   Gaze Patterns Between Groups

Social interaction patterns can be very complex. For example, imitation, conflict, arguments, and collaboration include sequences of complex cognitive processes that are not completely observable. In this study, we start with very basic events that are almost always present in social interactions. We start with an indirect indicator of visual attention: the gaze between subjects. According to [4], faces regulate social interaction and face recognition is part of a specific mechanism evolved ancestrally for monitoring both human and non-human animals. Face detection and identification are very low-level social interaction events. They only capture who is looking at whom. Even though these are very basic interactions, they can give us a powerful picture of the nature of social interaction in the wild; particularly in the classroom. For example, it can give us information about gaze patterns between the teacher and students or among students according to gender, subject, time of day, grade point average, etc. These low-level interactions are very important and form the building blocks of more complex interactions. The situation is similar to that of medicine, where there are basic, low-level physiological measurements such as temperature, pressure and heart beats. While the thermometer, stethoscope and sphygmomanometer may only measure basic events, they are essential for detecting high-level patterns that are typical of different diseases. Similarly, in social interaction, patterns of high-level interactions such as conflicts or team work can be guessed from low-level events. For this reason, it is highly desirable to detect and correctly identify gaze patterns.

We are interested in identifying gaze patterns and detecting possible differences between groups. For example, gaze patterns between male students, gaze patterns between female students, gaze patterns from male to female students, and gaze patterns from female to male students. These patterns provide important input for comparing the social impact of different explicit teaching strategies such as class layout, changes in seating, short quizzes, collaborative learning, and project-based teaching. They also provide information about implicit teaching practices such as gender discrimination and student segregation. By doing so, we hope to develop tools that will help measure the influence of certain teaching practices on social interaction patterns in the classroom.

a)   b)

Fig. 1. a) Student wearing eyeglasses with mini video camera mounted on the frame and with the lenses taken out. After a couple of minutes they completely forgot they were wearing them. b) Classroom where the teacher and a sample of 3 students are wearing the eyeglasses.

3   Methods

In this paper, we report the low-level social interaction patterns provided by the gaze between subjects in the classroom. These are computed from first-person video recordings taken from mini video cameras mounted on eyeglass frames worn by the teacher and students. These are very cheap eyeglasses sold in consumer electronic stores for about USD 60. We adapted these slightly in order to record longer videos. The original batteries were removed and replaced with rechargeable, longer-lasting, lithium batteries so that they could be worn for the whole morning (from 08:15 am to 13:15 pm) without changing the battery. An extra 8Mb memory card was also added in order to be able to record the whole day. Additionally, as shown in fig.1 a, the lenses were removed from the eyeglasses in order to facilitate the view (the original lenses were not high quality) and also to minimize their weight. The recordings were manually downloaded at the end of each day.

The information corresponds to a fourth grade class from an elementary school in Santiago, Chile. This is a school with a student population that is qualified by the government as being low socio-economic status (SES). There are 36 students: 21 boys and 15 girls. The average age of the students was 10.5 years, with standard deviation of 214 days. All classes were taught by the same male teacher. The videos were recorded inside the classroom from September 26th, 2012, to November 27th, 2012, from 08:15 am to 15:40 pm. Lunch break was from 13:15 pm to 14:10 pm. The students' parents signed a consent agreement for their children to wear the video cameras and to record the information for the purposes of research. The subjects wore the eyeglasses for the whole school day, but only during class time. They took the eyeglasses off during breaks and lunch time, as well as to go to the bathroom. A total of 12,133 minutes of video were recorded, 2,600 coming from the teacher camera and the rest from student cameras. Every 30 seconds, a frame was sampled from the videos. The pictures that were obtained were then processed with the OpenCV software to detect the presence of faces. A total of 26,936 faces were detected. Each face was saved as an image file. Image files with faces were subsequently processed semi-automatically using the Google Picasa software in order to identify the subject in each file. Picasa identified a large proportion of faces, but many low-resolution images of faces were subsequently identified manually.

We define subject 1 as looking at subject 2 if the video frame taken at a given time interval reveals the face of subject 2. With this definition, a subject can be considered to be looking at several subjects at any given moment. In this paper, we do not look to identify the main subject of the observer's gaze. We simply count the number of faces in each screen shot. Each day, the teacher and a sample of 3 or 4 students wore the mini video cameras, as shown in fig. 1.b..

There are many more male students than female students, and only one teacher. The time that the subjects wore the eyeglasses also varies considerably from one user to another. It is therefore necessary to normalize the information. The idea is to estimate the intrinsic tendency to look at a certain group. This is defined as the proportion an observer looks at the group in the ideal case that each group comprises the same number of subjects. Two normalization algorithms were used. The first method assumes that the presence of each group is constant throughout the different sessions and days across the whole two-month period. The second normalization is a maximum likelihood estimation of the tendencies. These are parameters that have to be found by the CCCP [19] algorithm that tries to maximize the likelihood of the observed data. The likelihood function comes from a multinomial distribution. More precisely it is a Fisher's noncentral hypergeometrical distribution. The probability distribution assumes that the observations at different moments are independent. It also assumes that the probability of looking at any member of the group is the same for every member of the group. At moment n for a given observer, the probability distribution for the K groups is given as

$$X_n \sim Multinomial\left(\frac{M_{n,1}p_1}{\sum_{k=1}^{K} M_{n,k}p_k}, \frac{M_{n,2}p_2}{\sum_{k=1}^{K} M_{n,k}p_k}, \dots, \frac{M_{n,K}p_K}{\sum_{k=1}^{K} M_{n,k}p_k}\right)$$

Where $M_{n,k}$ is the number of members of the kth group at the moment n. The Maximum likelihood algorithm finds the tendencies p_k that maximizes the probability of the observed data. p_k represents the intrinsic tendency to look at members of the group k when all groups have the same size, and $M_{n,k}$ adjust the probability according to the sizes of the groups at moment n.

If there is high variation in the size of the groups over time then the estimation from both algorithms can be very different. There is some random chronic absenteeism within the school and therefore there is some variation in the classroom composition from one day to another. We used both algorithms and the results are very similar. We report the tendencies estimated with the simpler method that assumes a constant group size. Finally, to compute the tendency of a group of observers we average the tendencies of the individual observers.

In addition to the video recordings, a sociometric survey was given to the students where they had to assess the popularity of each classmate on a scale from 1 (least popular) to 7 (most popular). The students' overall grade point average (GPA) was also obtained. Additionally, three independent teachers of the school and that know the students, rated the students' upper-body strength and physical attractiveness in scale from 1(low) to 7(high) using a digital survey showing the name and two pictures of each student. The average rates were computed for each student for both features.

4   Results

All of the results obtained correspond to just one fourth grade class. Even though the videos were taken over two months and produced huge amounts of data, it is still just one class.

Since there was only one teacher, it is not possible to infer from the data any patterns that could be typical of every teacher, nor of fourth grade teachers or teachers from a particular kind of elementary school. The patterns detected with the teacher are only valid for this particular teacher. While they may be very interesting and could be very typical, they only illustrate the power of observation instruments based on first-person videos. However the student patterns that were obtained can be valid for students from other classes and schools. The big difference is that they come from 35 students; not just a single subject. There are several other important features of the data. The patterns are obtained from hundreds of hours of video recordings. This is much more data than is typically gathered for social studies. The data corresponds to several months of student behavior; not just a couple of sessions. In addition, this is not data that is gathered from surveys; they are video recordings that in principle can illustrate conscious and unconscious, as well as unintended behavior. Another important feature is that it is data in the wild. This is a big difference to the usual data obtained in laboratory conditions. Subjects can answer and behave very differently in laboratory conditions than in the classroom. The student patterns could therefore provide interesting findings that are present in typical fourth grade classrooms and are not otherwise detected.

First let us review only the data regarding the observers. There are some interesting patterns in the proportion of the recorded time that students look at other students. Male students tend to have more moments looking at others (39.4%) than female students (32.4%). Every time they look at a face, on average male students see 5.2 faces, while female students see 4.7 faces. These rates are much lower than the corresponding rate for the teacher. Every time the teacher looks at a face, he sees an average of 13.3 faces. From table 1 it can be seen that students tend to have more moments looking at other students during the first two sessions in the morning (46.9 %) than during the other two sessions (29.0%, 38.7% and 34.8%, respectively).

Table 1. Distribution of student video recordings throughout the day.

Block	Recorded time (min)	Time with face (min)	% time looking at others
08:15 to 09:45	2202	1032	46.9
10:00 to 11:30	3409	990	29.0
11:45 to 13:15	1608	623	38.7
14:10 to 15:40	2309	804	34.8

Students look at more people in math, language and technology classes (44.0%, 48.9%, 50.5%, respectively) than in the other subjects, such as computer lab (15.8%) and science (33.3%). More popular students (above average on the sociometric scale used) tend to have more moments looking at others (39.4%) than less popular students (31.9%).

Let us now review the data that includes both the observer and the observed. This is data about the subjects that the different groups of students are looking at. Remember that the tendencies defined and computed correspond to rates based on the ideal case that all of the groups were of equal size. It can be seen that the tendency for male students to look at the teacher (0.415) is higher than for female students (0.306) (Table 2). Both male and female students have the tendency to look at students of their own gender more than students of the opposite gender. Males have a tendency to look more at other males than at females (p-value is 0.01412). Females have a tendency to look more at other females than at males but it is not statistically significant (p-value is 0.05253). On the other hand, the tendency for males to look at other males is higher than the tendency for females to look at males (p-value is 0.002). Similarly, the tendency for females to look at other females is also higher than the tendency for males to look females (p-value is 0.002).

Table 2. Tendencies to look at classmates by gender.

Group	Tendency to look at male students	Tendency to look at female students	Tendency to look at the teacher
Male Students	0.346	0.239	0.415
Female students	0.266	0.428	0.306
Teacher	0.504	0.496	0

During the first four morning sessions, the tendencies for both male and female students to look at the teacher (0.453 for males and 0.355 for females) are much higher than for the remaining four sessions of the day (0.211 for males and 0.148 for females). Both differences are statistically significant, with p-values 0.006 and 0.035, respectively. Both males and females have the tendency to look at the teacher a lot more in math classes (0.463 for males and 0.621 for females) than in other classes. For females, the difference between math and language is statistically significant with p-value = 0.002. Interestingly, in art classes female students almost never look at the teacher (tendency is 0.116), instead mainly look at female classmates (tendency is 0.640). The tendency for females to look at other females is much higher in art than in language (p-value = 0.07) and math (p-value = 0.0087). This is also the case in language classes (tendency is 0.056), although they increase their tendency to look at their male classmates (tendency is 0.567). The tendency for females to look at males is higher in language than in math classes (p-value = 0.000008), and is also higher than in art classes (p-value = 0.00097).In contrast, the tendency for males to look at other males does not change much from one subject to another (all p-values are above 0.18). Similarly, the tendency for males to look at females does not differ between art and math classes, nor between math and language classes (p-values are 0.44, and 0.25, respectively).

According to grade point average (GPA), above-average students have a higher tendency to look at the teacher (0.417) than below-average students (0.324), but it is not statistically significant (p-value is 0.27). However, in math sessions (Table 3), above-average students in terms of GPA have a much higher tendency to look at the teacher (tendency is 0.593) than in other classes (e.g. 0.246 for art class), which is statistically significant with p-value = 0.004.

Table 3. Tendency to look at classmates, by academic performance (GPA) for different subjects.

Session – group	Tendency to look at high GPA students	Tendency to look at low GPA students	Tendency to look at the teacher
Art sessions – high GPA	0.473	0.281	0.246
Math sessions – high GPA	0.250	0.157	0.593
Language sessions – high GPA	0.309	0.283	0.407
Art sessions – low GPA	0.530	0.311	0.159
Math sessions – low GPA	0.298	0.241	0.461
Language sessions – low GPA	0.373	0.401	0.225

This pattern for above-average GPA students is also true for all the sessions of academic subjects (math, language, science and history). The tendency to look at the teacher is 0.474, which is higher than the tendency to look the teacher during the other sessions (tendency is 0.166). The difference is statistically significant with p-value = 0.0006. Popular students (above-average according to the sociometric scale used) also have a higher tendency to look at the teacher in math classes (tendency is 0.533) than in other classes (e.g. tendency for art is 0.242, the difference between math and art is statistically significant, with p-value =0.038). Interestingly, this teacher's tendency to look at above-average GPA students is 0.580, which is much higher than his tendency to look at the rest of the students (0.420). The same is true for his tendency to look at popular students (tendency is 0.585) than the others (tendency is 0.415).

Studies in evolutionary psychology have found interesting patterns of social interaction according to physical features in male and female adults. Upper-body strength in males and visual attractiveness in females provides better bargaining positions in conflicts and regulates emotions such as anger [15],[16]. There are also studies with a K12 student population. For example, according to [6], "the core of the self-schema is predicted, from an evolutionary perspective, to be referenced in terms of one's standing vis-à-vis peers and particularly for traits that have an evolutionary history, including physical abilities (greater importance for boys than girls), physical attractiveness (greater importance for girls than boys), social influence, and family status…the best predictor of global self-esteem from childhood to adulthood is perceived physical attractiveness [7] and not, for instances, grades in high school mathematics classes". To investigate the impact of these physical features on gaze patterns, three independent teachers subjectively scored the upper-body strength and visual attractiveness of the 36 students. By averaging these scores, an upper-body strength index and attractiveness index were computed for each student. It can be seen (Table 4) that the tendency for above-average upper-body strength students to look at classmates with above-average strength (0.574) is much higher than their tendency to look at below-average upper-body strength classmates (0.426). However p-value is only 0.093.

Table 4. Tendency to look at classmates according to upper-body strength.

Upper-body Strength	Tendency to look at students with high upper-body strength	Tendency to look at students with low upper-body strength
High	0.574	0.426
Low	0.497	0.503

The tendencies for both above- and below-average attractive students to look at attractive classmates (0.558 and 0.568) are higher than for below-average students (0.442 and 0.432, respectively), both findings are statistically significant, with p-values less than 0.001. In the case of male students, the difference in tendency is also pronounced. The tendency for male students to look at attractive female classmates is 0.537, which is higher than chance (0.5) with p-value less than 0.001. It is also much higher than their tendency to look at below-average attractive female classmates (0.463).

5   Conclusions

We have analyzed two months of first-person video recordings. Every day, the teacher and a sample of 3 students wore eyeglasses with mounted mini video cameras. By doing so, a big amount of first-person visual data was recorded. The huge data sets obtained offer unique possibilities to do reality mining [14] to find very valuable face-to-face social patterns that are very difficult to obtain by other means. The gaze patterns that are reported are very interesting. Some of them are specific to this particular teacher and classroom. Others are more general findings that help us understand what happens at the classroom level with fourth graders. First-person video analysis is an interesting social mining tool. We have seen how low-level gaze patterns inform interesting social interaction patterns in the classroom. We have explored and discovered different gaze patterns according to the time of day, subject, academic performance, social popularity, gender, and certain physical features of the students.

There are several extensions that would be very important to carry out in the near future. First, there is the need to record and analyze video data in other classrooms. The statistics obtained in this study correspond to a particular classroom. It is necessary to record interactions with other classes in order to claim general patterns that are independent of specific situations. It is crucial to test the current findings on larger populations of students and teachers. Another extension to be carried out in the future is to analyze the gaze in terms of the main face that appears on the screen capture. In this paper we have counted all of the faces that are captured at each moment. Identifying the main face and analyzing interactions with the main subject could lead to some other patterns that can complement the ones reported here. Another extension for future work is to control and measure the impact of different teaching strategies. This requires examining basic strategies like classroom layout, position of whiteboards or other materials and changes to seating strategies, to more complex strategies such as teaching a whole class or team work. Another important issue is to measure and research the synchronization of gaze interactions. What happens just after the teacher's gaze is directed at a student? Will she direct her gaze at the teacher? What happens just after a student gaze is directed at a classmate? Another extension of this work is the use of first-person videos to measure student learning in interpersonal skills. Educators [12] criticize standardized tests such as PISA, which exclude assessment of the learning of social interactions in the classroom, such as power and position, leadership, submission, and social rules. These types of learning are very important. However, they are very difficult to measure with typical standardized tests and extremely slow and laborious to do with traditional ethnographic and class observation visits. However, analysis of interactions from first-person videos opens up tremendous opportunities for measuring progress in social learning.

The first-person instrument used in this study provides a new methodology for observing classroom dynamics that could lead to the discovery of important classroom realities that are hard to detect for the naked eye. First-person video analysis can help to open the black box of classroom practices. It can also help teachers share their pedagogical capital [5].

The invention of the microscope in the 17th century caused a revolution in biology by revealing structures that could not be observed with the naked eye [3]. Similarly, the ability to record first-person videos and thereby to detect gaze patterns between groups can reveal unsuspected social patterns and significantly improve our understanding of the dynamics of classroom practices.

Acknowledgments. This work was possible thanks to the support of Jorge Morkoff, principal of Santa Rita de Casia School, and the enthusiasm and collaboration of Stenio Morales, the fourth grade teacher studied in this paper. We also thank Paulina Sepúlveda, Josefina Hernandez and Luis Fredes for software developed for this project; Avelio Sepúlveda for preliminary statistical analysis; Manuela Guerrero for analyzing videos; Marylen Araya and Manuela Guerrero for manual classification of faces obtained from the videos; and Basal Funds for the Centers of Excellence Project FB 0003 from the Associative Research Program of CONICYT.

References

1. Araya, R., Dartnell, P.: Video study of mathematics teaching in Chile. In: Procceedings 11th Internacional Conference on Mathematics Education Conference, Monterrey, Mexico. (2008)
2. Araya, R., Plana, F., Dartnell, P., Soto-Andrade, J., Luci, G., Salinas, E., Araya, M.: Estimation of teacher practices based on text transcripts of teacher speech using a support vector machine algorithm. British Journal of Educational Technology (2012)
3. Cohen, J.E.: Mathematics Is Biology's Next Microscope, Only Better; Biology Is Mathematics' Next Physics, Only Better. Only Better. PLoS Biol. **2**(12), e439 (2004). doi:10.1371/journal.pbio.0020439
4. Cosmides, L., Tooby, J.: Evolutionary Psychology: New Perspectives on Cognition and Motivation. Annu. Rev. Psychol. **64**, 201–229 (2013)
5. Cuban, L.: Inside the Black Box of Classroom Practice. Harvard Education Press. Change without Reform in American Education (2013)
6. Geary, D.: Educating the evolved mind: conceptual foundations for an evolutionary educational psychology. In: Carlson, J.S., Levin, J.R. (eds.) Psychological Perspectives on Contemporary Educational Issues. Information Age Publishing, Greenwich (2007)
7. Harter, S.: The development of self representations. In: Eisenberg, N. (Vol. ed.) Social, emotional, and personality development, vol. 3, pp. 1017–1095. Damon, W. (Gen. ed.) Handbook of child psychology, 5th edn. John Wiley & Sons, New York (1998)
8. Hektner, J., Schmidt, J., Csikszentmihalyi, M.: Experience Sampling Method. Measuring the Quality of Everyday Life. Sage Publications (2007)
9. Hiebert, J., Gallimore, R., Garnier, H., Givvin, K.B., Hollingsworth, H., Jacobs, J., et al.: Teaching mathematics in seven countries: Results from the TIMMS 1999 video study. U.S. Department of Education, National Centre for Educational Statistics (2003)
10. Kane The Reliability of Classroom Observations by School
11. LessonLab, Inc.: TIMSS-R Video Math Coding Manual (2003)
12. Labaree, D.: Let's measure what no one teaches: PISA, NCLB, and the shrinking aims of education. Teachers College Record (2013). http://web.stanford.edu/~dlabaree/publication 2013/Let's_Measure.pdf
13. Park, H., Jain, E., Sheikh, Y.: 3D Social Saliency from Head-Mounted Cameras (2012). http://www.cs.cmu.edu/~hyunsoop/nips/NIPS12.pdf
14. Pentland, A.: Social Physics. The Pinguin Press, New York (2014)
15. Sell, A., Tooby, J., Cosmides, L.: Formidability and the logic of human anger. PNAS **106**(35), 15073–15078 (2009)
16. Sell, A., Tooby, J., Cosmides, L.: Human adaptations for the visual assessment of strength and fighting ability from the body and face. Proc. R. Soc. London Ser. B **276**, 575–584 (2009)
17. Stiegler, J., Hiebert, J.: The Teaching Gap. The Free Press (1999)
18. TIMSS: Video Study of Eighth-Grade, Mathematics Teaching. National Center for Statistics. Department of Education. Institute of Educational Sciences (1999)
19. Yuille, A.L., Rangarajan, A.: The Concave-Convex Procedure (CCCP). Neural Comput. **15**(4), 915–936 (2003)

Using Concept Maps in Education of Immunological Computation and Computational Immunology

Martina Husáková[✉]

Faculty of Informatics and Management,
University of Hradec Králové, Hradec Králové, Czech Republic
martina.husakova.2@uhk.cz

Abstract. The paper presents development of the concept map and the concept map-based web portal for education in immunological computation and computational immunology. The concept map should help with learning of students who do not study medicine, pharmacy or similar study programs and support active learning on the basis of collaborative development of the concept map by students with the assistance of a teacher.

Keywords: Concept mapping · Immunological computation · Artificial immune system · Computational immunology · Immunology · Topic maps

1 Introduction

Concept maps are used for knowledge representation and organization in visual manner. This approach was proposed by J. D. Novak in the research program at the Cornell University in 1972 [1], [2], [3]. The research program had the aim to understand how the knowledge is changed during study of children. Concept maps are based on the meaningful learning theory of D. Ausubel [4] claiming that learning process and understanding of new knowledge is mainly based on the idea what person already knows about studied topic. Concept maps help with comprehension of complex topics [15]. They are used during brainstorming [13], [14], [15]. These maps are applied for knowledge sharing [12] and capturing tacit knowledge of experts for expert systems development [13], [15]. The paper uses concept maps for speeding-up learning in multi-disciplinary topics - immunological computation and computational immunology. Concept map is then used as a backbone for topic map-based educational web portal development. Collective development of the concept map by students and a teacher is going to be realized. The paper is organized as follows. The section 2 introduces the steps for educational concept map development. The section 3 presents the application of the approach for representation of the educational concept map. The section 4 describes the usage of the concept map in building the topic map playing the role of the digital repository of information and knowledge sources. The section 5 evaluates developed topic map together with future directions. The section 6 concludes the paper.

© Springer International Publishing Switzerland 2015
M. Núñez et al. (Eds.): ICCCI 2015, Part II, LNCS 9330, pp. 461–470, 2015.
DOI: 10.1007/978-3-319-24306-1_45

2   Approach for Concept Map Building

Concept map is very simple tool consisting of concepts (representing entities of application domain) and linking words (representing relations between concepts of the same or different domain (cross-links)). Relations specify context used for comprehension of concepts meaning. There is not pre-defined sequence of steps for concept maps development. General approach for concept mapping is mentioned in various papers and books [3], [13], [14]. There has to be some reason why the concept map is developed, i. e. the concept map should help to answer a question ("focus question") or to help in problem solving. Core concepts of application domain are identified. These concepts can be ordered according to priority or generality. Firstly, it is useful to identify only limited collection of concepts because of simplification. Concept mapping is as a never-ending story due to the richness and complexity of our world. Preliminary concept map can be built with the usage of paper or software (e. g. CmapTools, VUE, LucidChart, MindMapper, SmartDraw or Microsoft Visio). Concept map structure is revised after that because new concepts or relations between them can appear, change their position or can be deleted because of their uselessness. Cross-links can be added after concept map revision. Re-positioning of all concepts, linking words or cross-links improve the overall clarity of the concept map.

Concept map ComImmuno-AIS presented in the paper is built similarly as was mentioned above. Process of development is divided into the following steps considering multi-disciplinary modelling:

1. *Educational context* mentions purpose of the concept map, information or knowledge sources used for concept map development, approach for concept mapping (paper-based, software support) and keywords emphasizing topics covered by the concept map.
2. *Domain(s) of interest*: Short written description of the application domain(s) can clarify some parts of the concept map or highlight the most important parts.
3. *Backbone of a concept map*: Key concepts are identified with limited collection of linking-words. Concepts (linking-words, cross-links) can be ordered according to some criteria (e. g. importance, generality/specificity). Only crucial cross-links are mentioned in this step.
4. *Backbone extension with core concepts and linking*-words: The main attention is paid to the concept map extension with concepts and linking-words for each application domain without cross-links.
5. *Refinement*: It is useful to check whether the information and knowledge is represented correctly. The help of the expert is beneficial.
6. *Backbone extension with more cross-links*: More concepts and cross-links of various domains are represented in this phase of concept mapping.
7. *Refinement*: This process integrates changes in concept map structure according to the knowledge model of the student (teacher) or help of expert. Overall structure of concept map is revised.

3   Concept Map Development

The following paragraphs introduce development of the ComImmuno-AIS concept map according to the steps proposed in the section 2. Graph-based representation of concept mapping is included in the paper.

3.1   Educational Context

Purpose of the ComImmuno-AIS concept map is to facilitate education and learning of the multi-disciplinary areas computational immunology and immunological computation. These topics are presented in the introductory optional course of the artificial intelligence that is taught for students of Applied Informatics and Information Management at the University of Hradec Králové – Faculty of Informatics and Management (Czech Republic). Immunological computation (artificial immune systems (AIS)) and computational immunology is strongly multi-disciplinary, because supposes knowledge of immunology, biology, computer science, mathematics or statistics. Students perceive these two topics as really interesting, but they are very difficult for them, especially because of the necessity to know body of knowledge of cell biology and immunology. Immune system is very complex system which is hardly understandable by students which do not study medicine or similar study programme. Students receive background in computer science (software engineering, knowledge engineering), data analysis, techniques of artificial and computational intelligence at the university, but majority of students are not familiar with immunology, biology, medicine or pharmacy. The concept map should ease the orientation in these two domains of interest. Main aim of the map is to offer the most important information and knowledge in visual and user friendly form. The concept map is distributed in electronic form, i. e. as the concept map developed in the VUE (Visual Understanding Environment) [5] and as a topic map visualized with the usage of the Ontopia open source environment [6].

3.2   Domains of Interest

Natural immune system (NIS) is complex and adaptive system ensuring homeostasis maintenance in the inner environment of the living system. Complexity of the NIS is mainly cause by the amount of "players" and relations between them. Properties and behaviour of the NIS attract among others the computer scientists. Two main branches of investigation of this highly complex system exist: computational immunology (CI) and immunological computation (artificial immune systems (AIS)). Main aim of the CI is to investigate different aspects of the NIS behavior that cannot be observed directly or it is difficult to predict its manifestations under different conditions. Modelling, simulation, physical, mathematical or statistical methods are used for deeper understanding of the amazing complexity of the NIS. Specific processes of the NIS became the inspiration for improvement of existing algorithms or for development of the new class of bio-inspired algorithms – immunity-based algorithms [8]. Usage of these specific algorithms gives rise the AIS that can be defined as [7]:

"The adaptive system inspired by theoretical immunology and observed immune functions, principles and models which can be used for problem solving." These algorithms are used in pattern recognition, machine learning, planning or scheduling, optimization, (multi-criteria) decision making or computer security [9].

3.3   Backbone of Concept Map – Raw Concepts Mapping

The concept map ComImmuno-AIS is developed with the VUE software tool [5]. Core concepts are identified for each area of interest with limited collection of linking words and cross-links. The most general concepts are not included during raw concepts mapping. Three segments are created as a basis for the next concept mapping, see the Fig. 1. The following core concepts are identified for each application domain:

- Immunity-based algorithms (immunological computation)
- Immune process and immune system (immunology)
- Modelling approach (computational immunology)

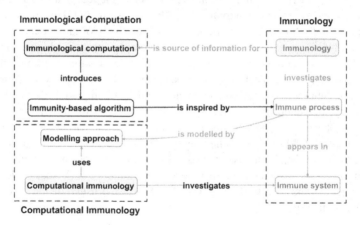

Fig. 1. The backbone of the concept map

3.4   Backbone Extension with Core Concepts and Linking-Words

Additional concepts and linking-words are added into the backbone of the concept map for each domain independently to other domains. Backbone extension with immunological concepts is depicted in the Fig. 2. Concept *Immune cell* is important especially because of the population-based algorithms dealing with decision whether the cell (pattern) has required properties or not. Concept *Immune organ* is included because some of organs are investigated by various computational techniques, i. e. in the computational immunology. *Immune processes* (positive/negative/clonal selection) are key concepts not only in the immunology, but also in the immunological computation (AIS). These immune processes were in the beginning of research in the AIS and became the inspiration for population-based algorithms.

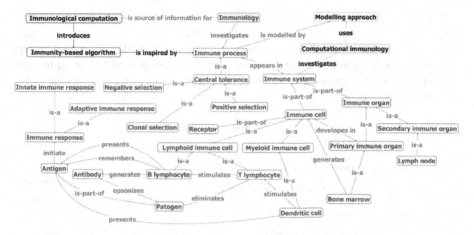

Fig. 2. Backbone extension – Immunology

Immunity-based algorithms are the core of the AIS research and education. Older network-based models are represented with younger population-based and dendritic cell-based algorithms in the concept map, see the Fig. 3. *Approach for modelling behavio*r occurring in the NIS is the key concept of the computational immunology, see the Fig. 3.

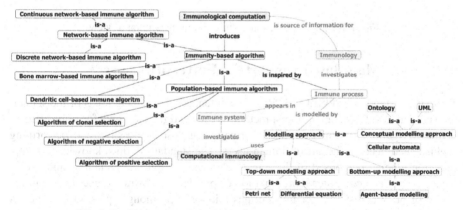

Fig. 3. Backbone extension – AIS and computational immunology

3.5   First Version of Concept Map

More multi-disciplinary concepts and cross-links are integrated after receiving sufficient backbone of the concept map ComImmuno-AIS (see section 2, step 6). The most important concepts and linking-words are represented for each application domain. Not all cross-links are mentioned in the Fig. 4 because of the clarity of the concept map. The Fig. 4 illustrates how three domains are interrelated. Revision of all elements is made in the concept map and the next multi-disciplinary concepts and relations between them are integrated.

The first version of the concept map is used for familiarization of students with multi-disciplinary topics during the introductory optional course of the artificial intelligence. During the lectures, students are going to actively participate in the concept map extension. They are going to study individually practical applications of the AIS and computational immunology, and integrate these facts into the ComImmuno-AIS concept map together a teacher.

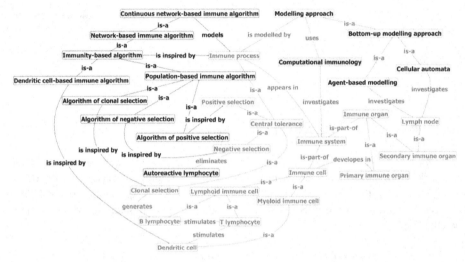

Fig. 4. Backbone extension – Multi-disciplinary point of view

4   Topic Map-Based Support for University Course

Topic maps standard ISO/IEC 13250 [10] is intended on realizing ideas of the semantic web – to develop efficient tools, methodologies and technologies for improvement of the navigation, finding relevant information as fast as possible and performing sophisticated tasks on behalf of human users. The standard defines knowledge representation schema composed of three main elements – topics, associations and occurrences. Topic is a computer representation of physical or abstract object that is in the centre of our interest. Association defines relationship among topics. Occurrence is the information source related to the topic. Information source can be internal (integral part of the topic map, e. g. definition, keywords, notes) or external (web document). Each of these elements has specific class – type. Having these elements, we can create topic map-based documents representing the metadata layer above the existing digital sources of different types. Developed multi-disciplinary concept map ComImmuno-AIS is used as a basis for development of an ontology that is a part of the topic map.

4.1   Topic Map Development

Topic map (ComImmuno-AIS-TM) development is divided on representation of the ontology, integration of instances and linking information sources with ontological concepts or instances. Representation of the ontology covers identification of topic types, association types and occurrence types on the basis of the concept map Com-Immuno-AIS. The following the most general topic types are distinguished: *Field of study*, *Approach*, *Process*, *Application*, *System*, *Person* and *Theory*, see the Fig. 5. Occurrence types are not included in the Fig. 5 because of the clarity of the ontology. Several revisions of the ontology were made before instances integration.

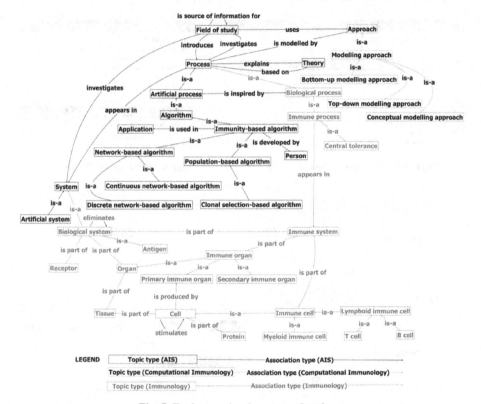

Fig. 5. Topic map development – Ontology

4.2   Topic Map as Digital Collection of Sources for Multi-disciplinary Topics

Source code of the multi-disciplinary topic map ComImmuno-AIS-TM is created in the LTM (The Linear Topic Map Notation) syntax [11]. The LTM is non-standardized syntax able to represent content of the topic map in much shorter piece of code in comparison to the standardized syntax XTM (XML Topic Maps). The Ontopia [6] open-source environment is used for management of the topic map ComImmuno-AIS-TM. The Ontopia offers the Omnigator that is used for verification of the

LTM-based topic map and browsing its content, see the Fig. 6. The Vizigator is applied for visualization the structure of the topic map as a graph, see the Fig. 7 where the graph-based view on the algorithm of the clonal selection and relating concepts is depicted.

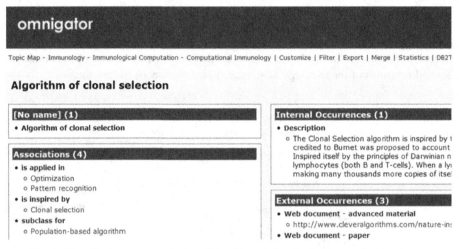

Fig. 6. Visualization of the topic type Algorithm of clonal selection in the Omnigator

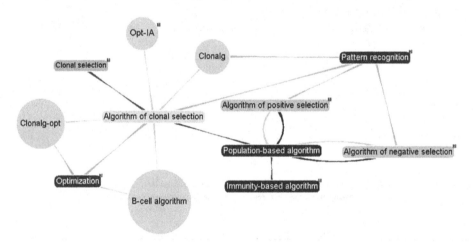

Fig. 7. Visualization of the topic type Algorithm of clonal selection in the Vizigator

5   Discussion

The developed topic map ComImmuno-AIS-TM is not in the final stage of the implementation. Building ontologies is iterative and never ending process because of the complexity of our world. Statistics of the ComImmuno-AIS-TM topic map is depicted on the Tab. 2. The topic map is going to be tested by students studying introductory course of the artificial intelligence. Comments and advices of students are going to be implemented into the next version of the topic map. Knowledge-based technologies II is the compulsory course for university students. The subject covers the semantic web technologies and topics of ontological engineering. The topic map ComImmuno-AIS-TM is used in this subject. It helps to explain how the ontology can be created and used in the practice. The Omnigator has limited possibilities in customization of the view on the content of the topic map. The Ontopia navigator framework is going to be used for web portal development based on the ComImmuno-AIS-TM topic map with the usage of the HTML, the Tolog query language, the JSP programming language and specific Tolog-based tags of the framework. The Tolog is used for extraction of required elements of the topic map. The elements are displayed with the Tolog-based tags of the framework in the web pages. The HTML and the JSP can be used for implementation of additional functionality for web pages.

Table 1. Statistics of the ComImmuno-AIS-TM topic map

Element of the topic map	Count
Topic types	17
Association types	16
External occurrence types	10
Internal occurrence types	8
Role types	29
Topics	152
Association	83
External occurrences	129
Internal occurrences	32

6   Conclusion

The paper presents the usage of concept mapping for representation multi-disciplinary topics immunological computation and computational immunology. The concept map supports education and learning of these areas presented in the introductory course of the artificial intelligence at the University of Hradec Králové – Faculty of Informatics and Management in the Czech Republic. The concept map ComImmuno-AIS is developed by systematic approach. Multi-disciplinary concept map is used for the topic map ComImmuno-AIS-TM development. The content of the LTM-based topic map is visualized with the Omnigator browser and the Vizigator plugin. The topic map is not in the final stage of its development. The collaborative development of the concept map is going to be realized where students are going to extend the concept map Com-Immuno-AIS together with a teacher during lectures.

Acknowledgements. The paper is supported by the project No. CZ.1.07/2.3.00/30.0052 - Development of Research at the University of Hradec Králové with the participation of postdocs, financed from EU (ESF project) and the project of specific science "Application of Artificial

Intelligence in Bioinformatics" (1/2015) at the University of Hradec Kralove, Faculty of Informatics and Management, Czech Republic.

References

1. Novak, J.D.: Concept maps and Vee diagrams: two meta-cognitive tools to facilitate meaningful learning. Instructional Science **19**, 29–52 (1990)
2. Novak, J.D.: Meaningful Learning: The Essential Factor for Conceptual Change in Limited or Inappropriate Propositional Hierarchies Leading to Empowerment of Learners. Science Education **86**, 548–571 (2002)
3. Novak, J.D., Cañas, A.J.: The Theory Underlying Concept Maps and How to Construct Them. Technical Report IHMC CmapTools2006-01, rev. 2008-01. Florida Institute for Human and Machine Cognition, Pensacola. http://cmap.ihmc.us/publications/researchpapers/theorycmaps/theoryunderlyingconceptmaps.htm
4. Ausubel, D.P.: The psychology of meaningful verbal learning. California Medicine **99**, 434 (1963)
5. Kumar, A., Saigal, R.: Visual understanding environment. In: Proc. of the 5th ACM/IEEE-CS Joint Conference on Digital Libraries, pp. 413–413. ACM New York (2005)
6. Ontopia: Ontopia. http://www.ontopia.net/ (cit. February 15, 2015)
7. De Castro, L.N., Timmis, J.: Artificial Immune Systems: A New Computational Intelligence Approach, 1st edn, p. 380. Springer-Verlag, London (2002)
8. De Castro, L.N.: Fundamentals of Natural Computing: Basic Concepts, Algorithms, and Applications, 1st edn., p. 696. Chapman and Hall/CRC (2006)
9. Dasgupta, D., Niño, L.F.: Immunological Computation: Theory and Applications, 1st edn., p. 277. CRC Press, Auerbach Publications (2009)
10. ISO: ISO 13250-2 Topic Maps – Data Model. http://www.isotopicmaps.org/sam/ (cit. February 20, 2015)
11. Garshol, L.M.: The Linear Topic Map Notation. Definition and introduction, ver. 1.3 (2006). http://www.ontopia.net/download/ltm.html (cit. February 4, 2015)
12. Hussain, H., Shamsuar, N.R.: Concept Map in Knowledge Sharing Model. The International Journal of Information and Education Technology **3**(3), 387–400 (2013)
13. Rachlis, B., et al.: Using concept mapping to explore why patients become lost to follow up from an antiretroviral therapy program in the Zomba District of Malawi. BMC Health Services Research **13**, 210 (2013). http://www.biomedcentral.com/1472-6963/13/210
14. Christodoulou K.: Collaborative On-line Concept Mapping (project report). School of Computer Science – The University of Manchester, p. 31 (2010). https://studentnet.cs.manchester.ac.uk/resources/library/thesis_abstracts/BkgdReportsMSc10/Christodoulou-Klitos.pdf
15. Lobb, R., Pinto, A.D., Lofters, A.: Using concept mapping in the knowledge-to-action process to compare stakeholder opinions on barriers to use of cancer screening among South Asians. Implementation Science **8**, 37 (2013) http://www.implementationscience.com/content/8/1/37

Framework of Design Requirements for E-learning Applied on Blackboard Learning System

Aneta Bartuskova[1(✉)], Ondrej Krejcar[1], and Ivan Soukal[2]

[1] Faculty of Informatics and Management, Center for Basic and Applied Research, University of Hradec Kralove, Rokitanskeho 62, 500 03 Hradec Kralove, Czech Republic
aneta.bartuskova@uhk.cz, ondrej@krejcar.org
[2] Faculty of Informatics and Management, Department of Economics, University of Hradec Kralove, Rokitanskeho 62, 500 03 Hradec Kralove, Czech Republic
ivan.soukal@uhk.cz

Abstract. Existing evaluations of learning systems are mostly based on a subjective feedback from end users, in order to measure their satisfaction and experience with the system. The results usually only indicate which areas are weaker than others; comments obtained from users are often casual and fuzzy. Therefore we propose a systematic review of usability and visual appeal by experts in HCI to complement current practice. Our solution of assessment is based on a framework of design requirements for e-learning purposes, which provides a concise easy to follow checklist, covering all significant areas of design assessment. This approach offers factual design issues which should be dealt with and can be especially useful in cases, where feedback from users is hard to obtain or it does not reveal any useful information.

Keywords: E-learning · LMS · Design evaluation · Usability · Blackboard

1 Introduction

Design of learning environment and content has a significant influence on learner's performance, experience, satisfaction etc. In this paper we present an application of framework of design requirements, which reflect two main design dimensions – usability and visual appeal. The evaluation by this framework was performed on selected computer science courses in Blackboard Learn with two main objectives: a) validate the proposed framework and show an application in real scenario, b) reveal potential design issues in Blackboard Learn, which is one of the most popular LMSs.

Existing evaluations of learning systems are mostly based on a subjective feedback from end users. These results however usually only indicate which areas are weaker than others; comments obtained from users are often casual and fuzzy. Our proposal is a systematic review of usability and visual appeal by experts in human-computer interaction to complement current practice. This solution provides a concise easy to follow checklist, covering all significant areas of design assessment, based on a framework of design requirements [1]. This approach takes use of collective intelligence and knowledge in the area of design and usability, since all web-based systems share common principles for information retrieval and interaction.

© Springer International Publishing Switzerland 2015
M. Núñez et al. (Eds.): ICCCI 2015, Part II, LNCS 9330, pp. 471–480, 2015.
DOI: 10.1007/978-3-319-24306-1_46

2   State of the Art

2.1   Usability and Visual Appeal

Usability and visual appeal (also aesthetics) are widely researched design constructs in human-computer interactions (HCI). Fee specified design as one of the three essential component parts of e-learning [2]. We can distinguish between classical and expressive design [3]. The former means orderly traditional design, which is directed by standards in web design, the latter is created creatively with a goal to break the rules. The relationship between classical aesthetics and functionality is being consistently established across studies [4]. There are many studies, which investigate a connection between usability and aesthetics in HCI [5].

Usability can be defined as a quality attribute that assesses how easy user interfaces are to use [6]. Usability measures are divided into the three groups: the measures of effectiveness, efficiency and satisfaction of the ISO 9241 standard for usability [7]. Most widely used method for measuring usability is user testing, also an inspection and an inquiry [8]. Usability can be in fact measured as both subjective and objective factor. Subjective evaluation can be realized by end-users with inquiries such as "how easy was working with the system", "would you use the system again", "do you feel that the system was easy to use" etc. Objective measurements are for example the task time, number of errors or severity of the errors during fulfilling tasks in the system. It was confirmed that providing web users with a usable environment can lead to significant savings and improved performances [9, 10].

Several terms for visual appeal are used alternatively in its research, e.g. aesthetics, attractiveness, beauty etc. Schenkman and Jönsson uncovered that beauty is the best predictor for an overall user judgement [11]. Research of van der Heijden introduced a perceived visual attractiveness of website, which influences usefulness and ease-of-use (in other words usability) [12]. Aesthetics in disciplines such as web design is a key predictor of an overall impression or user satisfaction [13], its role in e-learning is more of a backstage unobtrusive nature [1]. Therefore a classical design instead of expressive is usually utilized in learning systems. Good visual design is about solving problems, not drawing attention [14]. Visual aesthetics of computer interfaces in general is a strong determinant of users' satisfaction [3].

2.2   Learning Management Systems

A learning management system (LMS) is software used for delivering, tracking and managing education [15]. LMSs usually include a wide variety of features that can be utilized to support both distance and traditional teaching [16]. Learning management system can be interchanged for course management system (CMS). These systems are an increasingly important part of academic systems in higher education [9] and generally when conducting educational activities [17]. Cavus and Alhih discussed positive implications of LMSs in education [15]. Among the most popular LMSs belong commercial Blackboard and open source Moodle.

There are many studies which dealt with a comparison of LMSs. Unal and Unal conducted a comparative usability study on two different course management systems, BlackBoard and Moodle [9]. This study focused on a feedback from participants - their experience, satisfaction, what they liked and disliked. Similarly as the study of Machado and Tao, which compared Moodle and Blackboard from the view of a user experience [18]. Carvalho et al. criticised previous studies, which were comparing Blackboard and Moodle, that they were focused only on students' perceptions [19]. Therefore they assessed the extent and depth of use of the two LMSs based on students' perceptions and experience with both Blackboard and Moodle [19]. While some of the studies claimed that Moodle was preferred [9], other results were favourable for Blackboard [19]. Data and consequently also results of these studies were based on a subjective experience and evaluation by students.

Other studies are focused on a standalone assessment of one learning system rather than comparing several systems. Escobar-Rodriguez and Monge-Lozano researched the acceptance of Moodle technology by business administration students and provided insight on factors that contribute to intention to adopt this technology [20]. Oztekin et al. introduced a UseLearn method which supports the determination of usability problems by criticality metric analysis [21]. As in the previous studies, data were collected from the end-users by Likert-type evaluations. There are also studies which assessed LMSs objectively by quantifiable parameters. E.g. Caminero et al. compared LMSs according to their performance like memory or CPU usage [17].

3   Methodology

3.1   The Proposed Approach

The existing studies were mostly based on a subjective feedback from end users. This is of course a well-founded method for measuring user satisfaction, perceived ease of use etc., sometimes accompanied with open-ended inquiries, which can e.g. gather comments from users about what they liked or disliked. However evaluation from users on Likert scale only indicates which areas are weaker than others and comments from users are often casual and fuzzy. Most of all, results from such studies usually do not offer factual design issues which should be dealt with.

As a counterpart to the existing studies, we propose a systematic review of usability and visual appeal by experts in HCI. This type of assessment is very useful because it quickly reveals design issues based on well-founded recommendations and best practice. Our solution of assessment is based on a framework of design requirements for e-learning purposes, combining criteria of both usability and aesthetics [1]. In this paper we present an application of this framework on learning courses in one of the most popular LMSs - Blackboard Learn. Naturally the person that performs the evaluation has to have a good practice in HCI field of knowledge; in that case the presented framework provides a concise checklist, which is easy to follow and covers all significant areas of design assessment.

This approach could be especially useful in cases, where feedback from users is hard to obtain or it does not reveal any useful information. It can be also used repeatedly during the development cycle to save costly repairs of the final product.

3.2   Selection of Learning Courses

LMS Blackboard Learn was selected for an application of framework of design requirements for e-learning purposes. This system is primarily used at the Faculty of Informatics and Management of the University of Hradec Kralove in the Czech Republic for managing learning courses, currently the release 9.1. The assessment was performed on selected computer science courses. These courses are used at the faculty as complementary to traditional courses, either to function as a blended learning or just a repository of learning materials.

Out of all available courses in Blackboard Learn in the scope of the faculty, we have selected those which:

- were intended for Czech students in the full-time study program
- were administered for the current school term
- were categorized under computer science or similar sciences

Courses in this selection were the most complete and updated. The computer science topic was specified to narrow down the selection. The total of 18 learning courses was selected for the detailed assessment.

3.3   Design Requirements

For the purposes of this study we have used a list of design requirements of usability and aesthetics for e-learning [1]. Proposed requirements and recommendations are grouped into five categories – legibility, design consistency, visual presentation, content arrangement and content adjustment – each with a specified distinct purpose, summarized in [Table 1]. Dominating design dimensions (usability and aesthetics) are marked in the respective columns.

Table 1. Usability and aesthetics design requirements [1]

Usability	Aesthetics	Requirement	Purpose / effect
x		Legibility	Readability of content
x	x	Design consistency	Representation and recall based on mental models
	x	Visual presentation	Overall appeal and feeling
x		Content arrangement	Understanding and recognition
x	x	Content adjustment	Learnability of content

These design categories are influenced by both LMS environment and also particular implementation of the learning course. Legibility will be mostly influenced by the settings of LMS, because default values are rarely changed even if it is possible. However legibility of additional files (like presentations or PDFs) is a different matter and media legibility also depends on a quality of uploaded files. Visual presentation is also mostly dependent on LMS theme, but there are some limited choices for instructors to customize their course. Design consistency can be accessed in both levels – consistency across the courses (in LMS environment) and consistency within the particular course. Content arrangement is also dealt with at both levels – basic arrangement is provided by LMS (layout, navigation) and the content is further organized in the particular course. Finally, content adjustment primarily depends on the instructor skills and therefore is dealt with in the particular course; however the system needs to facilitate easy implementation of these principles. Individual attributes of design requirements are displayed in [Table 2], criteria for each attribute are discussed in more detail in [1].

Table 2. Individual attributes of design requirements, rearranged from [1]

Requirement	Individual attributes
Legibility	Typeface, Font size, Tonal contrast, Spacing, Alignment, Line length, Media legibility
Design consistency	Functional consistency, Aesthetic consistency, Consistency in layout and structure
Visual presentation	Aesthetic design, Colour, Colour contrast, Relevant graphics, Supportive graphics, Visual hierarchy
Content arrangement	Layout, Organization, Navigation mechanism, Multiple presentation media
Content adjustment	Chunking, White space, Gestalt Proximity, Emphasis mechanisms, Noise reduction

3.4 Evaluation Procedure

Although Blackboard Learn environment is well known to both authors, we have dedicated enough time to visual assessment and interaction with the system for this evaluation. As was already mentioned, both features on the system level and features implemented in individual learning courses were analyzed, as they both participate on the final design. The distinctions will be commented in respective sections. Discovered issues apply to the version of the system which was evaluated in this paper and particular implementation in selected computer science courses. The system was evaluated with a regard to the end user – the learner.

Some design requirements could be analyzed through visual appearance, in case of e.g. functional consistency or navigation, the actual use of the system was performed in a sufficient amount and diversity of use. Any difference of opinion during the assessment was resolved through negotiation between the two authors similarly as in a research by Hew and Cheung [22].

4   Evaluation of LMS Blackboard Learn

4.1   Legibility

Legibility is the first category for evaluation. Recommended characteristics are: simple character shape (typeface), font size bigger than 10pt, high tonal contrast, line spacing 1/30 of line length, line length 40-60 characters and left-aligned text [14].

Blackboard Learn uses standard sans-serif typeface, spacing and left alignment, which complies with the recommendations. The smallest font used in this system is 10.6px (e.g. in top navigation bar), which is very near the recommended minimum, the default font size in the content area is set to 13px. These values were accessed by developer tools in a browser. Legibility is very important in e-learning, because reading on a computer screen is harder than reading paper [23]. Unfortunately the system does not facilitate font size customization, so users with problems reading on screen or visual impairment have to use a magnifier integrated within the browser, which often breaks the layout and requires additional scrolling. Instructors can change font size only in the text fields, but they cannot influence font size of e.g. headlines or navigation items. However increasing font size in the text fields should be avoided as well, because since line height cannot be adjusted, larger fonts result in crowded text. Font size, colours and other styles can be in fact changed in Blackboard theme by the instructors. However it requires familiarity with CSS programming and design [26] and it affects styles as a whole. There is no such customization option for the learner.

Web Content Accessibility Guidelines (WCAG) 2.0 is a technical standard with a goal to make web content more accessible to people with disabilities [24]. WCAG 2.0 level AA requires a contrast ratio of 4.5:1 for normal text (3:1 for large text), level AAA requires a contrast ratio of 7:1 for normal text (4.5:1 for large text) [25]. Colours and consequently contrast can be changed by customization of colour schemes, but in our experience the default scheme is usually kept. Therefore we have checked contrast of default colour scheme in headings and content by Color Contrast Checker [25]. While contrast ratio for content was computed as 21:1 (which is the highest possible ratio for combination black and white) and passed both WCAG level AA and AAA, contrast ratio for headings (white text on blue-gray background) was computed as 2.87:1. This value fails both levels for both standard and large text. This means that headings in default scheme have insufficient contrast and readability.

Line length is a variable characteristic, because width of the content area partially adapts to size of the screen. At the minimum width of content area, the line of text has around 60 characters without spaces, which is in order. However on the screen with width of 1920px, line length consists of around 200 characters, which is too much. Blackboard has partially fluid layout, not responsive, so the width of the content area is the only flexible dimension. Media legibility depends solely on the quality of uploaded content, so it cannot be evaluated as the characteristics of the system.

4.2   Design Consistency

Aesthetic consistency and consistency in layout and structure was assessed by visual appearance, functional consistency was tested by interaction with the system. Several issues were encountered during consistency analysis. One of the main reasons for these issues is that Blackboard offers more different ways of handling the content with similar result (e.g. Page, Content item or File content type, also Learning module or Content folder). The instructors usually do not spend time to learn the best ways to work with the system, they often use the first functionality which brings an acceptable result, and they are often not consistent with their choice next time. It was already researched that confusion about the uses of available features and the mismatch of course features to learning expectations can, and often do, impede learning [9]. This behaviour then causes both visual and functional inconsistencies not only across the courses (caused by diverse strategies of instructors), but also within one course.

We have checked in Blackboard Learn that before the user clicks on some learning item, he does not know if that link would lead to a new page within the LMS, to some file opened in a new window, to some file opened within the LMS, if it triggers downloading the file or some other behaviour. This is quite a serious usability issue, considering that the majority of learning materials in Blackboard is available through some kind of link, either in the navigation bar or in the content area (with the exception of content in LMS's text fields, which is displayed directly). We have also encountered the same icons used for different types of files (visual inconsistency).

LMSs like Blackboard make the inconsistencies possible and likely. Inconsistency in layout and structure arises mainly from diverse approaches of instructors and used functionalities; however this is a necessary consequence of instructors' freedom in managing their learning courses, so it cannot be regarded as a shortcoming.

4.3   Visual Presentation

Aesthetic design of user interface in Blackboard is classical and unobtrusive, default colours are moderate, graphics is used relevantly mostly in a form of icons. Visual hierarchy in a form of visual contrast between blocks of content is limited, but since there are only several main segments (primarily content area, local navigation and system navigation), the distinction between them is relatively maintained. This was mostly a functional evaluation of aesthetics, which can be performed reasonably objectively; otherwise visual appeal (as the only category which is determined mostly by aesthetics and not usability) is very subjective so it cannot be assessed credibly, although its influence in HCI is undisputed.

4.4   Content Arrangement

Content arrangement is the next category of requirements, which includes layout, organization and navigation. The basic layout of any learning course in Blackboard is: system navigation bar at the top, local navigation in the left column and course content in the remaining area. Navigation items, content and its organization naturally differ in individual courses. We however noticed a common issue of inefficient content arrangement in learning courses in Blackboard. When creating a new course, a group of navigation items is predefined. Many instructors keep these navigation items in their courses even when they are empty and/or not used.

Also functionalities of the default "homepage" were not very useful in the majority of analyzed courses, but they were present usually as the opening screen in most of these courses. This is of course not as much a problem of the system as of the attitude of instructors. However the system could prevent these issues e.g. by customized initial build of the course instead of packing the new course with features and expect the instructor to browse through them one by one and delete unnecessary items.

As the next issue we consider organization and navigation in the learning courses, which is controlled solely by the instructor. Users can neither sort the items nor search for specific text arrays or files. Sorting is possible only in the section "Course Files" (central file storage area for a single course), but sorting through all files from the course including duplicities from imports and excluding anything which is not a file (e.g. content in LMS's text fields) is not very useful, and again the search option by name or at least file type is missing.

Another arrangement issue which makes orientation in the system difficult is support of duplicities, e.g. users can find the same file in several places including "Learning modules", "Course files" or "Media library". However this replication of files was revealed to be often incomplete and inconsistent, increasing interaction cost for the student who tries to get access to all available materials in the particular topic.

4.5   Content Adjustment

Blackboard supports chunking (or rather progressive disclosure) in the sense of content division into navigation items and folders. Use of white space depends on the theme and amount of content. The default theme is rather packed to support high density of information, however if there is not enough content in a space dedicated for that particular type of content, then undesirable white space appears. This cannot be remedied because of the lack of customization and manifests in both levels of user interface (system and individual course). E.g. in the welcome screen of the system, you can rearrange modules into three columns, however you cannot delete one column if you don't need it and if it is not filled with modules, you get an ineffective layout and unwanted white space. In the content area within the particular learning course, default layout causes unbalanced combinations of long text lines and narrow lists of links with large unused areas, which is the most apparent on wider screens.

Gestalt Proximity to support learning can be utilized more in learning materials in a form of files than in scope of the system. Blackboard is used more as a repository of learning materials probably because it does not support advanced means of styling the content such as e.g. Microsoft Word or Powerpoint. Limited emphasis mechanisms are available, headlines and links are emphasised automatically in the predefined style. To conclude, since this category of requirements is about learnability of content, these principles will be utilized mostly in the learning materials (files) rather than within the system. Maybe because the LMS does not offer enough support for content adjustment, storing learning content in files is still preferred over text fields in LMS.

5   Results and Conclusions

The design issues revealed during the assessment and discussed in the previous section are summarized in [Table 3]. This table presents a list of factual design issues, which can be regarded as the result of the evaluation.

Table 3. Summarized results of an evaluation of LMS Blackboard Learn

Requirement	Usability Issues
Legibility	small font (10.6px) which cannot be customized
	insufficient contrast (2.87:1) in the default colour scheme
	oversized uncomfortable line length (200 chars)
Design consistency	inconsistent appearance of links within the course (visual)
	inconsistent behaviour of links within the course (functional)
Content arrangement	predefined navigation items, empty or without purpose
	the lack of navigation and organization options for users (sorting of items, searching for text or files within the course)
	support of duplicities which make the interaction cost higher
Content adjustment	unbalanced layout with large unused areas of white space
	the lack of advanced means of styling the content

The results indicate that several design issues were revealed by the assessment of Blackboard Learn, which can be performed better to comply with approved standards and best practice. Most of the encountered problems in the area of legibility and content adjustment could be solved by changes in default CSS styles. Design consistency could be generally solved by reducing available functions for content management and making the remaining ones more versatile and easy to use. Content arrangement would also benefit from reduction of complexity. Navigation would be enhanced by adding sorting and searching functions. Nearly all areas would benefit from customization in both visual styles and organization.

This paper presented a different approach to evaluation of web-based learning systems than usual method of inquiries on user satisfaction, based on a framework of design requirements. The performed assessment revealed a set of factual design issues, which can be used for enhancing both usability and visual design of the system in order to improve user (learner) performance, experience and satisfaction.

Acknowledgments. This paper was written with financial support from specific university research funds of the Czech Republic Ministry of Education, Youth and Sport allocated to the University of Hradec Kralove, Faculty of Informatics and management, project no. 2/2015, order 2102.

References

1. Bartuskova, A., Krejcar, O.: Design requirements of usability and aesthetics for E-learning purposes. In: Sobecki, J., Boonjing, V., Chittayasothorn, S. (eds.) Advanced Approaches to Intelligent Information and Database Systems. SCI, vol. 551, pp. 235–245. Springer, Heidelberg (2014)
2. Fee, K.: Delivering E-Learning: A Complete Strategy for Design, Application and Assessment. 4th edn. Kogan Page (2009). ISBN 978-0749453978
3. Lavie, T., Tractinsky, N.: Assessing dimensions of perceived visual aesthetics of web sites. International Journal of Human-Computer Studies **60**, 269–298 (2004)

4. Mbipom, G., Harper, S.: The interplay between web aesthetics and accessibility. In: ASSETS 2011, pp. 147–154. ACM Press, New York (2011)
5. Bartuskova, A., Krejcar, O.: Evaluation framework for user preference research implemented as web application. In: Bădică, C., Nguyen, N.T., Brezovan, M. (eds.) ICCCI 2013. LNCS, vol. 8083, pp. 537–548. Springer, Heidelberg (2013)
6. Nielsen, J.: Usability 101. http://www.nngroup.com/articles/usability-101-introduction-to-usability/ (retrieved March 20, 2015)
7. ISO: Ergonomic requirements for office work with visual display terminals (VDTs)-Part 11: guidance on usability - Part 11: guidance on usability (ISO 9241-11:1998)
8. Fernandez, A., Insfran, E., Abrahão, S.: Usability evaluation methods for the web: A systematic mapping study. Information and Software Technology 53(8), 789–817 (2011)
9. Unal, Z., Unal, A.: Evaluating and comparing the usability of web-based course management systems. Journal of Information Technology Education, 1019–1038 (2011)
10. Kibaru, F., and Dickson-Deane, C.: Model for training usability evaluators of E-learning. In: Proceedings of World Conference on E-learning in Corporate, Government, Healthcare, and Higher Education, pp. 517–522 (2010)
11. Schenkman, B., Jönsson, F.: Aesthetics and preferences of web pages. Behaviour and Information Technology 19(5), 367–377 (2000)
12. van der Heijden, H.: Factors influencing the usage of websites: the case of a generic portal in the Netherlands. Information & Management 40(6), 541–549 (2003)
13. Hassenzahl, M.: The interplay of beauty, goodness, and usability in interactive products. Human-Computer Interaction 19(4), 319–349 (2004)
14. Horton, W.: E-learning by Design. Pfeiffer (2006). ISBN 978-0787984250
15. Cavus, N., Alhih, M.S.: Learning Management Systems Use in Science Education. Procedia - Social and Behavioral Sciences 143, 517–520 (2014)
16. Islam, A.K.M.N.: Investigating E-learning system usage outcomes in the university context. Computers & Education 69, 387–399 (2013)
17. Caminero, A.C., et al.: Comparison of LMSs: Which is the Most Suitable LMS for my Needs? International Journal of Emerging Technologies in Learning 8(S2), 29 (2013)
18. Machado, M., Tao, E.: Blackboard vs. Moodle: Comparing user experience of learning management systems. In: Education, pp. 7–12 (2007)
19. Carvalho, A., Areal, N., Silva, J.: Students' Perception of Blackboard and Moodle in a Portuguese University. British Journal of Education Technology 42(5), 824–841 (2011)
20. Escobar-Rodriguez, T., Monge-Lozano, P.: The acceptance of Moodle technology by business administration students. Computers & Education 58(4), 1085–1093 (2012)
21. Oztekin, A., et al.: UseLearn: A novel checklist and usability evaluation method for eLearning systems by criticality metric analysis. International Journal of Industrial Ergonomics 40(4), 455–469 (2010)
22. Hew, K.F., Cheung, W.S.: Use of Web 2.0 technologies in K-12 and higher education: The search for evidence-based practice. Educational Research Review 9, 47–64 (2013)
23. Weinschenk, S.: 100 Things Every Designer Needs to Know About People. New Riders (2011). ISBN 978-0321767530
24. W3C: Web Content Accessibility Guidelines (WCAG) Overview. http://www.w3.org/WAI/intro/wcag (retrieved March 15, 2015)
25. WebAIM: Color Contrast Checker. http://webaim.org/resources/contrastchecker/ (retrieved March 18, 2015)
26. Edugarage by Blackboard: Themes Showcase. http://www.edugarage.com/display/BBDN/Themes+Showcase (retrieved March 18, 2015)

SIDATA: Special Session on Science Intelligence and Data Analytics

Expressing a Temporal Entity-Relationship Model as a Traditional Entity-Relationship Model

Hoang Quang[1(✉)] and Thuong Pham[2]

[1] Department of Information Technology, College of Sciences, Hue University, Hue, Vietnam
hquang@hueuni.edu.vn
[2] Department of Information Technology, Pham van Dong University, Quang Ngai, Vietnam
ptmthuong@pdu.edu.vn

Abstract. According to the approach of conceptual database design, database models are designed from the Entity-Relationship (ER) model. However, the real world of most database applications are related to the temporal aspects, so data modelling using the ER diagrams had encountered many difficulties. In an attempt to more naturally support the modeling of temporal aspects of information, many different temporal ER models have been proposed such as RAKE, TEER, TimeER... In order to design a logical data model, we have to develop a new conversion method. To solve the problem, this paper addresses expressing a temporal ER model as a traditional ER model. Thus, it provides a mapping algorithm to convert a temporal model to a traditional ER model. Hereby, we can inherit conversion method from a traditional ER model to a logical data model. In this paper, the TimeER model is used to represent temporal ER models, because it provide built-in support for capturing temporal aspects more sufficiently compared to other models.

Keywords: Temporal ER model · Timeer model · ER model · Temporal aspects

1   Introduction

The development of information systems leads to the coming of many different logical data models/schemas such as relational model, object-oriented model, XML schema,...To design the logical data models according to top-down approach (the approach of conceptual database design), a series of studies have been successfully developed to make conversion from ER model to these logical data models [8] [11] [12].

However, the real world of most database applications are related to the temporal aspects, so data modelling using the ER diagrams has encountered many difficulties. This is because some components of the model require supporting temporal aspects. Many IT workers even think that traditional ER model does not meet this requirement. In an attempt to more naturally support the modeling of temporal aspects of information, many different temporal ER models have been proposed such as RAKE, TEER, TERM, MOTAR, STEER, ERT, TimeER [1][2].

© Springer International Publishing Switzerland 2015
M. Núñez et al. (Eds.): ICCCI 2015, Part II, LNCS 9330, pp. 483–491, 2015.
DOI: 10.1007/978-3-319-24306-1_47

But it also arises an issue the research community need to solve. That is, if there is any proposed temporal model/schema (such as temporal relational model, temporal object-oriented model, temporal XML schema,...) and if it is necessary to design from the conceptual level, then the research community need to develop a conversion method from temporal ER model to the new proposed model. This is really complicated for each arising work.

Thus, an issue arising naturally from this is whether we will be able to express a temporal ER model as a traditional ER model. If the answer is yes, "arising" work mentioned above will not require a solution. That is because, if data designers want to make a temporal model/schema from the conceptual level, they just make three steps:

- Step 1: Design conceptual temporal ER model;
- Step 2: Express the temporal ER model defined in Step 1 as traditional ER model;
- Step 3: Use the conversion method to convert traditional ER model to output data model.

However, unfortunately, there has been no author studying this issue up to the present time. Thus, this paper will focus on the development of a conversion method from temporal ER model to ER model.

Actually, there are many different temporal ER models. In this paper, TimeER model is used to represent temporal ER models because it is relatively updated and and it provides built-in support for capturing temporal aspects more sufficiently compared to other models.

Thereby, this paper will be structured as follows. In Section 2 will give an overview of the components of the TimeER model. In Section 3 will provide mapping rules from TimeER model to ER model. Finally, in Section 4, a conclusion and a discussion of future work will be then given.

2   An Overview of the TimeER Model

The TimeER model is developed from the EER model [2]. This model provides built-in support for capturing the following temporal aspects: the lifespan of an entity, the valid time of a fact, and the transaction time of an entity or a fact.

As defined, temporal aspects of the entities in the database can be either the lifespan (LS), or the transaction time (TT), or both the lifespan and the transaction time (LT). The temporal aspects of the attributes of an entity can be either the valid time (VT), or the transaction time (TT), or both the valid time and the transaction time (BT). Moreover, because a relationship type can be seen as an entity type or an attribute, the designer then can define the temporal aspects supported with this relationship type if necessary.

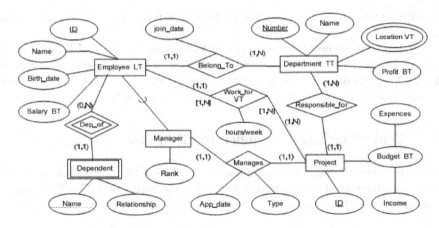

Fig. 1. TimeER diagram of a company [6]

Components of the TimeER Model

- **Entity types.** An entity type is graphically represented by a rectangle, while a weak entity type is by a double rectangle. If the lifespan, or the transaction time, or both of them of the entity type is captured, it is indicated by placing a LS, or a TT, or a LT, respectively, behind the entity type name. The semantics of this store is determined as follows.

If the lifespan LS (or the transaction time TT) of the entity type E is captured, the system can store the lifespan (or the transaction time) of each entity $e \in E$ on the set of time series $[t_1, t_1'] \cup [t_2, t_2'] \cup \ldots \cup [t_n, t_n']$, with $t_1 \leq t_1' < t_2 \leq t_2' < \cdots < t_n \leq t_n'$; where t_i and t_i' are respectively the corresponding values of the lifespan start (or the transaction time start) and the lifespan end (or the transaction time end) of each interval time $[t_i, t_i']$ ($i = \overline{1, n}$).

- **Attributes.** A single-valued attribute is represented by an oval, while a multi-valued attribute is by a double oval. Different from a simple attribute, a composite attribute is represented by an oval connected directly to the components of the composite attribute. We can give an assumption that each single-valued composite attribute is replaced with a set of simple attributes. Therefore, any attribute of an entity type or a composite attribute is the one of the following attribute types: single-valued simple attribute, or multi-valued simple attribute, or multi-valued composite attribute.

If the valid time, or the transaction time, or both of them of the attribute is captured, it is indicated by placing VT, or TT, or BT (BiTemporal), respectively, behind the attribute name. The semantics of this store is determined as follows.

If the valid time VT (or the transaction time TT) of the attribute A is captured, the system can store its different values associated with each interval time $[t, t']$; where t and t' are respectively the corresponding values of the valid time start (or the transaction time start) and the valid time end (or the transaction time end).

- **Relationship types.** A relationship type is represented by a diamond. For each relationship type it can be decided by the database designer whether or not to capture the temporal aspects of the relationships of the relationship type. If some temporal aspect is captured for a relationship type we term it temporal; otherwise, it is called non-temporal.

- **Superclass/subclass relations.** As in the EER model, the TimeER model offers support for specifying superclass/subclass relations. It is not possible to change the temporal support of the inherited attributes, but it is possible to add attributes and to further expand the inherited temporal support of the class itself.

3   Converting TimeER Models to ER Models

Procedure for converting a TimeER model to an ER model is presented in the form of 4 mapping rules to allow expressing temporal components in a TimeER model as components in an ER model.

3.1   Mapping Rules

Rule 3.1: Mapping rule of the temporal entities

If: E^* is the entity type E having the temporal support * and has the set of non-temporal attributes U,

then: E^* is mapped to the non-temporal entity type E which has the set of non-temporal attributes U. We then create the weak entity type $W(E)$ having the identifying relationship type S with the entity type E to represent the temporal support * for each entity of E. The weak entity type $W(E)$ has the set of attributes T and the partial key T' indicated in Table 1. The semantic of the identifying relationship type S is determined as follows.

$(e, w) \in S \iff$ The time $w \in W(E)$ is a temporal record of the entity $e \in E$

Table 1. Abbreviation used for temporal support of entity types and relationship types

(a)	If * = LS	then $T = \{LSs, LSe\}$ and $T' = \{LSs\}$
(b)	If * = TT	then $T = \{TTs, TTe\}$ and $T' = \{TTs\}$
(c)	If * = LT	then $T = \{TTs, TTe, LSs, LSe\}$ and $T' = \{TTs, LSs, LSe\}$

Proof:

In order to prove Rule 3.1 we need to prove: For each the entity $e \in E^*$, there exists an equivalent instance of e on the output of this rule.

To make it simple, , we only consider the case of * = LS (i.e. the temporal aspect of E is lifespan), we prove similarity in other cases.

Indeed, suppose lifespan of e in system is: $[t_1, t_1'] \cup [t_2, t_2'] \cup...\cup [t_n, t_n']$,

where: $t_1 \leq t_1' < t_2 \leq t_2' < \cdots < t_n \leq t_n'$ (1)

Then the equivalent instance of e on the output of this rule can be illustrated in the following figure:

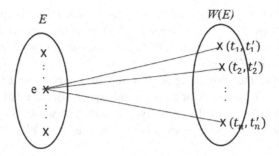

Fig. 2. The data illustrates the identifying relationship S between E and $W(E)$ on the entity $e \in E$

In the above figure, each tuple $(t_i, t_i') \in W(E)$ corresponds to an instance of an entity of $W(E)$, where: t_i and t_i' are respectively values of LSs and LSe.

From (1) lead to: $t_1 < t_2 < \cdots < t_n$

So we can choose $T' = \{LSs\}$ as the partial key of the weak entity type $W(E)$.

Rule 3.2: Mapping rule of temporal attributes of an entity type

If: The entity type E has attribute A^* (i.e. the attribute A having the temporal support *),

then: A^* is mapped to the weak entity type $W_A(E)$ having the identifying relationship type S_A with the entity type E to represent the temporal aspect * of the attribute A for each entity of E. The weak entity type $W_A(E)$ has the set of attributes $A \cup T$ and the partial key T', where T and T' are indicated in Table 2.

Table 2. Abbreviation used for temporal support of attributes and relationship types

(a)	If	* = VT	then $T = \{VTs, VTe\}$ and $T' = \{VTs\}$	
(b)	If * = TT		then $T = \{TTs, TTe\}$ and $T' = \{TTs\}$	
(c)	If	* = BT	then $T = \{TTs, TTe, VTs, VTe\}$ and $T' = \{TTs, VTs, VTe\}$	

Note that this mapping rule can be applied in both cases: A is a single-valued attribute or A is a multivalued attribute. In addition, it also can be applied in cases: the entity type E is non-temporal or temporal.

A general note to Rule 3.1 and Rule 3.2: Because the weak entity types $W(E)$ and $W_A(E)$ do not participate in any other relationships, so they can be represented as the composite multivalued attributes of the entity type E.

Proof: We prove similarly as Rule 3.1.

Rule 3.3: Mapping rule of temporal relationship types

If: R^* is the relationship type between two entity types E_1 and E_2 (i.e. the relationship type R having the temporal support *),

then:

— Firstly, the relationship type R^* is mapped to the relationship type $E(R)$. The relationship type $E(R)$ is also considered as an entity type to represent the relationship type R. In addition, we assign maximum cardinalities of lines connecting E_1 and E_2 with $E(R)$ to n.
— We then create the identifying relationship type S between $E(R)$ and the weak entity type $W(R)$ to represent the temporal aspect * of the relationship type R (i.e. the temporal aspect * of the entity type $E(R)$). The weak entity type $W(R)$ has the set of attributes T and the partial key T', where T and T' are indicated in Table 1 or Table 2.

We note that for the relationship type R^* with attributes, these attributes are also the attributes of the entity type $E(R)$ when we convert to ER model.

Proof: Because the relationship type R is considered as the entity type $E(R)$, so the temporal relationship type becomes temporal entity type. Thus, we have Rule 3.3 by applying Rule 3.1.

Rule 3.4: Mapping rule of temporal attributes of a relationship type
If: A^* is a temporal attribute of the relationship type R between two entity types E_1 and E_2 (i.e. the attribute A having the temporal support *),
then: Consider the following cases:

(a) If R is the non-temporal relationship type, then:
— The relationship type R is mapped to the relationship type $E(R)$ considered as an entity type.
— We then create the identifying relationship type S_A between $E(R)$ and the weak entity type $W_A(R)$ to represent the temporal aspect * of the attribute A. The weak entity type $W_A(R)$ has the set of attributes $A \cup T$ and the partial key T', where T and T' are indicated in Table 2.
(b) If R is the temporal relationship type, then we only create the identifying relationship type S_A like above case (a) (because according to Rule 3.3, the entity type $E(R)$ has existed).

Proof: Because the relationship type R is considered as the entity type $E(R)$, so a temporal attribute of a relationship type becomes the temporal attribute of the entity type. Thus, we have Rule 3.4 by applying Rule 3.2.

We note that Rule 3.3 and Rule 3.4 can be applied similarly to recursive relationship types and ternary relationship types.

4   An Example

In the above section, this paper has presented the mapping rules of the components in the TimeER model to the ER model. Thereby, we can solve the mapping of the TimeER model to the ER model. Figure 3 shows a TimeER model to be used in the example.

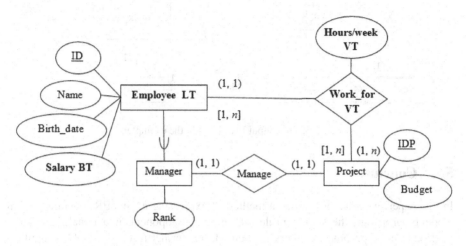

Fig. 3. The original relational model in the example

The temporal components in the above TimeER model are as indicated in Table 3.

Table 3. The temporal components of the illustrative example

NO.	TEMPORAL COMPONENT	NAME	APPLIED RULE
1	Entity type	Employee	3.1
2	Attribute of entity type	Salary (of the Employee entity type)	3.2
3	Relationship type	Work_for (binary relationship type between the Employee and the Project entity types)	3.3
4	Attribute of relationship type	Hours/week (of the Work_for relationship type)	3.4

By applying the rules in Section 3.1, we obtain the ER model given in Figure 4.

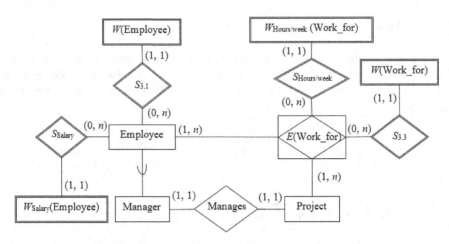

Fig. 4. The output ER model in the example

5   Conclusion

In this paper, we have developed a method of expressing a TimeER model as an ER model by proposing the mapping rules of temporal components in a TimeER model.

Based on the proposed conversion method, our further research would be applications of this result for developing direct conversion method from the TimeER model to other data models/schemas, such as object-oriented model, XML schema, logical description and others.

References

1. Jensen, C.S.: Temporal Database Management. Dr. techn. thesis by Christian S. Jensen. http://people.cs.aau.dk/~csj/Thesis/
2. Gregersen, H., Jensen, C.S.: Temporal Entity-Relationship Models-A Survey. IEEE Transactions on Knowledge and Data Engineering **11**(3), 464–497 (1999)
3. Quang, H., Thanh, H.T.: A mapping algorithm from TimeER model to relational model. In: ICT-Hanoi 2008: Proceedings of the 2nd Hanoi Forum on Information – Communication Technology, Hanoi, Vietnam, pp. 37–45 (2008)
4. Quang, H., Thanh, H.T.: Extension of Method for Converting TimeER Model to Relational Model. Journal of Computer Science and Cybernetics **25**(3), 246–257 (2009)
5. Hoang, Q., Van Nguyen, T.: Extraction of TimeER model from a relational database. In: Nguyen, N.T., Kim, C.-G., Janiak, A. (eds.) ACIIDS 2011, Part I. LNCS, vol. 6591, pp. 57–66. Springer, Heidelberg (2011)
6. Hoang, Q., Nguyen, V.T.: Extraction of a Temporal Conceptual Model from a Relational Database. Int. J. Intelligent Information and Database Systems **7**(4), 340–355 (2013)
7. Hoffer, J.A., George, J.F., Valacich, J.S.: Modern Systems Analysis and Design, 4th edn. Prentice Hall, USA (2005)

8. Franceschet, M., Gubiani, D., Montanari, A., Piazza, C.: A Graph-Theoretic Approach to Map Conceptual Designs to XML Schemas. ACM Transactions on Database Systems **38**(1), article 6 (2013)
9. Nguyen Viet Chanh: Hoang Quang: An Approach to Design Temporal Database. Journal of Science, Hue University **74A**(5), 99–107 (2012)
10. Chen, P.P.: The Entity-Relationship Model – Toward a Unified View of Data. ACM Transactions on Database Systems **1**(1) (1970)
11. Hoang, Q.: Translation from E-R Model to Object-Oriented Model. Journal of Computer Science and Cybernetics **17**(4), 78–86 (2001)
12. Elmasri, R., Navathe, S.B.: Fundamentals of Database Systems, 6th edn. Addison-Wesley, USA (2011)

A Probabilistic Ontology for the Prediction of Author's Interests

Emna Hlel[(✉)], Salma Jamoussi, and Abdelmajid Ben Hamadou

Miracl Laboratory, Technology Center of Sfax, BP 242 – 3021, Sakiet Ezzit, Sfax, Tunisia
{emnahlel,jamoussi,abdelmajid.benhamadou}@gmail.com

Abstract. The Bayesian network, a probabilistic model of knowledge representation, has the ability to represent and reason with uncertainty. It measures the
dependencies between a set of variables and infer new knowledge. In this paper,
we try to propose a method for building a probabilistic ontology, which models
a list of publications (dblp base). We have used for this aim a Bayesian Network to measure dependencies between different instances of ontology and to
infer new interests of authors from obtained Probabilistic Ontology.

Keywords: Classical ontology · Probabilistic ontology · Bayesian network

1 Introduction

Building a complete Probabilistic Ontology (PO) is a long and tedious task especially
the step of uncertain knowledge extraction (this step requires a very complex calculation to estimate the probabilities). That is why we try to use the Bayesian Network
(BN) as a way to easily measure the dependencies between a set of variables with a
probabilities. In this paper to build a PO from a list of publications dblp (Digital Bibliography & Library Project), we have combined the BN with the ontology while
retaining the advantages of each of them: Ontology is an explicit model which allows
to formally representing the knowledge of a domain and BN has the ability to represent and reason with uncertainty. Then, we have tried to infer new knowledge (the
interests of an author) from the obtained PO based on Bayesian inference.

This article is organized as follows. First, we begin with a state of the art in which
we describe the BN, PO, etc. Next, we explain our approach for building a PO from
XML documents. Then, we explain how we can infer the interests of a particular author from PO using Bayesian inference. Finally, we finish with a conclusion and
perspectives.

2 Related Works

2.1 Probabilistic Ontology Definition

According to [1], PO is "an explicit, formal knowledge representation that expresses
knowledge about a domain of application. This include : (1) Types of entities that

© Springer International Publishing Switzerland 2015
M. Núñez et al. (Eds.): ICCCI 2015, Part II, LNCS 9330, pp. 492–501, 2015.
DOI: 10.1007/978-3-319-24306-1_48

exist in the domain; (2) Properties of those entities; (3) Relationships among entities; (4) Processes and events that happen with those entities; (5) Statistical regularities that characterize the domain; (6) Inconclusive, ambiguous, incomplete, unreliable, and dissonant knowledge related to entities of the domain; and (7) Uncertainty about all the above forms of knowledge". It is an extension of the classical ontology (CO) by integrating uncertain knowledge. It can describe the knowledge of a particular domain ideally in a format that can be read and processed by a computer and incorporate uncertainty to this knowledge. We can define the PO simply as a CO enriched with uncertain knowledge. In other words, ontology is a PO only if it contains at least one probabilistic component. We can divide the PO components in precise components and probabilistic components. The precise components are: precise concepts, precise instances, precise properties and etc (proposed in [2]). They keep the same definitions and play the same roles as in CO. The probabilistic components (probabilistic relations (conceptual and semantic), probabilistic instances, etc.) are used to model the uncertain knowledge. We can define them as precise components attached by a particular values (probabilistic value) representing the uncertain aspect. This value can be specified manually by a domain expert or automatically using a learning process taking into consideration a set of observations. In this paper, we focused only on probabilistic conceptual relationship that is used to describe an uncertain conceptual relation between two instances. Some examples for this probabilistic relation are represented by: panel$_1$ (Panel) *is connected* with pipe$_1$ (Pipe) with 74% probability, John *is interested* in imagery (Theme) with 80% probability.

2.2 Bayesian Network Definition

BN is used to analyze large amounts of data to extract useful knowledge for decision making or predict the behaviour of a system. For example, we can make diagnoses depending on the symptoms of a patient. We can also estimate the probabilities of pathologies consistent with these symptoms. They are used in such fields as Medicine, computer Science and so on [3], [4], [5]. BN, which is a probabilistic graphical model, represents random variables in the form of a directed acyclic graph, with nodes expresses random variables; arcs indicate conditional dependencies between the nodes and the conditional probabilities measure dependencies between these nodes [3], [6], [7]. Mathematically, a Bayesian network is defined by B= (G, C):

— G=(X, E) is a directed acyclic graph whose nodes are associated with a set of random variables X={X_1, ..., X_n} and E is the set of arcs of G that represent dependencies between these nodes,
— C={P(X_i|Pa(X_i))} is a set of probabilities that each node X_i is conditionally dependent to the state of its parent Pa(X_i) in G.

2.3 Bayesian Network and Probabilistic Ontology

Various researchers in previous years have tried to present several works to represent uncertain knowledge [12], [2], [15], [20], [25], [11]. They have combined the BN with the ontology to model uncertain knowledge. They have proposed probabilistic extensions of ontology languages (PR-OWL, OntoBayes, etc) and probabilistic description logics in order to address the inability of classical description logics and ontology languages to represent and reason with uncertainty. But these works have not focused on the PO construction stages from data, its probabilistic and precise components and the difference between PO and CO. Yang Y. and Calmet J. have integrated the BN in OWL, they have defined the new OWL classes and OWL properties to represent uncertainty [10]. Essaid A. and Ben Yaghlane B. have proposed a probabilistic extension of OWL which is able to support the uncertainty thanks to new classes and new properties [8]. They are based on the theory of Dempster-Shafer [13] to represent uncertain knowledge in the form of a graph extracted from ontology. Once the graph is obtained, they have assigned masses a priori for each network node with the aid of a domain expert. Then they launched a probability propagation process to calculate conditional probabilities of network variables. The disadvantages of this work are: uncertainty is built only on ontology classes and probability assignment is made with aid by a domain expert by contrast there are several probabilistic machine learning methods from data that can be used. Ding Z. and Peng Y. have proposed a method for integrating uncertain knowledge in an ontology based on a BN [9]. They have converted all ontology concepts and relations into a graph of a BN to estimate the dependencies between them without distinguishing between probabilistic components and precise components of a PO. In other words, they have assumed that all the components of ontology are probabilistic components; by contrast a PO includes not only probabilistic components but also precise ones, similarly for [10], [8]. Octavian A. and al. also have tried to combine BN with the ontology to create an application that can be used by doctors to make diagnoses [11]. They have translated ontology instances as well as relationships between them of ontology, constructed with the aid of a domain expert, into a set of BN graph variables to measure dependencies between these instances. Carvalho R.N. has proposed a methodology to guide the expert in the development of PO (manual method) [14]. They also used the BN to represent the uncertainty characteristics. Their process of PO development consists of three major stages: Requirements, Analysis and Design, and Implementation. The first stage involves the definition of the objectives to be achieved. The Analysis and Design stage describes classes, relations, and rules (probabilistic and deterministic), etc. The implementation stage maps the entities, groups and rules (previously indicated) to a specific language that incorporates uncertainty in the Semantic Web (PR-OWL).

A reader interested in a CO can find various works describing in detail the lifecycle of CO, its construction steps, components, various automatic or manual construction methods of CO. But this is not much available for PO as the majority of researchers have focused only on the proposal of probabilistic description logic and ontology language to model the uncertain knowledge of a particular field. This paper is part of this framework in which we try to propose a PO construction method.

3   Our Methodology for Building PO

The PO aims to model certain and uncertain knowledge of a particular domain. In this work, we have tried to propose a method having as a goal the creation of PO which models a set of publications for an author a_i as well as the publications (base dblp) of his M principal co-authors. Indeed, we made a filtering of publications of an author a_i (using a threshold M which expresses the number of primary co-authors of a_i, who have the maximum of articles with this author): we leave only the publications of a_i with his M primary co-authors. The same procedure applied too for his M coauthors. So, we can say that the input of our method is a name of an author and the output is a probabilistic formal representation of knowledge (PO) modeling the publications of this author and his M coauthors. To build a PO, we have used a probabilistic model (BN) which allows to represent uncertain knowledge and easily to measure the dependencies between a set of variables with probabilities (conditional and marginal). We have represented the knowledge extracted from the base dblp as a formal model that includes a set of components: concepts (Author, Theme, etc), conceptual relations (co-author, be-interested, etc), etc. To show where is the uncertainty in this case, we take an example of a set of publications extracted from this base. Imagine this situation: A_1 is an author who has published two papers in the research theme th_1, three papers in the research theme th_2 and five papers in the research theme th_3. The calculation of the degree of the author's interest in these themes gives these results:

— A_1 is interested in th_1 with a degree of interest equals to 0.2: $P(th_1)$,
— A_1 is interested in th_2 with a degree of interest equals to 0.3: $P(th_2)$,
— A_1 is interested in th_3 with a degree of interest equals to 0.5: $P(th_3)$.

With $P(th_i)$ is calculated as the number of papers having the theme th_i divided by the total number of papers. These results are not certain and not complete, because we calculated these interest degrees from some of this author's publications (only from dblp base and there are other publications by this author saved in other bases as Springer). So the relationship be-interested is considered as incomplete and uncertain relationship between the concept "Author" and the concept "Theme". It is attached by a particular value (interest degree) to represent the uncertain aspect. In the following, we present our PO construction method which consists of two phases: Construction of CO from dblp base and uncertainty integration in the obtained CO (see figure 1).

3.1   Classical Ontology Construction

In this phase, we aim to formally represent the ontology components, assuming they are all classic or precise components. Indeed, the role of this phase is: the creation of terminological knowledge of our ontology and extraction of ontology instances. This phase contains three sub-steps. We begin with the creation of precise concepts: our CO is composed of concepts list which are: Author, Key-Word and Theme and its sub-concepts (as PCA, NLP, ONTOLOGY, etc). The creation of theme hierarchy is performed manually. Each author may be interested in one or more research themes that are characterized by a keyword list. Each author has from 1 to M coauthors. The second sub-step of this stage is the creation of conceptual relationships. Our ontology consists of three conceptual relationships that are have-word, co-author and be-interested. The relation "have-word" expresses that a Theme is characterized by a set of keywords. The relationship "co-author" indicates that an author can have a co-author list. The relation "be-interested" indicates that an author is interested in a research theme list. The final step of this phase is the population of ontology which consists to instantiate the ontology obtained.

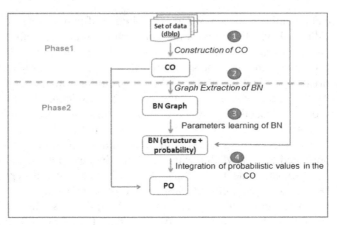

Fig. 1. The stages of our PO construction method.

3.2   Uncertainty Integration in CO

The aim of this phase is integrate uncertain knowledge in obtained CO (called COdblp) to create our PO (called POdblp). This phase consists of three stages which are: Extraction of BN graph from COdblp, Parameters learning of obtained BN and Integration of uncertainty in COdblp to obtain a PO.

BN Graph Extraction

In literature, various researchers have proposed approaches to build a BN graph from ontology to measure dependencies between its constituents [12], [9], [8] without distinguishing between precise components and probabilistic components of an ontology: they assumed that all elements of ontology are probabilistic components. By contrast, in our approach, we converted only ontology components that can support uncertainty in a BN graph (which are instances of Author and Theme concepts and the relationships between them) to estimate the dependencies between them. For building this graph, it is necessary to specify its nodes, the states of these nodes and the arcs between them. To perform this step, we have followed this approach:

— Author concept instances are converted into a set of BN nodes $A=\{a_1, a_2, ..., a_N\}$. Each node of A has two values {true, false}. If a_1 is co-author of a_2, then an arc between a_1 and a_2 will be added. The relationships of "co-authors" between Author instances that we provide a co-author network. The co-authors relationships can cause cycles in the obtained BN graph (BN graph is acyclic). To avoid these cycles, we have just a change in the arc orientation which poses the problem. For example, if a_2 is the co-author of a_1, a_1 is the co-author of a_3 and a_3 is the co-author of a_2. These relations of co-authors form a cycle. To remedy this problem, we have changed the orientation of one of the three arcs. Indeed, arcs orientation in BN meaning less and often just indicate that conditional dependency exists between two variables.

— The list of Theme concept instances will be converted in a set of nodes $T=\{t_1, ..., t_i, ..., t_Q\}$ with everyone has two values {true, false}.

— The relationship "be-interested" between an Author instance and a Theme instance will be translated in an arc between nodes coming from these two instances.
— The orientation of arcs between an instance of Theme and an instance of Author: We have sought to measure dependencies between Author instances and Theme instances to obtain the probability that an author a_j is interested in a theme t_i: $P(t_i=\text{true}|a_j=\text{true})$, with $i \in \{1, ..., Q\}$, $j \in \{1, ..., N\}$, $t_i \in T$ and $a_j \in A$. For this reason, the direction of arcs between a node a_j of A and a node t_i of T is from a_j to t_i ($a_j \rightarrow t_i$: a_j is the parent of t_i).

BN Parameter Learning

The parameter learning is an essential step in BN. It consists of estimating the probability distributions, assuming that the network structure has been fixed, from a base of examples. It can be performed with a simple statistical or Bayesian learning. For more information on parameter learning methods, we recommend reading [15], [16], [17]. In our case, the parameter learning consists in determining the best set of parameters of obtained BN (probability) taking into consideration the observed data of the dblp base. In other words, the goal of this step is the enrichment of the obtained structure from the CO by quantification of dependencies between different BN variables with probabilistic values in [0, 1] that can represent the uncertain aspect.

Integration of the Probabilistic Values in CO

To make the CO able to support uncertainty, we have changed the characteristics of the relationship "be-interested" by others: it becomes a probabilistic N-ary[1] associative relation [18], which is characterized by a data-Property, named ProbTA. This relationship is represented using a new concept "AUT-Theme" which is characterized by ProbTA data-Property and two ObjectProperty: "have-author" and "have-theme" (see figure 2). The ProbTA property expresses the probability that an author a_j is interested in a research theme t_i ($P(t_i=\text{true}|a_j=\text{true})$). The value of this property for each pair of instances (Author, Theme) is determined from the result of the learning process of BN parameters which is explained in the previous step.

```
<owl:Class rdf:ID="AUT-Theme"/>
<owl:ObjectProperty rdf:ID="have-author">
 <rdfs:range rdf:resource="#Author"/>
 <rdfs:domain rdf:resource="#AUT-Theme"/>
</owl:ObjectProperty>
<owl:ObjectProperty rdf:ID=have-theme">
 <rdfs:range rdf:resource="#Theme"/>
 <rdfs:domain rdf:resource="#AUT-Theme"/>
</owl:ObjectProperty>
<owl:DatatypeProperty rdf:ID="ProbTA">
 <rdfs:domain rdf:resource="#AUT-Theme"/>
 <rdfs:range rdf:resource="http://www.w3.org/2001/XMLSchema#float"/>
</owl:DatatypeProperty>
```

Fig. 2. OWL code that represents probabilistic relation "be-interested".

---

[1] N-ary relation: is a relationship having more than two arguments or additional attribute.

4   Inference of Author's Interests

A powerful technique in BN is Bayesian inference (or updated probability). It consists to propagate one or more certain information (values established by certain variables) to deduce how this intervenes on the probabilities of other network variables. Mathematically, the inference in a BN is the calculation of $P(U|\varepsilon)$ (with U is the set of variables and ε is the new information on one or more variables U), that is to say calculating the posterior probability of the network knowing that ε. This update is performed using Bayes theorem. Given two random variables A and B, Bayes theorem consists to determine the probability of A knowing that B [19]:

$$P(A|B) = \frac{P(B|A)P(A)}{P(B)} \tag{1}$$

The most popular inference algorithm for BN was proposed by Lauritzen and Speigelhalter [20]. This algorithm converts a BN into a junction tree and then performs inference on the junction tree. In this work, we used this algorithm to infer the new interests of a particular author, which are implicitly represented in our PO. The following steps show the procedure used to infer new interests of an author a_1:

1. Variable instantiation: That means forcing the value of a BN variable for example the value of author a_1 is set to true.
2. BN variable probability estimation: this is to calculate the probabilities of different BN variables knowing that a_1 equal to true. In fact, we are only interested in $P(t_i=true)$ knowing that a_1 is equals to true (with t_i belongs to T). This step provides us a set of probabilities Prob={P_1, P_2, ..., P_i, ..., P_Q}, with P_i belongs to [0, 1], Q is the number of Theme instances available in our PO and P_i is equal to $P(t_i="true")$ knowing that $a_1="true"$.
3. Selecting the new deduced interests of author a_1: After having ordered the probabilities of Prob, the new deducted interests of author a_1 from the PO are the research themes that have the highest probability values.

5   Results

In this Work, we try to create a PO from publications of "Walid Mahdi" and his 4 main co-authors: Akrout Belhassen, Abdelmajid_Ben Hamadou, Werda Salah and Zlitni Tarek. The total number of selected publications is 50 publications of 13 authors. The set of these publications is available at this link: *http://www.fichier-rar.fr/2015/03/08/dblpintf/*. The extraction of themes for each publication is made in a manual way (see this link *http://www.fichier-rar.fr/2015/03/15/theme4/*). Our PO includes 13 Author concept instances and 38 sub-concept of Theme and 173 instances of Key-Word. The implementation of our PO is written in Java. The result of our ontology construction method is a PO, which is saved in an OWL file (available for download at this link *http://www.fichier-rar.fr/2015/04/11/podblp/*). Note that this implementation has intermediate results as CO and obtained network structure from this ontology based on the set of conversion rules. The obtained BN graph, which is written in a format understandable by the R tool, is publicly available under this link *http://www.fichier-rar.fr/2015/03/16/bngraph1/*. To implement our application, we are based on a set of modules:

— Jena API[2]: This is a Java class library that facilitates the development of applications for the semantic web. We used it to implement Java classes to build our ontology.
— R tool[3]: this is a programming language for the statistical analysis of data. We used its packages for performing: learning of BN parameters (bnleran [21]), Bayesian inference (gRain [22]) and BN graph visualization (Rgraphviz[4]).

The second objective of this paper is the inference of new interests for a particular author from the obtained PO based on Bayesian inference. To validate the results obtained, we divided the set of publications selected from base dblp into two sets: E_1 includes publications of "Walid Mahdi" and his 4 main co-authors that were published before 2013 (this set contains 33 publications) and E_2 includes only publications of "Tarek Zlitni" which are published between 2013 and 2015 (this set contains 4 publications). The theme extraction step from E_2 is performed manually. It gives the set of themes $I_2 = \{t_1, ..., t_i, ..., t_{12}\} = \{$VIDEO_PROCESSING, LOCALIZATION, NLP, ANALYSIS_DATA, PROGRAM_IDENTIFICATION, PCA, GRAMMAR, MULTIMEDIA, VIDEO_GRAMMAR, MODELLING, CLASSIFICATION VIDEO_TEXT_EXTRACTION$\}$. The construction of PO, called POdblp1, is made from the E_1 (see this link *http://www.fichier-rar.fr/2015/04/11/podblp2/*). This ontology includes 29 instances of Theme $I_1 = \{t_1, ..., t_i, ..., T_{29}\}$. The inference step of preferences of "Tarek Zlitni", from POdblp1, provides us a set of probabilities Prob=$\{P_1, P_2, ..., P_i, ..., P_{29}\}$, with $P_i = P(t_i =$"true") knowing that "Tarek Zlitni"= "true". To check if the new interests deducted for this author from the POdblp1 are the same as interests extracted from the set of publications of this author (between 2013 and 2015: E_2), we made a comparison between I_2 and $I_3 = \{t_1, ..., t_i, ..., t_{12}\}$: the 12 themes of I_1 which have the highest probabilities. We noticed that the interests list deducted for "Tarek Zlitni" (I_3) from the PO based on Bayesian inference are very close to the reality. In other words, most of the research themes of this author extracted from E_2 (the set I_2) are present in the set I_3 with an error rate equals to 0,25 (see table 1). In addition, we can deduce the interests of an authors' group if they have decided to work together. For example, if "Tarek Zlitni" and "Bassem Bouaziz" decide to work together, then we can conclude that they will probably be interested in the list of themes NT_1 that are ordered by P_i (see table 1).

[2] http://jena.apache.org
[3] http://cran.r-project.org/
[4] http://www.bioconductor.org/packages/release/bioc/html/Rgraphviz.html

Table 1. Some result of our method interests' inference.

I_3	NT_1
ANALYSIS_DATA : 0.434018 GRAMMAR : 0.4560566 NLP : 0.4560566 LOCALIZATION : 0.5435524 MULTIMEDIA : 0.4517459 **TEMPORAL_ANALYSIS** : 0.5477035 VIDEO_GRAMMAR : 0.5058311 PROGRAM_IDENTIFICATION : 0.4273357 **WEB_APPLICATION** : 0.3043866 **FORMAL_ANALYSIS** : 0.2596237 VIDEO_PROCESSING : 0.228509 MODELLING : 0.2196281	VIDEO_GRAMMAR : 0.5061957 PROGRAM_IDENTIFICATION : 0.4227921 VISUAL_RECOGNITION : 0.333909 ANALYSIS_DATA : 0.4251515 GRAMMAR : 0.4279865 NLP : 0.4279865 LOCALIZATION : 0.5795182 MULTIMEDIA : 0.4114805 TEMPORAL_ANALYSIS : 0.6736602 WEB_APPLICATION : 0.2976109 VIDEO_PROCESSING : 0.2436057 LIP-READING : 0.2436057 FORMAL_ANALYSIS : 0.2470989 LINGUISTIC_KNOWLEDGE : 0.221843

6   Conclusion and Perspectives

In this paper, we have proposed a PO construction method from a publications list (base dblp) of a particular author and his main co-authors. This method includes these steps: Classical ontology construction, Conversion of uncertain components in a BN graph, Probability Distribution Estimation in the obtained graph based on learning of BN parameters and Conversion of the CO (COdblp) to a PO (POdblp). Then, we have tried to infer the new preferences for a particular author from our PO based on Bayesian inference.

As perspectives for this work, we will try to enrich the obtained PO by other new concepts and the use of standard ontologies (such as FOAF) to ensure interoperability between different systems. And, we will consider using the obtained PO as being a knowledge base in real-world applications to enhance their performances such as improving the results of a personalized search based on a PO.

References

1. da Costa, P.C.G., Laskey, K.B., Laskey, K.J.: PR-OWL: a bayesian ontology language for the semantic web. In: da Costa, P.C.G., d'Amato, C., Fanizzi, N., Laskey, K.B., Laskey, K.J., Lukasiewicz, T., Nickles, M., Pool, M. (eds.) URSW 2005 - 2007. LNCS (LNAI), vol. 5327, pp. 88–107. Springer, Heidelberg (2008)
2. Gruber, T.: Towards Principles for the Design of Ontologies Used for Knowledge Sharing. International Journal of Human and Computer Studies, 907–928 (1995)
3. Philippe, L.: Réseaux bayésiens: apprentissage et modélisation de systèmes complexes. Habilitation (2006)
4. Mechraoui, A., Thiriet, J.-M., Gentil, S.: Aide à la décision et diagnostic par réseaux bayésiens d'un robot mobile commandé en réseau. In: Actes de la Sixième Conférence Internationale Francophone d'Automatique (2010)
5. Onisko, A., Druzdzel, M.J., Wasyluk, H.: A probabilistic causal model for diagnosis of liver disorders. In: Proceedings of the 7 th Symposium on Intelligent Information Systems, pp. 379–387, June 1998

6. Naim, P., Wuillemin, P.H., Leray, P., Pourret, O., Becker, A.: Réseaux Bayésiens. Editions Eyrolles (2008)
7. Pearl, J.: Probabilistic Reasoning in Intelligent Systems: Networks of Plausible Inference, Morgan Kaufmnn, in Representation and Reasoning, CA (1988)
8. Essaid, A., Ben Yaghlane, B.: BeliefOWL: an evidential representation in OWL ontology. In: International Workshop on Uncertainty Reasoning for the Semantic Web, Washington DC, USA, pp. 77–80 (2009)
9. Ding, Z., Peng, Y.: A probabilistic extension to ontology language OWL. In: Proceedings HICSS, pp. 5–8 (2004)
10. Yang, Y., Calmet, J.: Ontobayes: an ontology-driven uncertainty model. In: Proceeding of the International Conference on Intelligent Agents, Web Technologies and Internet Commerce (IAWTIC 2005), pp. 457–464 (2005)
11. Arsene, O., Dumitrache, I., Mihu, I.: Medicine expert system dynamic Bayesian Network and ontology based. Expert Systems with Applications 38, 253–261 (2011)
12. Fabio, G., Polastroa, R.B., Takiyamaa, F.I., Revoredob, K.C.: Computing Inferences for Credal ALC Terminologies. In: URSW, pp. 94–97 (2011)
13. Shafer, G.: A Mathematical Theory of Evidence. Princeton University Press (1976)
14. Carvalho, R.N.: Probabilistic ontology: Representation and modeling methodology, Ph.D., George Mason University, Fairfax, VA, USA (2011)
15. Jordan, M.: Learning in Graphical Models. Kluwer Academic Publishers, Dordecht (1998)
16. Jacobs, R., Jordan, M., Nowlan, S., Hinton, G.: Adaptive mixtures of local experts. Neural Computation, 79–87 (1991)
17. Myung, J.: Tutorial Tutorial on maximum likelihood estimation. Journal of Mathematical Psychology, 90–100 (2002)
18. Hebeler, J., Fisher, M., Blace, R., Perez-Lopez, A.: Semantic web programming, Foreword by Mike Dean, Principal Engineer, BBN Technologies, livre (2009)
19. Bayes, T., Price, R.: An Essay towards Solving a Problem in the Doctrine of Chances, Philosophical Transactions of the Royal Society of London, pp. 370–418 (1763)
20. Lauritzen, S.L., Spiegelhalter, D.J.: Local computation with probabilities and graphical structures and their application to expert systems. J. Royal Statistical Society, 157–224 (1988)
21. Scutari, M.: Package 'bnlearn': Bayesian network structure learning, parameter learning and inference (2014). http://www.bnlearn.com
22. Højsgaard, S.: Graphical independence networks with the gRain package for R. Journal of Statistical Software 46(10), 1–26 (2012)

A Framework of Faceted Search for Unstructured Documents Using Wiki Disambiguation

Duc Binh Dang[1]([⊠]), Hong Son Nguyen[2],
Thanh Binh Nguyen[3], and Trong Hai Duong[1]

[1] School of Computer Science and Engineering, International University, VNU-HCMC,
Ho Chi Minh City, Vietnam
ducdb07@mp.hcmiu.edu.vn, haiduongtrong@gmail.com
[2] Faculty of Information Technology, Le Quy Don University, Hanoi, Vietnam
son_nguyenhong2002@yahoo.com
[3] International Institute for Applied Systems Analysis (IIASA), Schlossplatz 1, 2361,
Laxenburg, Austria
nguyenb@iiasa.ac.at

Abstract. According to literature review, there are two challenges in search engines: lexical ambiguity and search results filtering. Moreover, the faceted search usually applies for structured data, rarely uses for unstructured documents. To solve the aforementioned problems, we proposed an effective method to build a faceted search for unstructured documents that utilizes wiki disambiguation data to build the semantic search space; the search is to return the most relevant results with a collaborative filtering. The faceted search also can solve the lexical ambiguity problem in current search engines.

Keywords: Faceted search · Semantic search · Lexical ambiguity · Disambiguation · Wiki

1 Introduction

The major problem is shortage of specification of semantic heterogeneousness and ambiguity. We aim to look the *apple fruit* by putting the word *apple* in famous search engines including Google, Yahoo! and Bing. We could not find the page, which mentions *apple fruit* easily; most of the top results are about the *Apple Inc.*, an American multinational corporation.

Furthermore, searching with keyword *java* on the current search engines returns most of the results mentioning *java* as programming language, however, *java* is not only a programming language, *java* is the world's most populous island in Indonesia, located in Indonesia; it is also breed of chicken originating in the United States or also a brand of Russian cigarette. We only have the Java as an island when using the query "*Java Island*", as a chicken breed once the "*java chicken*" is entered. Almost of search-engines such as Google, Yahoo! or Bing are search engine, which are not knowledge engine. They are good at returning a small number of relevant documents from a tremendous source of webpages on the Internet; but they still experiences

M. Núñez et al. (Eds.): ICCCI 2015, Part II, LNCS 9330, pp. 502–511, 2015.
DOI: 10.1007/978-3-319-24306-1_49

the lexical ambiguity issue, the presence of two or more possible meanings within a single word.

In addition, the faceted search usually applies for structured data, rarely uses for unstructured documents. Dynamic queries defined as interactive user control of visual query parameters that generate a rapid animated visual display of database search results [2, 5-7, 13-16]. The authors emphasize the interface with outstanding speed and interactivities; in contrary to method of using database-querying language like SQL. This is one of outstanding works in faceted search in early phase. It is catalyst for interest of faceted search. Ahlberg and Shneiderman built Film Finder to explore the movie database [1]. The graphical design contains many interface elements; parametric search is also included in a faceted information space. However, the results of Film Finder returning to users are not proactive and users are still able to select unsatisfactory combination. Later on, Shneiderman and his colleagues addressed the above problem on query previews. Query Previews are prevent wasted steps by eliminating zero-hit queries. That is mean the parametric search is replace by faceted navigation. In general, it helps user to have an overview over the selected documents. In [12], both of the view-based search and query preview have same limitation that they do not support faceted search. The mSpace project [13] described as an interaction design to support user-determined adaptable content and describe three techniques, which support the interaction: preview cues, dimensional sorting and spatial context.

According to aforementioned study, there are two problems in today search engines:

• Lexical ambiguity: the question is that can the search engine returns the correct meaning of keyword when providing a limited information of search query?

• Search results filtering: can the search engine offer a better results filtering mechanism such as collaborative filtering or content-based filtering?

In addition, the faceted search usually applies for structured data, rarely uses for unstructured manner.

To address the mentioned problems, our approach is to build a faceted search for unstructured documents that utilizes wiki disambiguation data to build semantic search space; the search is to return the most relevant results with collaborative filtering. The faceted search can also solve the lexical ambiguity problem in current search engines.

The remainder of this paper is organized as follows. Section 2 provides methodology including disambiguation reference and semantic search space building using Wiki's disambiguation. Section 3 presents experiment with result evaluation, and Section 4 reports the conclusions.

2 Methodology

2.1 Disambiguation Reference Building Using Wikipedia

Wikipedia is a free access as an open internet-encyclopedia. Wikipedia is one of most popular websites, constitutes the Internet's largest, and most popular general reference work [11]. In Wikipedia, a page (article) is the basic entry, which describes entity or event. Any entities or events always contain hyperlinks, which guide users to other related pages.

Disambiguation of Wikipedia[1] is the process of resolving the conflicts that arise when a single term is ambiguous-when it refers to more than one topic covered by Wikipedia. For example, the word *Mercury* can refer to a chemical element, a planet, a Roman god, and many other things. Disambiguation is required whenever, for a given word or phrase in which a reader might search, there is more than one existing Wikipedia article to which that word or phrase might be expected to lead. In this situation, there must be a way for the reader to navigate quickly from the page that first appears to any of the other possible desired articles. The content of a disambiguation pages are created by a huge community - internet users, which make them, are very easy to understand. The facet and sub-facets are legible entities; common facets are *People, Places, Company, Computing* or *Music*, which make disambiguation data suitable for generic users.

In our database, all disambiguation entities are extracted from the articles in Wikipedia data; each disambiguation entity contains a list of all existing Wikipedia article of the given word; for example, *Java* is, an island, a programming language, an animal. Each meaning is categorized into facets and sub-facets.

In *Figure* 1, facets and sub-facets of *Java* disambiguation are:

• Facets: *Places, Animal, Computing, Consumables, Fictional characters, Music, People, Transportation*, Other uses.

• Sub-facets: *Indonesia, United States, Other* are sub-facets of facet Places.

In order to obtain the Wikipedia disambiguation data, we decide to download the Wikipedia dumps file, which contains Wikipedia contents, and load them into a MySQL database serve; the total size is more than 150 Gigabytes.

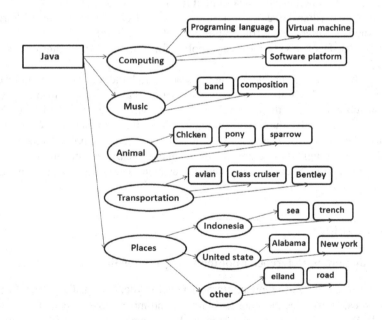

Fig. 1. Sample facet structure of Java

---

[1] https://en.wikipedia.org/wiki/Wikipedia:Disambiguation

After importing *Wikipedia*, all the disambiguation pages are extracted out from the database based on template of *Wiki Disambiguation*; for every disambiguation entity, an algorithm is applied to get the features of disambiguation documents. The whole process is as *Figures* 2:

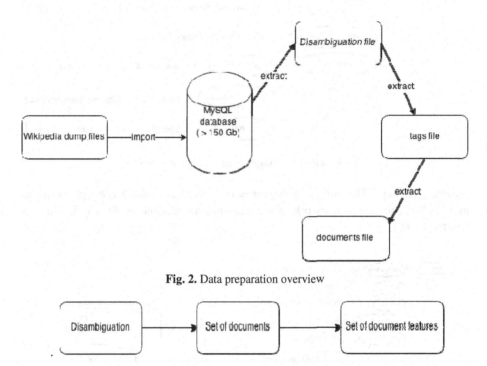

Fig. 2. Data preparation overview

Fig. 3. Document features extraction process

The document features are used as search data. There are two following main steps to construct the feature vector for each concept. First step, using the vector space model tf/idf to construct the generalization of the documents. Second step, for each leaf concept, the feature vector is calculated as the feature vector of set of documents associated with the concept. For each none-leaf concept, the feature vector is calculated by taking into consideration contributions from the documents directly associated with the concept and its direct sub-concepts. The detail of these two steps is presented in [3]. The result of this step is disambiguation reference model as shown in *Figure* 4:

2.2 Semantic Search Space Building Using Disambiguation Reference Model

This semantic space is constructed by organizing documents in the disambiguation reference model (see *Figure* 4). Each document is represented by a feature vector. The similarity between a document's feature vector and a reference concept's feature vector is considered as a relevant weight of the belong from the document to the

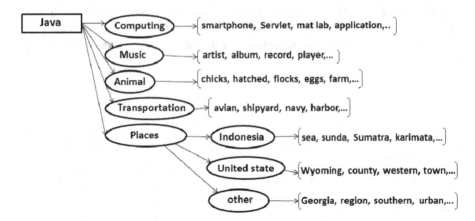

Fig. 4. A Part of Disambiguation Reference Model

reference concept. The similarity degree is used to assign which concept the document belongs to. The framework of the algorithm is depicted as *Figure* 5, which is described in [4].

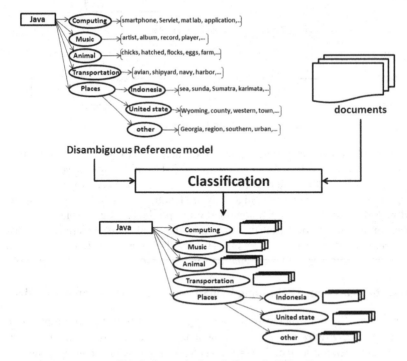

Fig. 5. Semantic Search Space Building Process

3   Experiment

3.1   Search Implementation

We illustrate our proposed system via a search situation. *Figure* 6 shows interface of our search program.

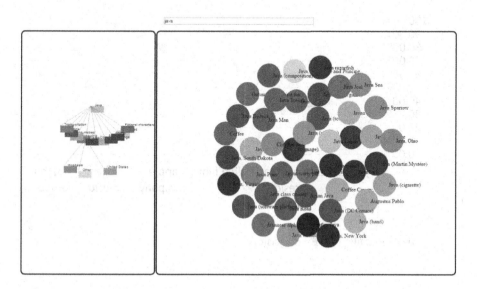

Fig. 6. Search user interface

The user interface of the faceted search is consisting of three UI components.

1)   The search input field (top): user will enter his/her query there. An on change event is triggered on that input. After query is provided, the program will start looking for the results.

2)   The search results section (left bottom): show all the relevant documents to the user. Each document is showed as a node. On double clicking on the node, search program will open the link to corresponding document.

3)   The Facet graph (right bottom): show facets and sub-facets of tree hierarchy (which is the disambiguation reference). Beside results in result graph, this facet graph helps obtaining the documents quickly: a).The tree gives user an overview about the returned results, what the current results are related to. b) When most weighted vertex in result graph is not user's intention, he can use facet graph to interactively refine the result.

3.2   Evaluation

Six tests are executed and the search program behaves differently for different profiles; in the search results, sizes of vertices for specific documents are varied for various users (Tom, Richard and Patricia).

- Accuracy evaluation

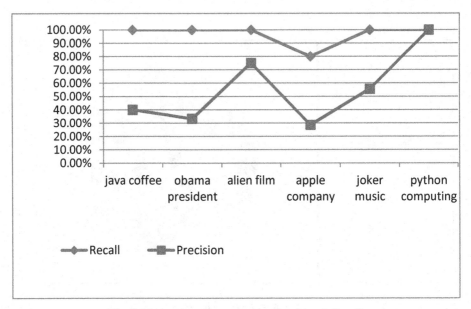

Fig. 7. Measure search result using Precision & Recall

In *Figure* 7, the effectiveness of the returned results of our program is evaluated. In most of cases, all the relevant documents are retrieved but the accuracy is not very good, many times the irrelevant documents are shown to user.

- User's satisfaction evaluation

Now we continue to look at the importance of vertices' size, normally the biggest vertex will draw user's attentions.

Table 1. Extracted facets/documents compared to original data

Query	Top weighted vertex	Satisfied
java	Java (programming language)	Yes
obama president	Barrack Obama	Yes
alien film	Alien Sun	No
apple company	Apple Inc.	Yes
joker music	Joker Phillips	No
python computing	Python (programming language)	Yes

- Filtering evaluation

Because of the problem when matching words together, there will be the cases that user is not able to get his interested results (Table 4); in these situation, user can refine search results by selecting facet or sub facets in the facet graph.

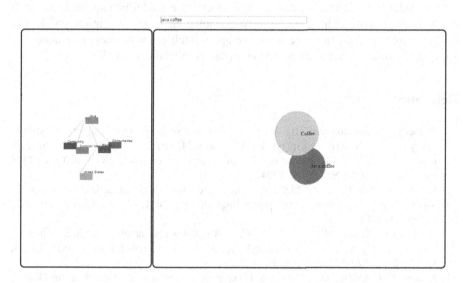

Fig. 8. Search results for *Java coffee* are filtered by *Consumables* facet

In a common flow, user to be able to select the correct information after two steps in our search program; the first is to enter search query, next step is to refine the search result based on facet value on facet graph.

- Efficiency evaluation

Now we evaluate the efficiency of the search program, the time consuming for sample queries are measured as follow:

Table 2. Query time consuming

Query	Time consumed(milliseconds)
Java	318
java coffee	47
Alien	316
Obama president	29
joker batman	77

According to the time consumed for each query, we observe that querying for generic query (java, alien) takes around 300 milliseconds, but querying for more specific query (java coffee) where users have knowledge about what they are searching for, the waiting time is negligible. The reason is that the time when search program calculates weight of each vertex, the more the documents are in the search result, the more time are needed.

4   Conclusion

We proposed a novel framework of faceted search for unstructured documents utilizing wiki disambiguation data. Wiki disambiguation is to learn a reference disambiguation model to build the semantic search space. The model is also used for result collaborative filtering. The faceted search can solve the lexical ambiguity problem in current search engines. In future work, we apply intelligence business techniques [9-10] to make visualization and interaction for facets to facilitate filtering.

References

1. Ahlberg, C., Shneiderman, B.: Visual information seeking: tight coupling of ynamic query filters with starfield displays. In: Adelson, B., Dumais, S., Olson, J. (eds.) Proceedings of the SIGCHI Conference on Human Factors in Computing Systems (CHI 1994), pp. 313–317. ACM, New York (1994)
2. Brunk, S., Heim, P.: tFacet: hierarchical faceted exploration of semantic data using well-known interaction concepts. In: proceedings of DCI 2011. CEUR-WS.org, vol. 817, pp. 31–36 (2011)
3. Duong, T.H., Uddin, M.N., Li, D., Jo, G.S.: A collaborative ontology-based user profiles system. In: Nguyen, N.T., Kowalczyk, R., Chen, S.-M. (eds.) ICCCI 2009. LNCS, vol. 5796, pp. 540–552. Springer, Heidelberg (2009)
4. Duong, T.H., Uddin, M.N., Nguyen, C.D.: Personalized semantic search using ODP: a study case in academic domain. In: Murgante, B., Misra, S., Carlini, M., Torre, C.M., Nguyen, H.-Q., Taniar, D., Apduhan, B.O., Gervasi, O. (eds.) ICCSA 2013, Part V. LNCS, vol. 7975, pp. 607–619. Springer, Heidelberg (2013)
5. Heim, P., Ertl, T., Ziegler, J.: Facet graphs: complex semantic querying made easy. In: Aroyo, L., Antoniou, G., Hyvönen, E., ten Teije, A., Stuckenschmidt, H., Cabral, L., Tudorache, T. (eds.) ESWC 2010, Part I. LNCS, vol. 6088, pp. 288–302. Springer, Heidelberg (2010)
6. Heim, P., Ziegler, J., Lohmann, S.: gFacet: a browser for the web of data. In: Proceedings of the International Workshop on Interacting with Multimedia Content in the Social Semantic Web (IMC-SSW 2008), vol. 417, pp. 49–58. CEUR-WS (2008)
7. Heim, P., Ziegler, J.: Faceted visual exploration of semantic data. In: Ebert, A., Dix, A., Gershon, N.D., Pohl, M. (eds.) HCIV (INTERACT) 2009. LNCS, vol. 6431, pp. 58–75. Springer, Heidelberg (2011)
8. Hostetter, C.: Faceted searching with Apache Solr. ApacheCon US (2006)
9. Nguyen, T.B., Schoepp, W., Wagner, F.: GAINS-BI: business intelligent approach for greenhouse gas and air pollution interactions and synergies information system. In: iiWAS 2008, pp. 332–338 (2008)
10. Nguyen, T.B., Wagner, F., Schoepp, W.: Cloud intelligent services for calculating emissions and costs of air pollutants and greenhouse gases. In: Nguyen, N.T., Kim, C.-G., Janiak, A. (eds.) ACIIDS 2011, Part I. LNCS, vol. 6591, pp. 159–168. Springer, Heidelberg (2011)

11. OECD Internet Economy Outlook 2012, 1st edn. [ebook] OECD (2012)
12. Pollitt, A.S., Smith, M., Treglown, M., Braekevelt, P.: View-based searching systems: progress towards effective disintermediation. In: proceedings of Online 1996 conference, London, England (1996)
13. Schraefel, M.C., Karam, M., Zhao, S.: mSpace: interaction design for user-determined, adaptable domain exploration in hypermedia. In: AH 2003: Workshop on Adaptive Hypermedia and Adaptive Web Based Systems, Nottingham, UK, pp. 217–235 (2003)
14. Shneiderman, B.: Dynamic queries for visual information seeking. IEEE Software 11(6), 70–77 (1994)
15. Tunkelang, D.: Faceted Search. Morgan & Claypool Publishers (2009)
16. Wagner, A., Ladwig, G., Tran, T.: Browsing-oriented semantic faceted search. In: Hameurlain, A., Liddle, S.W., Schewe, K.-D., Zhou, X. (eds.) DEXA 2011, Part I. LNCS, vol. 6860, pp. 303–319. Springer, Heidelberg (2011)

Current Derivative Estimation of Non-stationary Processes Based on Metrical Information

Elena Kochegurova[✉] and Ekaterina Gorokhova

Tomsk Polytechnic University, Lenin Avenue 30, Tomsk, Russian Federation, 634050
kocheg@tpu.ru, gorokhovaes@mail.ru

Abstract. Demand for estimation of derivatives has arisen in a range of some applied problems. One of the possible approaches to estimating derivatives is to approximate measurement data. The problem of real-time estimation of derivatives is investigated. A variation method of obtaining recurrent smoothing splines is proposed for estimation of derivatives. A distinguishing feature of the described method is recurrence of spline coefficients with respect to its segments and locality about measured values inside the segment. Influence of smoothing spline parameters on efficiency of such estimations is studied. Comparative analysis of experimental results is performed.

Keywords: Recurrent algorithm · Derivatives estimation · Variation smoothing spline

1 Introduction

Some basic information about a process or phenomenon is usually collected by an observed object on its own and recorded by measuring equipment within a certain interval. Any accompanying noise that distorts parameters of the object and non-stationary behavior of the object result in non-stationarity of the observed process. Such processes occur in every industrial sector.

Although many researchers investigate processing of non-stationary random processes, a wide range of problems still exists in this field. One of such problems is estimation of derivatives from experimental data.

Difficulty of obtaining estimations of derivatives is caused by incorrectness of the problem because a rate of change of a function is sensitive to drastic changes of error-containing data.

Having said that, methods of real-time recovery of a useful signal and its lower derivatives are of the utmost interest. Efficiency of any existing solutions is still not high and the problem remains relevant.

2 Method

This work is focused on the algorithms that are used in discrete-to-analog measuring equipment where at t_i, $i = 1, 2, ...,$ input sequence of data $y(t_i)$ with a random error $\xi(t_i)$ is recorded:

© Springer International Publishing Switzerland 2015
M. Nunez et al. (Eds.): ICCCI 2015, Part II, LNCS 9330, pp. 512–519, 2015.
DOI: 10.1007/978-3-319-24306-1_50

$$y(t_i) = f(t_i) + \xi(t_i),$$
$$M\{\xi(t)\} = 0,\ M\{\xi^2(t)\} = \sigma_\xi^2,\ M\{f(t), \xi(t)\} = 0 \tag{1}$$

A useful signal $f(t)$ is low-frequency and $\xi(t)$ denotes a wide-band noise.

The quality of the received information is mostly determined by methods and algorithms of processing metrical information.

One of the possible approaches to solving the problem of derivative estimation in observed processes is to smooth measurement results and their further usage in reverse tasks. Obtaining a smoothed analytical function that represents the major properties of the observed process allows calculating derivatives of any set order. Basic classes of approximation methods for experimental data are given and analyzed in [1].

Efficient approximation of metrical information has been achieved for interpolating and smoothing splines. [2, 3]

Global interpolating polynomial splines are the most developed. However, they associate with solving systems of algebraic equations, which causes enormous computational cost and makes this method improper for real-time estimations. Moreover, an interpolating spline cannot be used when there is much noise in measurement results.

The use of a minimal-norm-derivative spline allows giving a solution of required smoothness. Smoothing splines are based on optimization of special functions, which allows obtaining a model of a smoothed signal. [4, 5, 6]

The problem of derivation of measurement results using polynoms is traditionally considered in two variants. [7, 8, 9]

In the first case the target derivative is a solution of an integral equation of first kind: [7, 8]

$$\int_0^t \frac{(t-\tau)^{k-1}}{(k-1)!}(t-\tau)^{k-1}\varphi(\tau)d\tau = f(t) , \tag{2}$$

in the operational form

$$A \cdot \varphi = f \tag{3}$$

where φ denotes the k-th derivative of function $f(t)$.

The other way of derivation is related to application of a differential operator to the recovered function f(t):

$$A \cdot f = \varphi \tag{4}$$

The problem (2) is known as interpretation of indirect measurements while the problem (4) denotes interpretation of direct measurements.

In the problem (2), the Tikhonov regularization method has proved its efficiency. This method is characterized by strong dependence of obtained results from a regularization parameter value and absence of rigorous criteria for its use in case of an unknown measurement error. Furthermore, one regularization parameter is usually not sufficient for complex signals.

Solution of the problem (4) consists in derivation of an approximation function $f(t)$. A polynomial basis of high degree ensures computation of the k-th derivative. Because splines have some variable parameters – degree, defect, and nodes – the problem of successive derivation based on spline approximation may be considered constructive. Most attention in such solutions should be given to selection of a presentation form and a method for constructing a smoothing spline. Some fundamental results were obtained in a posteriori smoothing splines. In particular, in [9] there can be found an overview of estimated smoothing splines for precisely determined as well as for inaccurate data. Nevertheless, parametrization of a posteriori spline requires solving systems of linear equations (SLE), which leads to enormous calculations and makes real-time application of this method nearly impossible.

An appropriate approach to real-time estimation of parameters in an observed process is to construct recurrent calculating schemes. This paper proposes a calculating scheme for a recurrent smoothing spline where the number of measurements of every segment is higher than the number of nodes and spline coefficients are found with a variation method.

A recurrent smoothing spline (RSS) is based on the Tikhonov functional, which has been modified as: [6]

$$J(S) = (1-\rho)(h\Delta t)^2 \int_{t_0^i}^{t_h^i} [S''(t)]^2 \, dt + \rho \sum_{j=0}^{h} [S(t_j^i) - y(t_j^i)]^2 \tag{5}$$

where ρ denotes a weight coefficient, which determines a balance between smoothing and interpolative properties of the spline, $S(t)$, $\rho \in [0,1]$;

Δt is an interval of discretization for the observed process;

t_0^i, t_h^i are the first and the last abscises of the i-th segment of the spline;

$h = \dfrac{t_h^i - t_0^i}{\Delta t}$ denotes a number of measurements in the i-th segment of the spline;

later $h = \text{const}$ for every segment of the spline in the interval of the observed process; and $y(t_i) = f(t_i) + \xi(t_i)$ are measurements that represent an additive compound of the useful low-frequency signal $f(t)$ and a random error $\xi(t)$.

A cubic spline for the i-th segment has the following form:

$$S_i(\tau) = a_0^i + a_1^i \cdot \tau + a_2^i \cdot \tau^2 + a_3^i \cdot \tau^3, \\ -q \le \tau < h-q \tag{6}$$

In expression (6), τ is an index of a discrete sample of the i-th segment t_j^i and q denotes an index of a discrete sample of the i-th segment t_j^i where continuous derivatives of the spline are joined.

Conditions of joining spline segments determine continuous coefficients of the spline. Disconnected coefficients of the spline are found from the condition of minimizing the functional (5) and depend on the defect of the spline d. Thus, for defect $d = 2$ coefficients a_2^i and a_3^i are disconnected; for defect $d = 1$ only coefficient a_3^i is disconnected. Calculation formulas for spline coefficients are given in [6].

The other RSS parameter is the moment of joining spline segments q. There are several modes of using RSS depending on correlation between parameters q and τ. A current mode is shown in Figure 1, any other modes are described in [6].

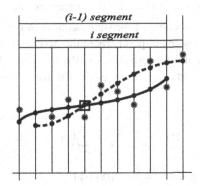

Fig. 1. Scheme of current RSS functioning. * - observed process; · - computed values; □ - moment of joining RSS segments

In a current mode of RSS estimation, a spline segment is calculated only after subsequent measurement is obtained from a previous $(h-1)$ measurement. In this mode joining happens at $t_{q+1}^{i-1} = t_q^i$ where q is the index of a discrete sample t_j^i of the i-th segment where continuous derivatives of the spline are joined.

This mode of using smoothing splines is especially valid for problems with minimum delays of the process estimation. It is also quite useful for reiterated real-time smoothings.

Let us now move on to calculation of a spline with the defect $d =2$ and $\Delta t = 1$.

Coefficients a_0^i, a_1^i are calculated from the condition $S^{(k)}(t_h^{i-1})_+ = S^{(k)}(t_0^i)_-$, ($k = 0,1$) and coefficients a_2^i, a_3^i are found from the condition of minimization (5) $\dfrac{\partial J(S)}{\partial a_2^i} = 0, \dfrac{\partial J(S)}{\partial a_3^i} = 0$.

Calculation formulas of spline coefficients for all modes are given in [6].

Expressions for estimation of smoothing spline parameters for defect $d =1$ may be found in a similar manner.

The last step in estimation of a derivate for an observed process is to analytically derivate the expression (6):

$$S_i'(\tau) = a_1^i + 2a_2^i \cdot \tau + 3a_3^i \cdot \tau^2, \quad -q \leq \tau < h-q \tag{7}$$

This method allows finding coefficients of a spline and estimating derivatives without using numerical algorithms. It reduces computation time for real-time estimation of derivatives.

3   Simulations

RSS parameters determine the quality of derivate estimation to a great extent. Such parameters are: a smoothing factor ρ that balances smoothing and interpolating properties of the spline, a number of measurements h in the observed segment of the spline, an index τ of the segment point where the derivative is estimated. The quality of recovering a function and its derivatives in the presence of an additive wide-band error with a determined σ_ξ has been estimated using root mean squared errors (RMSE).

$$RMSE\left(\hat{x}_j(t)\right) = \sqrt{\frac{1}{K}\sum_{j=1}^{K}\frac{1}{n}\sum_{i=1}^{n}\left(x(t_i) - \hat{x}_j(t_i)\right)^2}$$

$$RMSPE = \frac{RMSE}{x_{max} - x_{min}} \cdot 100 \ [\%]$$ (8)

where $\hat{x}_j(t_i)$ is the estimation of the function (first derivative) at time (t_i) in the j-th simulation; $x(t_i)$ is the true function (first derivative) at (t_i); $K = 200$ is a replicate simulations; $n = 100$ is the numbers of function values .

Optimal values of RSS customizable parameters have been found in a computational experiment when test functions given in Figure 3a and Figure 3c ($f_1(t)$ and $f_2(t)$ respectively) were used.

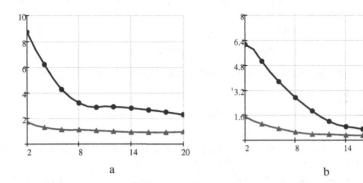

a   b

Fig. 2. Influence of the number of measurements in the spline segment h on a root mean square percentage error: (a) for the function $f_1(t) = 8 + 8.4 \cdot t - 0.2 \cdot t^2 + 1.33 \cdot 10^{-3} \cdot t^3$; (b) for the function $f_2(t) = 10 \cdot \sin\left(\frac{2\pi \cdot t}{100}\right)$

Length of a spline segment and a related parameter h have a dominant influence on the quality of recovering a function and its derivatives. Figure 2a and Figure 2b display dependence of RMSPE from the parameter h that changes in the range [0-20]. Widening the right range border at $\tau = 1$ in a current mode makes no sense, because it causes a critical delay, which is intolerable in real-time estimations. Error estimation function and the derivative is practically unchanged when length of a spline segment $h > 10$. Therefore $h = 10$ was selected for a further studies.

Figure 3 displays results of recovering functions $f_1(t)$ and $f_2(t)$ as well as their first derivatives. In this figure dots represent a useful signal; lines denote the recovered functions (3a and 3c) and their recovered derivatives (3b and 3d). The results are given for a particular case where $h = 10$ and $\sigma_\xi = 10\%$ from the maximum value of the useful signal. The graphs show that the derivatives have been recovered to a satisfying degree of visualization. However, although the recovery error is approximately the same (less than 2%), smoothness of the derivative in Figure 3d is much greater than that of the derivative in Figure 3b. Smoothness of derivative recovery can be regulated by the parameter ρ. Yet, improvement of smoothness leads to a systematic deviation, which is also an undesirable effect of recovery.

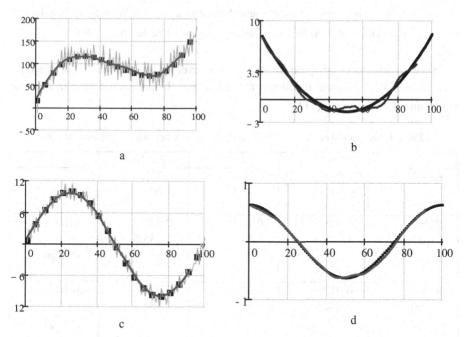

Fig. 3. Test functions and recovered derivatives: (a) useful signal $f_1(t) = 8 + 8.4 \cdot t - 0.2 \cdot t^2 + 1.33 \cdot 10^{-3} \cdot t^3$, signal-noise mixture, recovered function; (b) derivative $f_1'(t)$, recovered derivative; (c) useful signal $f_2(t) = 10 \cdot \sin\left(\dfrac{2\pi \cdot t}{100}\right)$, signal-noise mixture, recovered function; (d) derivative $f_2'(t)$, recovered derivative

A great amount of computations allowed estimating root mean squared errors in recovery of a function and its derivative depending on noise errors. Figure 4 demonstrates recovery errors in the error range (0-20)%. Throughout the estimated interval the value of the derivative recovery error is tolerable and can be used in engineering applications.

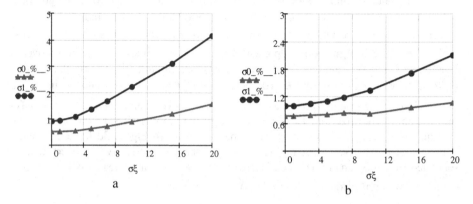

a   b

Fig. 4. Dependence between root mean squared percentage errors in recovery of function σ_0, derivative σ_1 and noise σ_ξ ; (a) for the function $f_1(t)$; (b) for the function $f_2(t)$;

In conclusion, let us compare the results of recovering a derivative by the proposed method (RSS) for the both functions $f_1(t)$, $f_2(t)$ and the parametric penalized spline smoothing (PPSS) given in [11].

Table 1. Estimation errors of a function derivatives when the parameter σ_ξ varies

	$\sigma_\xi = 3\%$	$\sigma_\xi = 7\%$	$\sigma_\xi = 10\%$	
PPSS	0.093	0.204	0.273	
	0.442 %	0.972 %	1.300 %	
RSS ($f_1(t)$)	0.108	0.168	0.220	h=10
	1.098 %	1.671 %	2.235 %	$\tau = 1$
				ρ - opt
RSS ($f_2(t)$)	0.013	0.015	0.017	h=10
	1.042 %	1.176 %	1.334 %	$\tau = 1$
				ρ - opt

Table 1 shows that PPSS provides a smaller error but RSS errors are the same degree for the chosen parameters of the spline. However only RSS allows real-time recovery of functions and their derivatives.

Acknowledgement. The work is performed out of the grant (agreement 14.575.21.0023, project unique identifier 2014-14-576-0052-044)

References

1. Golubinsky, A.N.: Approximation methods of experimental data and modeling. Herald of Voronezh Institute of MIA Russia **2**, 138–143 (2007). (In Russian)
2. Zavyalov, Y.S., Kvasov, B.I., Miroshnichenko, V.L.: Methods of spline functions. Nauka, Moscow (1980). (In Russian)
3. Vershinin, V.V., Zavyalov, Y.S., Pavlov, N.N.: Extreme properties of splines and smoothing problem, p. 102 c. Science, Novosibirsk (1988). (In Russian)
4. Rozhenko, A.I.: Theory and algorithms for variation spline approximation: Dr. Sci. Diss., January 1, 2007, Novosibirsk, p. 231 (2003). (In Russian)
5. Voskoboynikov, Y.E., Kolker, A.B.: Approximation of the contour image smoothing splines. Journal Avtometriya **39**(4), 3–12 (2003). (In Russian)
6. Ageev, U.M., Kochegurova, E.A.: Frequency properties of recurrent smoothing splines. Notify of High School, Instrumentmaking **3**, 3–8 (1990). (In Russian)
7. Dmitriev, V.I., Ingtem, J.G.: A two-dimensional minimum-derivative spline. Computational Mathematics and Modeling **21**, 206–211 (2010). Springer
8. De Brabanter, K., De Brabanter, J., De Moor, B.: Derivative Estimation with Local Polynomial Fitting. Journal of Machine Learning Research **14**, 281–301 (2013)
9. Ragozin, D.L.: Error bounds for derivative estimates based on spline smoothing of exact or noise data. Journal of Approximation Theory **37**, 335–355 (1983)
10. Kochegurova, E.A., Shebeko, E.V.: Usage of variation smoothing spline in short-term prediction problem. Notify of the Tomsk Polytechnic University **7**, 36–39 (2006). T.309, (In Russian)
11. Cao, J., Cai, J.: L. Wang: Estimating Curves and Derivatives with Parametric Penalized Spline Smoothing. Statistics and Computing **22**(5), 1059–1067 (2012)

An Intelligent Possibilistic Approach to Reduce the Effect of the Imperfection Propagation on Land Cover Change Prediction

Ahlem Ferchichi[(✉)], Wadii Boulila, and Imed Riadh Farah

Laboratoire RIADI, Ecole Nationale des Sciences de L'Informatique,
Manouba, Tunisia
ferchichi.ahlem@gmail.com, wadii.boulila@riadi.rnu.tn,
riadh.farah@ensi.rnu.tn

Abstract. The interpretation of satellite images in a spatiotemporal context is a challenging subject for remote sensing community. It helps predicting to make knowledge driven decisions. However, the process of land cover change (LCC) prediction is generally marred by imperfections which affect the reliability of decision about these changes. Propagation of imperfection helps improve the change prediction process and decrease the associated imperfections. In this paper, an imperfection propagation methodology of input parameters for LCC prediction model is presented based on possibility theory. The possibility theory has the ability to handle both aleatory and epistemic imperfection. The proposed approach is divided into three main steps: 1) an imperfection propagation step based on possibility theory is used to propagate the parameters imperfection, 2) a knowledge base based on machine learning algorithm is build to identify the reduction factors of all imperfection sources, and 3) a global sensitivity analysis step based on Sobol's method is then used to find the most important imperfection sources of parameters. Compared with probability theory, the possibility theory for imperfection propagation is advantageous in reducing the error of LCC prediction of the regions of the Reunion Island. The results show that the proposed approach is an efficient method due to its adequate degree of accuracy.

Keywords: LCC prediction · Parameter imperfection propagation · Sensitivity analysis · Possibility theory

1 Introduction

The interpretation of satellite images in a spatiotemporal context is a challenging subject for remote sensing community. It helps predicting to make knowledge driven decisions. However, the process of LCC prediction is generally marred by imperfections. Propagation of imperfection helps improve the change prediction process and decrease the associated imperfections. In general, imperfection in LCC prediction process can be categorized as both aleatory (randomness) and

© Springer International Publishing Switzerland 2015
M. Núñez et al. (Eds.): ICCCI 2015, Part II, LNCS 9330, pp. 520–529, 2015.
DOI: 10.1007/978-3-319-24306-1_51

epistemic (lack of knowledge). This imperfection, which can be subdivided into parameter and model structure imperfection. Only model parameters are used as factors in the imperfection propagation in this study. In LCC prediction process, the parameters imperfection is generally treated by probability theory. Moreover, numerous authors conclude that there are limitations in using probability theory in this context. So far, several non-probabilistic methods have been developed, which include Dempster-Shafer theory [6] and possibility theory [8]. These theories, in particular, may be the most attractive ones for risk assessment, because of their mathematical representation power. In our context of continuous measurements, the possibility theory is more adapted because it generalises interval analysis and provides a bridge with probability theory by its ability to represent a family of probability distributions. In summary, the possibility distribution has the ability to handle both aleatory and epistemic imperfection of pixel detection through a possibility and a necessity measures. On the other hand, the LCC prediction model involve a large number of input parameters. Thus, it is important to detect influential inputs, whose imperfections have an impact on the model output. Sensitivity analysis is the study of how imperfection in model output can be attributed to different sources of imperfection in the model input. Then, the objective of this study is to propose an imperfection propagation approach of parameters for the LCC prediction process. The proposed approach is divided into three main steps: 1) an imperfection propagation step based on possibility theory is used to propagate the parameters imperfection, 2) a knowledge base based on machine learning algorithm is build to identify the reduction factors of all imperfection sources, and 3) a global sensitivity analysis step based on Sobol's method is then used to find the most important imperfection sources of parameters. This paper is organized as follows: Section 2 describes the proposed approach. Section 3 presents our results. Finally, section 4 concludes the paper with summary and future works.

2   Proposed Approach

The proposed approach follows three main steps (Fig. 1): (1) propagation of parameter imperfection, (2) knowledge base is build to identify the reduction factors of all imperfection sources, and (3) global sensitivity analysis and step 1 is repeated only for those most important parameters in order to reduce the imperfections through LCC prediction model and the computational time.

2.1   Imperfection Propagation

This step is divided into three parts: (1) identification of uncertain input parameters for LCC prediction model, (2) identification of imperfection sources of parameters, and (3) propagation of parameter imperfection.

Identification of Uncertain Parameters. To better provide decisions about LCC, prediction model include several input parameters [3]. These parameters

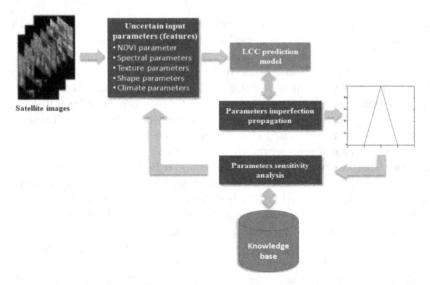

Fig. 1. General modelling proposed framework.

describe objects extracted from satellite images and which are subject of studying changes. In this study, objects are extracted from images coming from SPOT-4 image, including four multispectral bands and one monospectral band:

- 10 spectral parameters: Mean values and standard deviation values of green (MG, SDG), red (MR, SDR), NIR (MN, SDN), SWIR (MS, SDS) and monospectral (MM, SDM) bands for each image object;
- 5 texture parameters: homogeneity (Hom), contrast (Ctr), entropy (Ent), standard deviation (SD) and correlation (Cor).
- 7 shape and spatial relationships parameters: area (A), length/width (LW), shape index (SI), roundness (R), density (D), metric relations (MR) and direction relations (DR);
- 1 vegetation parameter: The NDVI (Normalized Difference Vegetation Index) is the ratio of the difference between NIR and red reflectance (VI).
- 3 climate parameters: temperature (Tem), humidity (Hum) and pressure (Pre).

Identification of Imperfection Sources. Imperfections related to the input parameters of LCC prediction model can be very numerous and affect model outputs. Therefore necessary to identify imperfections sources that should be considered for processing.
– *Spectral Parameters Imperfection Sources:* Several studies showed effects of spectral parameters[4]. Among these effects we list: spectral reflectance of the surface (S1), sensor calibration (S2), effect of mixed pixels (S3), effect of a shift in the channel location (S4), pixel registration between several spectral channels (S5), atmospheric temperature and moisture profile (S6), effect of haze particles (S7),

instrument's operation conditions (S8), atmospheric conditions (S9) and as well as by the stability of the instrument itself characteristics (S10).

− *Texture Parameters Imperfection Sources:* Those factors may produce a textural effect derived by the spatial interaction between the size of the object in the scene and the spatial resolution of the sensor (S11), a border effect (S12), and ambiguity in the object/background distinction (S13).

− *Shape Parameters Imperfection Sources:* Imperfection related to shape parameters can rely to the following factors [4]: accounting for the seasonal position of the sun with respect to the Earth (S14), conditions in which the image was acquired changes in the scene's illumination (S15), atmospheric conditions (S16) and observation geometry (S17).

− *NDVI Imperfection Sources:* Among factors that affect NDVI, we can list: variation in the brightness of soil background (S18), red and NIR bands (S19), atmospheric perturbations (S20), and variability in the sub-pixel structure (S21).

− *Climate Parameters Imperfection Sources:* Several studies showed effects of climate parameters [5]. These factors can be: atmospheric correction (S22), noise of the sensor (S23), land surface emissivity (S24), aerosols and other gaseous absorbers (S25), angular effects (S26), wavelength imperfection (S27), full-width half maximum of the sensor (S28) and band-pass effects (S29).

Propagation of Parameters Imperfection. The objective of this step is to propagate the imperfection of parameters using possibility theory. Using this theory to model imperfections has been studied extensively in several fields [6], [7].

a)Basics of Possibility Theory. The possibility theory developed by Dubois and Prade [2] handles imperfection in a qualitative way, but encodes it in the interval [0, 1] called possibilistic scale. The basic building block in the possibility theory is the notion of possibility distribution. A possibility distribution is defined as a mapping $\pi : \Omega \rightarrow [0,1]$. It is formally equivalent to the fuzzy set $\mu(x) = \pi(x)$. Distribution π describes the more or less plausible values of some uncertain variable X. To a possibility distribution are associated two measures, namely the possibility (Π) and necessity (N) measures, which read:

$$\Pi(A) = sup_{x \in A}\pi(x), \quad N(A) = inf_{x \notin A}(1 - \pi(x)) \tag{1}$$

The possibility measure indicates to which extent the event A is plausible, while the necessity measure indicates to which extent it is certain. They are dual, in the sense that $\Pi(A) = 1 - N(\overline{A})$, with $\overline{A}$ the complement of A. They obey the following axioms:

$$\Pi(A \cup B) = max(\Pi(A), \Pi(B)) \tag{2}$$

$$N(A \cap B) = min(N(A), N(B)) \tag{3}$$

An α-cut of π is the interval $[\underline{x}_\alpha, \overline{x}_\alpha] = \{x, \pi(x) \geq \alpha\}$. The degree of certainty that $[\underline{x}_\alpha, \overline{x}_\alpha]$ contains the true value of X is $N([\underline{x}_\alpha, \overline{x}_\alpha]) = 1 - \alpha$. Conversely, a collection of nested sets A_i with (lower) confidence levels λ_i can be modeled as a possibility distribution, since the α-cut of a (continuous) possibility distribution can be understood as the probabilistic constraint $P(X \in [\underline{x}_\alpha, \overline{x}_\alpha]) \geq 1 - \alpha$. In this setting, degrees of necessity are equated to lower probability bounds, and degrees of possibility to upper probability bounds.

b)Imperfection Propagation by Possibility Theory. In this section, the procedures of propagating the unified structures dealing with both imperfections of LCC prediction model will be addressed. Let us denote by $Y = f(X) = f(X_1, X_2, ..., X_j, ..., X_n)$ the model for LCC prediction with n uncertain parameters X_j, $j = 1, 2, ..., n$, that are possibilistic, i.e. , their imperfection is described by possibility distributions $\pi_{X1}(x_1), \pi_{X2}(x_2), ..., \pi_{Xj}(x_j), ..., \pi_{Xn}(x_n)$. In more detail, the operative steps of the procedure are the following:

1. set $\alpha = 0$;
2. select the α-cuts $A_\alpha^{X1}, A_\alpha^{X2}, , A_\alpha^{Xj}, ..., A_\alpha^{Xn}$ of the possibility distributions $\pi_{X_1}(x_1), \pi_{X_2}(x_2), ..., \pi_{X_j}(x_j), ..., \pi_{X_n}(x_n)$ of the possibilistic parameters X_j, $j = 1, 2, ..., n$, as intervals of possible values $\lfloor \underline{x}_{j,\alpha}, \overline{x}_{j,\alpha} \rfloor$ $j = 1, 2, ..., n$;
3. calculate the smallest and largest values of Y, denoted by $\underline{y}_\alpha$ and $\overline{y}_\alpha$, respectively, letting variables X_j range within the intervals $\lfloor \underline{x}_{j,\alpha}, \overline{x}_{j,\alpha} \rfloor$ $j = 1, 2, ..., n$; in particular, $\underline{y}_\alpha = inf_{j, X_j \in [\underline{x}_{j,\alpha}, \overline{x}_{l,\alpha}]} f(X_1, X_2, ..., X_j, ..., X_n)$ and $\overline{y}_\alpha = sup_{j, X_j \in [\underline{x}_{j,\alpha}, \overline{x}_{l,\alpha}]} f(X_1, X_2, ..., X_j, ..., X_n)$;
4. take the values $\underline{y}_\alpha$ and $\overline{y}_\alpha$ found in step 3. above as the lower and upper limits of the α-cut A_α^Y of Y;
5. if $\alpha < 1$, then set $\alpha = \alpha + \triangle\alpha$ and return to step 2. above; otherwise, stop the algorithm: the possibility distribution $\pi_Y(y)$ of $Y = f(X_1, X_2, ..., X_n)$ is constructed as the collection of the values $\underline{y}_\alpha$ and $\overline{y}_\alpha$ for each α-cut.

2.2   Knowledge Base

The objective of this knowledge base is to identify reduction factors of all parameters imperfection sources of satellite images objects which are then propagated. In general, the knowledge base stores the embedded knowledge in the system and the rules defined by an expert. In this study, we used an inductive learning technique to automatically build a knowledge base. Two main steps are proposed: training and decision tree generation. The learning step allows providing examples of the concepts to be learned. The second step is the decision tree generation. This step generates first decision trees from the training data. These decision trees are then transformed into production rules. Then, our knowledge base that contains all imperfection sources and their reduction factors is presented in Figure 2.

2.3   Sensitivity Analysis of Parameters

The main objective of the sensitivity analysis is to determine what are the input parameters of describing an object that contribute most to the imperfection of the prediction model output. In the proposed approach, a global sensitivity

If spectral parameters
   |Source = S1 and S2 : 'strict requirements for the instrument's design' and 'envisaging of
   appropriate procedures for on-board calibration' and 'choosing appropriate algorithms
   for radiometric correction'
   |Source = S3 and S4 and S5 : 'reducing the wavelength range of the irradiance or
   spectral response measurement' and 'reducing the cloud shadows and cloud
   contamination effects'
   |Source = S6 and S9 : 'choosing appropriate algorithms for atmospheric correction'
   |Source = S7 and S8 and S10 : 'reducing errors of sensor system itself'
If texture parameters
   |Source = S11 : 'using high spatial resolution'
   |Source = S12 and S13 : 'choosing appropriate methods for segmentation'
If shape parameters
   |Source = S14 and S15 : 'improving the platforms' stability and the carrier's velocity and
   'technological enhancement of the sensors themselves' and 'reducing effects of sun angles'
   |Source = S16 : 'reducing the effects of atmospheric conditions '
   |Source = S17 : 'improving the overall segmentation quality' and 'reducing the number of bad
   pixels and the size of bad areas and 'improvement the imperfection of pixels' response'
If NDVI parameter
   |Source = S18 and S20 : 'choosing appropriate algorithms for atmospheric correction'
   |Source = S19 : 'reducing errors in surface measurements for the NIR and red bands '
   |Source = S21 : 'reducing the temporal variations effects in the solar zenith and azimuth angles
   and 'reducing the sun angle effects and noise contamination'
If climate parameters
   |Source = S22 : 'choosing appropriate algorithms for atmospheric correction'
   |Source = S23 : 'reducing errors of sensor system itself'
   |Source = S24 : 'reducing the emissivity variations'
   |Source = S25 and S26 : 'reducing the sun angle effects and solar heating'
   |Source = S27 and S28 and S29 : 'reducing the errors of radiometer calibration and the errors
   of radiation' and 'reducing errors of spatial and temporal variability of clouds'

Fig. 2. Production rules generated from parameters imperfection sources.

analysis is used [11]. In this study, we focus more specifically on the Sobol's method for quantifying the influence of input parameters. This method is one of the most commonly used approaches to global sensitivity analysis. Let us denote by M the model for LCC prediction, $S = (A_1, ..., A_p)$ p independent input parameters with known distribution and Y the output of the prediction model. Y are decisions about LCC. Sobol indices (S_J), are used as an evaluation index for output parameters. The index S_J for output parameter Y is computed as follow:

$$S_J = \frac{V_J(Y_m)}{Var(Y)} \tag{4}$$

where $V_J(Y_m)$ is computed using equation (5) and $Var(Y)$ is computed using equation (6).

$$V_J(Y_m) = Var[E(Y_m|A_i)] = Var\{E[E(Y|A)A_i]\} \tag{5}$$

The variance of Y, $Var(Y)$, is the sum of contributions of all the input parameters $A = (A_i)_{i=1,...,p}$ and A. Here, A_ε denotes an additional input parameter, independent of A, named seed parameter.

$$Var(Y) = V_\varepsilon(Y) + \sum_{i=1}^{p} \sum_{|j|=i} [V_J(Y) + V_{J\varepsilon}(Y)] \tag{6}$$

$$V_\varepsilon(Y) = Var[E(Y|A_\varepsilon)], = Var[E(Y|A_iA_\varepsilon)] - V_i(Y) - V_\varepsilon(Y), ... \tag{7}$$

The interpretation of Sobol index is straightforward as their sum is equal to one. The larger an index value, the greater the importance of the variable or the group of parameters linked to this index. In the case of a model with p input parameters, the number of Sobol indices is $2^p - 1$.

3   Experimental Results

The aim of this section is to validate and to evaluate the performance of the proposed approach in imperfection propagation of LCC prediction model. This section is divided into three parts: Study area, validation and evaluation.

3.1   Study Area and Data

The study site is located in Reunion Island (Fig. 3), a small territory which is an overseas department of France in the Indian Ocean. Satellite images used for the experiments are coming from SPOT-4 satellite and acquired on 12 June 2007 and 09 June 2011 (Fig. 4).

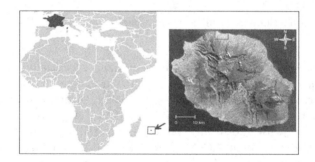

Fig. 3. The studied area.

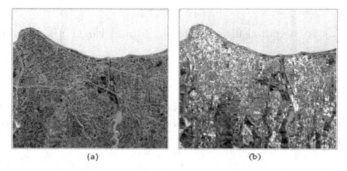

(a)   (b)

Fig. 4. Satellite images: (a) image acquired on 12 June 2007 and (b) image acquired on 09 June 2011.

3.2   Validation of the Proposed Approach

The validation section is divided into two main steps: 1) Propagation of parameters imperfection, and 2) Sensitivity analysis.

Propagation of Parameters Imperfection. In proposed approach, the propagating imperfections in the framework of possibility theory are realized. In fact,

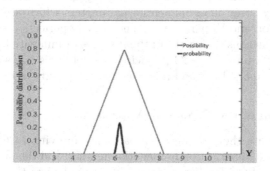

Fig. 5. The possibility distribution and its comparison with probability of the LCC.

Figure 5 shows the possibility distribution based on 10,000 samples. This distribution, as show in Figure 5, is compared with probability distribution.

Sensitivity Analysis. The second step consists to identify the most influential input parameters by sensitivity analysis. This helps determining input parameters that contribute most to the overall imperfection of prediction model. The results of the Sobol method are shown in Figure 6, and indicate that the MG, SDG, SDN, Ent, SD, Cor, and VI parameters are the most sensitive, while MR, SDR, MN, MS, A, MR and DR are relatively sensitive parameters and the parameters SDS, MM, SDM, Hom, Ctr, LW, SI, R, D, Tem, Hum and Pre are less sensitive for the first two objective function responses of interest. Before applying the sensitivity analysis, we can have the reduction factors of all imperfection

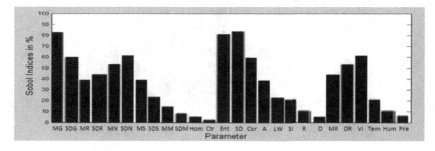

Fig. 6. Total sensitivity index of inputs parameters of LCC prediction model.

Table 1. Error for the prediction of LCCs between dates 2007 and 2011.

Approach	Predicting LCCs (%)
Approach in [10] *(possibilistic method without sensitivity analysis)*	0.32
Approach in [9] *(hybrid method without sensitivity analysis)*	0.28
Approach in [1] *(probabilistic method with sensitivity analysis)*	0.24
Proposed approach *(possibilistic method with sensitivity analysis)*	0.17

sources based on the knowledge base. After applying the sensitivity analysis, we will consider only seven most influential parameters which are the MG, SDG, SDN, Ent, SD, Cor, and VI parameters. Then, the imperfection propagation step can be repeated by considering the optimal values of influential parameters and we introduce them into the LCC prediction model. This helps obtaining more reliable decisions about LCC and reducing the complexity of the model.

3.3   Evaluation of the Proposed Approach

In order to evaluate the proposed approach in improving LCC prediction, we apply the proposed imperfection propagation method and the methods proposed in [1], [9], [10] for LCC prediction model. Table 1 depicts the error calculated between real LCCs, proposed method, and methods proposed in [1], [9], [10] between dates 2007 and 2011. As we note, the proposed approach provides a better results than the methods presented in [1], [9], [10] in predicting LCCs. This shows the effectiveness of the proposed approach in reducing imperfection related to the prediction process. Figure 7 illustrates percentages of change of the five land cover types (water, urban, bare soil, forest, and non-dense vegetation). Indeed, this Figure shows the difference between real changes and changes prediction for the proposed approach and the methods proposed in [1] [9] [10]. For example, as we note, the proposed approach predicts a change of 40,56% for urban object, while the real image shows a change of 42,89%. This provides a difference in the order of 2.33%. This result confirms the effectiveness of the proposed approach in improving LCC prediction.

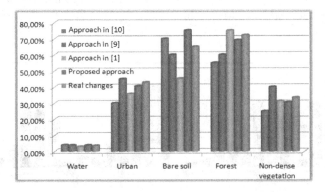

Fig. 7. Comparaison between real changes and changes prediction for the proposed approach and the methods proposed in [1], [9], [10].

4   Conclusion

This paper proposes the use of a possibility theory as the imperfection propagation method in LCC prediction model. The proposed approach is based on identification of imperfection sources of input parameters. After that, uncertain parameters are propagated with possibility theory. Then, we use a knowledge base to identify the reduction factors of all imperfection sources. To achieve this, we use a Sobol sensitivity analysis method. Finally, we repeat the imperfection propagation step by considering the optimal values of influential parameters and we introduce them into the LCC prediction model. This helps obtaining more reliable decisions about LCC. Proposed approach was compared on error prediction to existing propagation methods. Results show good performance of the proposed approach.

References

1. Boulila, W., Bouatay, A., Farah, I.R.: A Probabilistic Collocation Method for the Imperfection Propagation: Application to Land Cover Change Prediction. Journal of Multimedia Processing and Technologies **5**(1), 12–32 (2014)
2. Dubois, D., Prade, H.: When upper probabilities are possibility measures. Fuzzy Sets and Systems **49**, 65–74 (1992)
3. Boulila, W., Farah, I.R., Ettabaa, K.S., Solaiman, B., Ghézala, H.B.: A data mining based approach to predict Spatio-temporal changes in satellite images, International Journal of Applied. Earth **13**(3), 386–395 (2011)
4. Atanassov, V., Jelev, G., Kraleva, L.: Some Peculiarities of the Preprocessing of Spectral Data and Images. Journal of Shipping and Ocean Engineering **3**, 55–60 (2013)
5. Jimenez-Munoz, J.C., Sobrino, J.A.: Error sources on the land surface temperature retrieved from thermal infrared single channel remote sensing data. International Journal of Remote Sensing **27**(5), 999–1014 (2006)
6. Shafer, G.: A mathematical Theory of Evidence. Princeton University Press, Princeton (1976)
7. Tan, R.R., Culaba, A.B., Purvis, M.R.I.: Possibilistic Uncertainty Propagation and Compromise Programming in the Life Cycle Analysis of Alternative Motor Vehicle Fuels. Journal of Advanced Computational Intelligence and Intelligent Informatics **8**(1), 23–28 (2004)
8. Dubois, D., Prade, H.: Possibility theory, an approach to computerized processing of uncertainty. Plenum Press, New York (1988)
9. Aien, M., Rashidinejad, M., Firuzabadd, M.F.: On possibilistic and probabilistic uncertainty assessment of power flow problem: A review and a new approach. Renewable and Sustainable Energy Reviews **37**, 883–895 (2014)
10. André, J.C.S., Lopes, D.R.: On the use of possibility theory in uncertainty analysis of life cycle inventory. International Journal of Life Cycle Assessment **17**(3), 350–361 (2012)
11. Marrel, A., Iooss, B., Veiga, S., Ribatet, M.: Global sensitivity analysis of stochastic computermodels with joint metamodels. Statistics and Computing **22**(3), 833–847 (2012)

CIFM: Special Session on Computational Intelligence in Financial Markets

Measuring the Impact of Imputation
in Financial Fraud

Stephen O. Moepya[1,2]([✉]), Sharat S. Akhoury[1], Fulufhelo V. Nelwamondo[1,2],
and Bhekisipho Twala[2]

[1] CSIR Modeling and Digital Science, Pretoria, South Africa
{smoepya,sakoury,FNelwamondo}@csir.co.za
http://www.csir.co.za
[2] Faculty of Engineering and the Built Environment, University of Johannesburg,
Johannesburg, South Africa
btwala@uj.ac.za

Abstract. In recent years, data mining techniques have been used to
identify different types of financial frauds. In some cases, the fraud
domain of interest contains data with missing values. In financial statement fraud detection, instances which contain missing values are usually
discarded from the analysis. This may lead to crucial information loss.
Imputation is a technique to estimate missing values and is an alternative to case-wise deletion. In this paper, a study on the effectiveness
of imputation is taken using financial statement fraud data. Also, the
measure of similarity to the ground truth is examined using five distance
metrics.

Keywords: Financial statement fraud · Missing values · Imputation ·
Distance metrics

1 Introduction

Financial statement fraud (FSF) is a deliberate and wrongful act carried out
by public or private companies using materially misleading financial statements
that may cause monetary damage to investors, creditors and the economy. The
two most prominent examples of financial statement fraud are Enron Broadband
and Worldcom. In these two cases, wronged investors and the financial market
suffered as a result of this type of fraud. The collapse of Enron alone caused a
$70 billion market capitalization loss. The Worldcom scandal, caused by alleged
FSF, is the biggest bankruptcy in United States history [13]. In this light, the
application domain of financial statement fraud detection has attracted a keen
interest.

Public companies are required to issue audited financial statements at least
once a year. This requirement is to allow for *standardized* comparability between
companies. Financial statements contain information about the financial position, performance and cash flows of a company [10]. The statements also inform

M. Núñez et al. (Eds.): ICCCI 2015, Part II, LNCS 9330, pp. 533–543, 2015.
DOI: 10.1007/978-3-319-24306-1_52

readers about related party transactions. In practice, financial statement fraud (FSF) might involve [16]: the manipulation of financial records; intentional omission of events, transactions, accounts, or other significant information from which financial statements are prepared; or misapplication of accounting principles, policies, and procedures used to measure, recognize, report, and disclose business transactions.

Missing data, in general, is a common feature in many practical applications. In some cases, the data in FSF domain may contain missing values. Two approaches are taken when encountered with missing values: exclusion of all instances with missing values (casewise deletion) or an estimation of values for all missing items (imputation). Data can be missing completely at random (MCAR), missing at random (MAR) or missing not at random (MNAR) [14].

In this paper, the objective is to investigate the use of imputation techniques using authentic financial fraud data. Seven imputation techniques are investigated in order to test the significance of imputation. Imputed values will be evaluated via distance metric scores to attain a measure of similarity to known values ('ground truth'). The remainder of this paper is structured as follows. Section 2 reviews related research. Section 3 provides a brief description of the methodology. Sections 4 and 5 describe the sample data, experimental setup and details the results thereof. Finally, Section 6 presents the conclusion.

2   Related Work

In this section, a brief description of the impact of imputation in the finance domain follows.

Sorjamaa et al. [15] present a combination of Self-Organizing Maps (SOM) to treat missing values using corporate finance data. The study used data involving approximately 6000 companies listed on either the Paris or London Stock Exchange during the period 1999-2006. The authors concluded that using a combination of SOMs (instead of the traditional SOM imputation) yields better performance based on test error. Additional imputation schemes were not considered in this study to benchmark performance.

The use of multiple imputation (MI) on the consumer's choice between debit and credit cards is presented by King [9]. The dataset used in the experiment was from the 1998 Survey of Consumer Finances. The results suggest that the persistence of debit cards is due to the fact that even households that use credit cards without borrowing do not view credit as a substitute for debit. A limitation of the experiment was using only 5 imputed datasets for MI. Increasing the number of datasets used could have lead to a reduction in bias.

Fogarty [5] analyzed the importance of utilizing imputation to enhance credit score cards. The focus of this study was treating reject inference as a missing data problem. The data used was collected from a large German consumer finance company and includes 'accepts' and 'rejects' from the businesses application data and an associated credit performance variable from the behavioral data. The author concluded that model-based MI is an enhancement over traditional missing data approaches to reject inference. The effect of imputation was not tested against a benchmark dataset.

An interesting imputation approach on credit scoring data is presented by Paleologo et al. [11]. The authors' aim was to build and validate robust credit models on Italian company credit request data. Missing values were replaced with either the minimum or maximum value from the corresponding feature. There was no attempt to utilize standard imputation techniques to check predictive performance.

In light of the literature reviewed, the importance of imputing financial data has been highlighted. Imputation seems to be a viable option instead of case-wise deletion. Hence in this study, an investigation of several imputation methods using financial statement fraud data is undertaken. The similarity between a ground truth and imputed data will be computed using distance metrics.

3   Methodology

3.1   Imputation

The imputation of missing values can be broadly split into two categories: statistical and machine learning based imputation. Statistical imputation includes methods such as mean imputation, hot-deck and multiple imputation methods based on regression and the expectation maximization (EM) algorithm [6]. Machine learning approaches for imputing missing values create a predictive model to estimate missing values. These methods model the missing data estimation based on the available information in the data. For example, if the observed dataset contains some useful information for predicting the missing values, the imputation procedure can utilize this information and maintain a high precision. This section gives a description of both statistical and machine learning based imputation methods.

Mean imputation is one of the simplest methods to estimate missing values. Consider a matrix X containing a full data set. Suppose that the value x_{ij} belongs to the kth class C_k and it is missing. Mean imputation replaces x_{ij} with $\bar{x}_{ij} = \sum_{i:x_{ij} \in C_k} \frac{x_{ij}}{n_k}$, where n_k represents the number of non-missing values in the jth feature of the kth class.

In kNN imputation [8], missing cases are imputed using values calculated from corresponding k-nearest neighbors. The nearest neighbor of an arbitrary missing value is calculated by minimizing a distance function. The most commonly used distance function is the Euclidean distance between two instances y and z as $d(y, z) = \sqrt{\sum_{i \in D}(x_{yi} - x_{zi})^2}$, where D is a subset of the matrix X containing all instances without any missing values. Once k-nearest neighbors are computed, the mean (or mode) of the neighbors is imputed to replace the missing value.

Principal Component Analysis (PCA) imputation involves replacing missing values with estimates based on a PCA model. Suppose that the columns of matrix X are denoted by d-dimensional vectors $y_1, y_2, \cdots, y_n$. PCA imputation assumes that these vectors can be modeled as $y_j \approx W z_j + m$, where W is a

$d \times c$ matrix, z_j are the c-dimensional vectors of principal components and m is a bias vector. This imputation method iterates and converges to a threshold by minimizing the error $C = \sum_{j=1}^{n} \|y_j - W z_j - m\|^2$.

The Expectation-Maximization (EM) is an iterative procedure that computes the maximum likelihood estimator (MLE) when only a subset of the data is available. Let $X = (X_1, X_2, \cdots, X_n)$ be a sample with conditional density $f_{x|\Theta}(x|\theta)$ given $\Theta = \theta$. Assume that X has missing variables $Z_1, Z_2, \cdots, Z_{n-k}$ and observed variables $Y_1, Y_2, \cdots, Y_k$. The log-likelihood of the observed data Y is

$$l_{obs}(\theta; Y) = \log \int f_{X|\Theta}(Y, z|\theta) v_z(dz). \tag{1}$$

To maximize l_{obs} with respect to θ, the E-step and M-step routines are used. The E-step finds the conditional expectation of the missing values given observed values and current estimates of parameters. The second step, the M step, consists of finding maximum likelihood parameters as though the missing values were filled in [12]. The procedure iterates until convergence.

Singular Value Thresholding (SVT) [3] is a technique that has been used in Exact Matrix Completion (MC). The SVT algorithm solves the following problem:

$$\min_{X \in C} \tau \|X\|_* + \frac{1}{2} \|X\|_F^2, s.t. \; \mathcal{A}_{\mathcal{I}}(X) = \mathcal{A}_{\mathcal{I}}(M), \tag{2}$$

where $\tau \geq 0$ and the first and second norms are the nuclear and Frobenius norms respectively. M is an approximately low rank matrix. In the above equation, $\mathcal{A}$ is the standard matrix completion linear map where $\mathcal{A} : \mathbb{R}^{n_1 \times n_2} \to \mathbb{R}^k$. SVT is comprised of following two iterative steps:

$$\begin{cases} X_t = \mathcal{D}_t(A_{\mathcal{I}}^*(y_{t-1})) \\ y_t = y_{t-1} - \delta(A_{\mathcal{I}}(X_t) - b). \end{cases} \tag{3}$$

In the above equation, the shrinkage operator $\mathcal{D}_\tau$, also know as the soft-thresholding operator, is denoted as $\mathcal{D}_\tau = U \Sigma_\tau V^T$ where U and V are matrices with orthonormal columns and $\Sigma_\tau = diag(\max \{\sigma_i - \tau, 0\})$ with $\{\sigma_i\}_{i=1}^{\min\{n_1, n_2\}}$ corresponding to the singular values of the decomposed matrix. The step size of the iterative algorithmic process is given by δ.

Random Forests (RF), introduced by Breiman [1], is an extension of a machine learning technique named bagging which uses Classification and Regression Trees (CART) to classify data samples. Imputation via RF begins by imputing predictor means in place of the missing values. A RF is subsequently built on the data using roughly imputed values (numeric missing values are re-imputed as the weighted average of the non-missing values in that column). This process is repeated several times and the average of the re-imputed values is selected as the final imputation.

A brief explanation of Singular Value Decomposition (SVD) imputation follows. Consider the SVD of a matrix $X \in \mathbb{R}^{n_1 \times n_2}$ of rank r. In this instance, $X = U \Sigma V$, U and V are $n_1 \times r$ and $n_2 \times r$ orthogonal matrices respectively and $\Sigma = diag(\{\sigma_i\}_{1 \leq i \leq r})$. The σ_is are known as the positive singular values.

SVD imputation begins by replacing all missing values with some suited value (mean or random). The SVD is computed and missing values replaced with their prediction according to SVD decomposition. The process is repeated until the imputed missing data fall below some threshold.

3.2   Distance Metrics

In a formal sense, distance can be defined as follows. A distance is a function d with non-negative real values, defined on the Cartesian product $X \times X$ of a set X. It is termed a metric on X if $\forall\ x, y, z \in X$ it has the following properties:

1. $d(x,y) = 0 \iff x = y$;
2. $d(x,y) + d(y,z) \geq d(x,z)$; and
3. $d(x,y) = d(y,x)$.

Property 1 asserts that if two points x and y have a zero distance then they must be identical. The second property is the well known triangle inequality and states, given three distinct points x, y and z, the sum of two sides xy and yz will always be greater than or equal to side xz. The last property states that the distance measure is symmetrical.

Table 1. Distance/Similarity Metrics

Lorentzian	$d(x,y) = \sum_{i=1}^{n} \ln(1 + |x_i - y_i|)$
Minowski	$d(x,y) = \sqrt[p]{\sum_{i=1}^{n} |x_i - y_i|^p}$
Dice	$d(x,y) = \frac{\sum_{i=1}^{n}(x_i - y_i)^2}{\sum_{i=1}^{n} x_i^2 + \sum_{i=1}^{n} y_i^2}$
Squared Euclidean	$d(x,y) = \sum_{i=1}^{n}(x_i - y_i)^2$
Motyka	$d(x,y) = \frac{\sum_{i=1}^{n} \max(x_i, y_i)}{\sum_{i=1}^{n}(x_i + y_i)}$

Table 1 presents the definition of five distance metrics used in this experiment. The above similarity metrics each represent five of the eight types of similarity families. The data that will be used is not suitable for the Squared-chord, Shannon's entropy and combination families since it contains negative real values. The following section presents the results of the experimentation.

4   Data Description and Experimental Setup

4.1   Data

The dataset used in this experiment was obtained from INET BFA, one of the leading providers of financial data in South Africa. The data comprises publicly

Table 2. Summary statistics of the selected variables in the data

Variable	Min	1st Quartile	Median	Mean	3rd Quartile	Max
Assets-to-Capital Employed	-26.120	1.060	1.240	1.612	1.60	224.30
Book Value per Share	-966810	79	418	22279	1610	49596137
Cash flow per Share	-7198	8	84	3177	396	5159700
Current Ratio	0.010	1.00	1.410	3.283	2.251	726.130
Debt to Assets	0.0	0.270	0.4800	0.8804	0.680	1103.00
Debt to Equity	-182.370	0.350	0.820	2.516	1.730	760.940
Earnings per Share	-460338.5	2.3	44.5	391.6	218.2	825937.7
Inflation adjusted Profit per Share	-9232	1	43	2858	239	4898104
Inflation adjusted Return on Equity	-87472.97	3.11	13.45	-55.60	23.37	17063.16
Net Asset Value per Share	-373040	60	405	29438	1817	66658914
Quick Ratio	0.01	0.730	1.050	2.949	1.720	726.130
Retention Rate	-7204.35	57.67	89.39	70.97	100.00	5214.29
Return on Equity	-13600.00	4.045	14.830	-3.549	25.420	17063.160
Return on Capital Employed	-13600.00	1.500	8.700	-0.551	17.415	6767.330

listed companies on the Johannesburg Stock Exchange (JSE) between years 2003 and 2013. The different sectors for the listed companies on the JSE are: Basic Materials; Consumer Goods; Consumer Services; Financial; Health Care; Industrial; Oil and Gas; Technology; Telecommunications and Utilities. In the data, 123 (out of 3043) companies were known to have received a qualified financial report by an accredited auditor.

Each of the variables in Table 2 represent an aspect of measuring company performance. For example, the 'Quick Ratio' captures the amount of liquid assets per unit current liability. This ratio is a measure of how quickly a company can pay back its short-term debt. 'Earnings per share' (EPS) is a common metric for company valuation. It is a representation of a company's earnings, net of taxes and preferred stock dividends, that is allocated to each common stock. The other ratios fall into either the *profitability*, *solvency* or *leverage* category.

4.2 Experimental Setup

The experiments for this paper were conducted on a Intel(R) Core (TM) i5-3337U CPU @ 1.80 GHz with 6 GB memory. The implementation for algorithms are performed using the following R packages: 'imputation'[1] 'Amelia' [7], 'randomForest' [2] and 'yaImpute' [4].

The impact of imputation will be tested by running a Monte Carlo (MC) simulation (with 100 trails) as follows. In every MC trail:

1. Create 6 levels of missingness randomly from the 'ground truth' dataset using 1%, 2%, 5%, 10%, 15% and 20% as missing proportion;
2. Impute missing values on each missingness level using SVD, kNN, PCA, SVT, Mean, EM and RF imputation;
3. Compute the distance between the imputed values and the corresponding 'ground truth' values using the 5 distance/similarity measures (given in previous Section)

---

[1] This package has been now been archived in CRAN repository.

Once the Monte Carlo simulation is complete, compute the mean and standard deviations of the distances (from the 'ground truth') to evaluate the performance of the imputation techniques. The distances are evaluated using the five distance metrics.

The imputation schemes have parameters that needed to be selected. It was decided not to stray too far from the default parameter settings for the imputation methods. For PCA imputation: max iterations was set to 1000, number of principal components was set to 2 and a $1e - 06$ threshold. The rank approximation, r, was set the value 3 for SVD imputation. The Random Forest default parameters are as follows: $ntree = 300$ and $iter = 5$, these are the number of trees and iterations respectively. The default tolerance value for the EM imputation is $1e - 04$, and the value $k = 3$ was selected for kNN imputation.

5   Results

In this section, the results of the Monte Carlo simulation are presented. An analysis of the performance of the distance metrics will be undertaken.

Normalized distances of the 7 imputation methods using Lorentzian and Squared Euclidean distance

Table 3. Lorentzian

Method	1%	2%	5%	10%	15%	20%
EM	0.523	0.548	0.667	0.524	**0.487**	0.448
kNN	0.453	0.400	0.494	0.438	0.520	**0.395**
Mean	0.412	0.418	0.545	0.648	0.594	0.673
PCA	0.409	0.382	0.345	0.441	0.493	0.509
RF	0.486	0.399	0.437	0.491	0.560	0.533
SVD	**0.256**	**0.345**	**0.294**	**0.405**	0.504	0.536
SVT	0.445	0.412	0.457	0.492	0.550	0.412

Table 4. Squared Euclidean

Method	1%	2%	5%	10%	15%	20%
EM	0.018	**0.023**	**0.015**	**0.037**	**0.043**	**0.049**
kNN	0.019	0.035	0.067	0.168	0.191	0.277
Mean	**0.017**	0.032	0.050	0.147	0.165	0.239
PCA	0.025	0.039	0.069	0.165	0.201	0.285
RF	**0.017**	0.032	0.050	0.147	0.165	0.239
SVD	**0.017**	0.032	0.050	0.147	0.166	0.236
SVT	**0.017**	0.032	0.050	0.147	0.165	0.239

Table 3 presents the Lorentzian distance for each of the seven imputation techniques. Each column in the table represents the six levels of missingness that was randomly generated during the Monte Carlo simulation. Smaller values represent closer similarity to the ground truth. In this case, it can be seen that SVD imputation has the lowest normalized average distance for missingness levels $\leq 10\%$. For missingness above 10%, EM imputation and kNN produce the lowest average normalized distance with a score of 0.487 and 0.395 respectively. For all levels of missingness, Figure 1a shows a different imputation scheme attains the lowest standard deviation over the 100 Monte Carlo trails for the Lorentzia distance metric.

The Squared Euclidean distance results are presented in Table 4. At the 1% level of missingness, there is a four-way tie between the Mean, RF, SVD and SVT each achieving an average normalized distance of 0.17. For missingness greater than 1%, EM produced the lowest average distance values. These values seem to be substantially lower than distances of other imputation techniques, i.e., at the 20% missingness level EM achieves 0.049 and all other schemes have values

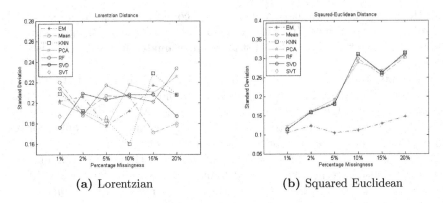

(a) Lorentzian   (b) Squared Euclidean

Fig. 1. Normalized standard deviation distances of the 7 imputation methods using Lorentzian and Squared Euclidean distance

greater than 0.2. Figure 1b shows that the minimum average standard deviation, for all levels of missingness, is produced by EM.

Normalized distances of the 7 imputation methods using Dice and Manhattan distance

Table 5. Dice

Method	1%	2%	5%	10%	15%	20%
EM	0.526	0.577	0.680	0.541	0.384	0.581
kNN	0.820	0.902	0.908	0.906	0.876	0.940
Mean	0.376	0.508	0.642	0.754	0.801	0.840
PCA	0.312	**0.297**	0.391	**0.391**	0.466	0.519
RF	**0.265**	0.467	**0.249**	0.526	**0.259**	0.834
SVD	0.297	0.352	0.420	0.440	0.417	**0.448**
SVT	0.821	0.766	0.835	0.792	0.649	0.864

Table 6. Manhattan

Method	1%	2%	5%	10%	15%	20%
EM	0.111	0.168	**0.121**	**0.201**	0.284	**0.212**
kNN	0.093	0.130	0.216	0.359	0.370	0.483
Mean	**0.065**	**0.085**	0.128	0.241	**0.248**	0.316
PCA	0.084	0.122	0.186	0.405	0.454	0.446
RF	0.072	0.100	0.163	0.298	0.275	0.390
SVD	0.073	0.108	0.170	0.325	0.346	0.418
SVT	0.068	0.101	0.153	0.283	0.282	0.365

Table 5 shows that RF achieves the lowest normalized average distance for missigness levels: 1%, 5% and 15%. PCA outperforms all other schemes for missingness level 2% and 10%. SVD attains the lowest value for the highest missingness level. At 20% missingness, EM, SVD and PCA attain average distances of lower than 0.6 whilst the other schemes score above 0.8. RF produces the lowest average standard deviation for most of the missingness levels. This is seen in Figure 2a where it achieves the least amount of variation in four out of the six levels using the dice distance metric.

The results for Manhattan distance are given in Table 6. These result bare some similarity to Table 5, where the similarity measure favors two imputation schemes. EM and Mean imputation achieve the lowest distance values levels 1%, 2% 15% and 5%, 10% and 20% respectively. Mean has increasing average distance values as the missingness increases. This is intuitive as the number of missing values increase, the mean of the remaining values are biased towards larger values. All imputation schemes exhibit this trend. Figure 2b mirrors results

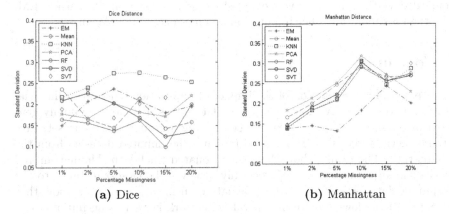

Fig. 2. Normalized standard deviation distances of the 7 imputation methods using Lorentzian and Squared Euclidean distance

in Figure 1b whereby the lowest values are produced by EM for all levels of missingness. Generally, it is seen that the greater the level of missingness, the larger the average standard deviation.

Table 7. Normalized distances of the 7 imputation methods using Motyka distance

Method	1%	2%	5%	10%	15%	20%
EM	**0.120**	0.850	**0.168**	0.495	0.901	**0.062**
kNN	0.659	0.623	0.425	0.507	0.514	0.522
Mean	**0.120**	**0.145**	0.186	**0.272**	**0.353**	0.407
PCA	0.178	0.276	0.532	0.860	0.428	0.469
RF	0.192	0.195	0.238	0.314	0.384	0.421
SVD	0.216	0.574	0.312	0.330	0.365	0.429
SVT	0.505	0.461	0.787	0.503	0.486	0.502

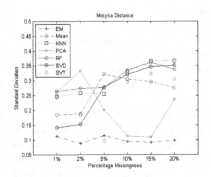

Fig. 3. Normalized standard deviation distances of the 7 imputation methods using Motyka distance

The results given by Table 7 slightly resemble those of Table 6 in that the lowest average distances are in favor of EM and Mean using the Motyka distance similarity. The minimum average distance (for EM and Mean) is 0.12 at the 1% level of missingness. The standard deviations of the Motyka similarity metric are presented in Figure 3. These values are similar to the Manhattan distance standard deviations (see Figure 2b) with the exception at the 5% level of missingness where the lowest standard deviation is produced by SVT. The average standard deviations for all levels of missingness are less than 0.14 using EM, while the maximum for other imputation schemes exceed 0.3 (Mean).

6 Conclusion

In this paper, an investigation of the impact of imputation schemes on financial data was undertaken. A Monte Carlo simulation was done and randomly generated missingness (1%-20%) was induced having a ground truth dataset. Five distance metrics were used in order to measure the imputed datasets from the ground truth. The results show that using Squared Euclidean, Manhattan and Motyka similarity measures, EM generally achieves the closest distance to the benchmark data. Also the standard deviations using those metrics show that EM obtained the lowest score in general. This work is seen as an initial study. Future possible extensions include parameter tuning for the imputation schemes. Another important contribution would be to test whether imputed datasets outperform the benchmark with respect to classification accuracy.

Acknowledgments. The current work is being supported by the Department of Science and Technology (DST) and Council for Scientific and Industrial Research (CSIR).

References

1. Breiman, L.: Random forests. Machine Learning **45**(1), 5–32 (2001)
2. Breiman, L.: randomforest: Breiman and cutlers random forests for classification and regression (2006)
3. Cai, J.F., Candès, E.J., Shen, Z.: A singular value thresholding algorithm for matrix completion. SIAM Journal on Optimization **20**(4), 1956–1982 (2010)
4. Crookston, N.L., Finley, A.O., et al.: yaimpute: An r package for knn imputation. Journal of Statistical Software **23**(10), 1–16 (2008)
5. Fogarty, D.J.: Multiple imputation as a missing data approach to reject inference on consumer credit scoring. Interstat (2006)
6. García-Laencina, P.J., Sancho-Gómez, J.L., Figueiras-Vidal, A.R.: Pattern classification with missing data: a review. Neural Computing and Applications **19**(2), 263–282 (2010)
7. Honaker, J., King, G., Blackwell, M., et al.: Amelia ii: A program for missing data. Journal of Statistical Software **45**(7), 1–47 (2011)
8. Jonsson, P., Wohlin, C.: An evaluation of k-nearest neighbour imputation using likert data. In: Proceedings of the 10th International Symposium on Software Metrics, 2004, pp. 108–118. IEEE (2004)
9. King, A.S., King, J.T.: The decision between debit and credit: finance charges, float, and fear. Financial Services Review **14**(1), 21–36 (2005)
10. Öğüt, H., Aktaş, R., Alp, A., Doğanay, M.M.: Prediction of financial information manipulation by using support vector machine and probabilistic neural network. Expert Systems with Applications **36**(3), 5419–5423 (2009)
11. Paleologo, G., Elisseeff, A., Antonini, G.: Subagging for credit scoring models. European Journal of Operational Research **201**(2), 490–499 (2010)

12. Regoeczi, W.C., Riedel, M.: The application of missing data estimation models to the problem of unknown victim/offender relationships in homicide cases. Journal of Quantitative Criminology **19**(2), 155–183 (2003)
13. Rezaee, Z.: Causes, consequences, and deterence of financial statement fraud. Critical Perspectives on Accounting **16**(3), 277–298 (2005)
14. Schafer, J.L., Graham, J.W.: Missing data: our view of the state of the art. Psychological Methods **7**(2), 147 (2002)
15. Sorjamaa, A., Lendasse, A., Severin, E.: Combination of soms for fast missing value imputation. In: Proceedings of MASHS, Models and Learnings in Human and Social Sciences (2010)
16. Zhou, W., Kapoor, G.: Detecting evolutionary financial statement fraud. Dezhoucision Support Systems **50**(3), 570–575 (2011)

An Improved Grey Forecasting Models: Case in China's Coal Consumption Demand

Chia-Nan Wang and Van-Thanh Phan[(✉)]

Department of Industrial Engineering and Management,
National Kaohsiung University of Applied Sciences, Kaohsiung, Taiwan
cn.wang@cc.kuas.edu.tw, thanhkem2710@gmail.com

Abstract. In order to improve the application area and the prediction accuracy of classical GM (1, 1) and Non Linear Grey Bernoulli Model (NGBM (1, 1)), a Fourier Grey model FRMGM (1, 1), and Fourier Non Linear Grey Bernoulli Model (abbreviated FRMNGBM (1, 1)) are proposed in this paper. These proposed models were built by using Fourier series to modify their residual values. To verify the effectiveness of these proposed models, the total coal consumption demand in China during period time from 1980 to 2012 was used to exam the forecast performance. The empirical results demonstrated that the accuracy of both GM (1, 1) and NGBM (1, 1) forecasting models after using Fourier series revised their residual error provided more accuracy than original ones. Furthermore, this paper also indicated that the FRMNGBM (1, 1) is the better model with MAPE=0.003.

Keywords: GM (1, 1) · Nonlinear grey bernoulli model · Fourier series · Coal consumption demand · China

1   Introduction

Grey system theory, founded by Prof. Deng in the 1980s [1], is a quantitative method for dealing with grey systems that are characterized by both partially known and partially unknown information [2, 3, 4]. As a vital part of Grey system theory, grey forecasting models with their advantages in dealing with uncertain information and using as few as four data points [5, 6]. Therefore, it has been successful in applied to various fields such as tourism [7, 8], sea transportation [9, 10], financial and economic [11, 12], integrated circuit industry [13, 14]. Especially, in the energy industry GM (1, 1) has also been used [15, 16].

In the recent years, many scholars have been proposed new procedures or new models to improve the precision of GM (1, 1) model and NGBM (1, 1), such as Wang and Phan [17], proposed FRMGM (1, 1) to forecast the GDP growth rate in Vietnam by used Fourier function. Wang et al [18] proposed optimized NGBM (1, 1) model for forecasting the qualified discharge rate of industrial wastewater in China by improved background interpolation value p and exponential value n is put forward in an NGBM (1, 1). Shang Ling [19] provided GAIGM (1, 1) by used genetic algorithm combined with improved GM (1, 1) to forecast the agriculture output from 1998 to 2010. Truong and Ahn [20] investigated a novel Grey model named "Smart Adaptive Grey Model, SAGM (1, 1)", etc.

© Springer International Publishing Switzerland 2015
M. Núñez et al. (Eds.): ICCCI 2015, Part II, LNCS 9330, pp. 544–553, 2015.
DOI: 10.1007/978-3-319-24306-1_53

In order to improve the quality of the forecasting models, in this study, a new approach to minimize the errors obtained from the conventional models is suggested by modifying the residual series with Fourier series. Practical example in coal consumption forecasting shows that the proposed FRMNGBM (1, 1) model has higher performance on model prediction. The remaining of this paper was organized as follows. In section 2, the modeling and methodology of GM (1, 1), NGBM (1, 1), and Fourier Residual Modification of their models was presented. Based on the fundamental function of GM (1, 1), NGBM (1, 1) and new two models FRMGM (1, 1) and FRMNGBM (1, 1) are created by using Fourier function to modified their residual , the case study and empirical results was shown in section 3. Finally, section 4 concluded this paper.

2   Modeling and Methodology

2.1   Classical GM (1, 1) Model

GM (1, 1) is the basic model of Grey forecasting modeling, a first order differential model with one input variable which has been successfully applied in many different researches. It is obtained as the following procedure.

Step 1: Let raw matrix $X^{(0)}$ stands for the *non-negative* original historical time series data

$$X^{(0)} = \left\{ x^{(0)}(t_i) \right\}, \; i = 1, 2, ..., \; n \tag{1.1}$$

Where $x^{(0)}(t_i)$ is the value at time t_i, and n is the total number of modeling data

Step 2: Construct $X^{(1)}$ by one time accumulated generating operation (1-AGO), which is

$$X^{(1)} = \left\{ x^{(1)}(t_i) \right\}, \; i = 1, 2, ..., n \tag{1.2}$$

$$\text{Where} \quad x^1(t_k) = \sum_{i=1}^{k} x^{(0)}(t_i), k = 1, 2, ..., \; n \tag{1.3}$$

Step 3: $X^{(1)}$ is a monotonic increasing sequence which is modeled by the first order linear differential equation:

$$\frac{dX^{(1)}}{dt} + aX^{(1)} = b \tag{1.4}$$

Where the parameter a is called the developing coefficient and b is the named the grey input.

Step 4: In order to estimate the parameter a and b, Eq. (1.4) is approximated as:

$$\frac{\Delta X^{(1)}(t_k)}{dt_k} + aX^{(1)}(t_k) = b \tag{1.5}$$

$$\text{Where} \quad \Delta X^{(1)}(t_k) = x^{(1)}(t_k) - x^{(1)}(t_{k-1}) = x^{(0)}(t_k) \tag{1.6}$$

$$\Delta t_k = t_k - t_{k-1} \tag{1.7}$$

If the sampling time interval is units, then let $\Delta t_k = 1$, Using

$$z^{(1)}(t_k) = px^{(1)}(t_k) + (1-p)x^{(1)}(t_{k-1}), \quad k = 2,3,..., n \tag{1.8}$$

To replace $x^{(1)}(t_k)$ in Eq. (1.5), we obtain

$$x^{(0)}(t_k) + az^{(1)}(t_k) = b, \quad k = 2,3,..., n \tag{1.9}$$

Where $z^{(1)}(t_k)$ in Eq. (1.8) is termed background value, and p is production coefficient of the background value in the range of (0, 1), which is traditionally set to 0.5.

Step 5: From the Eq. (1.9), the value of parameter a and b can be estimated using least-square method. That is

$$\begin{bmatrix} a \\ b \end{bmatrix} = (B^T B)^{-1} B^T Y_n \tag{1.10}$$

Where

$$B = \begin{bmatrix} -z^{(1)}(t_2) & 1 \\ -z^{(1)}(t_3) & 1 \\ ...... & ..... \\ -z^{(1)}(t_n) & 1 \end{bmatrix} \tag{1.11}$$

And $Y_n = \begin{bmatrix} x^{(0)}(t_2), & x^{(0)}(t_3),..., & x^{(0)}(t_n) \end{bmatrix}^T$ \tag{1.12}

Step 6: The solution of Eq. (1.4) can be obtained after the parameter "a" and "b" have been estimated. That is

$$\hat{x}^{(1)}(t_k) = \left[\left(x^{(0)}(t_1) - \frac{b}{a} \right) e^{-a(t_k - t_1)} + \frac{b}{a} \right], \quad k = 1,2,3,... \tag{1.13}$$

Step 7: Applying inverse accumulated generating operation (IAGO) to $\hat{x}^{(1)}(t_k)$, the predicted datum of $x^{(0)}(t_k)$ can be estimated as:

$$\begin{cases} \hat{x}^{(0)}(t_1) = x^{(0)}(t_1) & \tag{1.14} \\ \hat{x}^{(0)}(t_k) = \hat{x}^{(1)}(t_k) - \hat{x}^{(1)}(t_{k-1}) \end{cases}, \quad k = 2,3,... \tag{1.15}$$

2.2   Nonlinear-Grey Bernoulli Model "NGBM (1, 1)"

The procedures of deriving NGBM are as follows:

Step 1: Let raw matrix $X^{(0)}$ stands for the non-negative original historical time series data

$$X^{(0)} = \left\{ x^{(0)}(t_i) \right\}, \quad i = 1,2,..., n \tag{2.1}$$

Where $x^{(0)}(t_i)$ corresponds to the system output at time t_i, and n is the total number of modeling data.

Step 2: Construct $X^{(1)}$ by one time accumulated generating operation (1-AGO), which is

$$X^{(1)} = \left\{ x^{(1)}(t_i) \right\}, \; i = 1,2,..., \; n \tag{2.2}$$

Where $x^1(t_k) = \sum_{i=1}^{k} x^{(0)}(t_i), k = 1,2,..., \; n \tag{2.3}$

Step 3: $X^{(1)}$ is a monotonic increasing sequence which is modeled by the Bernoulli differential equation:

$$\frac{dX^{(1)}}{dt} + aX^{(1)} = b\left[X^{(1)} \right]^r \tag{2.4}$$

Where the parameter a is called the developing coefficient and b is the named the grey input and r is any real number excluding $r=1$.

Step 4: In order to estimate the parameter a and b, Eq. (2.4) is approximated as:

$$\frac{\Delta X^{(1)}(t_k)}{dt_k} + aX^{(1)}(t_k) = b\left[X^{(1)}(t_k) \right]^r \tag{2.5}$$

Where $\Delta X^{(1)}(t_k) = x^{(1)}(t_k) - x^{(1)}(t_{k-1}) = x^{(0)}(t_k)$

(2.6)

$$\Delta t_k = t_k - t_{k-1} \tag{2.7}$$

If the sampling time interval is units, then let $\Delta t_k = 1$, Using

$$z^{(1)}(t_k) = px^{(1)}(t_k) + (1 - p)x^{(1)}(t_{k-1}), \; k = 2,3,..., \; n \tag{2.8}$$

To replace $X^{(1)}(t_k)$ in Eq. (2.5), we obtain

$$x^{(0)}(t_k) + az^{(1)}(t_k) = b\left[z^{(1)}(t_k) \right]^r, \; k = 2,3,..., \; n \tag{2.9}$$

Where $z^{(1)}(t_k)$ in Eq. (2.8) is termed background value, and p is production coefficient of the background value in the range of (0, 1), which is traditionally set to 0.5.

Step 5: From the Eq. (2.9), the value of parameter a and b can be estimated using least- square method. That is

$$\begin{bmatrix} a \\ b \end{bmatrix} = (B^T B)^{-1} B^T Y_n \tag{2.10}$$

$$\text{Where } B = \begin{bmatrix} -z^{(1)}(t_2) & (z^{(1)}(t_2))^r \\ -z^{(1)}(t_3) & (z^{(1)}(t_3))^r \\ \cdots & \cdots \\ -z^{(1)}(t_n) & (z^{(1)}(t_n))^r \end{bmatrix} \tag{2.11}$$

$$\text{And } Y_n = \left[x^{(0)}(t_2), \, x^{(0)}(t_3), \ldots, \, x^{(0)}(t_n) \right]^T \tag{2.12}$$

Step 6: The solution of Eq. (2.4) can be obtained after the parameter a and b have been estimated. That is

$$\hat{x}^{(1)}(t_k) = \left[\left(x^{(0)}(t_1)^{(1-r)} - \frac{b}{a} \right) e^{-a(1-r)(t_k-t_1)} + \frac{b}{a} \right]^{\frac{1}{1-r}}, \, r \neq 1, \quad k = 1,2,3,\ldots \tag{2.13}$$

Step 7: Applying inverse accumulated generating operation (IAGO) to $\hat{x}^{(1)}(t_k)$, the predicted datum of $x^{(0)}(t_k)$ can be estimated as:

$$\begin{cases} \hat{x}^{(0)}(t_1) = x^{(0)}(t_1) \\ \hat{x}^{(0)}(t_k) = \hat{x}^{(1)}(t_k) - \hat{x}^{(1)}(t_{k-1}) \end{cases}, \quad k = 2,3,\ldots \quad \begin{matrix} (2.14) \\ (2.15) \end{matrix}$$

2.3   Fourier Residual Modification

In order to improve the accuracy of forecasting models, this paper proposed the Fourier series [7, 17] to modify the residuals in GM (1, 1) and NGBM (1, 1) models which reduces the values of MAPE. The overall procedure to obtain the modified model is as the followings:

Let x is the original series of n entries and v is the predicted series (obtained from GM (1, 1) or NGBM (1, 1). Based on the predicted series v, a residual series named $\mathcal{E}$ is defined as:

$$\mathcal{E} = \{\varepsilon(k)\}, \, k = 2,3,\ldots n \tag{16}$$

$$\text{Where } \varepsilon(k) = x(k) - v(k), \, k = 2,3,\ldots n \tag{17}$$

Expressed in Fourier series, $\varepsilon(k)$ is rewritten as:

$$\hat{\varepsilon}(k) = \frac{1}{2}a_{(0)} + \sum_{i=1}^{z} \left[a_i \cos\left(\frac{2\pi i}{n-1}(k) \right) + b_i \sin\left(\frac{2\pi i}{n-1}(k) \right) \right], \, k = 1,2,3,\ldots, n \tag{18}$$

Where $z = \left(\dfrac{n-1}{2} \right) - 1$ called the minimum deployment frequency of Fourier series [7] and only take integer number, therefore, the residual series is rewritten as:

$$\mathcal{E} = PC \tag{19}$$

Where

$$P = \begin{bmatrix} \frac{1}{2} & \cos\left(\frac{2\pi\times1}{n-1}\times2\right)\sin\left(\frac{2\pi\times1}{n-1}\times2\right) & \dots & \cos\left(\frac{2\pi\times Z}{n-1}\times2\right)\sin\left(\frac{2\pi\times Z}{n-1}\times2\right) \\ \frac{1}{2} & \cos\left(\frac{2\pi\times1}{n-1}\times3\right)\sin\left(\frac{2\pi\times1}{n-1}\times3\right) & \dots & \cos\left(\frac{2\pi\times Z}{n-1}\times3\right)\sin\left(\frac{2\pi\times Z}{n-1}\times3\right) \\ \dots & \dots & \dots & \dots \\ \frac{1}{2} & \cos\left(\frac{2\pi\times1}{n-1}\times n\right)\sin\left(\frac{2\pi\times1}{n-1}\times n\right) & \dots & \cos\left(\frac{2\pi\times Z}{n-1}\times n\right)\sin\left(\frac{2\pi\times Z}{n-1}\times n\right) \end{bmatrix} \tag{20}$$

And $C = [a_0, a_1, b_1, a_2, b_2, \dots, a_z, b_z]$ (21)

The parameter a_0, a_1, b_1, a_2, b_2... a_z, b_z are obtained by using the ordinary least squares method (OLS) which results in the equation of:

$$C = \left(P^T P\right)^{-1} P^T \left[\varepsilon\right]^T \tag{22}$$

Once the parameters are calculated, the modified residual series is then achieved based on the following expression:

$$\hat{\varepsilon}(k) = \frac{1}{2}a_{(0)} + \sum_{i=1}^{z}\left[a_i \cos\left(\frac{2\pi i}{n-1}(k)\right) + b_i \sin\left(\frac{2\pi i}{n-1}(k)\right)\right] \tag{23}$$

From the predicted series v and $\hat{\varepsilon}$, the Fourier modified series $\hat{v}$ of series v is determined by:

$$\hat{v} = \left\{\hat{v}_1, \hat{v}_2, \hat{v}_3, \dots, \hat{v}_k, \dots, \hat{v}_n\right\} \tag{24}$$

Where

$$\hat{v} = \begin{cases} \hat{v}_1 = v_1 \\ \hat{v}_k = v_k + \hat{\varepsilon}_k \quad (k = 2,3,\dots, n) \end{cases} \tag{25}$$

2.4   Evaluative Precision of Forecasting Models

In order to evaluate the forecast capability of these models, Means Absolute Percentage Error (MAPE) index was used to verify the performance and reliability of forecasting technique [21]. It is expressed as follows:

$$MAPE = \frac{1}{n}\sum_{k=2}^{n}\left|\frac{x^{(0)}(k) - \hat{x}^{(0)}(k)}{x^{(0)}(k)}\right| \times 100\ \% \tag{26}$$

Where $x^{(0)}(k)$ and $\hat{x}^{(0)}(k)$ are actual and forecasting values in time period k, respectively, and n is the total number of predictions. Wang et al [22] interprets the MAPE results as a method to judge the accuracy of forecasts, where more than 10% is an inaccurate forecast, 5%-10% is a reasonable forecast, 1%-5% is a good forecast, and less than 1% is a highly accurate forecast.

3   Simulation and Empirical Results

3.1   Data Analysis

The historical data of total coal consumption in China for 1980-2012 are obtained from the yearly statistical data published on the website of U.S. Energy Information Administration (EIA) [23]. All historical data was described in Figure 1, and the unit of coal consumption is measured in Quadrillion Btu (Q.T).

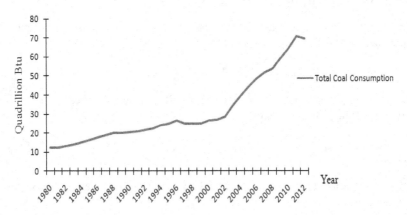

Fig. 1. Time series plots of China's coal consumption 1980-2012

Figure 1 clearly showed that the consumption demand of coal quickly increasing. The coal consumption reached to 70 Quadrillion Btu in 2012, increased 50 Quadrillion Btu compared with year of 1980.

To find out the parameters in GM (1, 1) and NGBM (1, 1) model as well as modified model of their models, computer software called Microsoft Excel is used. Beside a basic function in excel, Excel software also offers two useful functions named Mmult (array 1, array 2) to return the matrix product of two relevant arrays and Minverse (array) to return the inverse matrix. These two functions are of great help to find out the values of parameters in GM (1, 1), NGBM (1, 1) and Fourier residual modification.

3.2   GM (1, 1) Model for the Coal Consumption Demand

From historical data in the website of EIA [23] and based on the algorithm expressed in section 2.1, the coefficient parameter a and b in GM (1, 1) for the coal consumption is calculated as:

a= -0.0608, b= 8.7482 and the GM (1, 1) model for coal consumption demand are then:

$$\hat{x}^{(1)}(k) = 156.14 \times e^{-0.0608(k-1)} - 143.84$$

The evaluation indexes of GM (1, 1) model are also listed in Table 1. The residual series of GM (1, 1) is modified with Fourier series as illustrated in section 3.3

3.3   Modified GM (1, 1) Model by Fourier Series

The residual series of GM (1, 1) obtained in section 3.2 is now modified with Fourier series as per the algorithm stated in section 2.3. With this modified series, the forecasted values demand of coal consumption based on Fourier residual modified GM (1, 1) model FRMGM (1, 1) are calculated based on the equation (24). The evaluation index of FRMGM (1, 1) is summarized in Table 1.

3.4   NGBM (1, 1) Model for the Coal Consumption Demand

As the way to calculate and based on the mathematical algorithm expressed in section 2.2, the coefficient parameter a, b and the power of r in NGBM (1, 1) for the coal demand are calculated as: $a= -0.0659, b=25.1403$, and $r=-0.2374$

And the NGBM (1, 1) model for coal consumption is then:

$$\hat{x}^{(1)}(k)=\left[(403.7017) e^{0.06388(1+0.2374)(k-1)} -381.3896 \right]^{\frac{1}{1+0.2374}}$$

The evaluation index of NGBM (1, 1) model is also listed in Table 1. The residual series of NGBM (1, 1) is modified with Fourier series as illustrated in section 3.5

3.5   Modified NGBM (1, 1) Model by Fourier Series

The residual series of NGBM (1, 1) obtained in section 3.4 is now modified with Fourier series as per the algorithm stated in section 2.3. With this modified series, the forecasted values of coal consumption based on Fourier residual modified GM (1, 1) model FRMGM (1, 1) are calculated based on the equation (24). The evaluation index of FRMNGBM (1, 1) is summarized in Table 1.

Table 1. Summary of evaluation indexes of model accuracy

Model	MAPE (%)	Forecasting accuracy (%)	Performance
GM (1,1) with p =0.5	13.79	86.21	Good
NGBM(1, 1) with p=0.5 and r=-0.237	10.39	89.61	Good
FRMGM (1, 1) with p =0.5	**0.066**	**99.93**	**Excellent**
FRMNGBM (1 ,1) with p=0.5 and r=-0.237	**0.003**	**99.997**	**Excellent**

Table 1 shows all evaluation indexes of each model of GM (1, 1), FRMGM (1, 1), NGBM (1, 1) and FRMNGBM (1, 1) with its performance in forecasting coal consumption. The empirical results clearly show that the accuracy of both GM (1, 1) and NGBM (1, 1) forecasting models after using Fourier series revised their residual error provided more accuracy than original ones. Furthermore, this paper also indicated that the forecasting ability of FRMNGBM (1, 1) model is the better model with MAPE=0.003 (less much than FRMGM (1, 1) with MAPE =0.066) for forecasting the coal consumption demand in China. More detail was illustrated in figure 2.

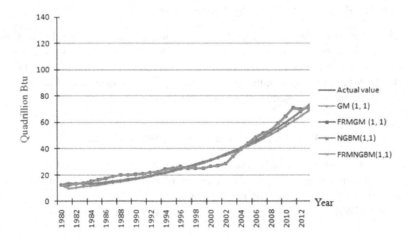

Fig. 2. Curves of actual values and simulated values using four models for coal consumption forecasting

4   Conclusion

Grey forecasting model is one of the most important parts of grey theory system. In the recent years, many scholars have been proposed new procedures or new models to improve the precision of GM (1, 1) model and NGBM (1, 1) with different ways, in this paper proposed an another way in order to improve the accuracy level of their performance by modify the residuals of GM (1, 1) and NGBM (1, 1) by Fourier series. This results displayed robust models in term of MAPE, compared with among them, FRMNGBM (1, 1) is the better model in forecast with the MAPE for forecasting the demand of cold consumption in China is 0.003%. Future researchers can be applied the proposed model on the other industries to forecast performance.

References

1. Deng, J.L.: Control problems of grey systems. Systems and Control Letters **5**, 288–294 (1982)
2. Emil, S., Camelia, D.: Complete analysis of bankruptcy syndrome using grey systems theory. Grey Systems: Theory and Application **1**, 19–32 (2011)
3. Hong, W., Fuzhong, C.: The application of grey system theory to exchange rate prediction in the post-crisis era. International Journal of Innovative Management **2**(2), 83–89 (2011)
4. Chen, H.J.: Application of grey system theory in telecare. Computers in Biology and Medicine **41**(5), 302–306 (2011)

5. Yi, L., Sifeng, L.: A historical introduction to grey system theory. In: IEEE International Conference on System, Man And Cybernetics, pp. 2403–2408 (2004)
6. Sifeng, L., Forrest, J., Yingjie, Y.: A brief introduction to grey system theory. In: 2011 IEEE International Conference on IEEE Grey Systems and Intelligent Services (GSIS) (2011)
7. Huang, Y.L., Lee, Y.H.: Accurately forecasting model for the stochastic volatility data in tourism demand. Modern economy 2(5), 823–829 (2011)
8. Chu, F.L.: Forecasting Tourism Demand in Asian-Pacific Countries. Annual of Tourism Research 25(3), 597–615 (1998)
9. Jiang, F., Lei, K.: Grey Prediction of Port Cargo Throughput Based on GM(1,1, a) Model. Logistics Technology 9, 68–70 (2009)
10. Guo, Z.J., Song, X.Q., Ye, J.: A Verhulst model on time series error corrected for port Cargo Throughput forecasting. Journal of the Eastern Asia Society for Transportation Studies 6, 881–891 (2005)
11. Kayacan, E., Ulutas, B., Kaynak, O.: Grey system theory-based models in time series prediction. Expert Systems with Applications 37, 1784–1789 (2010)
12. Askari, M., Askari, H.: Time Series Grey System Prediction-based Models: Gold Price Forecasting. Trends in Applied Sciences Research 6, 1287–1292 (2011)
13. Tsai, L.C., Yu, Y.S.: Forecast of the output value of Taiwan's IC industry using the Grey forecasting model. International Journal of Computer Applications in Technology 19(1), 23–27 (2004)
14. Hsu, L.C.: Applying the grey prediction model to the global integrated circuit industry. Technological forecasting and Social change 70, 563–574 (2003)
15. Hsu, C.C., Chen, C.Y.: Application of improved grey prediction model for power demand forecasting. Energy Conversion and management 44, 2241–2249 (2003)
16. Kang, J., Zhao, H.: Application of Improved Grey Model in Long-term Load Forecasting of Power Engineering. Systems Engineering Procedia 3, 85–91 (2012)
17. Wang, C.N., Phan, V.T.: An enhancing the accurate of grey prediction for GDP growth rate in Vietnam. In: 2014 International Symposium on Computer, Consumer and Control (IS3C), pp. 1137–1139 (2014). doi:10.1109/IS3C.2014.295
18. Wang, Z.X., Hipel, K.W., Wang, Q., He, S.W.: An optimized NGBM (1, 1) model for forecasting the qualified discharge rate of industrial wastewater in China. Applied Mathematical Modelling 35, 5524–5532 (2011)
19. Ou, S.L.: Forecasting agriculture output with an improved grey forecasting model based on the genetic algorithm. Computers and Electronics in Agriculture 85, 33–39 (2012)
20. Truong, D.Q., Ahn, K.K.: An accurate signal estimator using a novel smart adaptive grey model SAGM (1, 1). Expert Systems with Applications 39(9), 7611–7620 (2012)
21. Makridakis, S.: Accuracy measures: Theoretical and practical concerns. International Journal of Forecasting 9, 527–529 (1993)
22. Wang, C.N., Phan, V.T.: An improvement the accuracy of grey forecasting model for Cargo Throughput in international commercial Ports of Kaohsiung. International Journal of Business and Economics Research 3(1), 1–5 (2014). doi:10.11648/j.ijber.20140301.11
23. Website of the U.S. Energy Information Administration [Online]. http://www.eia.gov/

EL: Special Session on Ensemble Learning

Investigation of the Impact of Missing Value Imputation Methods on the k-NN Classification Accuracy

Tomasz Orczyk[✉] and Piotr Porwik

Institute of Computer Science, University of Silesia in Katowice,
Bedzinska 39, Sosnowiec, Poland
{tomasz.orczyk,piotr.porwik}@us.edu.pl

Abstract. Paper desribes results of an experiment where various scenarios of missing values occurrence in the data repository has been tested. Experiment was coducted on a publicly available database, containing complete, multidimensional continuous dataspace and multiple classes. Missing values were introduced using "completely at random" scheme. Tested scenarios were: training and testing using incomplete dataset, training on complete data set and testing on incomplete and vice versa. For comparison to data imputation methods also the ensemble of single-feature kNN classifiers, working withoud data imputation, has been tested.

Keywords: Missing values · Data imputation · Classification accuracy · Ensemble classification

1 Introduction

Classifiers are often used to build decision support systems, including medical support systems [11][13], identification and verification systems [12]. Such classifiers often are build and tested on archival data repositories, which may have some amount of missing values for some records. Problem of missing data is not new on the field of data analysis and statistics. Three categories of missing data have been defined in the literature [9][10]: NMAR (not missing at random), MAR (missing at random) and MCAR (missing completely at random). Most of known research about data imputation methods assume that data is missing completely at random. This means that the fact of the observation value is missing has no further informational meaning, i.e. does not correlate with the decision class.

Numerous methods exist for dealing with the problem of missing values [2]:

— Instances discarding (removing records with missing values)
— Feature discarding (removing features for which values are missing, or creating separate classifiers for different sets of features)
— Acquisition of missing values (repeating the observation to reacquire the missing value)
— Data imputation (calculate the missing value on the basis of remaining observations)

© Springer International Publishing Switzerland 2015
M. Núñez et al. (Eds.): ICCCI 2015, Part II, LNCS 9330, pp. 557–565, 2015.
DOI: 10.1007/978-3-319-24306-1_54

In the paper we have described single and multiple data imputation methods and an ensemble based modification of the idea of feature discarding [8] as an alternative to the data imputation [7].

The primary scenario for the experiment was to assess how the classifier will work on data with randomly missing values comparing to the same type of classifier trained and tested on an original (complete) data set. In practice this was a simulation of a common situation, where classification system is built and tested using repository with some values missing, but finally the system will work with a complete data, so it vital to determine system accuracy for complete data without having it. To do so, a degraded data sets have been prepared with different amount of missing (removed) values and they were imputed using different data imputation methods:

— single imputation methods:
  • global average value,
  • class average value,
— multiple imputation methods:
  • Amelia II [6] - bootstrap+EM algorithm to impute missing values from a dataset and produce multiple output datasets for analysis.

Multiple imputation involves imputing m values for each missing value in the data set and creating m completed data sets. Across these completed data sets, the observed (non-missing) values are the same, but the missing values are filled in with different imputations that reflect uncertainty about the missing data. The value of m for the experiment has been set to 10.

After imputation, the data sets were classified by a kNN classifier in a cross-validation process. Additionally the same data without imputation has been classified using the ensemble of single attribute based kNN classifier, as well as with the regular kNN. In all cases the value of k has been set to 3, on the basis of the previous experiments [7][14].

Some additional scenarios have also been tested in the same manner:

— classifier is trained on a complete data set but classifies data with missing values,
— classifier is trained on a data set with missing values but is tested on complete data.

Additionally some mixed scenarios have been tested.

2   Data Characteristics

Experiment has been performed on the publicly available UCI Wine Data Set [1]. Although experiment was designed mainly for medical databases, the Wine Data Set resembles such data sets, as it has many (namely thirteen) attributes (of real and integer type) and multiple (three) classes (see Table 1). What was also important, it has no missing values.

Table 1. Data characteristics of the UCI Wine Data Set

Parameter	Class 1 Mean (Std.dev.)	Class 2 Mean (Std.dev.)	Class 3 Mean (Std.dev.)	Total Mean (Std.dev.)
Alcohol	13.74 (0.46)	12.28 (0.54)	13.15 (0.53)	13.00 (0.81)
Malic acid	2.01 (0.69)	1.93 (1.02)	3.33 (1.09)	2.34 (1.12)
Ash	2.46 (0.23)	2.24 (0.32)	2.44 (0.18)	2.37 (0.27)
Alcalinity of ash	17.04 (2.55)	20.24 (3.35)	21.42 (2.26)	19.49 (3.34)
Magnesium	106.34 (10.50)	94.55 (16.75)	99.31 (10.89)	99.74 (14.28)
Total phenols	2.84 (0.34)	2.26 (0.55)	1.68 (0.36)	2.30 (0.63)
Flavanoids	2.98 (0.40)	2.08 (0.71)	0.78 (0.29)	2.03 (1.00)
Nonflavanoid phenols	0.29 (0.07)	0.36 (0.12)	0.45 (0.12)	0.36 (0.12)
Proanthocyanins	1.90 (0.41)	1.63 (0.60)	1.15 (0.41)	1.59 (0.57)
Color intensity	5.53 (1.24)	3.09 (0.92)	7.40 (2.31)	5.06 (2.32)
Hue	1.06 (0.12)	1.06 (0.20)	0.68 (0.11)	0.96 (0.23)
OD280/OD315 of diluted wines	3.16 (0.36)	2.79 (0.50)	1.68 (0.27)	2.61 (0.71)
Proline	1115.71 (221.52)	519.51 (157.21)	629.90 (115.10)	746.89 (314.91)
Samples count	59	71	48	178

3   Experiment Description

Experiment has been modeled in the KNIME [3] environment using WEKA [4] and R [5] components. For the purpose of the experiment six degraded datasets have been prepared. These were copies of original data set from which some feature values have been randomly removed. The amount of removed values in these sets were as follows: 9.7%, 14.3%, 18.0%, 21.7%, 25.3%, 29.4%. In the first step four reference experiments have been performed on these data sets (see Fig. 1):

— Training and testing using original (complete) dataset (we will call it FF experiment),
— Training classifier on complete data set, and testing on degraded data set (we will call it the FM experiment),
— Training classifier on degraded data set, and testing on complete data set (we will call it the MF experiment),
— Training and testing classifier on degraded data set (we will call it the MM experiment).

All experiments have been done by means of a 10 fold cross-validation.

In the next step common data imputation methods have been applied to the above-mentioned scenarios.

For experiments FM and MF data imputation has been done inside the cross-validation loop, while for MM experiment data has been imputed before the cross validation loop.

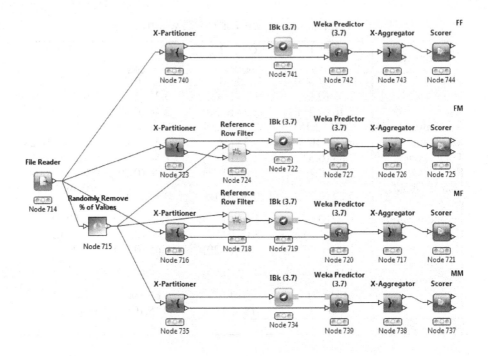

Fig. 1. Reference experiment setup in the KNIME Environment.

For the global average imputation and class average imputation the substitute (average) values were always computed according to the training set.

For the Amelia II algorithm the imputation has also been done inside the cross validation loop but imputation was done either on training or on testing data depending which one was containing missing values. For training data set, the class has been included   in the calculations of values to be imputed, while for testing set it was excluded.

For the last scenario the imputation has been done before the cross validation loop, and there were two variants for Amelia II: once the data has been imputed with class value inclusion and once two datasets were generated – for training with class included, and for testing with class excluded.

Two mixed scenarios were also tested. For single imputation, where missing values in the training set has been substituted using class average values from the testing set, and for testing set, the missing values has been substituted with global average values from training set. For Amelia II the training set with missing values has been imputed prior to cross-validation, while testing set with missing values has been imputed within the cross-validation loop on a concatenation of training and testing data. Training data rows were removed from testing set after the substitution values were calculated. Finally, above-mentioned scenarios were tested on the ensemble of single feature kNN classifiers (see Fig. 2).

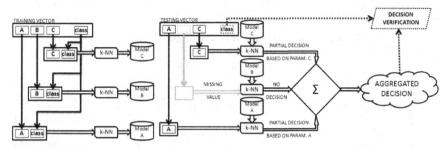

Fig. 2. General idea of the ensemble of a single feature based classifiers (*left*: training mode, *right*: classification/testing mode).

Proposed scenarios were designed to resemble situations where:

— we have incomplete data to test a classifier which will be finally working on a complete data (both training, and classified),
— we have repository of incomplete data to train a classifier, but classified data will have no missing values,
— we have repository of complete data, but classified data will likely be incomplete,
— classifier will finally work only on incomplete data (both training, and classified).

4   Results

Results will be presented in a graphical form (Fig. 3 – 6), where the description is as follows:

— NIm: No imputation,
— GAv: Single imputation using global average of a feature value,
— CAv: Single imputation using average of a feature value for each class,
— Ame: Multiple (10) imputation using Amelia II algorithm from the R package,
— AvMX: Single imputation with class average for training data, and global average for testing data,
— AmMB: Multiple imputation before cross validation loop (with class for training, and without class for testing data),
— AmMI: Multiple imputation inside cross validation loop.

Assuming that the classification accuracy should not be altered by missing values and especially by the imputation of missing values, both the positive and the negative difference of classification accuracy using incomplete data compared to the reference experiment should be treated as a problem. For the tested data base, the result of the MM experiment (see Fig. 5) shows that Amelia II has a tendency to heavily overestimate classifier's accuracy, also, what could be expected, the single imputation based on class average values overestimated classifier's accuracy. Both, the global average based imputation method and ensemble of single feature based classifiers, tend to underestimate accuracy in this experiment, and both have comparative results. For up to ~20% of missing values, their accuracy was close to the result of the reference experiment.

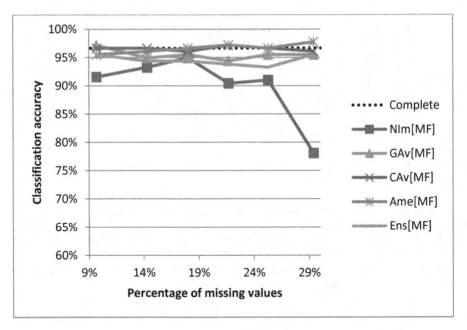

Fig. 3. Classification accuracy comparison for training on degraded data, and testing on complete data (MF).

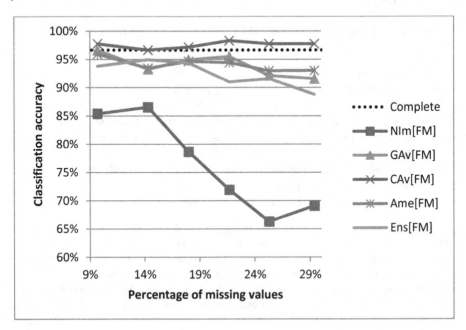

Fig. 4. Classification accuracy comparison for training on complete data, and testing on degraded data (FM).

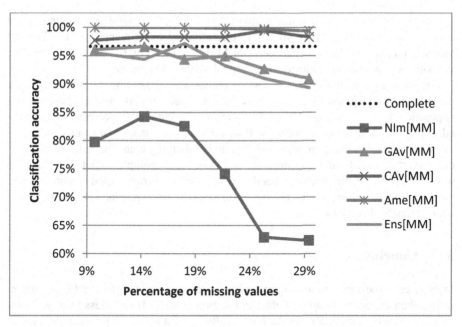

Fig. 5. Classification accuracy comparison for training and testing on degraded data (MM).

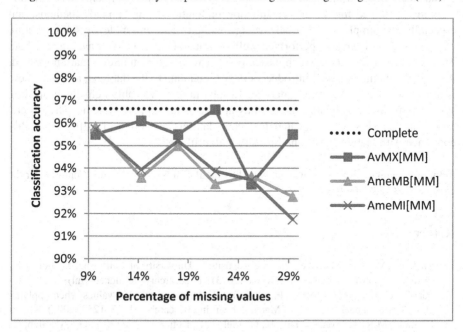

Fig. 6. Classification accuracy comparison for mixed methods on degraded data.

Results for "mixed" single imputation (class average for training data, and global average for testing data) were similar to the reference experiment's results. Two results of "mixed" Amelia II experiments of MM type (see Fig. 6) gave similar results, yet both were worse than "mixed" class/global average imputation.

For the simulated training on a complete data, and classifying degraded data (see Fig. 4), accuracy closest to the reference was obtained for class average based single imputation method (average 1% overestimation of accuracy). Three remaining methods gave comparative results with less than 4,5% underestimation of results.

For simulated training on degraded data and classifying complete data (see Fig. 3), class average based single imputation, and Amelia II multiple imputation methods were very close to the reference results, while global average based imputation and single feature based ensemble classifier gave slightly worse results with a tendency to underestimate classifier accuracy.

5   Conclusions

In each case the imputation and ensemble gave better (closer to the FF reference) results than experiments without data imputation. Amelia II and class average based substitution were giving always the best results (closest to the FF reference), but with a tendency to overestimate accuracy (especially in the MM experiment). Good results were also obtained for pre-cross validation imputation using "mixed" methods, and especially for simple class/global average variant. The ensemble of single feature based classifiers have also performed well – comparatively to global average method. This is good for a method which basically is a row based feature elimination method. As such the method doesn't introduce any artificial data to the datasets, and thus there is no risk of over fitting to the dataset. Results may be valuable help for projecting system tests for classification algorithms and decision support systems. The experiment can and should be repeated for other datasets and other classifiers, possibly combined with feature selection/elimination algorithms.

Acknowledgement. This work was supported by the Polish National Science Centre under the grant no. DEC-2013/09/B/ST6/02264.

References

1. Lichman, M.: UCI Machine Learning Repository. University of California, School of Information and Computer Science, Irvine (2013). (http://archive.ics.uci.edu/ml)
2. Saar-Tsechansky, M., Provost, F., Caruana, R.: Handling missing values when applying classification models. Journal of Machine Learning Research **8**, 1217–1250 (2007)
3. Berthold, M.R., Cebron, N., Dill, F., Gabriel, T.R., Kötter, T., Meinl, T., Ohl, P., Sieb, C., Thiel, K., Wiswedel, B.: KNIME: The Konstanz Information Miner. Studies in Classification, Data Analysis, and Knowledge Organization (GfKL 2007). Springer (2007)
4. Hall, M., Frank, E., Holmes, G., Pfahringer, B., Reutemann, P., Witten, I.H.: The WEKA Data Mining Software: An Update. SIGKDD Explorations **11**(1) (2009)

5. R Core Team: R: A Language and Environment for Statistical Computing, R Foundation for Statistical Computing, Vienna, Austria (2014)
6. Honaker, J., King, G., Blackwell, M.: Amelia II: A Program for Missing Data. Journal of Statistical Software **45**(7), 1–47 (2011)
7. Orczyk, T., Porwik, P., Bernas, M.: Medical Diagnosis Support System Based on the Ensemble of Single-Parameter Classifiers. Journal of Medical Informatics and Technologies **23**, 173–180 (2014)
8. Wozniak, M., Krawczyk, B.: Combined classifier based on feature space partitioning. International Journal of Applied Mathematics and Computer Science **22**(4), 855–866 (2012)
9. Little, R.J.A., Rubin, D.B.: Statistical Analysis with Missing Data. John Wiley & Sons, New York (1987)
10. Schafer, J.L.: Analysis of Incomplete Multivariate Data. Chapman and Hall/CRC (1997)
11. Porwik, P., Sosnowski, M., Wesolowski, T., Wrobel, K.: A computational assessment of a blood vessel's compliance: a procedure based on computed tomography coronary angiography. In: Corchado, E., Kurzyński, M., Woźniak, M. (eds.) HAIS 2011, Part I. LNCS, vol. 6678, pp. 428–435. Springer, Heidelberg (2011)
12. Doroz, R., Porwik, P.: Handwritten signature recognition with adaptive selection of behavioral features. In: Chaki, N., Cortesi, A. (eds.) CISIM 2011. CCIS, vol. 245, pp. 128–136. Springer, Heidelberg (2011)
13. Foster, K.R., Koprowski, R., Skufca, J.D.: Machine learning, medical diagnosis, and biomedical engineering research – commentary. Biomedical Engineering Online **13**, Article No. 94 (2014). doi: 10.1186/1475-925X-13-94
14. Bernas, M., Orczyk, T., Porwik, P.: Fusion of Granular Computing and k –NN Classifiers for medical data support system. In: Nguyen, N.T., Trawiński, B., Kosala, R. (eds.) ACIIDS 2015. LNCS, vol. 9012, pp. 62–71. Springer, Heidelberg (2015)

Enhancing Intelligent Property Valuation Models by Merging Similar Cadastral Regions of a Municipality

Tadeusz Lasota[1], Edward Sawiłow[1], Zbigniew Telec[2], Bogdan Trawiński[2(✉)],
Marta Roman[3], Paulina Matczuk[3], and Patryk Popowicz[3]

[1] Department of Spatial Management, Wrocław University of Environmental and Life Sciences,
ul. Norwida 25 50-375, Wrocław, Poland
{tadeusz.lasota,edward.sawilow}@up.wroc.pl
[2] Department of Information Systems, Wrocław University of Technology, Wybrzeże
Wyspiańskiego 27 50-370, Wrocław, Poland
{zbigniew.telec,bogdan.trawinski}@pwr.edu.pl
[3] Faculty of Computer Science and Management, Wrocław University of Technology,
Wybrzeże Wyspiańskiego 27 50-370, Wrocław, Poland

Abstract. A method for enhancing property valuation models consists in determining zones of an urban municipality in which the prices of residential premises change similarly over time. Such similar zones are then merged into bigger areas embracing greater number of sales transactions which constitute a more reliable basis to construct accurate property valuation models. This is especially important when machine learning algorithms are employed do create prediction models. In this paper we present our further investigation of the method using the cadastral regions of a city as zones for merging. A series of evaluation experiments was conducted using real-world data comprising the records of sales and purchase transactions of residential premises accomplished in a Polish urban municipality. Six machine learning algorithms available in the WEKA data mining system were employed to generate property valuation models. The study showed that the prediction models created over the merged cadastral regions outperformed in terms of accuracy the models based on initial component regions.

Keywords: Prediction models · Machine learning · Real estate appraisal · Trend functions · Merging cadastral regions

1 Introduction

Real estate appraisal is performed by licensed professionals and requires expert knowledge and enough amount of data. One of the most common approaches to determining the market value of a property is sales comparison method. To apply this approach it is necessary to possess transaction prices of the properties sold which attributes are similar to the one being appraised. If good comparable examples of transactions are available, then it is possible to attain reliable estimates. The professional standards demand for data of real estate located nearby the property being ap-

© Springer International Publishing Switzerland 2015
M. Núñez et al. (Eds.): ICCCI 2015, Part II, LNCS 9330, pp. 566–577, 2015.
DOI: 10.1007/978-3-319-24306-1_55

praised. However, the appraisers have often big problems acquiring appropriate data of similar premises because they are mostly sparse on the market or even lacking. To tackle this problem the authors proposed a method for determining and merging similar zones to obtain homogenous areas comprising greater number of samples [1]. Such uniform zones may be utilized to elaborate more reliable and accurate models for real property appraisal including automated valuation models.

Numerous approaches to real estate appraisal have been proposed for many years. They ranged from statistical methods through operational research to computational intelligence techniques. Various techniques can be found in rich literature in the field including models created using statistical multiple regression and neural networks [2], [3], [4], linear parametric programming [5], decision trees [6], rough set theory [7], fuzzy systems [8], and hybrid approaches [9].

Some hot machine learning approaches as ensemble methods [10], [11], [12], hybrid methods including evolutionary fuzzy systems [13] and evolving fuzzy systems [14], [15] have inspired us to explore them in the field of real estate appraisal.

For almost a decade we have been devising methods and algorithms for generating data driven regression models to assist with property valuation based on genetic fuzzy systems [16], [17]. We worked out also multiple models constructed using various resampling techniques including bagging with evolutionary fuzzy systems, decision trees, and neural networks [18], bagging with linear regression, decision trees, support vector machines, and artificial neural networks [19], bagging with 16 various fuzzy algorithms implemented in KEEL [20] as well as subagging, repeated holdout, and repeated cross validation employing genetic fuzzy systems [21]. A relatively good performance revealed evolving fuzzy models applied to cadastral data [22], [23]. We have also conducted research into application of ensembles of genetic fuzzy systems and neural networks to predict from a data stream of real estate sales transactions [24], [25].

In this paper we present our further investigation of the method for property valuation based on the concept of merging different areas of an urban municipality into uniform zones reflecting the dynamics of real estate market [1]. The cadastral regions of a city have been taken as zones for merging. The experiments were carried out using real-world data of sales transactions of residential premises accomplished in a Polish urban municipality. Six machine learning algorithms taken from WEKA [26] were employed to construct real estate appraisal models.

2   Method for Merging Similar Cadastral Regions of a Municipality

The approach to enhancing property valuation models consists in determining and merging zones within an urban municipality in which the prices of residential premises change similarly over time. Such similar zones are then merged into bigger areas embracing greater number of sales transactions which constitute a more reliable basis to construct accurate property valuation models. This is especially important when machine learning algorithms are used because they need numerous samples to train prediction models. Moreover, the trend functions are more reliable if they are determined using a larger number of samples. The area of a city considered in the paper is partitioned into 69 cadastral regions as shown in Figure 1. Within a city cadastral regions can be delimited by the boundaries of residential quarters and housing estates as well as by streets, railway lines, rivers etc. Therefore, the cadastral regions might be assumed to be uniform areas in respect of real estate appraisal.

The measure of similarity between two zones is based on determining the distance among trends of real estate price changes. A trend can be represented by a polynomial which degree is selected to reflect the nature of changes. The chart illustrating the change trend of average transactional prices per square metre modelled by the polynomial of degree three is given in Figure 2. It can be easily noticed that at the edges of time period considered the polynomial trend of degree three may produce unacceptable errors; the same applies to the polynomial trend of degree four. Therefore, instead of using higher degree polynomials we proposed to divide the period of time into a number of segments and apply linear trends for each of them. In Figure 3 three periods with different linear trends of price variability are illustrated.

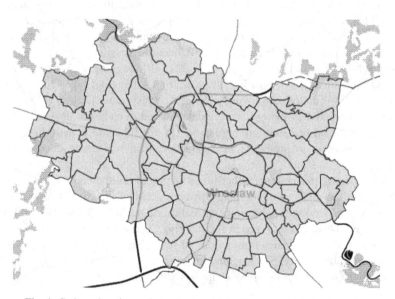

Fig. 1. Cadastral regions of an urban municipality considered in the paper

The similarity measures D_{ik} between a pair of zones i and k can be expressed by Formula 1, where: y_{ij}^{b} and y_{ij}^{e} denote the values of linear trend functions at the beginning b and end e of each *segment j* and m is the number of segments as shown in Figure 4. This relatively simple measure allows for taking into account similarity between the trends within all considered periods of time.

$$D_{ik} = \frac{1}{2m} \sum_{j=1}^{m} \left| y_{ij}^{b} - y_{kj}^{b} \right| + \left| y_{ij}^{e} - y_{kj}^{e} \right| \tag{1}$$

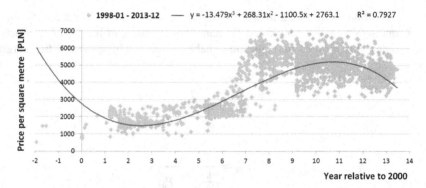

Fig. 2. Real estate price variability represented by a third degree polynomial trend

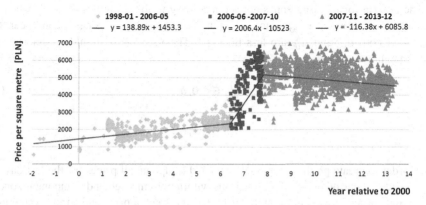

Fig. 3. Segments with three different linear trends of real estate price variability

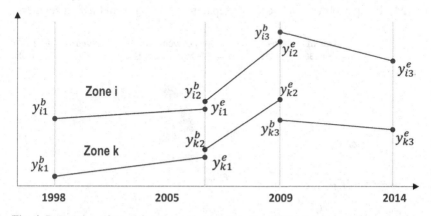

Fig. 4. Denotation of trend function values at the edges of three segments for two zones

Then the similarity measures D_{ik} are normalized by dividing them by the average price per square metre in the respective zones i and k applying Formulas 2 and 3. In Formula 2 N_i denotes the number of transactions in the *zone i* and P_{il} stands for the price of individual transaction in this zone. In turn, in Formula 3 N_k denotes the number of transactions in the *zone k* and P_{kl} stands for the price of individual transaction in this zone. This form of normalization enables us to establish criteria for merging zones.

$$d_{ik}^i = \frac{D_{ik}}{\frac{1}{N_i}\sum_{l=1}^{N_i} P_{il}} \qquad (2)$$

$$d_{ik}^k = \frac{D_{ik}}{\frac{1}{N_k}\sum_{l=1}^{N_k} P_{kl}} \qquad (3)$$

The criteria for merging zones are set according to the Formulas 4 and 5, where δ denotes the maximum acceptable deviation. For example the zones i and k can be merged if both d_{ik}^i and d_{ik}^k are less than 0.2. The value of δ should be determined experimentally.

$$d_{ik}^i \in < 0, \delta > \qquad (4)$$

$$d_{ik}^k \in < 0, \delta > \qquad (5)$$

In order to create prediction models we need to update all prices for the last day of the considered period using trend functions within both merged and component zones. The update procedure is based on the difference between a price and a trend value in a given time point (see Figure 5). It is very similar to the Delta method proposed in our work [24]. It is performed for each zone separately starting from the first segment.

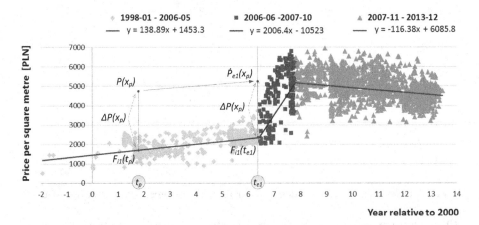

Fig. 5. Delta method for updating the premises prices for the end of a segment

As a result all the transaction prices are updated for the last day of the considered period. The prediction models for real estate appraisal are generated using revised prices together with the other attributes of the premises, such as usable area, age, number of storeys in a building, and distance from the city centre.

The detailed description of individual steps of the method for merging similar zones to improve intelligent models for real estate appraisal can be found in our previous work [1].

3   Setup of Evaluation Experiments

The goal of experiments was to show that the prediction models created over the merged cadastral regions reveal better accuracy than the models based on initial component regions. The course of experiment is shown in Figure 6 and consists of the following stages.

1) Preparing real-world datasets derived from local cadastral systems.
2) Comparing the change trends of real estate prices among cadastral regions.
3) Merging cadastral regions with similar trends.
4) Updating prices in individual regions using trend functions.
5) Creating property valuation models in individual regions.
6) Comparing the accuracy of models built over merged and component regions.

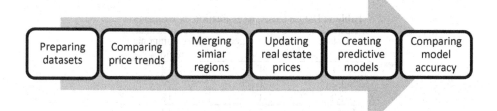

Fig. 6. Course of experiment to evaluate the method for merging cadastral regions

The experiments were conducted using real-world datasets comprising sales and purchase transactions derived from a cadastral system and a public registry of real estate transactions. The data used in experiments was taken from an unrefined dataset containing above 150 000 records of sales of residential premises realized in one of Polish cities with the population of 640,000 and area of about 300 square kilometres within 16 years from 1998 to 2013. The final dataset after cleansing contained 54,994 samples and consisted of sales transaction data of flats in buildings over land with rights of ownership and flats in buildings where the land was leased on terms of perpetual usufruct. Polish appraisers routinely differentiate the rights to land between ownership and perpetual usufruct arguing that they affect significantly the prices of premises. Each set of transaction records was then divided into four classes according to the year of a building construction as shown in Table 1. The professional appraisers argue that the properties falling to individual classes differ in construction technology and dynamics of relative market value. The differences in behaviour of property prices over time within individual classes are so big that the valuation of properties should be accomplished for individual classes separately.

The algorithm for merging similar regions was run for individual classes of properties presented in Table 1. Next, all transactional prices were updated for the last day of the considered timespan, i.e. for June 30, 2013 within both merged and component regions. For further study eight datasets comprising merged regions were employed. They are shown in Table 2 where the numbers of datasets 1, 2, ..., 8 correspond to the denotation of classes in Table 1.

Table 1. Classes of real estate considered in the paper

Class	Property type	Year of construction	No. of trans.
1	Flats in buildings over land with rights of ownership	before 1945	4,948
2	Flats in buildings over land with rights of ownership	1946-1970	1,301
3	Flats in buildings over land with rights of ownership	1971-2000	5,240
4	Flats in buildings over land with rights of ownership	2001-2013	11,952
5	Flats in buildings over land in perpetual usufruct	before 1945	13,178
6	Flats in buildings over land in perpetual usufruct	1946-1970	7,329
7	Flats in buildings over land in perpetual usufruct	1971-2000	4,017
8	Flats in buildings over land in perpetual usufruct	2001-2013	7,029
		Total	54,994

Table 2. Datasets used in evaluation experiments

Dataset	No. of merged regions	No of component regions
Set-1	4	19
Set-2	2	5
Set-3	3	18
Set-4	4	34
Set-5	2	15
Set-6	1	7
Set-7	2	8
Set-8	3	23
TOTAL	21	129

The prediction models were built over individual merged and component cadastral regions using six following machine learning algorithms available in the WEKA data mining system [26]:

REG – Linear Regression Model. The algorithm is a standard statistical approach to build a linear model which uses the least mean square method in order to adjust the parameters of the linear function.

M5R - M5Rules. The algorithm divides the parameter space into areas (subspaces) and builds in each of them a linear regression model. It is based on M5 algorithm. In each iteration a M5 Tree is generated and its best rule is extracted according to a given heuristic. The algorithm terminates when all the examples are covered.

M5P – Pruned Model Tree. The algorithm implements routines for generating M5 model trees. It is based on decision trees, however, instead of having values at tree's nodes, it contains a multivariate linear regression model at each node. The input space is divided into cells using training data and their outcomes, then a regression model is built in each cell as a leaf of the tree.

SMR – SMOreg. The algorithm implements the support vector machine for regression. It learns SVM for regression using RegSMOImproved, i.e. sequential minimal optimization with Shevade, Keerthi, et al. adaption of the stopping criterion.

MLR – MultilayerPerceptron for regression problems. Multi-layer Perceptron is a feed-forward neural network that uses the backpropagation learning algorithm. For regression, i.e. when the class is numeric, the output nodes become unthresholded linear units.

TRE – REPTree. Fast decision tree learner that builds a decision/regression tree using information gain/variance and prunes it using reduced-error pruning (with backfitting). It sorts values for numeric attributes only once.

Five following attributes of premises proposed by professional appraisers were applied to the *REG, M5R, M5P, SMR, MLR,* and *TRE* machine learning algorithms. As four input features: usable area of a flat (*Area*), age of a building construction (*Age*), number of storeys in a building (*Storeys*), the distance of the building from the city centre (*Centre*) were taken, in turn, price per square metre (*Price*) was the output variable.

In order to perform comparative analysis of the model accuracy 10-fold cross-validation method (10cv) was employed. Two performance measures *Root Mean Square Error (RMSE)* and *Relative Absolute Error (RAE)* were computed according to Formulas 6 and 7, where y_i^a and y_i^p denote the actual and predicted values respectively, $\overline{y^a}$ stands for the arithmetic mean of actual values, and n is the number of transactions in the test set.

$$RMSE = \sqrt{\frac{1}{n}\sum_{i=1}^{n}\left(y_i^p - y_i^a\right)^2} \tag{6}$$

$$RAE = \frac{\sum_{j=1}^{n}\left|y_i^p - y_i^a\right|}{\sum_{j=1}^{n}\left|\overline{y^a} - y_i^a\right|} \tag{7}$$

4   Analysis of Experimental Results

The comparative analysis was accomplished using 129 pairs of observations which constituted the results of 10cv for models over individual component regions and corresponding merged regions as shown in Table 2. According to Demšar [27] it was safer to use nonparametric statistical tests since they do not assume normal distributions. We employed the Friedman test followed by the paired Wilcoxon signed-ranks test [28]. Average rank positions of models determined during Friedman test for the models over merged and component regions for *RMSE* and *RAE* are presented in Table 3. The ranks produced by the Friedman test mean the lower rank value the better model. In turn, the results of the paired Wilcoxon test indicate statistically significant differences when p-value is lower than 0.05. The Wilcoxon test revealed that models constructed over merged cadastral regions surpassed the ones over component regions for both performance measures *RMSE* and *RAE* and each *REG, M5R, M5P, SMR, MLR,* and *TRE* machine learning algorithm.

Table 3. Results of Friedman and Wilcoxon tests for model accuracy comparison

Alg.	RMSE			RAE		
	Avg. rank for merged	Avg. rank for comp.	Wilcoxon p-value	Avg. rank for merged	Avg. rank for comp.	Wilcoxon p-value
REG	1.27	1.73	0.000000	1.15	1.85	0.000000
M5R	1.30	1.70	0.000000	1.16	1.84	0.000000
M5P	1.27	1.73	0.000000	1.16	1.84	0.000000
SMR	1.30	1.70	0.000000	1.16	1.84	0.000000
MLR	1.25	1.75	0.000000	1.16	1.84	0.000000
TRE	1.25	1.75	0.000000	1.09	1.91	0.000000

The performance of *REG, M5R, M5P, SMR, MLR,* and *TRE* models built over 21 merged cadastral regions is depicted in Figure 7, where the accuracy RMSE is expressed in terms of thousand PLN (Polish currency). It is clearly seen that the models constructed using *MLR* and *TRE* reveal worse accuracy for some regions. However, in order to detect whether the differences are significant statistical tests adequate for multiple comparisons over multiple datasets should have been accomplished. The statistical methodology including nonparametric tests followed by post-hoc procedures designed especially for multiple $N \times N$ comparisons [28], [29], was employed. First, the nonparametric Friedman test, which detect the presence of differences among all algorithms compared, was made. Next, the post-hoc procedures were applied in order to point out the particular pairs of algorithms which revealed significant differences. For $N \times N$ comparisons nonparametric Nemenyi's, Holm's, Shaffer's, and Bergmann-Hommel's procedures were employed.

Average rank positions of *RMSE* produced by the models constructed over merged regions which were determined by the Friedman test are shown in Table 4, where the lower rank value the better model. Adjusted p-values for Nemenyi's, Holm's, Shaffer's, and Bergmann-Hommel's post-hoc procedures for $N \times N$ comparisons for all possible pairs of machine learning algorithms are shown in Table 5. For illustration the p-values of the paired Wilcoxon are also given. The significance level considered for the null hypothesis rejection was 0.05. Following main observations can be done: there are not significant differences among *REG, M5R, M5P,* and *SMR* models. In turn, *MLR* and *TRE* models reveal significantly worse performance than any of first four models in the ranking.

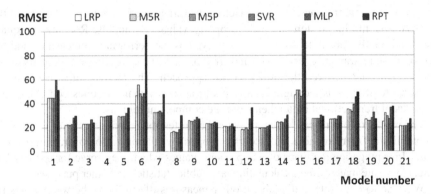

Fig. 7. Average *RMSE* values for 10-fold cross validation over merged cadastral regions

Table 4. Average rank positions of model performance in terms of *RMSE*

1st	2nd	3rd	4th	5th	6th
REG (1.45)	M5R (1.55)	M5P (3.00)	SMR (4.00)	MLR (5.09)	TRE (5.91)

Table 5. Adjusted p-values for $N \times N$ comparisons in terms of *RMSE*

Alg vs Alg	pWilcox	pNeme	pHolm	pShaf	pBerg
TRE vs REG	0.000069	0.000000	0.000000	0.000000	0.000000
MLR vs REG	0.000060	0.000002	0.000002	0.000001	0.000001
M5R vs TRE	0.000069	0.000002	0.000002	0.000001	0.000001
M5R vs MLR	0.000618	0.000017	0.000014	0.000011	0.000007
M5P vs TRE	0.000060	0.000127	0.000093	0.000084	0.000059
TRE vs SMR	0.000141	0.000389	0.000259	0.000259	0.000156
M5P vs MLR	0.000080	0.000797	0.000478	0.000372	0.000212
MLR vs SMR	0.000080	0.002223	0.001186	0.001038	0.000593
SMR vs REG	0.058187	1.000000	0.963521	0.963521	0.963521
M5P vs REG	0.192435	1.000000	1.000000	1.000000	0.963521
M5R vs SMR	0.244270	1.000000	1.000000	1.000000	1.000000
M5R vs M5P	0.821260	1.000000	1.000000	1.000000	1.000000
M5R vs REG	0.930756	1.000000	1.000000	1.000000	1.000000
TRE vs MLR	0.091846	1.000000	1.000000	1.000000	1.000000
M5P vs SMR	0.958420	1.000000	1.000000	1.000000	1.000000

5   Conclusions and Future Work

Our further study into the method for improving the procedures of real estate appraisal by merging zones of the city characterized by similar dynamics of premise
prices was presented in the paper. Cadastral regions of a city have been utilized as
zones for merging, i.e. bigger areas in comparison with zones determined based on a
land-use plan which were considered in [1]. Moreover, the cadastral regions are of
different nature because they include individual residential quarters or housing estates
and are delimited by streets, railway lines, and rivers.

We conducted a series of experiments using real-world data of sales and purchase
transactions of residential premises derived from a cadastral system and a public reg-

istry of real estate transactions. Six machine learning algorithms were taken from the WEKA data mining system to build property valuation models. Root mean square error and relative absolute error were employed as the performance measures and the statistical nonparametric Friedman and Wilcoxon tests were applied to analyse the results. The study proved the utility of our approach. The models built over merged cadastral regions revealed significantly better accuracy than the ones constructed on the basis of transactional data taken from the component regions.

The results of our research may provide the professional appraisers with the tool for accomplishing more reliable and accurate property valuations. Moreover, they may be utilized for preparing the maps of real property values which may serve for spatial planning, real estate management, calculating tax, public statistics and other purposes.

We plan to elaborate and evaluate other measures of similarity between zones reflecting the characteristics of premise price changes. Moreover, we intend to carry out further experiments to tune the parameters of the proposed algorithm using the other real-world datasets. The comparison with the heuristic premise estimation methods routinely employed by professional appraisers will be also performed.

References

1. Lasota, T., Sawiłow, E., Trawiński, B., Roman, M., Marczuk, P., Popowicz, P.: A method for merging similar zones to improve intelligent models for real estate appraisal. In: Nguyen, N.T., Trawiński, B., Kosala, R. (eds.) ACIIDS 2015. LNCS, vol. 9011, pp. 472–483. Springer, Heidelberg (2015)
2. Kontrimas, V., Verikas, A.: The mass appraisal of the real estate by computational intelligence. Applied Soft Computing 11(1), 443–448 (2011)
3. Zurada, J., Levitan, A.S., Guan, J.: A Comparison of Regression and Artificial Intelligence Methods in a Mass Appraisal Context. Journal of Real Estate Res. 33(3), 349–388 (2011)
4. Peterson, S., Flangan, A.B.: Neural Network Hedonic Pricing Models in Mass Real Estate Appraisal. Journal of Real Estate Research 31(2), 147–164 (2009)
5. Narula, S.C., Wellington, J.F., Lewis, S.A.: Valuating residential real estate using parametric programming. European Journal of Operational Research 217, 120–128 (2012)
6. Antipov, E.A., Pokryshevskaya, E.B.: Mass appraisal of residential apartments: An application of Random forest for valuation and a CART-based approach for model diagnostics. Expert Systems with Applications 39, 1772–1778 (2012)
7. D'Amato, M.: Comparing Rough Set Theory with Multiple Regression Analysis as Automated Valuation Methodologies. Int. Real Estate Review 10(2), 42–65 (2007)
8. Kusan, H., Aytekin, O., Özdemir, I.: The use of fuzzy logic in predicting house selling price. Expert Systems with Applications 37(3), 1808–1813 (2010)
9. Musa, A.G., Daramola, O., Owoloko, A., Olugbara, O.: A Neural-CBR System for Real Property Valuation. Journal of Emerging Trends in Computing and Information Sciences 4(8), 611–622 (2013)
10. Woźniak, M., Graña, M., Corchado, E.: A survey of multiple classifier systems as hybrid systems. Information Fusion 16, 3–17 (2014)
11. Krawczyk, B., Woźniak, M., Cyganek, B.: Clustering-based ensembles for one-class classification. Information Sciences 264, 182–195 (2014)
12. Burduk, R., Walkowiak, K.: Static classifier selection with interval weights of base classifiers. In: Nguyen, N.T., Trawiński, B., Kosala, R. (eds.) ACIIDS 2015. LNCS, vol. 9011, pp. 494–502. Springer, Heidelberg (2015)

13. Fernández, A., López, V.: José del Jesus, M., Herrera, F.: Revisiting Evolutionary Fuzzy Systems: Taxonomy, applications, new trends and challenges. Knowledge-Based Systems **80**, 109–121 (2015)
14. Lughofer, E.: Evolving Fuzzy Systems – Methodologies, Advanced Concepts and Applications. STUDFUZZ, vol. 266. Springer, Heidelberg (2011)
15. Lughofer, E., Cernuda, C., Kindermann, S., Pratama, M.: Generalized smart evolving fuzzy systems. Evolving Systems (2015). doi:10.1007/s12530-015-9132-6
16. Król, D., Lasota, T., Nalepa, W., Trawiński, B.: Fuzzy system model to assist with real estate appraisals. In: Okuno, H.G., Ali, M. (eds.) IEA/AIE 2007. LNCS, vol. 4570, pp. 260–269. Springer, Heidelberg (2007)
17. Król, D., Lasota, T., Trawiński, B., Trawiński, K.: Comparison of mamdani and TSK fuzzy models for real estate appraisal. In: Apolloni, B., Howlett, R.J., Jain, L. (eds.) KES 2007, Part III. LNCS (LNAI), vol. 4694, pp. 1008–1015. Springer, Heidelberg (2007)
18. Lasota, T., Telec, Z., Trawiński, B., Trawiński, K.: Exploration of bagging ensembles comprising genetic fuzzy models to assist with real estate appraisals. In: Corchado, E., Yin, H. (eds.) IDEAL 2009. LNCS, vol. 5788, pp. 554–561. Springer, Heidelberg (2009)
19. Lasota, T., Telec, Z., Trawiński, B., Trawiński, K.: A multi-agent system to assist with real estate appraisals using bagging ensembles. In: Nguyen, N.T., Kowalczyk, R., Chen, S.-M. (eds.) ICCCI 2009. LNCS, vol. 5796, pp. 813–824. Springer, Heidelberg (2009)
20. Krzystanek, M., Lasota, T., Telec, Z., Trawiński, B.: Analysis of bagging ensembles of fuzzy models for premises valuation. In: Nguyen, N.T., Le, M.T., Świątek, J. (eds.) Intelligent Information and Database Systems. LNCS, vol. 5991, pp. 330–339. Springer, Heidelberg (2010)
21. Lasota, T., Telec, Z., Trawiński, G., Trawiński, B.: Empirical comparison of resampling methods using genetic fuzzy systems for a regression problem. In: Yin, H., Wang, W., Rayward-Smith, V. (eds.) IDEAL 2011. LNCS, vol. 6936, pp. 17–24. Springer, Heidelberg (2011)
22. Lasota, T., Telec, Z., Trawiński, B., Trawiński, K.: Investigation of the eTS Evolving Fuzzy Systems Applied to Real Estate Appraisal. Journal of Multiple-Valued Logic and Soft Computing **17**(2–3), 229–253 (2011)
23. Lughofer, E., Trawiński, B., Trawiński, K., Kempa, O., Lasota, T.: On Employing Fuzzy Modeling Algorithms for the Valuation of Residential Premises. Information Sciences **181**, 5123–5142 (2011)
24. Trawiński, B.: Evolutionary Fuzzy System Ensemble Approach to Model Real Estate Market based on Data Stream Exploration. Journal of Universal Computer Science **19**(4), 539–562 (2013)
25. Telec, Z., Trawiński, B., Lasota, T., Trawiński, G.: Evaluation of neural network ensemble approach to predict from a data stream. In: Hwang, D., Jung, J.J., Nguyen, N.-T. (eds.) ICCCI 2014. LNCS, vol. 8733, pp. 472–482. Springer, Heidelberg (2014)
26. Witten, I.H., Frank, E., Hall, M.A.: Data Mining: Practical Machine Learning Tools and Techniques, 3rd edn. Morgan Kaufmann, San Francisco (2011)
27. Demšar, J.: Statistical comparisons of classifiers over multiple data sets. J. of Machine Learning Research **7**, 1–30 (2006)
28. Trawiński, B., Smętek, M., Telec, Z., Lasota, T.: Nonparametric Statistical Analysis for Multiple Comparison of Machine Learning Regression Algorithms. International Journal of Applied Mathematics and Computer Science **22**(4), 867–881 (2012)
29. García, S., Herrera, F.: An Extension on "Statistical Comparisons of Classifiers over Multiple Data Sets" for all Pairwise Comparisons. Journal of Machine Learning Research **9**, 2677–2694 (2008)

Blurred Labeling Segmentation Algorithm for Hyperspectral Images

Paweł Ksieniewicz[1][(✉)], Manuel Graña[2,3], and Michał Woźniak[1]

[1] Department of Systems and Computer Networks,
Wroclaw University of Technology, Wroclaw, Poland
`pawel.ksieniewicz@pwr.edu.pl`
[2] University of the Basque Country, Leioa, Bizkaia, Spain
[3] ENGINE Center, Wroclaw University of Technology, Wroclaw, Poland

Abstract. This work is focusing on the hyperspectral imaging classification, which is nowadays a focus of intense research. The hyperspectral imaging is widely used in agriculture, mineralogy, or food processing to enumerate only a few important domains. The main problem of such image classification is access to the ground truth, because it needs the experienced experts. This work proposed a novel three-stage image segmentation method, which prepares the data for the classification and employs the active learning paradigm which reduces the expert works on image. The proposed approach was evaluated on the basis of the computer experiments carried out on the benchmark hyperspectral datasets.

Keywords: Machine learning · Hyperspectral imaging · Image processing · Classification · Classifier ensemble

1 Introduction

The contemporary man lives is surrounded by the data coming from different sources and it is becoming more and more difficult to obtain the valuable knowledge from it. Additionally, data analyzing requires plenty of time and sometimes the results are disappointing. Therefore the efficient analytical tools which are capable to analyze the huge volume of multidimensional data are still focus of intense research.

One can say that *if a picture is worth 1000 words, a hyperspectral image is worth almost 1000 pictures*[1]. One of such a tool is hyperspectral imaging, which widely used in agriculture, mineralogy etc. Hyperspectral cameras are able to capture hundreds of monochrome images correlated with a particular spectrum, nevertheless they still need to be manually analyzed by human expert. Of course, it causes that labeling is very expensive and time consuming what makes such an approach impractical. Therefore methods which can work on the partially labelled data are desirable tool for future image classification [1]. One of

---

[1] J.P. Ferguson, An Introduction to Hyperspectral Imaging, Photonics and Analytical Marketing Ltd.

© Springer International Publishing Switzerland 2015
M. Núñez et al. (Eds.): ICCCI 2015, Part II, LNCS 9330, pp. 578–587, 2015.
DOI: 10.1007/978-3-319-24306-1_56

the very promising direction is of semi-supervised learning uses active learning paradigm [2–5], which employs an iterative data labeling and classifier training strategy to obtain an accurate classifier, with the smallest possible training set. Human expert introduces only the labels for the new selected examples, which can provide the greatest accuracy increase.

In this work we propose a method of hyperspectral image segmentation, using concept of blurred labeling. After selecting set of representative signatures, which are provided by a unsupervised pre-processing method (but could be also gained from signatures database, or selected by an expert) we are stacking similarity maps. Final labels come from calculating maximum with index, from anisotropic diffusion of achieved similarity stack. The labeled set of regions may be used to train a classifier. The proposed method was evaluated on the basis of experiments which answer the question about the possible classification accuracy.

The main contributions of the paper are as follows:

1. Proposition of region separation, from border detection, using local dynamics of image.
2. Noisy ranges filtering, using the measure of border information.
3. Implementation of blurred labeling for image segmentation.

In section 2 we will describe the proposed segmentation algorithm. Section 3 presents the experimental environment. At the end, in section 4 shows conclusions and possible usages of BLS algorithm.

2 Algorithm Description

The *Blurred Labeling Segmentation* algorithm consists of three consecutive stages. First one is detecting borders and filtering noised bands from spectral signature. Second uses filtered border map to collect a list of representative signatures from image. Third is an proposition of blurred labeling, which uses this list of signatures to finally segment an image.

Figure 1 shows the processing chain of proposed algorithm. Whole process is presented using popular benchmark image called Salinas C, which false-colored version is in Figure 2. Matlab implementation and documentation of code is available at the GIT repository[2].

2.1 Border Detection and Filtering

Procedure of border detection relies on calculation of local diversity of each image pixels signature band [6]. Using hyperspectral cube, we are constructing another one, called an *edge cube*, where every element equals a difference between minimal and maximal value in given radius. Figure 3 shows two example bands of the edge cube calculated with this method.

---

[2] https://github.com/xehivs/natix

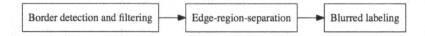

| Border detection and filtering | → | Edge-region-separation | → | Blurred labeling |

Fig. 1. Processing chain.

Fig. 2. False-color image.

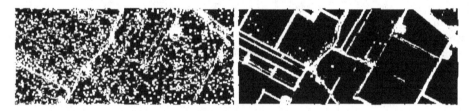

Fig. 3. Band 3 (left) and 100 (right) of edge cube.

Band 3 (on the left) is very noisy, so an edge mask is also unclear. band 100 (right side) is clean, so an edge mask is much cleaner. Basing on this we have made an assumption, that amount of information on the clean bands would be a stable value, which could easily separate them from noisy ones.

Amount of information can be measured in number of white pixels on band of the edge cube. It is calculated by dividing amount of all values (ρ) from every band by number of pixels per band (*ppl*):

$$\bar{\text{H}} = \frac{\sum \rho}{ppl} \tag{1}$$

Vector with normalized value of entropy is presented in Figure 4(a). We are using it to create a mean entropy filter (b).

Using the same method we are calculating dynamics of entropy changes (c).

$$\bar{\text{DH}} = |\bar{H} - \bar{H}'| \tag{2}$$

From dynamics we calculate mean dynamics filter (d). Union of those two filters (e) connects ability to filter noisiest parts (mean entropy filter) and ability to filter hills of entropy changes (mean dynamics filter).

Figure 4(f) shows ability of separation informations from noises.

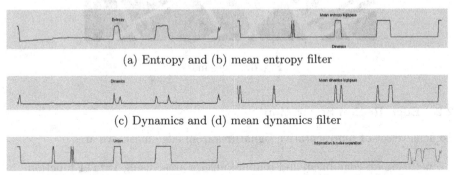

(a) Entropy and (b) mean entropy filter

(c) Dynamics and (d) mean dynamics filter

(e) Union of filters and (f) separation of noisy and informative bands.

Fig. 4. Filtering process.

Fig. 5. Flattered borders cube.

To construct an edges mask using the edges cube we are:

- filtering an edges cube, using an union filter,
- flattering an edges cube, using maximum function (Figure 5),
- blurring flat mask, using the anisotropic diffusion [7],
- converting flat mask into binary mask, using threshold of mean value of mask.

Resulting edge mask is presented in Figure 6.

Fig. 6. Edges mask.

2.2   Edge-Region Separation

To achieve set of representative signatures from edge mask, we are:

– negating an edge mask and refilling unsure areas using `imfill` function,
– assigning the label to every separated region of mask, using the `bwlabel()` function (Figure 7),

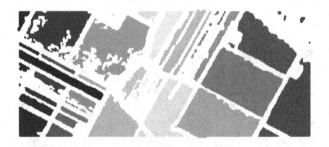

Fig. 7. Separated regions labeling.

– removing all regions smaller than given percentage of image,
– merging the labels for regions with similar signatures, using the mean value of signatures difference and given threshold.

Representative signatures set is an collection of mean signatures for every label achieved using this processing (Figure 8).

2.3   Blurred Labeling

Blurred labeling process is presented in Figure 9. We are taking every label generated in previous step (left column on image), and calculating the difference between its mean signature and whole, filtered, hyperspectral cube, which gives us a mask of similarity (center column). Mask of similarity is cleaned up, using the anisotropic diffusion [7] (right column).

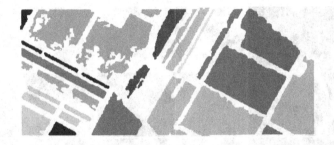

Fig. 8. Labels after signature merging.

All the diffused masks are stacked. Calculating maximum value of this stack gives us information about conviction of labeling (value) and assigned label (index). Result of maximum calculation is presented in the last row of Figure 10.

To achieve a final segmentation we are doing few more, post-processing steps.

- Every disjoint region of every label becomes a new label (Figure 11).
- Every label smaller than given percentage of image is removed.
- Every pair of neighbor labels with similar signatures are united.

Final segmentation of image is presented in Figure 12.

3   Experimental Evaluation

This section tries to experimentally analyze the usefulness of proposed segmentation method for classification task. We are using several state-of-the-art machine learning methods.

Three benchmark datasets of hyperspectral images was used to evaluate proposed approach[3].

- *Salinas C*, was collected in California, USA, over the Valley of Salinas. It contains 217 x 512 signatures over 224 bands in range 0.4—2.5 m, with nominal resolution of 10 nm.
- *Salinas A* is a segment of 86 x 83 signatures from the first image.
- *Indian Pines* dataset was collected over the Indian Pines in Indiana, USA. It contains 145 x 145 signatures over 224 bands in range 0.4—2.5 m, with nominal resolution of 10 nm. It's very noisy and unclear.

We use single-model based approaches, using four different popular classifiers: k-Nearest-Neighbors (k-NN), Neural Network (MLP), Decision Tree (DT), Naive Bayes (NB) and Support Vector Machine (SVM). As classifier ensemble, we selected the Random Forest. For the experiments we used their computer

---

[3] These datasets are standard benchmarking datasets according to the literature. A more detailed description of this datasets can be found in website http://www.ehu.eus/ccwintco/index.php?title=Hyperspectral_Remote_Sensing_Scenes

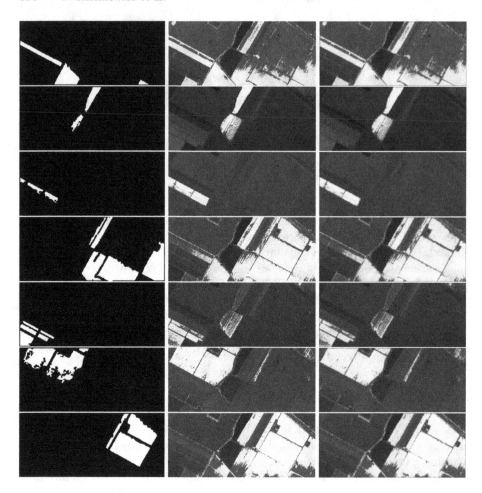

Fig. 9. Blurred labeling process. Rows presents chosen labels. First column is edge-region separation label. Second is a distance between mean signature from label and filtered image. Third is anisotropic filtered distance map.

implementations available in KNIME software. All experiments were carried out using 5x2 cross validation [8] and presented as averaged results.

Results of experiment for every dataset are presented in Figure 13.

The proposed method performs quite well comparing with other state-of-art approaches. The Random Forrest seems to be valuable classifier for this task. Admittedly, it is not significantly better than the minimal distance classifier, but its computational load is extremely smaller.

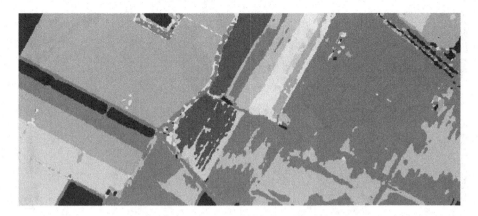

Fig. 10. Result of blurred labeling.

Fig. 11. Labels after separation of labels.

Table 1. Results of the experiment with respect to accuracy [%], extended by information from t-test. Values in table denote for how many datasets, algorithm from row is statistically better than algorithm from column.

	KNN	MLP	DT	SVM	Ens	NB
KNN	—	3	0	3	0	1
MLP	0	—	0	0	0	1
DT	0	0	—	1	0	1
SVM	0	1	1	—	0	1
ENS	2	0	2	0	—	1
NB	0	0	0	0	0	—

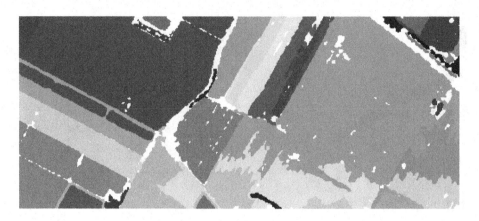

Fig. 12. Final segmentation.

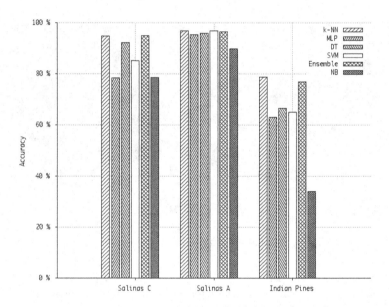

Fig. 13. Classification algorithms accuracy.

4 Conclusions

The work presents a novel method of segmentation dedicated to hyperspectral images. The results of the computer experiments carried out on several benchmark datasets seems to confirm it usefulness, especially in the case where an ensemble classifier is used on the classification step. Because the results are very promising, therefore authors are going to extend the computer experiments on the wider set of hyperspectral images and compare the proposed method with other well-known approaches as Cluster-based Ensemble Algorithm (CENA) [9], ensemble algorithm which combines generative (mixture of Gaussians) and discriminative (support cluster machine) models for classification [10], or classifier ensemble of Support Vector Machines (SVMs) [11].

Acknowledgments. The work was supported in part by the Polish National Science Center under the grant no. DEC-2013/09/B/ST6/02264, by EC under FP7, Coordination and Support Action, Grant Agreement Number 316097, ENGINE — European Research Centre of Network Intelligence for Innovation Enhancement (http://engine. pwr.wroc.pl/), and by the statutory funds of the Department of Systems and Computer Networks, Wroclaw University of Technology.

References

1. Bennett, K.P., Demiriz, A.: Semi-supervised support vector machines. In: Advances in Neural Information Processing Systems, pp. 368–374. MIT Press (1998)
2. Rajan, S., Ghosh, J., Crawford, M.M.: An active learning approach to hyperspectral data classification. IEEE T. Geoscience and Remote Sensing **46**(4), 1231–1242 (2008)
3. Li, J., Bioucas-Dias, J.M., Plaza, A.: Spectral-spatial classification of hyperspectral data using loopy belief propagation and active learning. IEEE T. Geoscience and Remote Sensing **51**(2), 844–856 (2013)
4. Krawczyk, B., Filipczuk, P.: Cytological image analysis with firefly nuclei detection and hybrid one-class classification decomposition. Engineering Applications of Artificial Intelligence **31**, 126–135 (2014)
5. Jackowski, K.: Multiple classifier system with radial basis weight function. In: Graña Romay, M., Corchado, E., Garcia Sebastian, M.T. (eds.) HAIS 2010, Part I. LNCS, vol. 6076, pp. 540–547. Springer, Heidelberg (2010)
6. Davies, E.R.: Machine Vision: Theory, Algorithms, Practicalities. Elsevier, December 2004
7. Perona, P., Malik, J.: Scale-space and edge detection using anisotropic diffusion. IEEE Transactions on Pattern Analysis and Machine Intelligence **12**(7), 629–639 (1990)
8. Bengio, Y., Grandvalet, Y.: No unbiased estimator of the variance of k-fold cross-validation. J. Mach. Learn. Res. **5**, 1089–1105 (2004)
9. Chi, M., Qian, Q., Benediktsson, J.A.: Cluster-based ensemble classification for hyperspectral remote sensing images. In: IGARSS, vol. 1, pp. 209–212. IEEE (2008)
10. Chi, M., Kun, Q., Benediktsson, J.A., Feng, R.: Ensemble classification algorithm for hyperspectral remote sensing data. IEEE Geoscience and Remote Sensing Letters **6**(4) (2009)
11. Ceamanos, X., Waske, B., Benediktsson, J.A., Chanussot, J., Fauvel, M., Sveinsson, J.R.: A classifier ensemble based on fusion of support vector machines for classifying hyperspectral data. Int. J. Image Graphics **1**(4), 293–307 (2010)

Keystroke Data Classification for Computer User Profiling and Verification

Tomasz Emanuel Wesołowski$^{(\boxtimes)}$ and Piotr Porwik

Institute of Computer Science, University of Silesia,
Bedzinska 39, 41-200 Sosnowiec, Poland
{tomasz.wesolowski,piotr.porwik}@us.edu.pl

Abstract. The article addresses the issues of behavioral biometrics. Presented research concerns an analysis of a user activity related to a keyboard use in a computer system. A method of computer user profiling based on encrypted keystrokes is introduced to ensure a high level of users data protection. User's continuous work in a computer system is analyzed. This type of analysis constitutes a type of free-text analysis. Additionally, an attempt to user verification in order to detect intruders is performed. Intrusion detection is based on a modified k-NN classifier and different distance measures.

Keywords: Behavioral biometrics · Data classification · Free text analysis · Keystroke analysis

1 Introduction

The most commonly used methods of computer user authentication and verification usually require elements such as passwords or tokens, which are vulnerable to loss or theft. The biometric methods are free from the dangers of this kind because the forgery of the characteristics of the person being verified related to its appearance or behavior is much more difficult. By means of biometrics the automatic recognition of individuals based on the knowledge of their physiological or behavioral characteristics (e.g., signature characteristics [11,14,16], computer user activity [12]) is possible.

Behavioral characteristics are related to the pattern of behavior of a person. In computer science, for the purpose of computer user verification, biometric methods are based on the analysis of user's activity when using different manipulators (e.g., computer mouse [20]) or a keyboard [1,3,15]. The analysis of the activity connected to a keyboard involves detection of a rhythm and habits of a computer user while typing on a keyboard [21]. The detection of these dependencies allows a recognition of so called user profile. This profile can then be used in the access authorization systems.

In the proposed approach the profile of a user is built from the information on the sequence of keys and on time dependencies that occur between the key events. Next, the computer user's profile can be used in a Host-based Intrusion Detection

© Springer International Publishing Switzerland 2015
M. Núñez et al. (Eds.): ICCCI 2015, Part II, LNCS 9330, pp. 588–597, 2015.
DOI: 10.1007/978-3-319-24306-1_57

System (HIDS). The task of HIDS is to analyze (preferably in real-time) the logs with registered activity of the user and to respond when an unauthorized access is detected. A verification of a user based on the analysis of his typing habits while using a keyboard can effectively prevent an unauthorized access when a keyboard is overtaken by an intruder (so called masquerader) [17,18].

The innovation of the presented method is that the process of collecting a user's activity data is performed in the background, practically not involving a user. A new method for creating a computer use's profile based on the partially encrypted keystrokes is proposed. The encryption of the alphanumeric characters entered via a keyboard is performed in order to prevent an access to user's private and sensitive information (passwords, PIN codes). Literature sources indicate a high efficiency of classifiers for different purposes (e.g., medical diagnosis, biomedical engineering [6] or biometrics [1,8,9,13,16,19]). For this reason, intrusion detection has been performed by means of the k-NN classifier.

2   Related Works

Most methods regarding dynamics of typing on a keyboard are limited to the study of a user's activity involving only the repeated typing of a fixed text [2,4,5,7,9,21]. It is difficult to compare with one another the results of individual studies. There are large differences in the formats of data associated with the activity of users, the number of users participating in the study, time period of a study and the size of the data sets. In most cases, the data sets used in the studies are not publicly available, effectively making it impossible to compare.

An interesting approach to authentication using dynamics of typing is described in [19]. A user's task was to type any text consisting of about 650 characters. Entered characters were stored as plain text. When analyzing, the keys were divided into groups that were organized in a hierarchical structure. However, in analysis assumptions typical for the English language were used. For this reason the authentication method was intended only for texts written in English. The authentication performance under non-ideal conditions was 87,4% in average and it decreased to 83,3% over two-week intervals. This method processes the plain text, which is a serious threat. Furthermore, it is limited to texts written in a determined language.

3   User's Activity Data Acquisition

The software used for collecting and saving information generated while using the keyboard is designed for MS Windows operating systems and does not require any additional libraries. It was designed to work continuously, recording the activity associated with a keyboard, mouse, and the use of popular programs. To ensure a user's private data protection the identifiers of alphanumeric keys are encoded using the MD5 hash function.

The program saves the data in real time during a user's activity. Data are stored in the text files. The first line of the data file contains the screen resolution. The next lines contain the sequence of events related to a user's activity. Each line starts with the prefix followed by the time of the event and additional information (Table 1).

Table 1. Event prefixes and data description for a recorded user's activity

Prefix	Event	Event data
K	key down	an encrypted code of an alphanumeric key
k	key up	or a code of a function key

4   Data Preprocessing

In the studies only the events related to the use of the keyboard are taken into consideration. The keyboard has been divided into groups of keys. The division of the keys is consistent with the following scheme for standard QWERTY keyboard layout:

- left function keys (with assigned identifiers $L1$– $L14$): $F1..F7$, Esc, Tab, *Caps lock, Left shift, Left ctrl, Windows, Left alt*;
- right function keys (with assigned identifiers $R1$ – $R25$): $F8..F12$, *PrtScr, Scroll lock, Pause, Insert, Delete, Home, End, PgUp, PgDown, NumLck, Backspace, Enter, Right Shift, Right Ctrl, Context, Right alt, Arrows (up, down, left, right)*;
- alphanumeric keys (with assigned identifiers $ID1$ – $ID64$);
- other keys.

Each keyboard event (consisting of pressing or releasing the key with the identifier id) is stored in the subsequent i-th row of the input data file as a following vector w_i:

$$w_i = [prefix, t_i, id] \qquad (1)$$

where:
$prefix$ – type of an event, $prefix \in \{'K','k'\}$, $'K'$ - key down, $'k'$ - key up,
t_i – timestamp of an event,
id – key identifier (e.g. $L1$, $L10$, $R25$, etc.).

In order to create computer user's profile time dependencies between the keyboard events are extracted. The text data file containing a set of vectors w_i is converted into a set of vectors v_{id} representing time dependencies between the keyboard events. The data file is searched for the consecutive pairs of rows with identical identifier id, and then each such a pair (containing one key down and following one key up event) is converted into a vector v_{id} according to the following formula:

$$\begin{cases} w_i = ['K', t_i, id] \\ w_j = ['k', t_j, id] \end{cases} \rightarrow v_{id} = [t_i, t_j], i < j. \tag{2}$$

Vectors w_i of the same type (with the same identifier id) should be present in the data file an even number of times. Otherwise, the vector, for which the pair was not found, will be considered as an artifact and will be removed.

5   Data Analysis and User Profiling

In the first stage of user profiling the set of vectors v_{id} is organized into two tree structures T_{keys} (Fig. 1) and T_{pairs} (Fig. 2). Each vector v_{id} containing the element id is assigned to the group G_{id} in a leaf of the T_{keys} structure. After enrollment, the same vector v_{id} is added to all the groups higher in the branches of the tree structure until reaching the root group G_{keys}. For example, if element id of a vector v_{id} is assigned an identifier $ID1$ than the vector v_{ID1} will be added to the groups G_{ID1}, $G_{alphanumeric}$ and finally to G_{keys}.

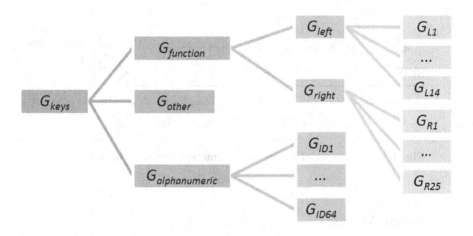

Fig. 1. Tree structure for organizing the groups of keys

By analogy, in the second tree structure T_{pairs} information associated with the occurrence of the events for a specific pair of keys is stored.

In our approach both tree structures include 113 different groups G_{id}.

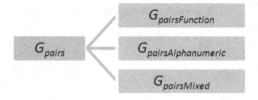

Fig. 2. Tree structure for organizing the groups of key pairs

5.1 Eliminating Outliers

To eliminate the outliers keystroke events are connected into sequences. The event is added to the sequence if the time that has elapsed since the previous event does not exceed the maximum allowed time between two events t_{max}. A sequence, in which a number of elements meeting the first condition does not reach the minimum number of elements c_{min} is omitted in further analysis. The values of parameters t_{max} and c_{min} have been determined experimentally.

During the preliminary analysis and the outliers elimination process only the time of key down event is taken into account. The reason for this is that the time of key up event may be in the wrong order in the data set due to the use of the function keys (Fig. 3).

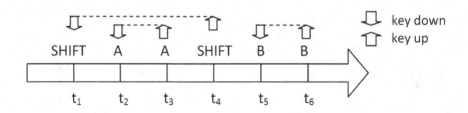

Fig. 3. The sequence of occurrence of function keys and character keys

5.2 Feature Vectors

The next step is to create feature vectors. The groups of keys G_{id} organized into structures T_{keys} and T_{pairs} created in the previous stage maintain information on the number of vectors v_{id} added to the particular group. If the number of vectors v_{id} in the group G_{id} reaches the maximum value specified by the parameter g_{max} the feature vector is created. The group G_{id}, which reached the number of vectors v_{id} equal to g_{max} is cleared and the process of adding further vectors v_{id} to the groups is resumed. The process ends when all the vectors v_{id} of a user have been processed or when the required number of feature vectors has been obtained. The value of the g_{max} parameter has been determined experimentally and applies to all the groups of keys in both tree structures.

The feature vector is calculated from the data collected in both structures T_{keys} and T_{pairs} consisting together of 113 different groups G_{id}. Two different methods for calculating the elements of the feature vector were examined. For each group G_{id} separately the average t_{id} (3) and standard deviation σ_{id} (4) of dwell times (for individual keys) and the time between keystrokes (for pairs of keys) was calculated.

Let the number of vectors v_{id} registered in the group G_{id} be N. Then:

$$t_{id} = \frac{1}{N}\sum_{k=1}^{N} t_k \tag{3}$$

$$\sigma_{id} = \sqrt{\frac{1}{N}\sum_{k=1}^{N}(t_k - t_{id})^2} \tag{4}$$

where:
t_k - dwell time of the k-th key belonging to the group G_{id} in T_{keys} or time between keystrokes for pairs of keys belonging to the group G_{id} in T_{pairs}.

Finally, each feature vector consists of 113 values (features). The process is repeated and for a given user the next feature vector is created. The process ends when the required number of feature vectors has been obtained or all user's data were processed. The required number of feature vectors was experimentally established. In our case, from a biometric point of view, the optimal number of feature vectors in a user's profile was equal to 1000.

A subset of the feature vectors of the given user constitutes its profile.

6   Intrusion Detection

The profile describes the activity of a user associated with the use of the keyboard on the computer workstation. The testing set consists of the feature vectors of another user, which will represent an intruder in the experiment. To decide whether the registered activity is derived from the intruder it is necessary to check how the tested set of the feature vectors differs from the profile.

First, the feature vectors are normalized to the range of [0,1]. The studies were verified using *leave-one-out* method for 4 users and additionally repeated 20

Table 2. Values of biometric system parameters used in the study

Parameter	Value
t_{max}	650 ms
c_{min}	5
g_{max}	15
k-NN	$k = 3$
Acceptance threshold	0.5

times for different subsets of feature vectors of an intruder. The results obtained from tests were averaged. The number of feature vectors generated for each user was for *user*1 to *user*4 equal 1470, 1655, 1798, 1057 respectively.

The testing procedure for intrusion detection took into account the parameters described in Table 2. The values of parameters have been determined experimentally in order to obtain the lowest values of the EER.

7   Results of Experiments

First, the experiments to select the optimal values of parameters of the biometric system based on the k-NN classifier were performed (Table 2). Separate experiments were performed to tune each of the parameters - the values of parameters have been determined experimentally in order to obtain the lowest values of the EER for the developed methods.

Next, two methods for calculating the distances were examined. Fig. 4 depicts an example of the distances obtained for *user*1 constituting the legitimate user. The first 100 samples represent data subsets of the legitimate *user*1. Samples no. 101-200 represent intruders. The method based on Euclidean Distance had an average EER value lower by 7.4 than the one based on Manhattan Distance. The values for Manhattan Distance were much more dispersed also for the testing sets belonging to the same user as the profile.

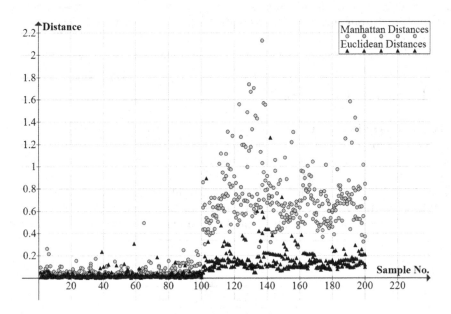

Fig. 4. The distances measured by two methods for legitimate *user*1

The final results of the study for feature vectors calculated as average (3) and standard deviation (4) are presented in Fig. 5 and Fig. 6 respectively.

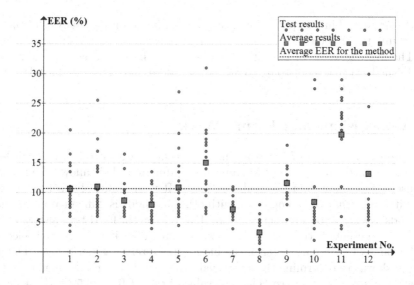

Fig. 5. Values of EER for the intrusion detection and profiling method based on the average values

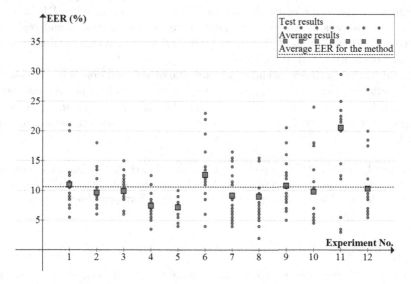

Fig. 6. Values of EER for the intrusion detection and profiling method based on the standard deviation

The charts in Fig. 5 and Fig. 6 should be interpreted as follows. The columns represent different pairs of a legitimate user and an intruder, for which the tests were performed. Each column indicates 20 tests of the same intruder for different subsets of input data (dots). Squares represent the average score of 20 attempts for each pair of users. The dashed line represents the average value of the EER for the presented in this paper biometric system based on the analysis of the use of a computer keyboard.

The average value of the EER for the studies based on average values was established at the level of 10.62% and for the studies based on standard deviation at the level of 10.59%.

8   Conclusions and Future Work

The proposed method of recording user;s activity data introduces a high level of security through the use of MD5 encoding function. This allows the analysis of user's continuous work in real conditions. It is an innovative solution, however, it causes that the comparison with other methods is difficult because most methods are based on an open text and limited length, fixed text analysis.

In the presented study, a series of experiments allowing an optimal selection of the value of each parameter of the biometric system was performed. The priority task was to determine the lowest value of the EER of the biometric system based on a use of a keyboard. The best value of the EER = 10.59% was achieved for the intrusion detection based on Euclidean Distances and standard deviation. The introduced approach is not only better than the one presented in [19], it eliminates also some of its limitations. The study described in [19] was limited to the analysis of English language, stored as an open text and registered in a controlled environment. The presented method is also comparable or better than the ones presented in [4,5,7,9].

In future, the authors intend to explore the other methods of data classification in order to detect intrusions and masqueraders in particular. An interesting approach to consider was described in [10]. Additional studies should consider merging user's activity data connected to the use of computer mouse and keyboard.

Acknowledgments. The research described in this article has been partially supported from the funds of the project "DoktoRIS – Scholarship program for innovative Silesia" co-financed by the European Union under the European Social Fund.

References

1. Alsultan, A., Warwick, K.: Keystroke Dynamics Authentication: A Survey of Free-text Methods. J. of Computer Science Issues **10**(4), 1–10 (2013). No. 1
2. Araujo, L.C.F., Sucupira Jr, L.H.R., Lizarraga, M.G., Ling, L.L., Yabu-Uti, J.B.T.: User Authentication Through Typing Biometrics Features. IEEE Transactions on Signal Processing **53**(2), 851–855 (2005)
3. Banerjee, S.P., Woodard, D.L.: Biometric Authentication and Identification Using Keystroke Dynamics: A survey. J. of Pattern Recognition Research **7**, 116–139 (2012)
4. Dowland, P.S., Singh, H., Furnell, S.M.: A preliminary investigation of user authentication using continuous keystroke analysis. In: The 8th Annual Working Conference on Information Security Management and Small Systems Security (2001)

5. Filho, J.R.M., Freire, E.O.: On the Equalization of Keystroke Timing Histogram. Pattern Recognition Letters **27**(13), 1440–1446 (2006)
6. Foster, K.R., Koprowski, R., Skufca, J.D.: Machine learning, medical diagnosis, and biomedical engineering research - commentary. Biomedical Engineering Online **13**, Article No. 94 (2014). doi:10.1186/1475-925X-13-94
7. Gunetti, D., Picardi, C., Ruffo, G.: Keystroke analysis of different languages: a case study. In: Famili, A.F., Kok, J.N., Peña, J.M., Siebes, A., Feelders, A. (eds.) IDA 2005. LNCS, vol. 3646, pp. 133–144. Springer, Heidelberg (2005)
8. Hu, J., Gingrich, D., Sentosa, A.: A k-nearest neighbor approach for user authentication through biometric keystroke dynamics. In: IEEE International Conference on Communications, pp. 1556–1560 (2008)
9. Killourhy, K.S., Maxion, R.A.: Comparing anomaly-detection algorithms for keystroke dynamics. In: International Conference on Dependable Systems & Networks (DSN 2009), pp. 125–134. IEEE Computer Society Press (2009)
10. Krawczyk, K., Woźniak, M., Cyganek, B.: Clustering-based ensembles for one-class classification. Information Sciences **264**, 182–195 (2014)
11. Kudłacik, P., Porwik, P.: A new approach to signature recognition using the fuzzy method. Pattern Analysis & Applications **17**(3), 451–463 (2014). doi:10.1007/s10044-012-0283-9
12. Kudłacik, P., Porwik, P., Wesołowski, T.: Fuzzy Approach for Intrusion Detection Based on User's Commands. Soft Computing. Springer-Verlag, Heidelberg (2015). doi:10.1007/s00500-015-1669-6
13. Lopatka, M., Peetz, M.: Vibration sensitive keystroke analysis. In: Proc. of The 18th Annual Belgian-Dutch Conference on Machine Learning, pp. 75–80 (2009)
14. Płys, M., Doroz, R., Porwik, P.: On-line signature recognition based on an analysis of dynamic feature. In: IEEE International Conference on Biometrics and Kansei Engineering, pp. 103–107. Tokyo Metropolitan University Akihabara (2013)
15. Teh, P.S., Teoh, A.B.J., Yue, S.: A Survey of Keystroke Dynamics Biometrics. The Scientific World Journal **2013**, Article ID 408280, 24 (2013) . doi:10.1155/2013/408280
16. Porwik, P., Doroz, R., Orczyk, T.: The k-NN classifier and self-adaptive Hotelling data reduction technique in handwritten signatures recognition. Pattern Analysis and Applications. doi:10.1007/s10044-014-0419-1
17. Raiyn, J.: A survey of cyber attack detection strategies. International J. of Security and Its Applications **8**(1), 247–256 (2014)
18. Salem, M.B., Hershkop, S., Stolfo, S.J.: A survey of insider attack detection research. In: Advances in Information Security, vol. 39, pp. 69–90. Springer, US (2008)
19. Tappert, C.C., Villiani, M., Cha, S.: Keystroke biometric identification and authentication on long-text input. In: Wang, L., Geng, X. (eds.) Behavioral Biometrics for Human Identification: Intelligent Applications, pp. 342–367 (2010). doi:10.4018/978-1-60566-725-6.ch016
20. Wesolowski, T., Palys, M., Kudlacik, P.: Computer user verification based on mouse activity analysis. In: Barbucha, D., Nguyen, N.T., Batubara, J. (eds.) New Trends in Intelligent Information and Database Systems. SCI, vol. 598, pp. 61–70. Springer, Heidelberg (2015)
21. Zhong, Y., Deng, Y., Jain, A.K.: Keystroke dynamics for user authentication. In: IEEE Computer Society Conference on Computer Vision and Pattern Recognition Workshops, pp. 117–123 (2012). doi:10.1109/CVPRW.2012.6239225

The New Multilayer Ensemble Classifier for Verifying Users Based on Keystroke Dynamics

Rafal Doroz[(✉)], Piotr Porwik, and Hossein Safaverdi

Institute of Computer Science, University of Silesia,
Ul. Bedzinska 39, 41-200 Katowice, Sosnowiec, Poland
{rafal.doroz,piotr.porwik,hossein.safaverdi}@us.edu.pl
http://www.biometrics.us.edu.pl, http://www.zsk.us.edu.pl

Abstract. In this work we proposed the new multilayer ensemble classifier which can be applied in many domains, especially in the biometric systems. Proposed classifier works on database which comprises data from keystroke dynamics. Such kind of data allows us to recognize computer users who use password. It is a typical case among the users every day work. Obtained results confirm that proposed multilayer ensemble classifier gives the high security level. For this reason our method can be used to protect computer resources against forgers and imposters.

Keywords: Keystroke dynamics · Ensemble classifiers · Biometrics

1 Introduction

The increasing use of internet and computers makes that in practice our work can be observed by illegitimate users. Today internet service is widely available, hence many, even remotely, users have access to computers and resources [7]. For example, almost everyday we need to check our bank account, e-mail account or we have to fill out online forms, so we need to use our personal information. There are several ways to protect the sensitive information. Biometric techniques are, where we analysis voice samples, fingerprint, iris, lip prints [18,19], digital signature, gait, etc. allow us to increase the security of the computer system. Keystroke dynamic is one of the important features in biometrics because it does not need any additional devices to install on computer and it could be fully implemented by means of the software [8], [12–14], [17]. A password is also one of the best solutions to increase the computer system security [4]. Although the password has a lot of benefits, in some cases it raises the danger. When someone illegally gets the password, the system cannot correctly recognize the person, then impostor can obtain unauthorized access to computer resources.

One way to overcome this problem is behavioral biometric of keystroke dynamics [4]. The keystroke rhythms of a user are measured to develop a unique biometric template of the user's typing pattern for future authentication [7].

© Springer International Publishing Switzerland 2015
M. Núñez et al. (Eds.): ICCCI 2015, Part II, LNCS 9330, pp. 598–605, 2015.
DOI: 10.1007/978-3-319-24306-1_58

Keystroke dynamics allows to measure timing patterns which could be used to recognize individuals. In this technique time durations between two keystrokes and pressure on the key can be measured [8], [12–14]. It allows us to build a unique pattern for the individuals. Keystroke dynamics is developed [1], [9,10] and accuracy of this biometric is improved. In our paper we used a machine learning technique to improve the recognition of authorized (legitimated) and unauthorized (illegitimated) users [9–11].

2 Ensemble Classifier

To make a pattern for each user, each of them types by means of the computer keyboard the password "try-mbs" 10 times [5,6]. In practice password is represented by the vector $\mathbf{v}_j^k = [\,156\ 188\ 266\ 375\ 343\ 219\ 203\ k\,]$, where numbers indicate time between two consecutive pressed keys, k is a label of a given user, and $j = 1, ..., 10$ is number of pattern. Mentioned above vectors will be supplied to an input of classifiers. By means of computer keyboard, user registers the password which was mentioned above, the password consists of eight letters and it ends by label. Such data forms a vector that is parallelly supplied to the inputs of four classifiers (see Fig. 1).

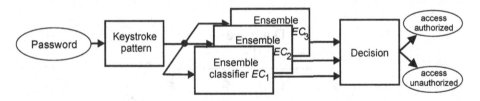

Fig. 1. The general structure of proposed classifier devoted to computer user's recognition.

It could be seen that proposed classifier consists of three ensemble-based sub-classifiers. The main part of a system is ensemble of the classifiers EC_i, $i = 1, ..., 3$ which consists of four single classifiers: c_1, c_2, c_3 and c_4. The ensemble classifiers EC_i work parallelly and each of them has the same structure.

In practice different types of classifiers can be applied, but we propose the following classifiers: Kstar(c_1), BayesNet(c_2), LibSVM(c_3) and HoeffdingTree(c_4).

Kstar: is an instance-based classifier. The classifier works on the training instance dataset. And for classification some similarity functions are used. It differs from other instance-based learners in that it uses entropy as distance function. The fundamental assumption of such a classifier is that similar instances will have similar classifications.

BayesNet: Bayesian networks (BNs), belong to the family of probabilistic models. This classifier is used to represent knowledge about an uncertain domain.

In particular, each node in this network represents a random variable, while the connections between network nodes represent probabilistic dependencies between corresponding random variables. These conditional dependencies are often estimated by using known statistical methods. Hence, BNs combine principles from network, and probability theory, as well as computer science, and statistics.

LibSVM: implements algorithm for kernelized support vector machines (SVMs). This classifier utilized well known support vector classification. It is special library which allow us to accelerate computation.

HoeffdingTree: Hoeffding tree is an incremental, anytime decision tree induction algorithm that is capable of learning from massive data streams, assuming that the distribution generating examples does not change over time. Hoeffding trees exploit the fact that a small sample can often be enough to choose an optimal splitting attribute. This idea is supported mathematically by the Hoeffding bound, which quantifies the number of observations needed to estimate some statistics within a prescribed precision [3].

This selection follows from the fact that mentioned classifiers give the best accuracy level compared to other classifiers. Each the i'th classifier c_i on its output calculates probability $P_i \in [0, 1]$ that a given individual is legitimated or not. The EC_i structure is presented in Fig. 2. The local IF THEN rule creates a local EC_i ensemble decision.

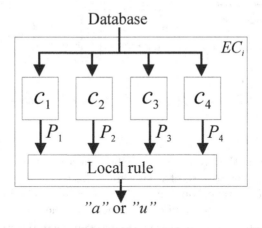

Fig. 2. Single ensemble classifier structure (one out of three).

Let S_g be a total probability generated by all classifiers. Let S_f be a probability that a given user password is forged. For such assumptions, we have:

$$S_g = \sum_{i=1}^{4} P_i, \tag{1}$$

$$S_f = \sum_{i=1}^{4} (1 - P_i), \tag{2}$$

where P_i is probability of legitimate user password produced by the i'th classifier.

The classifiers c_1, c_2, c_3 and c_4 are single classifiers and local rule produce results of classification - password was authorized ("a") or not ("u"). For such assumption, the local rule can be formulated as follow:

$$
\begin{aligned}
if \quad & S_g > S_f \ then \ \text{user is legitimated ("a")} \\
otherwise \quad & \text{user is illegitimated ("u")}
\end{aligned}
\tag{3}
$$

3   Classifier in the Training Mode

Learning set for a given user consists of 10 genuine and 10 forged passwords [5,6]. In practice password is changing to a vector form, which has been shown in the previous section. Our database comprises only original passwords [5,6], therefore forged examples have to be formed from the genuine passwords of other users.

Let $O^k = \{\mathbf{v}_1^k, \mathbf{v}_2^k, ..., \mathbf{v}_{10}^k\}$ be a set of genuine passwords of the person k. The database W includes passwords of 100 users and each of them typed the same password 10 times, therefore $W = \{O^k\}$, $k = 1, ..., 100$. Each ensemble classifier EC_i, $i = 1, ..., 3$ (see Fig. 2) is learned separately by means of the three kinds of learning sets DS_1^k, DS_2^k and DS_3^k, respectively. For such assumptions the forged ($F^a \subset W$) and genuine ($O^k \subset W$) passwords of legitimated user k are marked below as follows:

$$
\begin{aligned}
DS_1^k = O^k \cup F^a, \quad DS_2^k = O^k \cup F^b, \quad DS_3^k = O^k \cup F^c \\
a \neq b \neq c \neq k \quad \text{and} \quad k = 1, ..., 100
\end{aligned}
\tag{4}
$$

It should be noticed that password F^a is a genuine password of the person a but it is treated as a forged password for the person k.

The classifiers c_1, c_2, c_3 and c_4 work in supervised mode. After training, classifier can be switched into verified mode.

4   Classifier in the Verify Mode

Unlike the other approaches [1], [4], [16] instead of mixing biometric features we used different machine learning algorithms only. As was explained above, the global classifier has a multi-layer structure which brightly follows from Fig. 1.

During verification procedure three ensembles of classifiers form the answer in the majority voting scheme, which is presented synthetically in Fig. 3. The voting scheme uses IF THEN rule to build ultimate decision whether the password of a given user is authorized ("a") or unauthorized ("u") [9–11]:

$$
\begin{aligned}
if \quad & \bigcup_{i=1}^{3} \{a\}_{EC_i} > \bigcup_{i=1}^{3} \{u\}_{EC_i} \ then \ \text{user is legitimated ("a")} \\
otherwise \quad & \text{user is illegitimated ("u")}
\end{aligned}
\tag{5}
$$

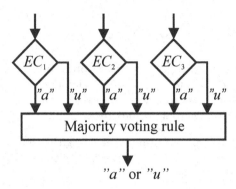

Fig. 3. Ensemble of classifiers work in the password verification mode.

5   Results Obtained

Experiments carried out allow us comparing the obtained results with other methods. In the consecutive experiments, the best structure of classifiers has been established. The results have been gathered in Table 1. In this experiment instead of ensemble, various single classifiers have been checked. Table 1 presents accuracy of the best five single classifiers from many others which were tested.

Table 1. Computer users recognizing accuracy of the best single classifiers.

Classifier	Accuracy [%]
NaiveBayesUpdatable	84.85
RBFClassifier	84.56
NaiveBayesSimple	84.25
HoeffdingTree	84.22
NaiveBayes	82.98

In the next investigation we built a different ensemble classifiers. Instead of single classifier, the one-layer ensemble with different members (c_1,c_2,c_3 and c_4) have been tested. Table 2 presents the accuracy results for the best two one-layer ensemble classifiers.

From Table 2 follows that accuracy of ensemble classifiers is still not enough in professional biometric systems. Such results should be improved. To realize this task we proposed the new type of multilayer ensemble classifiers. The general structure of these ensembles have been already presented in Fig 1. In the long time investigation carried out, we found the best classifier members of ensemble classifiers. Members of the three ensemble classifiers have been presented in Table 3 as well as the best accuracy which such classifiers reached.

Additionally we built confusion matrix for the ensemble which worked in the computer users verification mode. It presented in Table 4. Confusion matrix has

Table 2. Different ensemble classifiers (like as in Fig 2) and their accuracy for the users recognition by means of keystroke dynamics.

Ensemble Classifiers	Accuracy [%]	
c_1	BayesNet	
c_2	NaiveBayesSimple	
c_3	IBK	88.7
c_4	RandomForest	
c_1	Kstar	
c_2	HoeffdingTree	
c_3	NaiveForest	87.7
c_4	IsolationForest	

Table 3. The best selection of the ensemble classifier members (like as in Fig 3) for keyboard dynamics - based biometric system.

Members of ensemble of ensemble classifiers	Accuracy [%]	
c_1	Kstar	
c_2	BayesNet	
c_3	LibSVM	98.4
c_4	HoeffdingTree	

Table 4. Confusion matrix for the best ensemble classifier selection.

TP = 971	FN = 29
FP = 3	TN = 997

Table 5. Accuracy comparison with different methods.

Method	Accuracy [%]	Database	Used feature(s)	Nr of users
Proposed method	98.40	try4-mbs	latency	100
Loy [5]	96.14	try4-mbs	latency	100
Monrose [7]	92.14	private	variuos	63
Guven [2]	95.00	private	variuos	12
Sung [15]	98.13	practice	variuos	100

been built for ensemble with members from Table 3. Our results were compared with other investigations, which is presented in Table 5. Form this table follows that proposed multilayer classifier structure gives the best accuracy level compared to state of the art proposition announced in literature. It means that in some cases biometric systems can be simply improved.

6   Conclusions

From obtained results follow that new type of multilayer classifier gives the highest accuracy level compared to single classifier approaches and for other single layer ensemble classifier structures. Accuracy of 98.4% is promising result and can be treated as very high biometric factor. Hence our method can be included

into the professional biometric systems based on the keystroke dynamics. The
newest literature announced that it is possible to obtain the highest accuracy
level compare to our proposition, but it needed more sophisticated method what
causes that system works slowly.

In future we will try to investigate other ensemble classifiers with various
members.

Acknowledgments. This work was supported by the Polish National Science Centre
under the grant no. DEC-2013/09/B/ST6/02264.

References

1. Doroz, R., Porwik, P.: Handwritten signature recognition with adaptive selection of behavioral features. In: Chaki, N., Cortesi, A. (eds.) Computer Information Systems – Analysis and Technologies. CCIS, vol. 245, pp. 128–136. Springer, Heidelberg (2011)
2. Guven, A., Sogukpinar, I.: Understanding users' keystroke patterns for computer access security. Computers Security **22**(8), 695–706 (2003)
3. Kirkby R.: Improving Hoeffding Trees, PhD thesis, University of Waikato (2007)
4. Kang, P., Cho, S: Keystroke dynamics-based user authentication using long and free text strings from various input devices, Information Sciences **308**, 72–93 (2015)
5. Loy, C.C., Lim, C.P., Lai, W.K.: Pressure-based typing biometrics user authentication using the fuzzy ARTMAP neural network. In: International Conference on Neural Information Processing, Taiwan (2005)
6. Loy, C.C., Lai, W.K., Lim, C.P: Keystroke patterns classification using the ARTMAP-FD neural network. In: International Conference on Intelligent Information Hiding and Multimedia Signal Processing, Taiwan (2007)
7. Monrose, F., Rubin, A.D.: Keystroke dynamics as a biometric for authentication. Future Generation Computer Systems **16**(4), 351–359 (2000)
8. Panasiuk, P., Saeed, K.: Influence of database quality on the results of keystroke dynamics algorithms. In: Chaki, N., Cortesi, A. (eds.) CISIM 2011. CCIS, vol. 245, pp. 105–112. Springer, Heidelberg (2011)
9. Porwik, P., Doroz, R., Wrobel, K.: A new signature similarity measure. In: World Congress on Nature Biologically Inspired Computing, NaBIC 2009, pp. 1022–1027. IEEE (2009)
10. Porwik, P., Doroz, R., Orczyk, T.: The k-NN classifier and self-adaptive Hotelling data reduction technique in handwritten signatures recognition. Pattern Analysis and Applications, 1–19 (2015)
11. Porwik, P., Doroz, R.: Self-adaptive biometric classifier working on the reduced dataset. In: Polycarpou, M., de Carvalho, A.C.P.L.F., Pan, J.-S., Woźniak, M., Quintian, H., Corchado, E. (eds.) HAIS 2014. LNCS, vol. 8480, pp. 377–388. Springer, Heidelberg (2014)
12. Rybnik, M., Panasiuk, P., Saeed, K.: User authentication with keystroke dynamics using fixed text. In: International Conference on Biometrics and Kansei Engineering, ICBAKE 2009, pp. 70–75. IEEE (2009)
13. Rybnik, M., Panasiuk, P., Saeed, K., Rogowski, M.: Advances in the keystroke dynamics: the practical impact of database quality. In: Cortesi, A., Chaki, N., Saeed, K., Wierzchoń, S. (eds.) CISIM 2012. LNCS, vol. 7564, pp. 203–214. Springer, Heidelberg (2012)

14. Rybnik, M., Tabedzki, M., Saeed, K.: A keystroke dynamics based system for user identification. In: 7th Computer Information Systems and Industrial Management Applications, CISIM 2008, pp. 225–230. IEEE (2008)
15. Sung, K., Cho, S.: GA SVM Wrapper Ensemble for Keystroke Dynamics Authentication Department of Industrial Engineering, Seoul National University, San 56–1, Shillim-dong, Kwanak-gu, Seoul, 151–744, Korea (2005)
16. SZS, I., Cherrier, E., Rosenberger, C., Bours, P.: Soft biometrics for keystroke dynamics: Profiling individuals while typing passwords. Computers Security **45**, 147–155 (2014)
17. Teh, P.S., Teoh, A.B.J., Tee, C., Ong, T.S.: Keystroke dynamics in password authentication enhancement. Expert Systems with Applications **37**(12), 8618–8627 (2010)
18. Wrobel, K., Doroz, R., Palys, M.: A method of lip print recognition based on sections comparison. In: IEEE Int. Conference on Biometrics and Kansei Engineering, pp. 47–52. Tokyo Metropolitan University Akihabara, Tokyo, Japan (2013)
19. Wrobel, K., Doroz, R., Palys, M.: Lip print recognition method using bifurcations analysis. In: Nguyen, N.T., Trawiński, B., Kosala, R. (eds.) ACIIDS 2015. LNCS, vol. 9012, pp. 72–81. Springer, Heidelberg (2015)

Application of Evolving Fuzzy Systems to Construct Real Estate Prediction Models

Martin Grześlowski[1], Zbigniew Telec[1], Bogdan Trawiński[1(✉)],
Tadeusz Lasota[2], and Krzysztof Trawiński[3]

[1] Department of Information Systems, Wrocław University of Technology,
Wybrzeże Wyspiańskiego 27 50-370, Wrocław, Poland
{zbigniew.telec,bogdan.trawinski}@pwr.edu.pl
[2] Department of Spatial Management, Wrocław University of Environmental and Life Sciences,
ul. Norwida 25/27 50-375, Wrocław, Poland
tadeusz.lasota@up.wroc.pl
[3] European Centre for Soft Computing, Edificio Científico-Tecnológico, 3ª Planta,
C. Gonzalo Gutiérrez Quirós S/N 33600, Mieres, Asturias, Spain
krzysztof.trawinski@softcomputing.es

Abstract. Four variants of the *eTS* algorithms (evolving Takagi-Sugeno fuzzy systems) were implemented and examined in respect of their usefulness for the intelligent system of real estate market. The *eTS* algorithms were compared as regards their predictive accuracy with the *Flexfis* algorithm and ensembles employing general linear models (*Glm*) devoted to predict from a data stream of real estate sales transactions. The experiments were conducted in Matlab environment using real-world data taken from a dynamically changing real property market. The analysis of the results was performed using statistical methodology including nonparametric tests followed by post-hoc procedures designed especially for multiple $N \times N$ comparisons. The models produced by two versions of *Simple_eTS* and *Flexfis* algorithms and as well as ensembles composed of *Glm* models revealed statistically similar performance.

Keywords: Evolving fuzzy systems · *eTS* · *Flexfis* · General linear models · Ensemble models · Data stream · Property valuation

1   Introduction

For several years we have been exploring techniques for constructing an intelligent model of real estate market to assist with of property valuation. It is devoted to a broad spectrum of users involved in the real property management ranging from individual citizens and real estate appraisers to banks, insurance companies, and local self-governments. The architecture of the system to be exploited on a cloud computing platform is outlined in Figure 1. Various public registers and cadastral systems establish a manifold data source for the system. The essential part of the system are property valuation models including models built in accordance with the professional standards as well as automated data-driven models constructed with machine learning algorithms.

© Springer International Publishing Switzerland 2015
M. Núñez et al. (Eds.): ICCCI 2015, Part II, LNCS 9330, pp. 606–616, 2015.
DOI: 10.1007/978-3-319-24306-1_59

Two classes of computer systems including Automated Valuation Models (AVM) and Computer Assisted Mass Appraisal (CAMA) are intensively developed to fulfil the needs of numerous users interested in the premises values. These systems combine models generated using statistical multiple regression [1], neural networks [2], rough set theory [3], and hybrid approaches [4]. The requirement for informative and interpretable models gave rise to fuzzy and neuro-fuzzy systems [5], [6] as alternative solutions. Incremental online techniques are especially suitable for this purpose because each day dozens of the records of real estate sales-purchase transactions income to cadastral information centres to be registered. The stream of transactional data reflect also the changes of real estate market over time. This motivated us to employ evolving fuzzy systems to process streams of data utilizing their ability to learn, update, expand their memory on-line on demand.

For almost a decade we have been working out and exploring techniques for constructing data driven regression models to assist with property valuation based on genetic fuzzy systems [7], [8]. We devised also ensemble models built employing different resampling techniques including bagging together with evolutionary fuzzy systems, decision trees, and neural networks [9], bagging comprising linear regression, artificial neural networks, decision trees, and support vector machines [10], bagging encompassing 16 various fuzzy algorithms implemented in KEEL [11] as well as subagging, repeated holdout, and repeated cross validation using genetic fuzzy systems [12]. A relatively good performance provided evolving fuzzy systems applied to a data stream formed from cadastral data [13], [14]. We have also studied the application of ensembles of genetic fuzzy systems and neural networks to predict from a data stream of real estate sales transactions [15], [16].

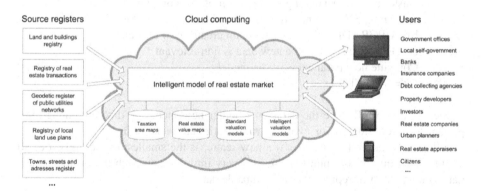

Fig. 1. General schema of the intelligent system of real estate market

The goal of the study reported in the paper was to implement four variants of the *eTS* evolving fuzzy systems devised by Angelov and Filev [17], [18] with modifications proposed by Victor and Dourado [19], [20] and examine from the point of view of their usefulness for the intelligent system of real estate market. The eTS algorithms were compared in respect of predictive accuracy with the *Flexfis* algorithm based on evolving fuzzy systems for incremental learning from data streams worked out by

Lughofer et al. [21], [22] and ensembles employing general linear models devoted to predict from a data stream of real estate sales transactions [23], [24]. The experiments were conducted in Matlab environment using using real-world data taken from a dynamically changing real property market.

2 Implemented Evolving Fuzzy Algorithms

For the purpose of the study reported in this paper we implemented four versions of the evolving Takagi-Sugeno system, proposed by Angelov and Filev in [17] (*eTS* algorithm) and [18] (*simple_eTS* algorithm), incorporating modifications based on ideas introduced by Victor and Dourado in [19], [20]. In the eTS model, its rule-base and parameters continuously evolve by adding new rules with more summarization power, and/or modifying existing rules and parameters. The core of the systems is keeping up to date Takagi-Sugeno consequent parameters and scatter antecedents derived from a fuzzy clustering process. The flow of the *simple_eTS* algorithm [18] can be outlined as follows:

1. Read input vector x_k
2. Predict output y'_k
3. Read output y_k
4. Compute scatter S_k for a new data point ($z_k = [x_k \ y_k]$)
5. Update centres of existing clusters by S_k
6. Compare S_k with potentials of existing clusters to check if the new data point ($z_k = [x_k \ y_k]$) is enough important to modify or upgrade an existing cluster center. Depending on the result of this comparison:
   a) MODIFY – the new data point replaces existing cluster
   b) UPGRADE – the new data point creates new cluster
   c) DO NOTHING – the new data is non-relevant
7. Compute new parameters of the model
8. $k \mathrel{+}= 1$; GOTO 1.

Steps 4 and 6 influence the decision whether new incoming data are enough important to be considered for evolving rules. Original decision in [18] about using a new data is based on the criterion if a new scatter is the smallest or the biggest of all scatters according to Formula 1. This is a very intuitive approach to consider a new data sample only if it represents some valuable data.

$$S_k\big(z(k)\big) < minS_k\big(z^i\big) \vee S_k\big(z(k)\big) > maxS_k\big(z^i\big), i \in \big(1, R\big) \qquad (1)$$

In this paper we used approach proposed in [19] which introduced threshold ε in between which new data will be relevant (2). In consequence the new significant data can be detected earlier and in consequence the clusters can evolve faster. As proven in

[19] and [20] using threshold provides lower performance measures RMSE (*root mean square error*) and NDEI (*non-dimensional error index*) compared to the original *eTS* [17] without threshold (3).

$$S_k(z(k)) > maxS_k(z^i) \lor 0.5\varepsilon < S_k(z(k)) < 0.675\varepsilon, i \in (1, R) \tag{2}$$

$$(z(k)) > maxS_k(z^i) \tag{3}$$

Second modification changes the way how scatter value is computed (4). Scatter is an average distance from a data sample to all other samples [18]. The base *eTS* algorithm uses potentials instead of the scatter. The potential is a monotonic function inversely proportional to the distance which enables recursive calculation what is important for online implementation of the learning algorithm (6) [17].

$$S_k(z(k)) = \frac{1}{(k-1)(n+m)} \left((k-1)\sum_{j=1}^{n+m} z_j^2(k) - 2\sum_{j=0}^{n+m} z_j(k)\beta_j(k) + \gamma(k) \right) \tag{4}$$

where

$$\beta_j(k) = \sum_{l=1}^{k-1} z_j(l); \gamma(k) = \sum_{l=1}^{k-1}\sum_{j=1}^{n+m} z_j^2(l) \tag{5}$$

$$P_k(z_k) = \left(1 + \frac{1}{k-1}\sum_{l=1}^{k-1}\sum_{j=1}^{n+1}(d_{lk}^j)^2 \right)^{-1} \tag{6}$$

where

$$d_{lk}^j = z_l^i - z_k^j \tag{7}$$

The goal of this modification is to examine if simpler scatter function, in terms of computational complexity, will give better or equal results compared to the version of the algorithm using the potential (6). In evaluation experiments we used four versions of the eTS algorithm denoted as follows:

- *evTS-1* – modified *simple_eTS* algorithm using condition (2) and the scatter function (4);
- *evTS-2* – original *simple_eTS* algorithm using condition (1) and the scatter function (4);
- *evTS-3* – modified *eTS* algorithm using condition (1) and the potential (6);
- *evTS-4* – original *eTS* algorithm using condition (2) and the potential (6).

3   Evolving Fuzzy System *Flexfis*

The *Flexfis* algorithm, short for FLEXible Fuzzy Inference Systems, incrementally evolves clusters associated with rules and performs a recursive adaptation of consequent parameters by employing local learning approach. The *Flexfis* method, was first proposed in [21] and significantly extended in [22]. It was devised for incremental learning of Takagi-Sugeno fuzzy systems from data streams in a sample-wise single-pass manner. This signifies that always only one sample is loaded into the *Flexfis* learning engine where the model is updated and immediately discarded, afterwards. In consequence the method needs low resources both with respect to computational complexity and with respect to virtual memory and therefore is feasible for on-line modelling applications. In this sort of applications models should be kept-up-to-date as early as possible in order adapt to new operating conditions, systems states etc. and to prevent extrapolation situations when performing predictions on new samples. The basic steps in *Flexfis* technique can be outlined as follows [25]:

1. Rule evolution and updating of antecedent parameters in the cluster space with the help of an incremental evolving clustering variant.
2. Recursive adaptation of consequent parameters exploiting the local learning approach (parameters are updated for each rule separately).
3. Balancing out a non-optimal situation by adding a correction vector to the vector of consequent parameters.
4. In the extended version: Detecting redundant rules with the help of specific overlap measures (two variants: one-dimensional intersection points of fuzzy sets and inclusion metric) and performing on-line rule merging/pruning after each incremental update cycle.

The detailed description of *Flexfis* can be found in [22], [25].

4   Ensemble Approach to Predict from a Data Stream

For comparative experiments we employed our method to predict from a data stream of real estate sales transactions based on ensembles of regression models [15]. The method was worked out as the result of exploring the intermediate solution between evolving fuzzy systems and evolutionary fuzzy systems built over stationary datasets.

Our approach consists in the utilization of aged models to compose ensembles and correction of the output provided by component models by means of trend functions reflecting the changes of prices in the market over time. The idea of our method is illustrated in Figure 2.

The data stream is segmented into data chunks according to the periods of a constant length t_c. Each time interval determines the shift of a sliding time window which encompasses training data to create regression (*RM*) models. The length of the sliding window t_w is equal to the multiple of t_c so that $t_w=jt_c$, where $j=1,2,3,...$. The window determines the scope of training data to generate a model RM_i. from scratch. It is assumed that a model generated over a given training dataset is valid for the next interval which specifies a test dataset. Similarly, the interval t_t which delineates a test dataset is equal to the multiple of t_c so that $t_t=kt_c$, where k=1,2,3,.. . The sliding window is shifted step by step of a period t_s in the course of time, and likewise, the interval t_s is equal to the multiple of t_c so that $t_s=lt_c$, where l=1,2,3,... .

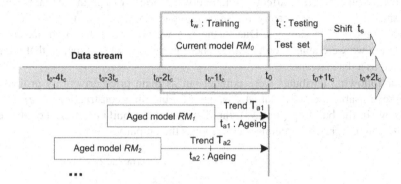

Fig. 2. Idea of an ensemble approach to predict from a data stream

We consider in Figure 2 a point of time t_0 at which the current model RM_0 was built over data that came in between time t_0-2t_c and t_0. The models created earlier, i.e. RM_1, RM_2, etc. aged gradually and in consequence their accuracy deteriorated. However, they are neither discarded nor restructured but utilized to compose an ensemble so that the current test dataset is applied to each component RM_i. However, in order to compensate ageing, their output produced for the current test dataset is updated using trend functions $T(t)$. As the functions to model the trends of price changes the polynomials of the degree from one to five can be employed. For the *Ens-glm* algorithm used in this paper the trends were determined over a time period: encompassing the length of a sliding window plus model ageing intervals, i.e. t_w plus, t_{ai} for a given aged model RM_i. The polynomial of the degree was used as trend function. The method for updating the prices of premises according to the trends of the value changes over time was based on the difference between a price and a trend value in a given time point [15].

5   Experimental Setup and Results

The goal of experiments was to examine the performance of four variants of the eTS algorithm: *evTS-1, evTS-2, evTS-3, evTS-4*. We compared also the results obtained with the outcome of the *Flexfis* algorithm and ensembles composed of general linear models models which output was updated using a second-degree polynomial trend function (*Ens-glm*). The latter two algorithms provided the best accuracy in our previous research [23]. All algorithms were implemented in Matlab environment.

Real-world dataset used in experiments was derived from an unrefined dataset containing above 100 000 records referring to residential premises transactions accomplished in one Polish big city with the population of 640 000 within 14 years from 1998 to 2011. First of all, transactional records referring to residential premises sold at market prices were selected. Then, the dataset was confined to sales transaction data of residential premises (apartments) where the land was leased on terms of perpetual usufruct. The other transactions of premises with the ownership of the land were omitted due to the conviction of professional appraisers stating that the land ownership and lease affect substantially the prices of apartments and therefore they should be used separately for sales comparison valuation methods. The final dataset counted 9795 samples. Due to the fact we possessed the exact date of each transaction we were able to order all instances in the dataset by time, so that it can be regarded as a data stream.

Four following attributes were pointed out as main price drivers by professional appraisers: usable area of a flat (*Area*), age of a building construction (*Age*), number of storeys in the building (Storeys), the distance of the building from the city centre (*Centre*), in turn, price of premises (*Price*) was the output variable.

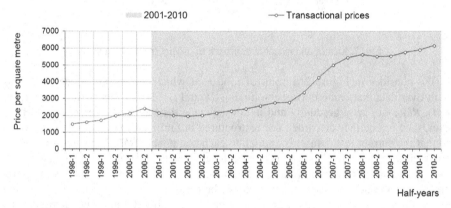

Fig. 3. Change trend of average transactional prices per square metre over time

The evaluating experiments were conducted for 108 points of time from 2002-01-01 to 2010-12-01. Single models were generated over chunks of a data stream delineated by a sliding window which was 12 months long. The window was shifted by one month. Each step, the test datasets were determined by the interval of 3 months, current for a given time point. 108 datasets, obtained in this way, ensured the variability of individual points of observation due to the dramatic rise of the prices of residential premises during the worldwide real estate bubble (see Figure 3). In the case of *evTS-1*, *evTS-2*, *evTS-3*, *evTS-4*, and *Fexfis* algorithms the accuracy of single models was determined the root mean square error (*RMSE*). In turn, for *Ens-glm* average value of *RMSE* provided by 24 component models was computed.

The analysis of the results was performed using statistical methodology including nonparametric tests followed by post-hoc procedures designed especially for multiple *N*×*N* comparisons [26], [27], [28]. The idea behind statistical methods applied to analyse the results of experiments was as follows. The commonly used paired tests i.e. parametric t-test and its nonparametric alternative Wilcoxon signed rank tests are not appropriate for multiple comparisons due to the so called family-wise error. The proposed routine starts with the nonparametric Friedman test, which detect the presence of differences among all algorithms compared. After the null-hypotheses have been rejected the post-hoc procedures should be applied in order to point out the particular pairs of algorithms which produce significant differences. For $N \times N$ comparisons nonparametric Nemenyi's, Holm's, Shaffer's, and Bergamnn-Hommel's procedures are recommended.

6   Statistical Analysis of Experimental Results

The goal of the statistical analysis was to compare the accuracy of the predictive models generated by *evTS-1*, *evTS-2*, *evTS-3*, *evTS-4*, *Flexfis* and ensembles composed of *Glm* models which output was updated using trend functions *T2* for *Age Trends*. Nonparametric tests of statistical significance adequate for multiple comparisons were conducted for all years together. Within this period 108 values of *RMSE*, constituted the points of observation.

Average rank positions of RMSE produced by the models constructed over merged regions which were determined by the Friedman test are shown in Table 1, where the lower rank value the better model. However, please note, it does not mean that each model placed in the lower position significantly outperforms the ones ranked higher on the list.

Table 1. Average rank positions of model performance in terms of *RMSE*

1st	2nd	3rd	4th	5th	6th
Flexfis (2.71)	evTS-2 (2.87)	Ens-glm (3.17)	evTS-1 (3.26)	evTS-3 (4.13)	evTS-4 (4.87)

Table 2. Adjusted p-values for $N \times N$ comparisons in terms of *RMSE*

Alg vs Alg	pWilcox	pNeme	pHolm	pShaf	pBerg
evTS-4 vs Flexfis	0.000000	4.15E-16	4.15E-16	4.15E-16	4.15E-16
evTS-2 vs evTS-4	0.000000	6.89E-14	6.43E-14	4.59E-14	4.59E-14
evTS-4 vs Ens-glm	0.000000	3.74E-10	3.24E-10	2.49E-10	1.74E-10
evTS-1 vs evTS-4	0.000000	4.18E-09	3.35E-09	2.79E-09	1.95E-09
evTS-3 vs Flexfis	0.000000	4.37E-07	3.21E-07	2.92E-07	2.92E-07
evTS-2 vs evTS-3	0.000000	1.25E-05	8.30E-06	8.30E-06	4.98E-06
evTS-3 vs Ens-glm	0.000000	0.002506	0.001503	0.001169	6.68E-04
evTS-1 vs evTS-3	0.000000	0.010087	0.005380	0.004707	0.002690
evTS-3 vs evTS-4	0.000000	0.054289	0.025335	0.025335	0.025335
evTS-1 vs Flexfis	0.067267	0.478323	0.191329	0.191329	0.191329
Ens-glm vs Flexfis	0.151434	1.000000	0.373654	0.298923	0.224193
evTS-1 vs evTS-2	0.067267	1.000000	0.506522	0.506522	0.379891
evTS-2 vs Ens-glm	0.288900	1.000000	0.733480	0.733480	0.379891
evTS-2 vs Flexfis	0.254190	1.000000	1.000000	1.000000	1.000000
evTS-1 vs Ens-glm	0.084960	1.000000	1.000000	1.000000	1.000000

The post-hoc procedures for $N \times N$ comparisons should be applied to point out the pairs of models between which statistically significant differences occur. Adjusted p-values for Nemenyi's, Holm's, Shaffer's, and Bergmann-Hommel's post-hoc procedures for $N \times N$ comparisons for all possible pairs of machine learning algorithms are shown in Table 2. For illustration the p-values of the paired Wilcoxon test are also given. The significance level considered for the null hypothesis rejection was 0.05. The p-values indicating the statistically significant differences between given pairs of algorithms are marked with italics.

Following main observations can be done: there are not significant differences among *evTS-1*, *evTS-2*, *Flexfis*, and *Ens-glm* models. In turn, *evTS-3* and *evTS-4* models reveal significantly worse performance than any of first four models in the ranking. This means that the simpler scatter function incorporated into the *simple_eTS* algorithm (4) provides better performance compared to the version of the algorithm using the potential (6). Moreover, employing modified conditions to modify and upgrade the fuzzy rules (2) does not result in significantly bigger accuracy of the *eTS* models.

7   Conclusions and Future Work

In this paper we examined the application of incremental data-driven fuzzy modelling techniques for the purpose to estimate the future prices of residential premises based on the past sales-purchase transactions.

Following main conclusions can be drawn from our study. The simpler scatter function incorporated into the *simple_eTS* algorithm provides better performance compared to the original version of the *eTS* algorithm using the potential. Moreover, there were not significant differences in accuracy between algorithms using original conditions to modify and upgrade the fuzzy rules and the ones with conditions employing threshold values. Finally, the models produced by two versions of *Simple_eTS* and *Flexfis* algorithms and as well as ensembles composed of *Glm* models revealed statistically similar performance.

We can conclude that evolving fuzzy modelling methods constitute a reliable and powerful alternative for valuation of residential premises and can be successfully utilized in the AVM systems to support appraisers' work.

Acknowledgments. Many thanks also to Edwin Lughofer for granting us his FLEXFIS algorithm code, thereby allowing us to conduct the experiments.

References

1. Nguyen, N.: Cripps. A.: Predicting housing value: A comparison of multiple regression analysis and artificial neural networks. Journal of Real Estate Research **22**(3), 313 (2001)
2. Selim, H.: Determinants of house prices in Turkey: Hedonic regression versus artificial neural network. Expert Systems with Applications **36**, 2843–2852 (2009)
3. D'Amato, M.: Comparing Rough Set Theory with Multiple Regression Analysis as Automated Valuation Methodologies. Int. Real Estate Review **10**(2), 42–65 (2007)
4. Kontrimas, V., Verikas, A.: The mass appraisal of the real estate by computational intelligence. Applied Soft Computing **11**(1), 443–448 (2011)
5. González, M.A.S., Formoso, C.T.: Mass appraisal with genetic fuzzy rule-based systems. Property Management **24**(1), 20–30 (2006)

6. Guan, J., Zurada, J., Levitan, A.S.: An Adaptive Neuro-Fuzzy Inference System Based Approach to Real Estate Property Assessment. Journal of Real Estate Research **30**(4), 395–421 (2008)
7. Król, D., Lasota, T., Nalepa, W., Trawiński, B.: Fuzzy system model to assist with real estate appraisals. In: Okuno, H.G., Ali, M. (eds.) IEA/AIE 2007. LNCS (LNAI), vol. 4570, pp. 260–269. Springer, Heidelberg (2007)
8. Król, D., Lasota, T., Trawiński, B., Trawiński, K.: Comparison of mamdani and TSK fuzzy models for real estate appraisal. In: Apolloni, B., Howlett, R.J., Jain, L. (eds.) KES 2007, Part III. LNCS (LNAI), vol. 4694, pp. 1008–1015. Springer, Heidelberg (2007)
9. Lasota, T., Telec, Z., Trawiński, B., Trawiński, K.: Exploration of bagging ensembles comprising genetic fuzzy models to assist with real estate appraisals. In: Corchado, E., Yin, H. (eds.) IDEAL 2009. LNCS, vol. 5788, pp. 554–561. Springer, Heidelberg (2009)
10. Lasota, T., Telec, Z., Trawiński, B., Trawiński, K.: A multi-agent system to assist with real estate appraisals using bagging ensembles. In: Nguyen, N.T., Kowalczyk, R., Chen, S.-M. (eds.) ICCCI 2009. LNCS, vol. 5796, pp. 813–824. Springer, Heidelberg (2009)
11. Krzystanek, M., Lasota, T., Telec, Z., Trawiński, B.: Analysis of bagging ensembles of fuzzy models for premises valuation. In: Nguyen, N.T., Le, M.T., Świątek, J. (eds.) Intelligent Information and Database Systems. LNCS, vol. 5991, pp. 330–339. Springer, Heidelberg (2010)
12. Lasota, T., Telec, Z., Trawiński, G., Trawiński, B.: Empirical comparison of resampling methods using genetic fuzzy systems for a regression problem. In: Yin, H., Wang, W., Rayward-Smith, V. (eds.) IDEAL 2011. LNCS, vol. 6936, pp. 17–24. Springer, Heidelberg (2011)
13. Lasota, T., Telec, Z., Trawiński, B., Trawiński, K.: Investigation of the eTS Evolving Fuzzy Systems Applied to Real Estate Appraisal. Journal of Multiple-Valued Logic and Soft Computing **17**(2–3), 229–253 (2011)
14. Lughofer, E., Trawiński, B., Trawiński, K., Kempa, O., Lasota, T.: On Employing Fuzzy Modeling Algorithms for the Valuation of Residential Premises. Information Sciences **181**, 5123–5142 (2011)
15. Trawiński, B.: Evolutionary Fuzzy System Ensemble Approach to Model Real Estate Market based on Data Stream Exploration. Journal of Universal Computer Science **19**(4), 539–562 (2013)
16. Telec, Z., Trawiński, B., Lasota, T., Trawiński, G.: Evaluation of neural network ensemble approach to predict from a data stream. In: Hwang, D., Jung, J.J., Nguyen, N.-T. (eds.) ICCCI 2014. LNCS, vol. 8733, pp. 472–482. Springer, Heidelberg (2014)
17. Angelov, P., Filev, D.: An approach to online identification of Takagi-Sugeno fuzzy models. IEEE Trans. on Systems, Man and Cybernetics, part B **34**(1), 484–498 (2004)
18. Angelov, P., Filev, D.: Simpl_eTS: a simplified method for learning evolving Takagi-Sugeno fuzzy models. In: The 2005 IEEE International Conference on Fuzzy Systems FUZZ-IEEE, Reno, Las Vegas, USA, 22–25 May 2005, pp. 1068–1073 (2005)
19. Victor, J., Dourado, A.: Evolving Takagi-Sugeno Fuzzy Models. Technical Report, CISUC (2003). http://cisuc.dei.uc.pt/acg/view_pub.php?id_p=760
20. Angelov, P., Victor, J., Dourado, A., Filev, D.: On-line evolution of Takagi-Sugeno fuzzy models. In: Proceedings of the 2nd IFAC Workshop on Advanced Fuzzy/Neural Control (AFNC 2004), Oulu, Finland (2004)
21. Lughofer, E., Klement, E.P.: FLEXFIS: a variant for incremental learning of Takagi-Sugeno fuzzy systems. In: Proc. of FUZZ-IEEE 2005, Reno, USA, pp. 915–920 (2005)
22. Lughofer, E.: FLEXFIS: A robust incremental learning approach for evolving TS fuzzy models. IEEE Transactions on Fuzzy Systems **16**(6), 1393–1410 (2008)

23. Telec, Z., Trawiński, B., Lasota, T., Trawiński, K.: Comparison of evolving fuzzy systems with an ensemble approach to predict from a data stream. In: Bădică, C., Nguyen, N.T., Brezovan, M. (eds.) ICCCI 2013. LNCS, vol. 8083, pp. 377–387. Springer, Heidelberg (2013)
24. Telec, Z., Trawiński, B., Lasota, T., Trawiński, G.: Evaluation of neural network ensemble approach to predict from a data stream. In: Hwang, D., Jung, J.J., Nguyen, N.-T. (eds.) ICCCI 2014. LNCS, vol. 8733, pp. 472–482. Springer, Heidelberg (2014)
25. Lughofer, E.: Evolving Fuzzy Systems – Methodologies, Advanced Concepts and Applications. Studies in Fuzziness and Soft Computing, vol. 266. Springer Verlag, Heidelberg (2011)
26. Demšar, J.: Statistical comparisons of classifiers over multiple data sets. Journal of Machine Learning Research 7, 1–30 (2006)
27. García, S., Herrera, F.: An Extension on "Statistical Comparisons of Classifiers over Multiple Data Sets" for all Pairwise Comparisons. Journal of Machine Learning Research 9, 2677–2694 (2008)
28. Trawiński, B., Smętek, M., Telec, Z., Lasota, T.: Nonparametric Statistical Analysis for Multiple Comparison of Machine Learning Regression Algorithms. International Journal of Applied Mathematics and Computer Science 22(4), 867–881 (2012)

BigDMS: Special Session on Big Data Mining and Searching

What If Mixing Technologies for Big Data Mining and Queries Optimization

Dhouha Jemal[1(✉)] and Rim Faiz[2]

[1] LARODEC, ISG Tunis, University of Tunis, Tunis, Tunisia
dh.jemal@gmail.com
[2] LARODEC, IHEC Carthage, University of Carthage, Carthage, Tunisia
rim.faiz@ihec.rnu.tn

Abstract. Data is growing at an alarming speed in both volume and structure. The data volume and the multitude of sources have an exponential number of technical and application challenges. The classic tools of data management became unsuitable for processing and unable to offer effective tools to deal with the data explosion. Hence, the imposition of the Big Data in our technological landscape offers new solutions for data processing. In this work, we propose a model that integrates a Big Data solution and a classic DBMS, in a goal of queries optimization. Then, we valid the proposed optimized model through experiments showing the gain of the execution cost saved up.

Keywords: Big data · Optimization · Mapreduce · Information retrieval · Integration · OLAP · Cost · Performance

1 Introduction

The data is currently in the middle of themes and its amount is increasing dramatically. It's not easy to measure the total volume of data generation and processing. Data generation is estimated of 2.5 trillion bytes of data every day[1]. This impressive figure masks even more important evolutions. First, unstructured data will grow faster than structured data. Moreover, beyond the storage, challenges will focus on the capacity to process the data and make it available to users. As described in [16], with the worldwide volume of data which does not stop growing, the classical tools for data management have become unsuitable for processing. Therefore a need of defining and developing new technologies and solutions for access to the massive quantities of data in order to extract meaningful information and knowledge.

In this context, Big Data is a popular phenomenon which aims to provide an alternative to traditional solutions database and analysis. It leads to not only a technology revolution but also a business revolution. It is not just about storage of and access to data, Big Data solutions aim to analyze data in order to make sense of that data and exploiting its value. Thus, MapReduce [3] is presented as one of the most efficient

---

[1] http://www-01.ibm.com/software/fr/data/bigdata/

M. Núñez et al. (Eds.): ICCCI 2015, Part II, LNCS 9330, pp. 619–627, 2015.
DOI: 10.1007/978-3-319-24306-1_60

Big Data solutions. As presented in [4], it is emerging as an important programming model for large-scale data-parallel applications.

Data is the most precious asset of companies and can be mainspring of competitiveness and innovation. As presented in [15], many organizations noticed that the data they own and how they use it can make them different than others. This explains that most organizations try to collect and process as much data as possible in order to find innovative ways to differentiate themselves from competitors. IDC[2] estimates that enterprise data doubles every 18 months. Otherwise, business intelligence and analytics are important in dealing with the magnitude and impact of data driven problems and solutions. According to the IDC[3] Research, the worldwide market for business analytics software grew 8.2% to reach $37.7 billion in 2013. That is why organizations need to be able to rapidly respond to market needs and changes, and it has become essential to have efficient and effective decision making processes with right data to make the decision the most adapted at a given moment. Although, according to Salesforce research, 89% of business leaders believe Big Data will revolutionize business operations in the same way the Internet did, and 83% have pursued Big Data projects in order to seize a competitive edge[4]. The question is how organizations should prepare for these developments in the big data ecosystem, what technology to use for data analysis in such an environment.

Conscious of the need to a powerful and optimized tools to verify and analyze data in order to support the decision-making process, the aim of this work is to optimize queries to meet new organizations needs by determining the most efficient way to execute a given query. We propose to integrate two main categories of data management systems: classic Database management systems (DBMS) and NoSQL DBMSs. The idea is to integrate the ORDBMS [17] PostgreSQL [6] as the most notable research project in the field of object relational database systems, and MapReduce to perform OLAP queries in a goal of minimizing Input/Output costs in terms of the amount of data to manipulate, reading and writing throughout the execution process. The main idea of integration leans on the comparison of query execution costs by both paradigms with the aim of minimizing the Input/Output costs. We valid the proposed approach through experiments showing the significant gain of the cost saved up compared to executing queries independently on MapReduce and PostgreSQL.

The remainder of this paper is organized as follows. In section 2, we give an overview of related work addressing data management systems integration search. In section 3, we present our proposed approach for optimizing the online analytical processing (OLAP) queries Input/Output execution cost, then in section 4 we discuss the obtained results. Finally, section 5 concludes this paper and outlines our future work.

---

[2] http://www.infoworld.com/article/2608297/infrastructure-storage/how-to-survive-the-data-explosion.html

[3] http://www.sas.com/content/dam/SAS/en_us/doc/analystreport/idc-ba-apa-vendor-shares-excerpt-103115.pdf

[4] http://www.forbes.com/sites/louiscolumbus/2015/05/25/roundup-of-analytics-big-data-business-intelligence-forecasts-and-market-estimates-2015/

2   Related Work

Both MapReduce and Parallel DBMS provide a means to process large volumes of data. However, with exponential generation, sources of data changes producing new types of data and content that gave rise to new challenges in the data treatment and analysis. Researchers asked as to whether the parallel DBMS paradigm can scale to meet needs and demands. Several existing studies have compared MapReduce and Parallel DBMS. A strong interest towards this field is arising in the literature actually.

A good amount of literature has been dedicated to provide a broad comparison of the two technologies. In [7], McClean et al. provide a highlevel comparison between MapReduce and Parallel DBMS. The authors provide a selection of criteria that can be used to choose between MapReduce and Parallel DBMS for a particular enterprise application and to find the best choice for different scenarios. However, in [8], Nance et al. offer a summary of the different arguments for a mix of both traditional RDBMS and NoSQL systems, as they are often designed for and solve different problems. Pavlo et al., to evaluate both parallel DBMS and the MapReduce model, conduct the experiments in [12]. This work It has been noted that the MapReduce model lacks of computation time. However, it showed the superior performance of MapReduce model relative to parallel database systems in scalability and fault tolerance.

In many research works, MapReduce has been presented as a complement to the DBMS technology. Stonebraker et al. in [5], compare MapReduce with the parallel DBMS. This work argues that the parallel DBMS showed clear advantages in efficient query processing and high-level query language and interface, whereas MapReduce excelled in ETL and analytics for "read only" semi-structured data sets. In this work, MapReduce is considered as a complement to the DBMS technology rather than competitor with it, since databases are not designed to be good at ETL tasks. Also, in [9] Gruska and Martin have been working to integrate strengths of the two technologies and avoid weaknesses of them. Researchers consider the two systems RDBMSs and MapReduce as complimentary and not competitors. This work aims to present different types of integration solutions between RDBMS and MapReduce, including loosely and hightly integration solutions.

Many research works aim to use the two approaches together. Yui and Kojima propose in [10] propose a purely relational approach that removes the scalability limitation of previous approaches based on user-defined aggregates. A database-Hadoop hybrid approach to scalable machine learning is presented, where batch-learning is performed on the Hadoop platform, while incremental-learning is performed on PostgreSQL. HadoopDB, presented in [11] by Abouzeid et al., attempted to bridge the gap between the two technologies and is adopted for illustrating the performance of MapReduce on processing relational data stored in database systems. It benefits from DB indexes by leveraging DBMS as a storage in each node.

Many research work are conducted to compare MapReduce and database systems. Some presented the use of DBMSs and MapReduce as complementary paradigms, others as competing technologies.

Several studies underlined the performance limitations of the MapReduce model, in terms of computation time, and explained this lack by the fact that the model was not originally designed to perform structured data analysis. Other works discuss the case of many data analysis tasks, where the algorithm is complex to define. The application requires multiple passes over the data, and operations are dependents and complementary where the output from one is the input to the next. For these complex applications, MapReduce provides a good alternative since it allows the user to build complex computations on the data, without the limitation and the difficulty of the SQL language to build such algorithms.

In the other hand, several research studies have been conducted between MapReduce and RDBMSs in the goal of integration, but all these works evaluate the model having as a metric the computation time and efficiency. No attempt was carried out to analyse the I/O execution cost.

3   MapReduce-PostgreSQL Integration Model

Organizations look to choose the technology to be used to analyze the huge quantity of data, and look for best practices to deal with data volume and diversity of structures for best performance and optimize costs. The question is about the selection criteria to be considered in processing the data while meeting the objectives of the organization.

Data processing needs are changing with the ever increasing amounts of both structured and unstructured data. While the processing of structured data typically relies on the well developed field of relational database management systems (RDBMSs), MapReduce is a programming model developed to cope with processing immense amounts of unstructured data.

For this purpose, we suggest a model to integrate the RDBMS PostgreSQL and the MapReduce framework as one of the Big Data solutions in order to optimize the OLAP queries Input/Output execution cost. We aim to integrate classic data management systems with Big Data solutions in order to give businesses the capability to better analyze data with a goal of transforming it into useful information as well as minimizing costs.

3.1   Inspection

The basic idea behind our approach is based on the cost model to approve execution and selectivity of solutions based on the estimated cost of execution. To support the decision making process for analyzing data and extracting useful knowledge while minimizing costs, we propose to compare the estimates of the costs of running a query on Hadoop MapReduce compared to PostgreSQL to choose the least costly technology. For a better control, we will proceed to a thorough query analysis.

The detailed analysis of the queries execution costs showed a gap mattering between both paradigms. Hence the idea of the thorough analysis of the execution process of each query and the implied cost. To better control the cost difference between costs of Hadoop MapReduce versus PostgreSQL on each step of the query's execution process, we propose to dissect each query to a set of operations that demonstrates the process of executing the query. In this way, we can check the impact of the execution of each operation of a query on the overall cost and we can control the total cost of the query by controlling the partial cost of each operation in the information retrieval process.

In this context, we are inspired from a work done in [13] to provide a detailed execution plan for OLAP queries. This execution plan zooms on the sequence of steps of the process of executing a query. It allows detailing the various operations of the process highlighting the order of succession and dependence.

In addition to dissect the implementation process, the execution plan details for each operation the amount of data by the accuracy of the number of records involved and the dependence implemented in the succession of phases. These parameters will be needed to calculate the cost involved in each operation. After distinguishing the different operations of the query, the next step is to calculate the unit cost of each operation. As part of our approach we aim to dissect each query and focus on each separate operation. That way we can control the different stages of the execution process of each query with the aim of calculate the cost implied by each operation as well as its influence on the total cost of the query. Therefore we can control the cost of each query to support the decision making process and the selectivity of the proposed solutions based on the criterion of cost minimization.

3.2   Queries Cost Estimation

Having identified all operations performed during the query execution process the next step is then to calculate the cost implied in each operation independently, in both paradigms PostreSQL and MapReduce with the aim of controlling the estimated costs difference according to the operations as well as the total cost of query execution. At this stage we consider each operation independently to calculate an estimate of its cost execution on PostreSQL on one hand then on MapReduce on the other hand.

Cost estimation on MapReduce. In a MapReduce system, a query star join between F (fact table) and n dimension tables, runs a number of phases, each phase corresponds to a MapReduce job. So, for MapReduce paradigm we have first to extract the number of jobs that will be run for executing the query. The MapReduce job number depends on the number of joint and the presence or absence of aggregation and sorting data. There are three cases of figure: The request contains only n successive joint operations between F and n dimension tables; Join operations are followed by a process of grouping and aggregation on the results of the joint; Sort is applied to the results. In this context, we propose to rely on the equation (1) presented below, and inspired from [14]. This equation allows to determine the number of MapReduce jobs implied in the execution of a given OLAP query. This number can be estimated by the following formula:

$$Nbrjobq = n + x \tag{1}$$

The "n" refers to the number of dimension tables. The x can be equal to: 0, if the query involves only join operations; 1, if the query contains grouping operations and aggregation; 2, if the results are sorted. After identifying all the jobs of query execution, the next step is to calculate the Input/Output cost implicated in each job. In this stage, we relied on the mentioned work presented in [BOK 14] to extract the amount of data and the number of records involved in each operation. These two parameters will be used in a MapReduce cost model that we implemented to approve our proposed approach, in order to calculate the Input/Output cost of each job.

Cost estimation on PostgreSQL. PostgreSQl provides the possibility of itemize each operation by an incremental value of the Input/Output cost implied in each step.

In PostgreSQL platform, the command "explain" shows the execution plan of a statement. This command displays the execution plan that the PostgreSQL planner generates for the supplied statement. Besides the succession of the executed operations, the most critical part of the display is the estimated statement execution cost (measured in cost units that are arbitrary, but conventionally mean disk page fetches). It includes information on the estimated start-up and total cost of each plan node, as well as the estimated number of rows. Actually two numbers are shown: the start- up cost before the first row can be returned, and the total cost to return all the rows. Therefore for each operation, the cost should be the difference between these two values.

3.3 Analysis

The analysis of the results of the estimated costs independently for each operation showed the high cost of the first join operation executed on PostgreSQL, and a noticeable difference for the Hadoop MapReduce paradigm. This can be explained by the fact that in the case of data warehouses, the fact table is still the largest table in terms of number of tuples, which explains the high cost of its analysis. The figure 1 illustrates the process of an OLAP query in our proposed smart model. Based on the previous ascertainment, we propose performing the first joint operation that integrates fact table on MapReduce framework which proves competence for heavy processing to be performed on a large volume of data. In this way we try to minimize the cost of the query execution by minimizing the cost of the most expensive operation. Then the output of the first sub process will be the input of the subsequent operations. Thus, other operations required by the query such as aggregation, sorting, in addition to other join operations will be passed on PostgreSQL.

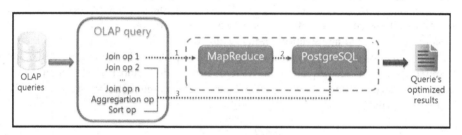

Fig. 1. MapReduce-DBMS smart model

4   Experimental Results

In order to validate our proposed approach, we present in this section the results of experiments conducted to evaluate the proposed model performance as well as gain cost compared with the cost required by each platform independently.

4.1   Experimental Environment

The experiments involve two DBMSs: an ORBMS PostgreSQL and a NoSQL DBMS Hadoop MapReduce. To test and compare our theoretical expectations with the real values, we set up a cluster consisting of one node. For all the experiments, we use the version 9.3 of PostgreSQL. For Hadoop, we used the version 2.0.0 with a single node as worker node hosting DataNode, and as the master node hosting NameNode.

We worked on a workload of 30 queries OLAP. The training data consisted of a data warehouse of 100GB of data with a fact table and 4 dimension tables.

4.2   Discussion

Our approach proposes an hybrid model between ORDBMS and Hadoop MapReduce, based on the comparison of Input/Output costs on the both paradigms. The results of the application of the proposed approach are presented in the table 1. It shows the total Input/Output cost of running the workload on Hadoop MapReduce (Tot_cost_MR), the total Input/Output cost of running the workload on ORDBMS PostgreSQL (Tot_cost_PG), the total Input/Output cost of the workload by applying our proposed approach (Tot_cost_intg), the total Input/Output cost of the workload by choosing for each operation of each query the lowest cost between Hadoop MapReduce and PostgreSQL (Best_Cost_MR_PG).

Table 1. Proposed approach results

Tot_cost_MR	Tot_cost_PG	Tot_cost_intg	Best_Cost_MR_PG
505579305,9	786297800,8	369473436,7	365360413

Observing the values presented in table 1 illustrate the difference of the cost saved up by the application of the proposed approach. The total cost of running all the workload under the proposed approach is only slightly over predicted compared to the cost estimated cost to run the workload by choosing for each operation of each query the lowest cost between the two proposed paradigm (Hadoop MapReduce and PostgreSQL).

The presented values shows the gain of the Input/Output cost by running the workload independently on Hadoop MapReduce and on postgreSQL. In addition, we can compare the Input/Output cost of the workload obtained by applying our proposed approach to the Input/Output the cost obtained if we choose the lower cost for each operation contained in the execution plan of each query.

The table illustrates the notable difference of the Input/Output cost of running the workload by applying the proposed approach compared to the Input/Output running cost on the two paradigms PostgreSQL and Hadoop MapReduce separately.

The gain estimated of the Input/Output cost saved up by running the workload thanks to applying the proposed approach is of 27% compared to Hadoop MapReduce and 53% compared to PostgreSQL. This percentage gain proves the performance of our proposed approach. In addition, the cost returned by our proposed approach is 98.8% of the optimal cost obtained in the case of choosing for each operation of each query the lowest cost between Hadoop MapReduce and PostgreSQL.

5   Conclusion

Given the exploding data problem, the world of databases has evolved which aimed to escape the limitations of data processing and analysis. There has been a significant amount of work during the last two decades related to the needs of new supporting technologies for data processing and knowledge management, challenged by the rise of data generation and data structure diversity.

Both technologies, MapReduce model and relational database systems present strengths and weaknesses. To provide an effective and efficient data management and analysis system, the author believes that Combining MapReduce and RDBMS technologies has the potential to create very powerful systems. In this paper we have proposed a smart model integrating the MapReduce model and a RDBMS, in a goal of optimization of OLAP queries I/O execution cost.

We expect future releases to enhance performance. We plan to deal with OLAP queries with many imbrications and to highlight their influence on the cost.

References

1. Sagiroglu, S., Sinanc, D.: Big data: a review. In: Collaboration Technologies and Systems (CTS) 2013 International Conference on IEEE, pp. 42–47 (2013)
2. Narasimhan, R., Bhuvaneshwari T.: Big Data - A Brief Study. International Journal of Scientic and Engineering Research 5(9) (2014)
3. Dean, J., Ghemawats, S.: Mapreduce : Simplified data processing on large clusters. Communications of the ACM 51(1), 107–113 (2008)
4. Zaharia, M., Konwinski, A., Joseph, A.D., Katz, R., Stoica, I.: Improving MapReduce performance in heterogeneous environments. In: Proceedings of the 8th USENIX Conference on Operating Systems Design and Implementation OSDI, vol. 8, no 4, pp. 29–42 (2008)
5. Stonebraker, M., Abadi, D., Dewitt, D., Madden, S., Paulsone, E., Pavlo, A., Rasin, A.: Mapreduce and parallel dbmss: friends or foes? Communications of the ACM 53(1), 64–71 (2010)
6. Douglas, K., Douglas, S.: PostgreSQL: A comprehensive guide to building, programming, and administring PostreSQL databases, 1st edn. Sams Publishing (2003)
7. McClean, A., Conceicao, R., O'halloran, M.: A comparison of MapReduce and parallel database management systems. In: ICONS 2013 The Eighth International Conference on Systems, pp. 64–68 (2013)

8. Nance, C., Losser, T., Iype, R., Harmon, G.: NOSQL vs RDBMS – why there is room for both. In: Proceedings of the Southern Association for Information Systems Conference Savannah USA, pp. 111–116 (2013)
9. Gruska, N., Martin, P.: Integrating MapReduce and RDBMSs. In: Proceedings of the 2010 Conference of the Center for Advanced Studies on Collaborative Research, IBM Corp., pp. 212–223 (2010)
10. Yui, M., Kojima, I.: A database-hadoop hybrid approach to scalable machine learning. In: IEEE International Congress on Big Data 2013, pp. 1–8. IEEE (2013)
11. Abouzeid, A., Pawlikowski, K., Abadi, D., Silberschatz, A., Rasin, A.: Hadoopdb: an architectural hybrid of mapreduce and dbms technologies for analytical workloads. In: Proceedings of the VLDB Endowment, pp. 922–933 (2009)
12. Pavlo, A., Rasin, A., Madden, S., Stonebraker, M., Dewitt, D., Paulson, E., Shrinivas, L., Abadi, D.: A comparison of approaches to large scale data analysis. In: Proceedings of the 2009 ACM SIGMOD International Conference on Management of data, pp. 165–178 (2009)
13. Boukorca, A., Faget, Z., Bellatreche, L.: What-if physical design for multiple query plan generation. In: Decker, H., Lhotská, L., Link, S., Spies, M., Wagner, R.R. (eds.) DEXA 2014, Part I. LNCS, vol. 8644, pp. 492–506. Springer, Heidelberg (2014)
14. Brighen, A.: Conception de bases de données volumineuses sur le cloud. In: Doctoral dissertation, Université Abderrahmane Mira de Béjaia (2012)
15. Demirkan, H., Delen, D.: Leveraging the capabilities of service-oriented decision support systems: Putting analytics and big data in cloud. Decision Support Systems 55(1), 412–421 (2013)
16. Ordonez, C.: Can we analyze big data inside a DBMS? In: Proceedings of the Sixteenth International Workshop on Data Warehousing and OLAP, pp. 85–92. ACM (2013)
17. Brown, P.G.: Object-Relational Database Development: A Plumber's Guide. Prentice Hall PTR, USA (2000)
18. Chandrasekar, S., Dakshinamurthy, R., Seshakumar, P.G., Prabavathy, B., Babu, C.: A novel indexing scheme for efficient handling of small files in hadoop distributed file system. In: 2013 International Conference on Computer Communication and Informatics (ICCCI), pp. 1–8 (2013)

Parallel K-prototypes for Clustering Big Data

Mohamed Aymen Ben HajKacem$^{(\boxtimes)}$, Chiheb-Eddine Ben N'cir,
and Nadia Essoussi

LARODEC, Université de Tunis, Institut Supérieur de Gestion de Tunis,
41 Avenue de la Liberté, Cité Bouchoucha, 2000 Le Bardo, Tunisia
`medaymen.hajkacem@gmail.com`, {`chiheb.benncir,nadia.essoussi`}`@isg.rnu.tn`

Abstract. Big data clustering has become an important challenge in data mining. Indeed, Big data are often characterized by a huge volume and a variety of attributes namely, numerical and categorical. To deal with these challenges, we propose the parallel k-prototypes method which is based on the Map-Reduce model. This method is able to perform efficient groupings on large-scale and mixed type of data. Experiments realized on huge data sets show the performance of the proposed method in clustering large-scale of mixed data.

Keywords: Big data · K-prototypes · Map-reduce · Mixed data

1 Introduction

Given the exponential growth and availability of data collected from different resources, analyzing these data has become an important challenge referred to as Big data analysis. Big data usually refer to three mains characteristics also called the three Vs [4] which are respectively Volume, Variety and Velocity. Volume refers to the increased quantity of generated data. Variety indicates the many different data types and formats in which data is produced. Velocity refers to the speed at which data should be analyzed. Analyzing Big data usually requires powerful methods and tools since traditional ones are not suitable for processing large and heterogeneous amount of data. For example, existing methods which are used to organize data into groups of similar objects, fail to deal with large-scale of data. The fail is explained by the high computational costs of the existing clustering methods which require unrealistic time to build the grouping.

To obviate this problem, several parallel clustering methods [1,2,6,8,10,11, 15] have been proposed in the literature. Most of these methods use the Map-Reduce [3], which is a parallel programming model to process large-scale data sets. For example, Zaho et al. [15] have proposed a parallel k-means based on Map-Reduce. This method first computes locally the centers of clusters in the map phase. Then, the reduce phase updates the global centers. Kim et al. [8] have introduced an implementation of DBSCAN method using Map-Reduce model. This method first partitions the data to find the clusters in each partition via map phase. Then, it merges the clusters from every partition in reduce phase.

© Springer International Publishing Switzerland 2015
M. Núñez et al. (Eds.): ICCCI 2015, Part II, LNCS 9330, pp. 628–637, 2015.
DOI: 10.1007/978-3-319-24306-1_61

Recently, a parallel implementation of fuzzy c-means clustering algorithm using Map-Reduce model is presented in [11]. The proposed method consists in two Map-Reduce jobs. The first one calculates the centroid matrix, and the second one is devoted to generate the membership matrix.

The later discussed methods offer for users an efficient analysis of a huge amount of data through a Map-Reduce model. Nevertheless, these methods can not handle different types of data and are limited to only numerical attributes. Since Big data are also characterized by the variety of attributes' type, including numerical and categorical attributes, we focus our study on the challenge of clustering large amount of mixed data. We propose in this paper the parallel k-prototypes method called PKP, which is based on the Map-Reduce model to cluster large-scale of mixed data. To the best of our knowledge, this is the first work that deals with numerical and categorical large amount of data.

The rest of this paper is organized as follows: Section 2 presents some existing clustering methods for mixed data, with an emphasis on the k-prototypes method. Then, Section 3 describes the Map-Reduce model while Section 4 describes our proposed method. After that, Section 5 presents experiments that we have realized to evaluate the performance of the proposed method. Finally, we conclude this paper in Section 6.

2   Clustering Methods for Mixed Data

Clustering is an important task in data mining that aims to organize data into groups of similar observations. Given that data are often described by different types of attributes such as, numerical and categorical, a pre-processing step is usually required to transform data into a single type since most of proposed clustering methods deal with only numerical or categorical attributes. However, few clustering methods have been proposed to deal with mixed types of data. For instance, Li and Biswas [9] introduced the Similarity-Based Agglomerative Clustering called SBAC, which is a hierarchical agglomerative algorithm for mixed data. Huang [7] proposed k-prototypes method which integrates k-means and k-modes methods to cluster numerical and categorical data. Ji et al. [5] proposed an improved k-prototypes to deal with mixed type of data. This method introduced the concept of the distributed centroid for representing the prototype of categorical attributes in a cluster. Among the later discussed methods, k-prototypes remains the most popular method to deal with mixed data because of its efficiency [5]. In the following, we present the k-prototypes method.

– K-prototypes method

Given a data set X containing n data objects described by m_r numerical attributes and m_c categorical attributes, the aim of k-prototypes [7] is to find k groupings where the following objective function is minimized:

$$J = \sum_{l=1}^{k} \sum_{i=1}^{n} p_{il} d(x_i, Q_l),$$ (1)

Where $p_{il} \in \{0,1\}$ is a binary variable indicating the membership of data object x_i in cluster l, Q_l is the prototype of the cluster l and $d(x_i, Q_l)$ is the dissimilarity measure which is defined as follows:

$$d(x_i, Q_l) = \sum_{r=1}^{m_r}(x_{ir} - q_{lr})^2 + \gamma_l \sum_{c=1}^{m_c} \delta(x_{ic}, q_{lc}), \qquad (2)$$

Here x_{ir} represents the values of numeric attributes and x_{ic} represents the values of categorical attributes for each data object x_i. $Q_l = \{q_{l1}, \ldots q_{lm}\}$ represents the cluster centers for cluster l, where q_{lr} is the mean of numeric attribute r and cluster l, q_{lc} is the most common value (mode) for categorical attributes c and cluster l. For categorical attributes, $\delta(p,q)=0$ for $p \equiv q$ and $\delta(p,q) = 1$ for $p \neq q$. γ_l is a weight for categorical attributes for each cluster l. The main algorithm of k-prototypes method is described in Algorithm 1.1.

Algorithm 1.1. k-prototypes method

Data: Data set X=$\{x_1 \ldots x_n\}$; number of clusters k

Result: Cluster centers Q=$\{Q_1 \ldots Q_k\}$;

begin

    Choose k cluster centers from X

    repeat

        Compute distance between data objects and cluster centers using Equation 2

        Update the cluster centers Q (Save the previous cluster centers as $Q^\wedge$ to analyze the convergence)

    until $Q^\wedge = Q$;

end

Although the interesting results shown by the k-prototypes method with small and medium mixed data sets, it fails to deal with large-scale data sets (from millions of instances) [13]. Since, it is necessary to compute the distance between each data object to each center, in each iteration. Here, the distance computation is the time consuming step, especially when the size of data sets increases highly. To overcome this weakness, we propose the parallel k-prototypes method which is based on the Map-Reduce model to handle large-scale of mixed data. We present in the following section the Map-Reduce model.

3   Map-Reduce Model

Map-Reduce [3] is a parallel programming model designed to process large-scale data sets among cluster nodes (i.e machines). It is characterized by the fact that the user is oblivious of the details about the data storage, distribution and replication. This model specifies the computation as two functions namely, map and reduce. Each one has $< key/value >$ pairs as input and output. The map function takes each $< key/value >$ pair and generates a set of intermediate

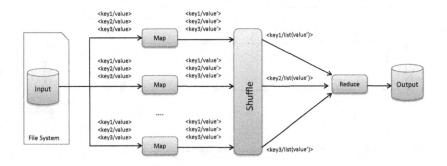

Fig. 1. Map-Reduce flowchart

$< key/value >$ pairs. Then, shuffle phase is devoted to merge all the values associated with the same intermediate key as a list. The reduce function takes this list as input for creating the final values. Fig. 1. illustrates the flowchart of the Map-Reduce model. In a Map-Reduce job, all map and reduce functions are executed in parallel way. All map (resp. reduce) functions are independently run by mapper (resp. reducer) node. The inputs and outputs of a Map-Reduce job are stored in an associated distributed file system that is accessible from any machine of the cluster nodes. The implementation of the Map-Reduce model is available in Hadoop[1]. Hadoop provides a distributed file system named Hadoop Distributed File System, (HDFS) that stores data on the nodes.

4   Parallel k-Prototypes Based on Map-Reduce

To offer for users the possibility to perform clustering of mixed type of large-scale data, we propose the parallel k-prototypes method based on the Map-Reduce model (PKP). As stated before, the most intensive calculation to occur in the k-prototypes method is the calculation of distances, which decreases its performance when dealing with large data sets. Indeed, the distance computations between one object with the cluster centers is independent to the distance computations between another object with the corresponding cluster centers. Thus, the distance computation between different objects and cluster centers can be executed in parallel. In each iteration, the new cluster centers, which are used in the next iteration, should be updated. Hence, this operation will be executed serially.

As shown in Fig. 2., the proposed method mainly consists into three functions: *map function* which performs the assignment of each data object to the nearest cluster, *combine function* which is based on calculating the local cluster centers, and *reduce function* which is devoted to compute the new cluster centers.

---

[1] http://hadoop.apache.org/

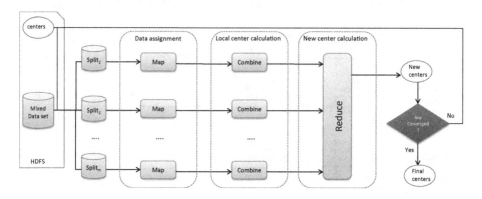

Fig. 2. Parallel k-prototypes method based on Map-Reduce

Suppose an input mixed data set stored in HDFS as sequence file of $< key/value >$ pairs, each of which represents a data object in the data set. The key is an index in bytes and the value represents the data object values. The input data set is composed by h blocks that are accessible from any machine of the cluster nodes. First, the PKP partitions input data set into m splits (i.e $Split_j \ldots Split_m$) where m is a user-defined parameter and $1 \leq j \leq m$. Each $Split_j$ is associated to map task j. We can mention that this partitioning process is performed sequentially means that the map task j corresponds to the $Split_j$ data chunk of h/m blocks. Then, the PKP creates a *centers* variable that contains the centers of the cluster, which is stored in HDFS.

4.1   Map Function

When each mapper receives the associated split, it first calculates the distance between each data object and the cluster centers using Equation 2. Given this information, the mapper then assigns the data object to its corresponding nearest cluster. Finally, the map function outputs the intermediate key and value pairs which are composed respectively of the nearest cluster index and the data object values.

Let Compute-Distance(x_i,$center_j$) be a function which returns the distance between data object x_i and $center_j$. The map function is outlined in Algorithm 1.2.

4.2   Combine Function

In this function, we combine the intermediate data produced from each map task, in order to reduce the amount of data transferred across to the reducers. To this end, we compute a partial information about the cluster centers denoted by the local cluster centers. First, we sum the values of the numerical attributes of the data objects assigned to the same cluster. Second, we compute the frequencies of different values of categorical attributes relative to the data objects for each

Algorithm 1.2. Map function

Data: $< key/value >$; where key: position and $value$: data object values, centers

Result: $< key_1/value_1 >$; where key_1: nearest cluster index and $value_1$: data object values

begin

> minDis← 0
>
> index← 0
>
> $x_i \leftarrow value$
>
> **foreach** $center_j \in centers$ **do**
>
> > $\lfloor$ dis= Compute-Distance$(x_i, center_j)$
>
> **if** $dis < minDis$ **then**
>
> > minDis ← dis
> >
> > $\lfloor$ index ← j
>
> $key_1 \leftarrow index$
>
> $value_1 \leftarrow x_i$

end

cluster. Third, we record the number of data objects assigned in the same cluster. As each combine function finishes its processing, the intermediate data which consists into the local cluster centers, are forwarded to a single reduce task.

Let $\text{Dom}[j] = \{a_j^1, a_j^2, \ldots a_j^t\}$ be the domain of categorical attribute j. Let $Sum_i[j]$ be the sum of values of numeric attribute j for the data objects assigned in cluster i, $Freq_i[j, a_j^k]$ be the frequency of value a^k of categorical attribute j for the data objects assigned in cluster i, and Num_i be the number of data objects of cluster i. The combine function is outlined in Algorithm 1.3.

4.3   Reduce Function

In this function, the local cluster centers are merged in order to compute the new cluster centers. So, we first sum the numeric values and the total number of data objects assigned to the same cluster. Then, we compute the global frequencies for different values of categorical attributes relative to the data objects. Given these information, we can compute both the mean and mode value of the new center. Once the new centers are obtained, the PKP moves to the next iteration until the convergence. It is important to mention that the convergence is achieved when the cluster centers become stable for two consecutive iterations.

Let $Newcenter_i$ be the new center values of the cluster i and Highest-Freq(Freq[j,.]) be a function which returns the most common value (i.e mode) of the categorical attribute j from Freq[j,.] variable. The reduce function is outlined in Algorithm 1.4.

5   Experiments and Results

In this section, we evaluate the performance of PKP using speedup and scaleup measures [14]. We run the experiments on the Amazon Elastic MapReduce

Algorithm 1.3. Combine function

Data: $< key/V >$; where key: cluster index and V: list of data objects assigned in the same cluster

Result: $< key_1/value_1 >$; where key_1: cluster index and $value_1$: local cluster center

begin
    $Sum_{key} \leftarrow \emptyset$
    $Freq_{key} \leftarrow \emptyset$
    $Num_{key} \leftarrow 0$
    while $V.HasNext()$ **do**
        % *Get a data object*
        $x_i \leftarrow V.next()$
        for $j \leftarrow 1 \ldots m_r$ **do**
           $Sum_{key}[j] \leftarrow Sum_{key}[j] + x_{ij}$
        for $j \leftarrow m_r + 1 \ldots m_c$ **do**
           for $k \leftarrow 1 \ldots t$ **do**
               if $x_{ij} = a^k$ **then**
                   $Freq_{key}[j, a_j^k] \leftarrow Freq_{key}[j, a_j^k] + 1$
        $Num_{key} \leftarrow Num_{key} + 1$
    $key_1 \leftarrow key$
    $value_1 \leftarrow Sum_{key} \cup Freq_{key} \cup Num_{key}$
end

(Amazon EMR)[2] which is a web service for processing huge amounts of data by exploiting the parallelism on a cluster of machines, each of which has single core 2.6 GHz cores and 1 GB of memory.

For the following evaluations, we utilize three data sets generated from KDD CUP 1999[3] named Data set1, Data set2, Data set3, which are listed in Table 1.

Table 1. The description of data sets

Data set	No. of data objects	No. of numerical attributes	No. of categorical attributes	No. of clusters
Data set1	$4 * 10^6$	37	4	23
Data set2	$8 * 10^6$	37	4	23
Data set3	$16 * 10^6$	37	4	23

To evaluate the speedup, we keep the size of the data set constant and increase the number of the nodes in the system. Speedup given by the larger system with m nodes is defined as follows [14]:

$$Speedup(m) = \frac{T_1}{T_m}, \tag{3}$$

[2] http://aws.amazon.com/elasticmapreduce/
[3] http://kdd.ics.uci.edu/

Algorithm 1.4. Reduce function

Data: $< key/V >$; where key: cluster index and V: list of local cluster centers

Result: $< key_1/value_1 >$; where key_1: cluster index and $value_1$: new center values

begin

$\quad Sum_{key} \leftarrow \emptyset$

$\quad Freq_{key} \leftarrow \emptyset$

$\quad Num_{key} \leftarrow 0$

$\quad Newcenter_{key} \leftarrow 0$

$\quad$ **while** $V.HasNext()$ **do**

$\qquad$ % *Get a local cluster center*

$\qquad x_i \leftarrow V.next()$

$\qquad$ **for** $j \leftarrow 1 \ldots m_r$ **do**

$\qquad\quad \lfloor \; Sum_{key}[j] \leftarrow Sum_{key}[j] + x_i.Sum_{key}[j]$

$\qquad$ **for** $j \leftarrow m_r + 1 \ldots m_c$ **do**

$\qquad\quad$ **for** $k \leftarrow 1 \ldots t$ **do**

$\qquad\qquad \lfloor \; Freq_{key}[j, a_j^k] \leftarrow Freq_{key}[j, a_j^k] + x_i.Freq_{key}[j, a_j^k]$

$\qquad\quad \lfloor \; Num_{key} \leftarrow Num_{key} + x_i.Num_{key}$

$\quad$ **for** $j \leftarrow 1 \ldots m_r$ **do**

$\qquad \lfloor \; Newcenter_{key}[j] \leftarrow Sum_{key}[j]/Num_{key}$

$\quad$ **for** $j \leftarrow m_r + 1 \ldots m_c$ **do**

$\qquad \lfloor \; Newcenter_{key}[j] \leftarrow \text{Highest-Freq}(Freq_{key}[j,.])$

$\quad key_1 \leftarrow key$

$\quad value_1 \leftarrow Newcenter_{key}$

end

Where T_1 is the execution time on one node and T_m is the execution time on m nodes. The efficient parallel method gives linear speedup. A system with m times the number of nodes yields a speedup of m. Meanwhile, linear speedup is difficult to achieve since the cost of network communication increases with the number of nodes becomes large. We perform the speedup evaluation on the three data sets. The number of nodes varied from 1 to 4. Fig. 3.(a) shows the speedup results. As the size of the data set increases, the speedup of PKP becomes approximately linear. From Fig. 3.(a), we can conclude that the PKP handles large amount of mixed types data efficiently.

Scaleup is defined as the ability of a m-times larger system to perform a m-times larger job in the same execution time [14].

$$Scaleup(DS, m) = \frac{T_{DS_1}}{T_{DS_m}}, \qquad (4)$$

Where DS is the data set, T_{DS_1} is the execution time for DS on one node and T_{DS_m} is the execution time for m*DS on m nodes. To show how well PKP can treat large-scale of mixed data sets when more nodes are available, we have realized scaleup experiments where we increase the size of the data sets in direct proportion to the number of machines in the cluster nodes. Roughly speaking, the data sets' size of $4*10^6$, $8*10^6$ and $16*10^6$ instances are executed on 1,2 and 4 machines respectively. Fig. 3.(b) shows the performance results of the data sets. Apparently, the PKP scales well when dealing with large-scale of data.

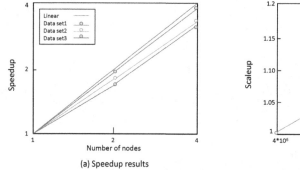   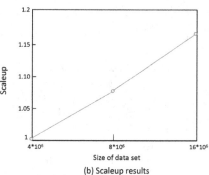

(a) Speedup results   (b) Scaleup results

Fig. 3. Experimental evaluation

6   Conclusion

Since traditional clustering methods can not deal with Big data, there is a need for scalable solutions. Furthermore, Big data are often characterized by the mixed type of attributes such as numerical and categorical. This paper have investigated the parallelization of k-prototypes method using Map-Reduce model to cluster large-scale of mixed data. We have used speedup and scaleup measures to evaluate the performances of the proposed method. Experimental results on large data sets show the performance of our method. A proper initialization of k-prototypes method is crucial for obtaining a good final solution. Thus, it should be interesting to parallelize the initialization step of k-prototypes using Map-reduce model, in order to create an initial set of cluster centers that is probably close to the optimum solution.

References

1. Bahmani, B., Moseley, B., Vattani, A., Kumar, R., Vassilvitskii, S.: Scalable k-means++. Proceedings of the VLDB Endowment **5**(7), 622–633 (2012)
2. Cui, X., Zhu, P., Yang, X., Li, K., Ji, C.: Optimized big data K-means clustering using MapReduce. The Journal of Supercomputing **70**(3), 1249–1259 (2014)
3. Dean, J., Ghemawat, S.: MapReduce: simplified data processing on large clusters. Communications of the ACM **51**(1), 107–113 (2008)
4. Gorodetsky, V.: Big data: opportunities, challenges and solutions. In: Ermolayev, V., Mayr, H.C., Nikitchenko, M., Spivakovsky, A., Zholtkevych, G. (eds.) ICTERI 2014. CCIS, vol. 469, pp. 3–22. Springer, Heidelberg (2014)

5. Ji, J., Bai, T., Zhou, C., Ma, C., Wang, Z.: An improved k-prototypes clustering algorithm for mixed numeric and categorical data. Neurocomputing **120**, 590–596 (2013)
6. Hadian, A., Shahrivari, S.: High performance parallel k-means clustering for disk-resident datasets on multi-core CPUs. The Journal of Supercomputing **69**(2), 845–863 (2014)
7. Huang, Z.: Clustering large data sets with mixed numeric and categorical values. In: Proceedings of the 1st Pacific-Asia Conference on Knowledge Discovery and Data Mining, pp. 21–34 (1997)
8. Kim, Y., Shim, K., Kim, M.S., Lee, J.S.: DBCURE-MR: An efficient density-based clustering algorithm for large data using MapReduce. Information Systems **42**, 15–35 (2014)
9. Li, C., Biswas, G.: Unsupervised learning with mixed numeric and nominal data. Knowledge and Data Engineering **14**(4), 673–690 (2002)
10. Li, Q., Wang, P., Wang, W., Hu, H., Li, Z., Li, J.: An efficient K-means clustering algorithm on mapreduce. In: Bhowmick, S.S., Dyreson, C.E., Jensen, C.S., Lee, M.L., Muliantara, A., Thalheim, B. (eds.) DASFAA 2014, Part I. LNCS, vol. 8421, pp. 357–371. Springer, Heidelberg (2014)
11. Ludwig, S.A.: MapReduce-based fuzzy c-means clustering algorithm: implementation and scalability. International Journal of Machine Learning and Cybernetics, 1–12 (2015)
12. MacQueen, J.: Some methods for classification and analysis of multivariate observations. Proceedings of the fifth Berkeley symposium on mathematical statistics and probability **14**(1), 281–297 (1967)
13. Vattani, A.: K-means requires exponentially many iterations even in the plane. Discrete Computational Geometry **45**(4), 596–616 (2011)
14. Xu, X., Jger, J., Kriegel, H.P.: A fast parallel clustering algorithm for large spatial databases. High Performance Data Mining, 263–290 (2002)
15. Zhao, W., Ma, H., He, Q.: Parallel k-means clustering based on mapreduce. Cloud Computing, 674–679 (2009)

Big Data: Concepts, Challenges and Applications

Imen Chebbi[(✉)], Wadii Boulila, and Imed Riadh Farah

Laboratoire RIADI, Ecole Nationale des Sciences de L'Informatique,
Manouba, Tunisia
ichebbi88@gmail.com, wadii.boulila@riadi.rnu.tn, riadh.farah@ensi.rnu.tn

Abstract. Big data is an evolving term that describes any voluminous
amount of structured, semi-structured and unstructured data that has
the potential to be mined for information.Big Data it's varied, it's growing, it's moving fast, and it's very much in need of smart management In
this paper, we first review a literature survey of big data, second review
the related technologies, such as could computinq and Hadoop, We then
focus on the five phases of the value chain of big data, i.e.,data sources,
data collection, data management, data analysis and data visualization.
And we finally examine the several representative applications of big
data.

Keywords: Big data · Cloud computing · Hadoop · Big data analytic ·
Mapreduce

1 Introduction

Today the term big data [1] [2] draws a lot of attention, but behind the hype
there's a simple story. For decades, companies have been making business decisions based on transactional data stored in relational databases. Beyond that
critical data, however, is a potential treasure trove of non-traditional, less structured data: weblogs, social media, email, sensors, and photographs that can be
mined for useful information. Decreases in the cost of both storage and compute power have made it feasible to collect this data which would have been
thrown away only a few years ago. As a result, more and more companies are
looking to include non-traditional yet potentially very valuable data with their
traditional enterprise data in their business intelligence analysis. Every day, we
create 2.5 quintillion bytes of data so much that 90By now, the term of big
data is mainly used to describe enormous datasets, a new fact of business life,
one that requires having strategies in place for managing large volumes of both
structured and unstructured data. This data comes from everywhere: sensors
and videos, purchase transaction record, and cell phone GPS signals to name
a few [3]. Compared with traditional datasets, big data [5] typically includes
masses of unstructured data that need more real-time analysis. In addition, big
data also brings about new opportunities for discovering new values, helps us to
gain an in-depth understanding of the hidden values, and also incurs new challenge, e.g, how to effectively organize and process such datasets (e.g. datasets

© Springer International Publishing Switzerland 2015
M. Núñez et al. (Eds.): ICCCI 2015, Part II, LNCS 9330, pp. 638–647, 2015.
DOI: 10.1007/978-3-319-24306-1_62

of satellite images) The rest of this paper is organized as follows. In Section 2, we present the definition of big data and its four features. Then, in Section 3 we introduce the related technologies. Section 4 focuses on the big data value chain (which is composed of five phases).Section 5 examine the several representative applications of big data. A brief conclusion with recommendations for future studies is presented in Section 6.

2   The era of Big Data

In this section, we first present a list of popular definitions of big data, followed by a definition of the 4V.

2.1   What Is Big Data

There is no universal definition of what constitutes "Big Data". In fact, several definitions for big data are found in the literature:

- IDC [4] Big data technologies describe a new generation of technologies and architectures, designed to economically extract value from very large volumes of a wide variety of data, by enabling high-velocity capture, discovery, and/or analysis.
- McKinsey Global Institute [6] "Big data" refers to datasets whose size is beyond the ability of typical database software tools to capture, store, manage, and analyse. This definition is intentionally subjective and incorporates a moving definition of how big a dataset needs to be in order to be considered big data–i.e., we don't define big data in terms of being larger than a certain number of terabytes (thousands of gigabytes). We assume that, as technology advances over time, the size of datasets that qualify as big data will also increase. Also note that the definition can vary by sector, depending on what kinds of software tools are commonly available and what sizes of datasets are common in a particular industry. With those caveats, big data in many sectors today will range from a few dozen terabytes to multiple petabytes (thousands of terabytes).
- Gartner [7] Big Data in general is defined as high volume, velocity and variety information assets that demand cost-effective, innovative forms of information processing for enhanced insight and decision making.
- Oracle [8] big data is the data characterized by 4 key attributes: volume, variety, velocity and value.
- IBM [9] big data is the data characterized by 3 attributes: volume, variety and velocity. Big data is an abstract concept. Apart from masses of data, it also has some other features, which determine the difference between itself and "massive data" or "very big data."

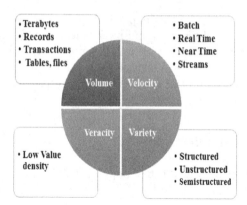

Fig. 1. The 4Vs features of Big Data

2.2   Features of Big Data

Big data is characterized by 4Vs: the extreme volume of data, the wide variety of types of data and the velocity at which the data must be must processed. Although big data doesn't refer to any specific quantity, the term is often used when speaking about petabytes and exabytes of data, much of which cannot be integrated easily [9]. IBM [4] offers a good and simple overview for the four critical features of big data :

- Big data is always large in volume. It actually doesn't have to be a certain number of petabytes to qualify. If your store of old data and new incoming data has gotten so large that you are having difficulty handling it, that's big data.
- Velocity or speed refers to how fast the data is coming in, but also to how fast you need to be able to analyze and utilize it. If we have one or more business processes that require real-time data analysis, we have a velocity challenge. Solving this issue might mean expanding our private cloud using a hybrid model that allows bursting for additional compute power as-needed for data analysis.
- Variety points to the number of sources or incoming vectors leading to your databases. That might be embedded sensor data, phone conversations, documents, video uploads or feeds, social media, and much more. Variety in data means variety in databases.
- Veracity is probably the toughest nut to crack. If we can't trust the data itself, the source of the data, or the processes you are using to identify which data points are important, we have a veracity problem. One of the biggest problems with big data is the tendency for errors to snowball. User entry errors, redundancy and corruption all affect the value of data.

2.3 Big Data Challenges

As big data wends its inextricable way into the enterprise, information technology (IT)practitioners and business sponsors alike will bump up against a number of challenges that must be addressed before any big data program can be successful.Five of those challenges are [10] [11]:

- Uncertainty of the Data Management : There are many competing technologies, and within each technical area there are numerous rivals. Our first challenge is making the best choices while not introducing additional unknowns and risk to big data adoption.
- Transparency: Making data accessible to relevant stakeholders in a timely manner.
- Experimentation: Enabling experimentation to discover needs, expose variability, and improve performance. As more transactional data is stored in digital form, organizations can collect more accurate and detailed performance data.
- Decision Support: Replacing/supporting human decision making with automated algorithms which can improve decision making, minimize risks, and uncover valuable insights that would otherwise remain hidden.
- Innovation: Big Data enables companies to create new products and services, enhance existing ones, and invent or refine business models.

3 Related Technologies

In order to gain a deep understanding of big data, this section will introduce several fundamental technologies that are closely related to big data, including cloud computing, MapReduce and Hadoop. Big data analytics is often associated with cloud computing because the analysis of large data sets in real-time requires a platform like Hadoop to store large data sets across a distributed cluster and MapReduce to coordinate, combine and process data from multiple sources.

3.1 Cloud Computing for Big Data

Cloud computing [12] is closely related to big data. Big Data is all about extracting value out of "Variety, Velocity and Volume" from the information assets available, while Cloud focuses on On-Demand, Elastic, Scalable, Pay-Per use Self Service models .Big Data need large on-demand compute power and distributed storage to crunch the 3V data problem and Cloud seamlessly provides this elastic on-demand compute required for the same. With the "Apache Hadoop", the de-facto standard for Big Data processing, the big data processing has been more batch oriented in the current state. The burst workload nature of the Big Data Computing Infrastructure makes it a true case for the Cloud. Big data is the object of the computation-intensive operation and stresses the storage capacity of a cloud system. The main objective of cloud computing [9] is to use huge computing and storage resources under concentrated management, so as to provide big data applications with fine-grained computing capacity. The development of cloud computing provides solutions for the storage and processing of big data. On the other hand, the emergence of big data also accelerates the development of cloud computing. The distributed storage technology based on cloud computing can effectively manage big data; the parallel computing capacity by virtue of cloud computing can improve the efficiency of acquisition and analyzing big data.

3.2   Hadoop/MapReduce for Big Data

Presently, Hadoop [13] is widely used in big data applications in the industry, e.g., spam filtering, network searching, clickstream analysis, and social recommendation. In addition, considerable academic research is now based on Hadoop. Some representative cases are given below. As declared in June 2012, Yahoo runs Hadoop in 42,000 servers at four data centers to support its products and services, e.g., searching and spam filtering, etc. At present, the biggest Hadoop cluster has 4,000 nodes, but the number of nodes will be increased to 10,000 with the release of Hadoop 2.0. In the same month, Facebook announced that their Hadoop cluster can process 100 PB data, which grew by 0.5 PB per day as in November 2012. In addition, many companies provide Hadoop commercial execution and/or support, including Cloudera, IBM, MapReduce, EMC, and Oracle. The Apache Hadoop software library is a framework that allows for the distributed processing of large data sets across clusters of computers using simple programming models. It is designed to scale up from single servers to thousands of machines, each offering local computation and storage. Rather than rely on hardware to deliver high-availability, the library itself is designed to detect and handle failures at the application layer, so delivering a highly-available service on top of a cluster of computers, each of which may be prone to failures.

4   Big Data Architecture

In this section we will focus on the value chain of big data [14] [9], which can be generally divided into five phases: data sources, data collection, data management, data analysis and data visualization.

4.1   Data Sources

Data sources is the first step of big data pipeline.Success for a big data strategy lies in recognizing the different types of big data sources [3], using the proper mining technologies to find the treasure within each type, and then integrating and presenting those new insights appropriately according to your unique goals, to enable the organization to make more effective steering decisions. The different sources of information includes [15] network usage, sensors, connected devices, mobile devices, worldwideweb applications and social networks.

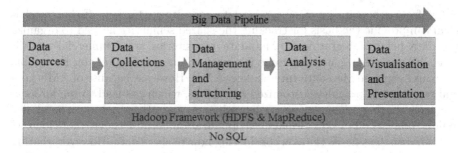

Fig. 2. The Big Data Value Chain

4.2   Data Collection

The second phase of the big data system is big data collection. Data collection refers to the process of retrieving raw data from real-world objects. The process needs to be well designed.Otherwise,inaccurate data collection would impact the subsequent data analysis procedure and ultimately lead to invalid results. At the same time, data collection methods not only depend on the physics characteristics of data sources, but also the objectives of data analysis. As a result, there are many kinds of data collection methods. In the subsection, we will focus on two common methods for big data collection.

– Log File: Log files [16], one of the most widely deployed data collection methods, are generated by data source systems to record activities in a specified file format for subsequent analysis. Log files are useful in almost all the applications running on digital devices. There are three main types of web server log file formats available to capture the activities of users on a website: Common Log File Format (NCSA), Extended Log Format (W3C), and IIS Log Format (Microsoft). All three log file formats are in the ASCII text format. Alternatively, databases can be utilized instead of text files to store log information to improve the querying efficiency of massive log repositories.Otherexamples of log file-based data collection include stock ticks in financial applications, performance measurement in network monitoring, and traffic management.
– Sensors [17] are used commonly to measure a physical quantity and convert it into a readable digital signal for processing. Sensor types include acoustic, sound, vibration, automotive, chemical, electric current, weather, pressure, thermal, and proximity. Through wired or wireless networks,this information can be transferred to a data collection point. Wired sensor networks leverage wired networks to connect a collection of sensors and transmit the collected information.

4.3   Data Management and Structuring

The data source developer makes deliberate organizational choices about the datas syntax, structure, and semantics and makes that information available either from schemata or from a metadata repository [15] [18]. Either mechanism can provide the basis for tracking the shared semantics needed to organize the data before integrating it. Metadata repositories are commercially available, and numerous generic metamodels exist, many of which rely on Extensible Markup Language Metadata Interchange (XMI). However, because of XMIs generality, each tool provides customized extensions, which can lead to vendor lock, problems sharing schemata among participants, and other tool-interoperability issues. Analysts often skip formal data organization because theyre more focused on their own data needs than on considering how to share data. However, sharing knowledge about internal data organization can enable more seamless integration with data providers environments (upstream) and data consumers environments (downstream).

4.4 Data Analysis

The most important stage of the big data value chain is data analysis [18]. Data sources are then ready for analysis, which includes maintaining the provenance between the input and results and maintaining metadata so that another analyst can recreate those results and strengthen their validity. Popular data analysis techniques, such as MapReduce, enable the creation of a programming model and associated implementations for processing and generating large datasets. Big data can be analyzed with the software tools commonly used as part of advanced analytics disciplines such as predictive analytics, data mining, text analytics and statistical analysis. Mainstream BI software [19] and data visualization tools can also play a role in the analysis process. But the semi-structured and unstructured data may not fit well in traditional data warehouses based on relational databases. Furthermore, data warehouses may not be able to handle the processing demands posed by sets of big data that need to be updated frequently or even continually for example, real-time data on the performance of mobile applications or of oil and gas pipelines. As a result, many organizations looking to collect, process and analyze big data have turned to a newer class of technologies that includes Hadoop [13] and related tools such as YARN, MapReduce, Spark, Hive and Pig as well as NoSQL databases. Those technologies form the core of an open source software framework that supports the processing of large and diverse data sets across clustered systems.

4.5 Data Visualization

Visualization [20] involves presenting analytic results to decision makers as a static report or an interactive application that supports the exploration and refinement of results. The goal is to provide key stakeholders with meaningful information in a format that they can readily consume to make critical decisions. Industries, such as media and training, have a wealth of data visualization techniques, which others could adopt. Virtual and augmented realities, for example, enhance the user experience and make it easier to grasp information thats elusive in two-dimensional media. Although this technology has promising implications, virtual and augmented reality systems continue to be viewed as only suitable for training, education, and other highly customized uses. Big data environments help organizations capture, manage, process and analyze large amounts of data from both new data formats, as well as traditional formats, in real time. When managing the data lifecycle of big data, organizations should consider the volume, velocity and complexity of big data.

5   Big Data Applications

In this section we present a diverse and a representative set of applications dealing with big data.

5.1   Social Media Applications

IBM estimates that 2.5 quintillion bytes of new data are created every day. To put this into perspective, social media alone generates more information in a short period of time than existed in the entire world just several generations ago. Popular sites [21] like Facebook, Instagram, Foursquare, Twitter, and Pinterest create massive quantities of data that if translated properly by large-scale applications would be any brands golden ticket into the minds of its consumers. Unfortunately, the data produced by social media is not only enormous its also unstructured. The task of capturing, processing, and managing this data is unquestionably beyond human scale. In fact, its beyond the scale of most common software. Because of this, a glass wall exists between marketers and the data they can see it, but they cant harness it. Its easy to see how Big Data fits into the picture. The Big Data industry deals in sets of data that range from a few dozen terrabytes to many hundreds of petabytes. A slew of Big Data applications have been created specifically to make sense of social media data.

5.2   Enterprise Applications

Theres no doubt Big Data is changing how companies are able to process and use data.The primary goal of big data analytics is to help companies make more informed business decisions by enabling data scientists, predictive modelers and other analytics professionals to analyze large volumes of transaction data, as well as other forms of data that may be untapped by conventional business intelligence (BI) [19] programs. That could include Web server logs and Internet clickstream data, social media content and social network activity reports, text from customer emails and survey responses, mobile-phone call detail records and machine data captured by sensors connected to the Internet of Things. Some people exclusively associate big data with semi-structured and unstructured data of that sort, but consulting firms like Gartner Inc. and For rester Research Inc. also consider transactions and other structured data to be valid components of big data analytics applications [22].

5.3 Science Applications

The sciences have always been big consumers of Big Data, with the added challenge of analyzing disparate data from multiple instruments and sources. For example, astronomy involves continuously collecting and analyzing huge amounts of image data, from increasingly sophisticated telescopes. The next big science astronomy project is the Large Synoptic Survey Telescope (LSST). The telescope will ultimately collect about 55 Petabytes of raw data. There is a streaming pipeline where software looks for patterns in the images (e.g., stars and other celestial objects) and then looks for the same object in different telescope images to obtain trajectory information. All of the data will be stored, and astronomers want to reprocess the raw imagery with different algorithms, as there is no universal image-processing algorithm that pleases all astronomers. New Big Data technologies can provide the capacity to store, process, analyze, visualize and share large amounts of image data, as well as remote sensing data from satellites [23]. Astronomy looks out to the sky from telescopes; remote sensing looks in towards the earth from satellites. The two are roughly mirror images of each other, presenting an opportunity for new approaches to analyzing both kinds of data simultaneously.

6 Conclusion

The Big data concept has progressively become the next evolutionary phase in batch processing, storing, manipulation and relations visualization in vast number of records. The era of big data is upon us, bringing with it an urgent need for advanced data acquisition, management, and analysis mechanisms. In this paper, we have presented the concept of big data, challenges, related technologies, applications and highlighted the big data value chain, which covers the entire big data lifecycle. The big data value chain consists of five phases: data sources, data collection, data management, data analysis and data visualization. We regard Big Data as an emerging trend in all science and engineering domains. With Big Data technologies, we will hopefully be able to provide most relevant and most accurate social sensing feedback to better understand our society at realtime. We can further stimulate the participation of the public audiences in the data production circle for societal and economical events. The era of Big Data has arrived.

References

1. Tsuchiya, S., Sakamoto, Y., Tsuchimoto, Y., Lee, V.: Big data processing in cloud environments. FUJITSU Science and Technology J. 48(2), 159–168 (2012)
2. Big Data: Science in the Petabyte Era: Community Cleverness Required. Nature 455(7209) (2008)
3. What is Big Data. IBM, New York (2013). http://www-01.ibm.com/software/data/bigdata/

4. Gantz, J., Reinsel, D.: The digital universe in 2020: big data, bigger digital shadows, and biggest growth in the far east. In: Proc. IDC iView, IDC Anal. Future (2012)
5. Mayer-Schonberger, V., Cukier, K.: Big data: a revolution that will transform how we live, work, and think. Eamon Dolan/Houghton Mifflin Harcourt (2013)
6. Manyikaetal, J.: Bigdata: The Next Frontier for Innovation, Competition, and Productivity, p. 1137. McKinsey Global Institute, San Francisco (2011)
7. Gartner Group, Inc. http://www.gartner.com/it-glossary/big-data/
8. Oracle Big Data. https://www.oracle.com/bigdata/index.html
9. Chen, M., Mao, S., Liu, Y.: Big Data: a survey. MONET **19**(2), 171–209 (2014). Spring science business Media, New York
10. Xindong, W., Zhu, X., Gong-Qing, W., Ding, W.: Data Mining with Big Data. IEEE Trans. Knowl. Data Eng. **26**(1), 97–107 (2014)
11. Kong, W., Wu, Q., Li, L., Qiao, F.: Intelligent data analysis and its challenges in big data environment. In: 2014 IEEE International Conference on System Science and Engineering (ICSSE), pp. 108–113 (2014)
12. Gupta, R., Gupta, H., Mohania, M.: Cloud computing and big data analytics: what is new from databases perspective? In: Srinivasa, S., Bhatnagar, V. (eds.) BDA 2012. LNCS, vol. 7678, pp. 42–61. Springer, Heidelberg (2012)
13. The Apache Software Foundation (2014). http://hadoop.apache.org/
14. Han, H., Wen, Y., Chua, T.-S., Li, X.: Toward Scalable Systems for Big Data Analytics: A Technology Tutorial. IEEE Access **2**, 652–687 (2014)
15. Evans, D., Hutley, R.: The explosion of data. white paper (2010)
16. Wahab, M.H.A., Mohd, M.N.H., Hanafi, H.F., Mohsin, M.F.M.: Data pre-processing on web server logs for generalized association rules mining algorithm. World Acad. Sci. Eng. Technol. **48**, 36 (2008)
17. Chandramohan, V., Christensen, K.: A first look at wired sensor networks for video surveillance systems. In: Proceedings LCN 2002, 27th annual IEEE conference on local computer networks, pp. 728–729. IEEE
18. Bhatt, C.A., Kankanhalli, M.S.: Multimedia data mining: state of the art and challenges. Multimedia Tools Appl. **51**(1), 35–76 (2011). [184] G. Blackett (2013)
19. Richardson, J., Schlegel, K., Hostmann, B., McMurchy, N.: Magic quadrant for business intelligence platforms (2012). [Online]
20. Friedman, V.: Data visualization and infographics (2008). http://www.smashingmagazine.com
21. Li, Y., Chen, W., Wang, Y., Zhang, Z.-L.: Influence diffusion dynamics and influence maximization in social networks with friend and foe relationships. ACM, pp. 657–666 (2013)
22. Agrawal, D., Bernstein, P., Bertino, E.: Challenges andopportunities withbigdata. Acommunity white paper (2012)
23. Ma, Y., Wu H., Wang, L.: Remote sensing big data computing: challenges and opportunities. Future Generation Computer Systems (2014)

Improving Information-Carrying Data Capacity in Text Mining

Marcin Gibert[✉]

Department of Information Systems Engineering, Faculty of Computer Science,
West Pomeranian University of Technology in Szczecin, Szczecin, Poland
mgibert@wi.zut.edu.pl

Abstract. In this article the relation between the selection of textual data representation and text mining quality has been shown. Due to this, the information-carrying capacity of data has been formalized. Then the procedure of comparing information-carrying data capacity with different structures has been described. Moreover, the method of preparing the γ-gram representation of a text involving machine learning methods and ontology created by the domain expert, has been presented. This method integrates expert knowledge and automatic methods to develop the traditional text-mining technology, which cannot understand text semantics. Representation built in this way can improve the quality of text mining, what was shown in the test research.

Keywords: Text mining · Information-carrying data capacity · Vector space model · Text documents representation

1 Introduction

Data mining is the automatic or semi-automatic process of exploration and analysis of large quantities of data in order to discover meaningful patterns and rules [13]. Depending on the type of analyzed data, proper methods and techniques of data mining are used. In case of data such as text documents, which are usually unstructured or semi structured, the following methods are used, whose main tasks are:

- development of replacement representation (data structuring) of text documents, which gives the possibility of using different techniques of exploration;
- main exploration of textual data with a new structure.

In general, to achieve this goal, the Vector Space Model, in which every text document is represented by a vector of features, is used [2]. These features are the component elements of a text, e.g. single words, called terms. A set of terms is the transformation of an original form of textual data to a more useful text representation for the exploration task. Due to a properly selected structure of textual data, exploration is possible only when concentrating on the construction of a text or on semantic information, which the text carries. It can be a quantity of specified words from a text or precise extraction of semantic information, e.g. a fragment of stock company news,

© Springer International Publishing Switzerland 2015
M. Núñez et al. (Eds.): ICCCI 2015, Part II, LNCS 9330, pp. 648–657, 2015.
DOI: 10.1007/978-3-319-24306-1_63

which provides help for investors in purchasing stocks – *investor related to the board* [16].

Depending on the adopted strategy of textual data representation, except the structure of data, the quantity and quality of important information, which is carried by these data, are also changed. Improper selection of data structure can cause:

- redundancy of representation and subsequently information noise, both of which complicate identifying important information;
- difficulty in extracting semantic information;
- fewer interpretation possibilities of data in the exploration process.

Every element mentioned above can have a negative influence on the result of text mining. Therefore, the relations between the selection of textual data and quantity and quality of important information in the text mining has been characterized and the factor of the information-carrying capacity of data has been defined. Then, three different models of text representation have been described. One of them is γ –gram representation, which can be created by using the machine-learning methods and the information patterns based on the domain ontology. Information patterns are the formal model of factual information, which occurs in the text [10]. Therefore, this approach develops the traditional text-mining technology, which cannot understand text semantics. Based on this solution, the proper structure of data (terms) can be prepared by using ontology in the context of the exploration task. Representation built in this way can improve the information-carrying data capacity, what was shown in the test research.

2   General Rules of Text Mining

Text mining is based on a structure called the semiotic triangle [18]. This is a set of relations between the term, object and concept. The term is a form of linguistic expression, e.g. a single word. The object is part of the reality indicated by the term. The concept is the meaning, which constitutes an image (mapping) of the object in the mind. Due to above relations, text mining can include an analysis on the level of the text construction or semantic information. Depending on the exploration task, the new text representation, which, to a greater extent allows us to focus exploration activities on text construction or on semantic information, is prepared. The Vector Space Model is usually used for preparing this structured representation of a text. Data exploration based on this model is performed in two steps:

- Step I – exploring components of text documents to prepare the proper representation of textual data. This step precedes the main textual data exploration which is called the data preparation step.
- Step II – general exploration of new textual data representation, which is prepared in the previous step.

The role of data preparation in step I, in which methods of data exploration are also used, is eliminating redundant elements occurring in the text, i.e. information noise. This is because information noise could have a negative influence on the selection of proper representation of textual data and then on the exploration result. In this step, the text document, which is usually expressed in natural language form is mapped by replacement representation in the form of terms vector. Preparing a structured representation of textual data in step I usually consists of three main parts [6]:

- tokenization and segmentation;
- lemmatization and stemming;
- reduction and selection of data representation.

Segmentation is the transformation of a text from a linear form to a set of semantic lingual units, from graphemes, trough words, to whole sentences[9]. Sometimes segmentation is understood as tokenization, which is the process of dividing a text into a series of single tokens, usually single words [20].

Lemmatization is the process of reducing words from their original to basic form, called lemma. Stemming is a similar process, which relies on extracting identical part of words, called stems. Because of lemmatization or stemming processes, distinct grammatical forms are regarded as one word, which allows to identify the same word, located in different positions of the text. In practice, there is a possibility of using lemmatization in relation to recognized words in a text and stemming for unrecognized words [7].

In the next step, unimportant words are removed from a large set of words extracted from the text. This reduction is usually achieved by using the stop lists, containing words with little value to the exploration process. These words do not directly affect the meaning of a text, but they shape the course of the utterance. In general, these are words, which are most commonly used in texts of a given language, e.g. conjunctions, pronouns. Sometimes stop rules are used interchangeably, most commonly based on Zipf's law [14]. These rules, using a given algorithm, specify a set of uninformative words for analyzing text sets. Sometimes, the so-called thesaurus e.g. WordNet is used to remove synonyms from a text [14].

Subsequently, text representation proper to the exploration task is created [21]. There are a few proposals of text representation, which have been described in the literature [5]:

- unigram model of representation – based on single words;
- n-gram model of representation – based on equal lengths of word sequences;
- γ -gram model of representation – based on different lengths of word sequences.

The unigram model of text representation is based on the notation of individual words occurring in a text. This notation builds a vector, which represents textual data. The elements of the vector r are calculated by formula [5]:

$$r = \begin{cases} 1 \ when \ \exists j; \ w_j = v_i, v_i \in V \\ 0 \ otherwise \end{cases}, \ i \leq n, j \leq k \qquad (1)$$

where:
V – lexicon;
v_i – i^{th} word in lexicon;
n – number of word in lexicon;
w_j – j^{th} word extracted from the text t;
k – number of word extracted from the text.

The n-gram model of text representation is based on notation of the constant length combination of words, which occur successively in a text. This model of text representation is a matrix M created according to the formula [5]:

$$Mx, y = \begin{cases} 1 \ when \ (w_j, w_{j+1}, \dots, w_{j+l-1}) = r_x \wedge w_{j+l} \in v_y \\ 0 \ otherwise \end{cases} ; \ 1 \le j + l - 1 \le k \quad (2)$$

where:

$M_{x,y}$ – matrix representing text document, rows x of matrix corresponding to combination of words r_x;
r_x- variation including l words form lexicon V;
w_j – j^{th} word extracted from the text t;
v_y- words from lexicon V corresponding to column y of matrix.

The γ-gram model of the text representation is built by using function $\gamma(w_1,\dots,w_k)$, whose values respond to the usefulness of given word sequences $w_1,\dots,w_k$ in analyzing a text document [5]. Determining the most useful sequence of different lengths is done by using a proper algorithm.

After the step of data representation, a standard procedure of exploration is done. Appropriate weights are calculated for elements of text representations (terms), by using the selected function of importance e.g. tf, idf, tf-idf [15]. Optionally, Latent Semantic Indexing, which detects sematic relations between terms by using algebraic calculation, could be used [8]. Finally, by using the proper measure, e.g. the cosine measure, the similarity between text documents is calculated [22].

3   Preparing Text Representation Based on Machine Learning and on Domain Expert Knowledge

While preparing text representation two approaches should be used as text document analysis is carried out on text construction or semantic information. They are as follows:

- machine learning;
- expert knowledge.

The former, machine learning, is automatic and largely based on statistical-mathematical methods [12]. The latter, expert knowledge, concerns the techniques of knowledge management and is associated strongly with semantic analysis [24] [16].

This approach uses lexical and syntax rules of a given language and takes into account the meaning of analyzing words and phrases. In this solution it is important to have   knowledge of   analytical language, grammar and specification of utterance, which is related to the used vocabulary. Machine-learning is the most popular way of acquiring it in terms of simplicity of practical application. This is mainly because there is no expert at hand. Therefore, many commercial systems are based on automatic methods. However, it does not mean that using such methods is the most effective approach. P. Gawrysiak wrote [5]: *Rather, it appears that future NLP systems will benefit both expert knowledge of linguists, stored in the form of a knowledge base, and from the systems of automatic analysis too, as it will be able to modify and update this knowledge.* Therefore the optimal solution seems to be integrating machine learning methods and the base of knowledge defined by the expert e.g. in the form of domain ontology.

In the literature, few exemplary methods, which use the base of knowledge defined by the domain expert as a solution to this kind of approach, has been described [16] [19]. For example it is possible to select the proper terms in the Vector Space Model, which are important for exploration purposes, on the basis of domain knowledge [23]. In this case, terms are usually in the form of word sequences of different lengths, which have semantic relations with each other. In fact, this is the γ -gram model of representation, which extracts and effectively uses semantic information carried by the text. This method improves the traditional text-mining technology, which cannot understand text semantics.

The solution proposed by the author is selecting terms in the Vector Space Model based on the information patterns defined by the expert. Information patterns are a general model of semantic information, which are specified by the relation of proper words [24]. There are few proposals of formally defining this kind of information pattern [22][23][24]. In the test research, the Web Ontology Language - OWL - the standard of data description in ontology form, has been chosen for this task [17]. The definition of information patterns by using ontology is natural because of relations occurring between the object, the term and the concept , which are expressed by the semiotic triangle. Thus, every term corresponds to some concept ,, which is identical to the concept used in ontology creation. Moreover, the relations between concepts, included in the information patterns, are mapped in the ontology model.

Factual information, which can be extracted from a text based on a specific algorithm, which uses the information pattern defined by OWL, is for example *investor related to the board*. The definition of the information pattern based on OWL tags for the above factual information is:

```
<owl:Class rdf:ID="investor" />
<owl:Class rdf:ID="related" />
<owl:Class rdf:ID="board" />

<owl:ObjectProperty rdf:ID="whichOne">
<rdfs:domain rdf:resource="#investor" />
<rdfs:range rdf:resource="#related" />
</owl:ObjectProperty>
```

```
<owl:ObjectProperty rdf:ID="withWhom">
<rdfs:domain rdf:resource="#related" />
<rdfs:range rdf:resource="#board" />
</owl:ObjectProperty>
```

The pattern definition by using OWL facilitates the use of proper patterns for a specified task of data exploration. The set of patterns may form the basis of knowledge, which can be used as factual information, which is important for the exploration process. Furthermore, because of the use of an OWL construction it is easy to extend several patterns with new elements. This operation improves the construction process of the total model of knowledge, which is necessary to extract proper factual information for a given exploration task.

4   Rating Text Document Exploration Process with Use of Information-Carrying Capacity

The information-carrying capacity of data depends on all-important and interpretable information, which is carried by data D with structure s and concerns the considered task of exploration. Information-carrying capacity has been defined by the author's formula:

$$N_{inf}(D, M) = f(W, R, \zeta) \tag{3}$$

where:

N_{inf} – information capacity of data D;
D – textual data;
M – is the set of selected quality measures;
W – quality of data exploration result expressed by the adopted measures;
R – representation of text document;
ζ – is an unknown and unidentified set of information, which has an influence on information capacity of data D and on the result of the decision process;
$f()$ – unknown function, whose value is specifying information capacity N_{inf}.

In the absence of knowledge about the function $f()$ and the set ζ from formula (3) and in the absence of methods for their identification, the information-carrying capacity N_{inf} of data D with structure s_1 can be determined as the rank of information-carrying capacity of data with structure s_2 regarding structure s_2, based on the quality of result W of the data exploration process. This is a kind of categorical rating expressing the rank of values. Therefore, the information-carrying capacity of data D with structure s_1 is greater than the information-carrying capacity of data D with structure s_2 in case of achieving the result in the exploration process, which is characterized by higher quality according to the adopted measure W, e.g. precision, recall, accuracy, specificity, fall-out [1]. The adopted quality measure should be related to the defined aim of the exploration process [25].

The whole set of quality measures is also possible to take into account in the rating of information-carrying data capacity. Therefore, comparison of the information-carrying data capacity based on an adopted set of different quality measures, is performed by using one of the multi-criteria decision methods., e.g. Electre I [3].

There are many multi-criteria decision methods e.g. AHP, ANP, Electre I Promethee. Each of them works differently and is used for different tasks. The comparison of multi-criterion decision methods and a detailed description of them is presented in the literature[4]. For the decision problem, which includes selecting the best decision variant, the Electre I method has been adopted for the exemplary research. This method is used for problems of selecting the best decision variant. Electre I allows every decision variant to specify the weight of importance and the searching direction of the best decision. Subsequently, the relations between decision variants are appointed. The graph of superiority relations specifies their levels from the best to the worst variants. The highest value of the exploration quality measure combination could be obtained by using the best variant of data structure from this graph.

The quality rating of textual data exploration is made on selected levels of recall. Depending on the exploration task it can be either on one level or a whole set of selected levels of recall, e.g. above value 0.5. Therefore, the final rating of decision variants by using for example the Electre I method is made on the selected levels of recall. The final superiority factor of information-carrying capacity, is calculated, according to formula:

$$W_{N_{inf}} = \sum_{o=1}^{u} \begin{cases} 1 \; when \; p_o = 1 \\ 0 \; otherwise \end{cases} \tag{4}$$

where:

po- level of variant achieved in Electre I method, for recall o;
u – maximum level of recall;
po=1 – means that for recall o, the variant is on the first level of superiority graph.

5   Research on the Influence of Different Textual Data Structures on Quality of Exploration Process – Case Study

The test research was carried out on the corpus of 200 texts, derived from information for Polish stock company investors. The task of exploration was to identify company news, which gives the latter signals to buy stocks. For example this is news, in which information is given about the purchase of company stocks by people, who are in relation with members of the Board. In the first step, there is a set of 10 news in the analyzing corpus, which should be returned in ranking. This news   determines 10 levels of recall. Next, those textual data structures were selected, which would be ranked in respect to the information-carrying capacity in the identical exploration process. The γ -gram model of representation has been selected on the basis of information patterns created by the expert by using the Web Ontology Language, according to the procedure described in chapter 3.

Finally, the set of quality measures, which was taken into account on the specified level of recall in rating, has been defined. For research purposes, three measures of quality have been selected, namely precision, negative prediction value and recall. Recall has values equal or higher than *0.5*. The example, taken into account in the selection of the most favorable ones, in respect of expert preferences, variant in Electre I method, has been presented in table 1.

Table 1. The data, which were taken into account in the rating of the most favourable decision variant for level of recall equal 0.8

For level of recall = 0,8			
criteria	precision	negative prediction value	Final rank (after normalization)
Direction of optimization	max	max	
weight	5	4	
Unigram model of representation	0,42	0.99	0,6
2-gram model of representation	0.12	0.98	0,4
γ -gram model of representation	0.62	0.99	1

Finally, according to the procedure, which has been described in chapter 4, the rank of data information capacity for individual structures has been calculated. All calculations in the test research have been made by using the author`s software, based on JAMA - basic linear algebra package for Java. On analyzing the results of the calculation it can be said that the most favorable data for the exemplary research has a γ-gram structure s_3, the second in order is a unigram structure s_1 and the last one is a 2-gram structure s_2.

6   Summary

In this article the importance of the correct choice of data structures in the text mining process has been described. The possibility of extracting more valuable knowledge was indicated, while taking into account the selected data structures in order to consider the problem of exploration. The relation between selecting terms and text mining quality has been shown. The method, which allows to select the proper term in the Vector Space Model has been presented. This method uses the information patterns defined by the Web Ontology Language to create better model of text representation for considering the exploration task. Moreover, by using the information-carrying data capacity factor, the procedure, which allows to compare the different data structure influence on a specified set of text mining quality measures, has been presented. Finally, in the article comparative research results of text mining quality has been presented with exemplary textual data of different structures. One of these structures, γ –gram representation, which is prepared by using machine-learning methods and ontology created by the domain expert, has proved to be capable of improving the information-carrying capacity of data, shown in this research paper.

References

1. Amiri, I.S., Akanbi, O.A., Fazeldehkordi, E.: A Machine-Learning Approach to Phishing Detection and Defense, p. 27 (2014)
2. Mbarek, R., Tmar, M., Hattab, H.: A New Relevance Feedback Algorith Based on Vector Space Basis Change. Computational Linguistics and Intelligent Text Processing **2**, 355–356 (2014)
3. Munier, N.: A Strategy for Using Multicriteria Analysis in Decision-Making: A Guide for Simple and Complex Environmental Projects, pp. 59–65 (2011)
4. Velasquez, M., Hester, P.T.: An Analysis of Multi-Criteria Decision Making Methods. International Journal of Operations Research **10**(2), 56–66 (2013)
5. Gawrysiak, P.: Automatyczna kategoryzacja dokumentów, pp. 36–45 (2001)
6. Ramasubramanian, C., Ramya, R.: Effective Pre-Processing Activities in Text Mining using Improved Porter's Stemming Algorithm. International Journal of Advanced Research in Computer and Communication Engineering, Volume **2**(12), 4537 (2013)
7. Weiss, S.M., Indurkhya, N., Zhang, T.: Fundamentals of Predictive Text Mining, pp. 17–19 (2010)
8. Landauer, T.K., McNamara, D.S., Dennis, S., Kintsch, W.: Handbook of Latent Semantic Analysis, p. 10 (2013)
9. Dale, R., Moisl, H., Somers, H.: Handbook of Natural Language Processing, p. 11 (2000)
10. Luna Dong, X., Gabrilovich, E., Murphy, K., Dang, V., Horn, W., Lugaresi, C., Sun, S., Zhang, W.: Knowledge_based Trust: Estimating the Trustworthiness of Web Sources. Computer Science Database (2015)
11. Gentile, A.L., Basile, P., Iaquinta, L., Semeraro, G.: Lexical and semantic resources for NLP: from words to meanings. In: Lovrek, I., Howlett, R.J., Jain, L.C. (eds.) KES 2008, Part III. LNCS (LNAI), vol. 5179, pp. 277–284. Springer, Heidelberg (2008)
12. Kononenko, I., Kukar, M.: Machine Learning and Data Mining, p. 17 (2007)
13. Berry, M., Linoff, G.: Mastering Data Mining: The Art and Science of Customer Relationship Management, p. 7 (2004)
14. Esposti, M.D.: Mathematical Models of Textual Data: A short Review, pp. 100–102 (2014)
15. Sabbah, T., Selemat, A.: Modified Frequency-Based Term Weighting Scheme for Accurate Dark Web Content Classification, pp. 185–187 (2014)
16. Jackson, P., Moulinier, I.: Natural Language Processing for Online Applications: Text Retrieval, Extraction, and Categorization, Amsterdam, vol. 5, pp. 125–126 (2007)
17. Bechhofer, S., Harmelen, F., Hendler, J., Horrocks, I., McGuinness, D.L., Patel-Schneider, P.F., Stein, L.A.: OWL Web Ontology Language (2015). http://www.w3.org/TR/owl-ref/
18. Merkelis, R.: Philosophy and Linguistics, p. 12 (2013). http://www.slideshare.net/robertasmerkelis/philosophy-and-linguistics-28940425
19. Jiang, L., Zhang, H.-b., Yang, X., Xie, N.: Research on semantic text mining based on domain ontology. In: Li, D., Chen, Y. (eds.) Computer and Computing Technologies in Agriculture VI, Part I. IFIP AICT, vol. 392, pp. 336–343. Springer, Heidelberg (2013)
20. Chakraborty, G., Pagolu, M., Satshi, G.: Text Mining and Analysis: Practical Methods, Examples, and Case Studies Using SAS, p. 70 (2014)
21. Sanders, T., Schilperoord, J., Spooren, W.: Text Representation: Linguistic and Psycholinguistic Aspects, pp. 1–19 (2001)
22. Śmiałkowska, B., Gibert, M.: The classification of text documents by using Latent Semantic Analysis for extracted information. Ekonomiczne Problemy Usług No. 106, Zeszyty Naukowe Uniwersytetu Szczecińskiego No. 781, pp. 345–358 (2013)

23. Smialkowska, B., Gibert, M.: The classification of text documents in Polish language by using Latent Semantic Analysis for extracted information. Theoretical and applied informatics **25**, 239–250 (2013)
24. Lubaszewski, W.: Słowniki komputerowe i automatyczna ekstrakcja informacji z tekstu (2009)
25. Wang, R.Y., Strong, D.M.: What data quality means to data consumers. Journal of Management Information Systems **12**(4), 7 (1996)

Author Index

Printed in the United States
By Bookmasters